Rehabilitation Robots for Neurorehabilitation in High-, Low-, and Middle-Income Countries

Current Practice, Barriers, and Future Directions

Rehabilitation Robots for Neurorehabilitation in High-, Low-, and Middle-Income Countries

Current Practice, Barriers, and Future Directions

Edited by

Michelle J. Johnson
Departments of Physical Medicine and Rehabilitation, Bioengineering, and Mechanical Engineering and Applied Mechanics, University of Pennsylvania, Philadelphia, PA, United States

Rochelle J. Mendonca
Department of Rehabilitation and Regenerative Medicine, Programs in Occupational Therapy, Columbia University, New York, NY, United States

ACADEMIC PRESS

An imprint of Elsevier

Academic Press is an imprint of Elsevier
125 London Wall, London EC2Y 5AS, United Kingdom
525 B Street, Suite 1650, San Diego, CA 92101, United States
50 Hampshire Street, 5th Floor, Cambridge, MA 02139, United States
The Boulevard, Langford Lane, Kidlington, Oxford OX5 1GB, United Kingdom

Notices
Knowledge and best practice in this field are constantly changing. As new research and
experience broaden our understanding, changes in research methods, professional practices, or
medical treatment may become necessary.

Practitioners and researchers must always rely on their own experience and knowledge in
evaluating and using any information, methods, compounds, or experiments described herein.
In using such information or methods they should be mindful of their own safety and the safety
of others, including parties for whom they have a professional responsibility.

To the fullest extent of the law, neither the Publisher nor the authors, contributors, or editors,
assume any liability for any injury and/or damage to persons or property as a matter of
products liability, negligence or otherwise, or from any use or operation of any methods,
products, instructions, or ideas contained in the material herein.

ISBN: 978-0-323-91931-9

For Information on all Academic Press publications
visit our website at https://www.elsevier.com/books-and-journals

Publisher: Mara Conner
Acquisitions Editor: Sonnini Yura
Editorial Project Manager: Emily Thomson
Production Project Manager: Fizza Fathima
Cover Designer: Mark Rogers

Typeset by MPS Limited, Chennai, India

Working together
to grow libraries in
developing countries

www.elsevier.com • www.bookaid.org

Contents

Part I
Fundamentals

Part II
Background on country healthcare systems, rehabilitation standards of care, stroke rehabilitation, and rehabilitation and assistive robotics in selected HICs

Rochelle J. Mendonca, Carol A. Wamsley, Chung-Ying Owen Tsai, Hao Su and Michelle J. Johnson

Amanda McIntyre, Javad K. Mehr, Marcus Saikaley, Mahdi Tavakoli, Dalton L. Wolfe and Ana Luisa Trejos

Part III
Background on healthcare systems, rehabilitation, stroke rehabilitation, and rehabilitation robotics in selected LMICs

Daniel Comadurán Márquez, Pavel Loeza Magaña and Karla D. Bustamante Valles

18. Europe region: Serbia   283

Ljubica M. Konstantinovic, Andrej M. Savic, Aleksandra S. Vidakovic,
Olivera C. Djordjevic and Sindi Z. Mitrovic

19. Asia Pacific region: India   293

Sivakumar Balasubramanian, Aravind Nehrujee, Abha Agrawal,
Guruprasad V., Shovan Saha and Sujatha Srinivasan

23.   Middle East region: Turkey   353

Duygun Erol Barkana, Ismail Uzun, Devrim Tarakci, Ela Tarakci,
Ayse Betul Oktay and Yusuf Sinan Akgul

24.   Africa region: Nigeria   367

Morenikeji A. Komolafe, Kayode P. Ayodele, Matthew O.B. Olaogun,
Philip O. Ogunbona, Michael B. Fawale, Abiola O. Ogundele,
Akintunde Adebowale, Oluwasegun T. Akinniyi,
Sunday O. Ayenowowan, Abimbola M. Jubril, Ahmed O. Idowu,
Ahmad A. Sanusi, Abiodun H. Bello and Kolawole S. Ogunba

List of contributors

Ebenezer Ad Adams National Stroke Support Services, Accra, Ghana

Akintunde Adebowale Department of Medicine, Obafemi Awolowo University, Ile-Ife, Osun, Nigeria

Abha Agrawal A4 Clinics, Indore, Madhya Pradesh, India

Javier Sánchez Aguilar European Neurosciences Center, Madrid, Spain

Yusuf Sinan Akgul Department of Computer Engineering, Gebze Technical University, Kocaeli, Turkey

Oluwasegun T. Akinniyi Department of Electronic and Electrical Engineering, Obafemi Awolowo University, Ile-Ife, Osun, Nigeria

Asongafac Center for Promotion of Rehabilitation Medicine and Disability Research, Yaoundé, Cameroon; Physiotherapy Unit, Buéa Regional Hospital, Buéa, Cameroon

Sunday O. Ayenowowan Department of Medical Rehabilitation, Obafemi Awolowo University Teaching Hospitals Complex, Ile Ife, Osun, Nigeria

Kayode P. Ayodele Department of Electronic and Electrical Engineering, Obafemi Awolowo University, Ile-Ife, Osun, Nigeria

Suman Badhal Vardhman Mahavir Medical College, Safdarjung Hospital, New Delhi, Delhi, India

Sivakumar Balasubramanian Department of Bioengineering, Christian Medical College, Bagayam, Vellore, Tamil Nadu, India

Jakita Baldwin Columbia University Irving Medical Center, New York, NY, United States

Santosh Balivada Andhra Pradesh Med Tech Zone, Visakhapatnam, Andhra Pradesh, India

Cori Barger Department of Rehabilitation Medicine, NewYork-Presbyterian Brooklyn Methodist Hospital, Brooklyn, NY, United States

Duygun Erol Barkana Department of Electrical and Electronics Engineering, Yeditepe University, İstanbul, Turkey

Saeed Behzadipour Department of Mechanical Engineering, Sharif University of Technology, Tehran, Iran; Djavad Mowafaghian Research Center in Neuro-rehabilitation Technologies, Sharif University of Technology, Tehran, Iran

Abiodun H. Bello Department of Medicine, University of Ilorin Teaching Hospital, Ilorin, Kwara, Nigeria

Paolo Boldrini Italian Society of Physical and Rehabilitation Medicine (SIMFER), Rome, Italy

Donatella Bonaiuti Italian Society of Physical and Rehabilitation Medicine (SIMFER), Rome, Italy

Kevin Doan-Khang Bui Selera Medical, Inc., Portola Valley, CA, United States

Karla D. Bustamante Valles Research Department, Centro de Investigación en Bioingeniería A.C., Chihuahua, Mexico; Orthopaedic & Rehabilitation Engineering Center, Marquette University, Milwaukee, WI, United States

Ning Cao Department of Physical Medicine and Rehabilitation, Johns Hopkins University School of Medicine, Baltimore, MD, United States

Arys Carrasquilla Batista Department of Mechatronics Engineering, Instituto Tecnológico de Costa Rica, Cartago, Costa Rica

Zhen Chen Neurorehabilitation Center, The First Rehabilitation Hospital of Shanghai, Shanghai, P.R. China; Tongji University School of Medicine, Shanghai, P.R. China; Tenth People's Hospital of Tongji University, Shanghai, P.R. China

Carlos A. Cifuentes Bristol Robotics Laboratory, University of the West of England, Bristol, United Kingdom; School of Engineering, Science and Technology, Universidad del Rosario, Bogotá, Colombia

Daniel Comadurán Márquez Research Department, Centro de Investigación en Bioingeniería A.C., Chihuahua, Mexico; Cumming School of Medicine, University of Calgary, Calgary, AB, Canada

Beatriz Coto-Solano Rehabilitation Service, Hospital R. A. Calderón Guardia, Caja Costarricense de Seguro Social, San José, Costa Rica

Vincent Crocher Department of Mechanical Engineering, University of Melbourne, Parkville, VIC, Australia

Timothy Dillingham Departments of Physical Medicine and Rehabilitation, Bioengineering, and Mechanical Engineering and Applied Mechanics, University of Pennsylvania, Philadelphia, PA, United States

Olivera C. Djordjevic Faculty of Medicine, University of Belgrade, Belgrade, Serbia; Clinic for Rehabilitation "Dr Miroslav Zotović", Belgrade, Serbia

Michael B. Fawale Department of Medicine, Obafemi Awolowo University, Ile-Ife, Osun, Nigeria

Ronit Feingold-Polak Department of Physical Therapy, Ben-Gurion University, Be'er Sheva, Israel

Justin Fong Department of Mechanical Engineering, University of Melbourne, Parkville, VIC, Australia

Nicolás García-Aracil Department of Control and Systems Engineering, Miguel Hernandez University, Alicante, Spain; Robotics and Artificial Intelligence Unit, Bioengineering Institute, Spain

Andrea Garzón School of Medicine and Health Sciences, Universidad del Rosario, Bogotá, Colombia

Carly Goldberg Programs in Occupational Therapy, Columbia University, New York, NY, United States; Department of Rehabilitation, New York Presbyterain, New York, NY, United States

Abderrazak Hajjioui Clinical Neuroscience Laboratory, Faculty of Medicine, Pharmacy and Dentistry, University Sidi Mohamed Ben Abdellah, Fez, Morocco; Department of Physical & Rehabilitation Medicine, University Hospital Hassan II, Fez, Morocco

Taya Hamilton Massachusetts General Hospital, Boston, MA, United States; Perron Institute of Neurological and Translational Science, University of Western Australia, Perth, WA, Australia

Fazah Akhtar Hanapiah Faculty of Medicine, Universiti Teknologi MARA, Sungai Buloh, Selangor, Malaysia

Yasuhisa Hirata School of Engineering, Department of Robotics, Tohoku University, Sendai, Japan

Koh Teck Hong (Zen) MotusAcademy, Zürich, Switzerland

Zikai Hua Shanghai University, Shanghai, P.R. China

Ahmed O. Idowu Department of Medicine, Obafemi Awolowo University, Ile-Ife, Osun, Nigeria

Gabriel Iturralde-Duenas University of Texas at Austin, Mechanical Engineering Department, Austin, TX, United States

Andrea Blanco Ivorra Robotics and Artificial Intelligence Unit, Bioengineering Institute, Spain

Rodrigo S. Jamisola, Jr Faculty of Engineering & Technology, Botswana International University of Science & Technology, Palapye, Botswana

Michelle J. Johnson Departments of Physical Medicine and Rehabilitation, Bioengineering, and Mechanical Engineering and Applied Mechanics, University of Pennsylvania, Philadelphia, PA, United States

Abimbola M. Jubril Department of Electronic and Electrical Engineering, Obafemi Awolowo University, Ile-Ife, Osun, Nigeria

Bong-Keun Jung Department of Mechanical Engineering, Seoul National University, Kwanak-gu, Seoul, Republic of Korea

Kavita Kachroo Kalam Institute of Health Technology (KIIT), Visakhapatnam, Andhra Pradesh, India

Sinforian Kambou Institute of Applied Neurosciences and Functional Rehabilitation, Yaoundé, Cameroon; Center for Promotion of Rehabilitation Medicine and Disability Research, Yaoundé, Cameroon

Maikutlo Kebaetse Faculty of Medicine, University of Botswana, Gaborone, Botswana

Shafagh Keyvanian Department of Mechanical Engineering and Applied Mechanics, University of Pennsylvania, Philadelphia, PA, United States

Kang Xiang Khor Techcare Innovation Sdn. Bhd., Taman Perindustrian Ringan Pulai, Skudai, Johor, Malaysia

Yvonne Y.W. Khor YK Natural Physio & Academy, Taman Austin Perdana, Johor Bahru, Johor, Malaysia

JiHyun Kim Department of Occupational Therapy, Far East University, Eumseung-gun, Chungcheongbuk-do, Republic of Korea

Sumudu Perera Kimmatudawage Department of Medicine, @AgeMelbourne, Royal Melbourne Hospital (Melbourne Health), University of Melbourne, Melbourne, VIC, Australia; Western Health, Melbourne, VIC, Australia; London School of Tropical Medicine, London, United Kingdom

Marlena Klaic Melbourne School of Health Sciences, University of Melbourne, Parkville, VIC, Australia

Venkata P. Kommula Faculty of Engineering & Technology, University of Botswana, Gaborone, Botswana

Morenikeji A. Komolafe Department of Medicine, Obafemi Awolowo University, Ile-Ife, Osun, Nigeria

Ljubica M. Konstantinovic Faculty of Medicine, University of Belgrade, Belgrade, Serbia; Clinic for Rehabilitation "Dr Miroslav Zotović", Belgrade, Serbia

Hermano I. Krebs Department of Mechanical Engineering, Massachusetts Institute of Technology, Cambridge, MA, United States; School of Medicine, University of Maryland, Baltimore, MD, United States; School of Medicine, Fujita Health University, Nagoya, Aichi Prefecture, Japan; Department of Mechanical Science and Bioengineering, Osaka University, Osaka, Osaka Prefecture, Japan; The Wolfson School of Engineering, Loughborough University, Loughborough, Leicestershire, United Kingdom

Eric Gueumekane Bila Lamou Department of Internal Medicine, Gynaeco-Obstetric and Paediatric Hospital, Douala, Cameroon

Shelly Levy-Tzedek Department of Physical Therapy, Ben-Gurion University, Be'er Sheva, Israel

Pavel Loeza Magaña Rehabilitation Medicine, Sports Science, Universidad Nacional Autónoma de México, Ciudad de México, Mexico

Stefano Mazzoleni Department of Electrical and Information Engineering, Politecnico di Bari, Bari, Italy and The BioRobotics Institute, Scuola Superiore Sant'Anna, Pisa, Italy

Lingani Mbakile-Mahlanza Faculty of Social Sciences, University of Botswana, Gaborone, Botswana

Amanda McIntyre Arthur Labatt Family School of Nursing, Western University, London, ON, Canada; Parkwood Institute Research, Lawson Health Research Institute, London, ON, Canada

Javad K. Mehr Department of Electrical and Computer Engineering, University of Alberta, Edmonton, AB, Canada; Department of Medicine, University of Alberta, Edmonton, Edmonton, AB, Canada

Rochelle J. Mendonca Department of Rehabilitation and Regenerative Medicine, Programs in Occupational Therapy, Columbia University, New York, NY, United States

Sindi Z. Mitrovic Faculty of Medicine, University of Belgrade, Belgrade, Serbia; Clinic for Rehabilitation "Dr Miroslav Zotović", Belgrade, Serbia

Ichiro Miyai Neurorehabilitation Research Institute, Morinomiya Hospital, Osaka, Japan

Ntsatsi Mogorosi Botswana-UPenn Partnership, Gaborone, Botswana

Natiara Mohamad Hashim Faculty of Medicine, Universiti Teknologi MARA, Sungai Buloh, Selangor, Malaysia

Inhyuk Moon Department of Robots and Automation Engineering, Dong-Eui University, Busanjin-gu, Busan, Republic of Korea

Marcela Múnera Bristol Robotics Laboratory, University of the West of England, Bristol, United Kingdom; Biomedical Engineering Department, Colombian School of Engineering Julio Garavito, Bogotá, Colombia

Said Nafai School of Health Sciences, Department of Occupational Therapy, American International College, Springfield, MA, United States

Imama A. Naqvi Columbia University Irving Medical Center, New York, NY, United States

Kagiso Ndlovu Faculty of Computer Science, University of Botswana, Gaborone, Botswana

Aravind Nehrujee Department of Bioengineering, Christian Medical College, Bagayam, Vellore, Tamil Nadu, India; Department of Mechanical Engineering, Indian Institute of Technology, Chennai, Tamil Nadu, India

Dawn M. Nilsen Programs in Occupational Therapy, Columbia University, New York, NY, United States; Department of Rehabilitation, New York Presbyterain, New York, NY, United States

Tomoyuki Noda Computational Neuroscience Laboratories (CNS), Advanced Telecommunication Research Institute (ATR), Kyoto, Japan

Justus Mackenzie Nthitu Department of Occupational Therapy, Mahalapye District Hospital, Mahalapye, Botswana

Frank Kwabena Afriyie Nyarko Department of Mechanical Engineering, Kwame Nkrumah University of Science and Technology, Kumasi, Ghana

Cassandra Ocampo Faculty of Medicine, University of Botswana, Gaborone, Botswana

Kolawole S. Ogunba Department of Electronic and Electrical Engineering, Obafemi Awolowo University, Ile-Ife, Osun, Nigeria

Philip O. Ogunbona School of Computing and Information Technology, University of Wollongong, Wollongong, NSW, Australia

Abiola O. Ogundele Department of Medical Rehabilitation, Obafemi Awolowo University Teaching Hospitals Complex, Ile Ife, Osun, Nigeria

Ayse Betul Oktay Department of Computer Engineering, Yildiz Technical University, İstanbul, Turkey

Matthew O.B. Olaogun Faculty of Medical Rehabilitation, University of Medical Sciences, Ondo, Ondo, Nigeria

Esteban Ortiz-Prado Universidad de las Americas, Facultad de Medicina, De los Colimes, Quito, Ecuador

José María Catalán Orts Robotics and Artificial Intelligence Unit, Bioengineering Institute, Spain

Chung-Ying Owen Tsai Department of Rehabilitation and Human Performance, Icahn School of Medicine at Mount Sinai, New York, NY, United States

Jin-Hyuck Park Department of Occupational Therapy, Soonchunhyang University, Asan, Chungcheongnam-do, Republic of Korea

Angie Pino Biomedical Engineering Department, Colombian School of Engineering Julio Garavito, Bogotá, Colombia

Federico Posteraro Rehabilitation Department, Versilia Hospital, AUSL Toscana Nord Ovest, Lucca, Italy

Benedict Okoe Quao Ankaful Leprosy & General Hospital, Ankaful, Ghana; National Leprosy Control Programme, Disease Control & Preventive Department, Ghana Health Service Public Health Division, Korle-Bu, Accra, Ghana

Narges Rahimi Department of Neurology, Jefferson-Einstein Medical Center, Philadelphia, PA, United States

Rakesh Srivastava WISH Foundation, Indian Council of Medical Research, Medical Council of India, NBE, Ministry of Health and Family Welfare, New Delhi, Delhi, India

Shovan Saha Department of Occupational Therapy, Manipal College of Health Professions, MAHE, Manipal, Karnataka, India

Marcus Saikaley Parkwood Institute Research, Lawson Health Research Institute, London, ON, Canada

Jose López Sánchez European Neurosciences Center, Madrid, Spain; Spanish Association for Intensive Therapy in Neurorehabilitation, Spain

Ahmad A. Sanusi Department of Medicine, Obafemi Awolowo University, Ile-Ife, Osun, Nigeria

Andrej M. Savic School of Electrical Engineering, Science and Research Centre, University of Belgrade, Belgrade, Serbia

Won-Kyung Song Department of Rehabilitation and Assistive Technology, National Rehabilitation Center, Gangbuk-gu, Seoul, Republic of Korea

Sujatha Srinivasan Department of Mechanical Engineering, Indian Institute of Technology, Chennai, Tamil Nadu, India

Eileen L.M. Su Faculty of Electrical Engineering, Universiti Teknologi Malaysia, Johor Bahru, Johor, Malaysia

Hao Su Mechanical and Aerospace Engineering Department, Lab of Biomechatronics and Intelligent Robotics, North Carolina State University, Raleigh, NC, United States

Abena Yeboaa Tannor Department of Health Promotion and Disability, School of Public Health, College of Health Sciences, Kwame Nkrumah University of Science and Technology, Kumasi, Ghana

Devrim Tarakci Department of Occupational Therapy, Istanbul Medipol University, İstanbul, Turkey

Ela Tarakci Department of Physiotherapy and Rehabilitation, Istanbul University-Cerrahpaşa, İstanbul, Turkey

Mahdi Tavakoli Department of Electrical and Computer Engineering, University of Alberta, Edmonton, AB, Canada

Michael Temgoua Institute of Applied Neurosciences and Functional Rehabilitation, Yaoundé, Cameroon

Philomène Synthia Tonye Physiotherapy Unit, Nkongsamba Regional Hospital, Nkongsamba, Cameroon

Ana Luisa Trejos Department of Electrical and Computer Engineering, Western University, London, ON, Canada; School of Biomedical Engineering, Western University, London, ON, Canada

Billy Tsima Faculty of Medicine, University of Botswana, Gaborone, Botswana

Valery Labou Tsinda Neuro-rehabilitation and Movement Disorders Society, Yaoundé, Cameroon

Ismail Uzun İNOSENS, Kocaeli, Turkey

Guruprasad V. Department of Occupational Therapy, Manipal College of Health Professions, MAHE, Manipal, Karnataka, India

Aleksandra S. Vidakovic Faculty of Medicine, University of Belgrade, Belgrade, Serbia; Clinic for Rehabilitation "Dr Miroslav Zotović", Belgrade, Serbia

Carol A. Wamsley Therapy Division, Good Shepherd Penn Partners, Penn Institute for Rehabilitation Medicine, Philadelphia, PA, United States

Patrice L. Weiss Pediatric and Adolescent Rehabilitation Center, ALYN Hospital, Jerusalem, Israel; Department of Occupational Therapy, University of Haifa, Haifa, Israel

Lauren Winterbottom Programs in Occupational Therapy, Columbia University, New York, NY, United States; Columbia Department of Rehabilitation and Regenerative Medicine, New York, NY, United States

Dalton L. Wolfe Parkwood Institute Research, Lawson Health Research Institute, London, ON, Canada

Dixon Yang Columbia University Irving Medical Center, New York, NY, United States

Che Fai Yeong Faculty of Electrical Engineering, Universiti Teknologi Malaysia, Johor Bahru, Johor, Malaysia

Amin Zammouri Engineering Department, EPF Graduate School of Engineering, Cachan, France

Loredana Zollo Advanced Robotics and Human-Centred Technologies Laboratory, University Campus Bio-Medico of Rome, Rome, Italy

Healthcare disparities and access to rehabilitation robots for neurorehabilitation

Our goal in editing this book was to draw attention to healthcare disparities in stroke neurorehabilitation worldwide with specific attention to the use of high-tech and robotic technologies to assist in stroke care from both the clinical and the engineering perspectives. Levels 2 and 3 and some Level 1 evidence support the use of a wide range of assistive technologies, including robotic ones during the assessment of stroke impairment, the rehabilitation of stroke impairment, and the maximization of function in daily life after stroke. Unfortunately although multiple high-income countries (HICs) and some low- and middle-income countries (LMICs) have independently identified the potential of rehabilitation robots and have begun early-to-mid-stage development efforts worldwide, there is a need study these efforts and to distill from them lessons that can be used to guide designers and clinicians in countries at the beginning of their assistive technology (AT) journey. This book describes the state of the art of stroke rehabilitation using robot and mechatronic systems in select HICs and LMICs and highlights potential solutions to enable these technologies to be available to clinicians worldwide no matter the country or economics. Whenever one of the selected country reported little to no adoption of robotic solutions in the care of patients who need neurorehabilitation, including stroke, we asked them the contributing authors to report on what exists and to speculate on the reasons for the lack of uptake. Fig. 1 shows a worldwide map highlighting the locations of countries covered in this book. Table 1 lists the countries surveyed in this book and provides an overview of their demographic and economic information and stroke statistics.

According to the World Health Organization, 15 million people suffer from stroke worldwide each year. Approximately, a third of those affected will suffer from lifelong functional disabilities, including problems with performing activities of daily living and regaining independence [1]. Stroke is the third leading cause of worldwide disability in adults [2]. Up to 75% of stroke survivors experience hemiparesis, and up to 53.4% of stroke survivors experience poststroke cognitive impairments within 1.5 years poststroke [3–5]. In 2019, there were about 101 million people in the world living with impairments due to a stroke

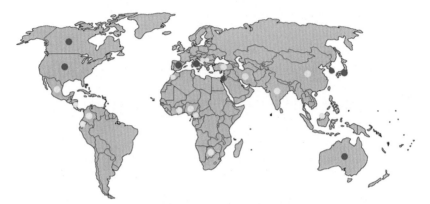

FIGURE 1 Preface description: This figure shows the countries reviewed in this book on the world map. The 8 HICs are located by blue markers—USA, Canada, Italy, Spain, Japan, Australia, Republic of Korea and Israel. The 15 LMICs are located by yellow markers—Mexico, Costa Rica, Columbia, Ecuador, Serbia, India, Malaysia, China, Iran, Turkey, Nigeria, Botswana, Ghana, Cameroon, and Morocco. *HICs*, High-income countries; *LMICs*, low- and middle-income countries.

with 89% of stroke deaths and disability living in LMICs [6]. Robot-assisted neurorehabilitation is one example of a technology-based approach to stroke rehabilitation that can help bridge healthcare gaps that may exist worldwide. This approach typically involves the use of robots in a rehabilitation context to provide repetitive and game-based or task-based exercise training to improve functional outcomes. These therapy robots can be used in a variety of hospital settings, but they are especially used in inpatient and outpatient rehabilitation. Depending on cost, some are found in subacute skill nursing facilities, community rehabilitation centers, and primary care facilities, but rarely are they found in low-resource settings whether in HICs or LMICs.

A country's percentage of the gross domestic product (GDP) per capita spent on healthcare dollars can influence the availability of medical facilities, rehabilitation facilities, and schools for training medical and rehabilitation professionals. In general, developing countries categorized as LMICs will use a much smaller percentage of their GDP per capita for healthcare expenditures, which can lead to the lack of skilled doctors, nurses, and therapists in sufficient numbers to treat their population inside or outside of cities, a fact which is supported by the strong correlation ($R^2 = 0.58$) between the healthcare access and quality (HAQ) index (out of 100) and healthcare expenditure (% of GDP/per capita) (Table 1, Fig. 2).

As a result of decreased access, we should expect that functional outcomes for someone in a HIC after a stroke will differ from those in an LMIC, especially since rehabilitation access, technology availability, and resources also differ. The correlation in the stroke deaths rate per million when plotted against the percentage of people over 65 years of age is another telling relationship. We expect that the greater percentage of people over 65

TABLE 1 Overview of their demographic and economic information for countries in this book.

	Country	Population (million)	Population aged 65 + (%)	Life expectancy at birth (years)	Stroke deaths/ million	GDP/ capita	Health spending /capita	Expenditure on health (% of GDP)	Healthcare index	Healthcare access and quality (HAQ) index
HIC	The United States	332.92	16.63	78.8	2.227	65134	10921.01	16.9	69.03	91·5
	Canada	38.07	18.10	82.2	1.512	46550	5048.37	10.8	71.8	91·6
	Italy	60.37	23.30	83.3	2.515	33090	2905.5	8.7	66.77	88·1
	Spain	46.75	19.98	83.4	1.837	29816	2711.19	9	78.8	85·7
	Japan	126.05	28.40	84.4	2.321	40063	4360.47	11	80.68	90·4
	Australia	25.79	16.21	83.3	1.713	54763	5427.46	9.3	77.71	91·0
	Republic of Korea	51.31	15.79	82.8	2.548	32143	2624.53	7.6	80.68	89·0
	Israel	8.79	12.41	82.7	1.679	46376	3456.39	7.5	73.76	87·1

(Continued)

TABLE 1 (Continued)

	Country	Population (million)	Population aged 65 + (%)	Life expectancy at birth (years)	Stroke deaths/ million	GDP/ capita	Health spending /capita	Expenditure on health (% of GDP)	Healthcare index	Healthcare access and quality (HAQ) index
LMI	Mexico	130.26	7.62	75	3.208	9849	540.37	5.4	72.51	77·6
	Costa Rica	5.14	10.25	77	2.444	12238	921.59	7.6	62.92	78·0
	Colombia	51.27	9.06	80	2.957	6432	495.33	7.6	66.72	76·5
	Ecuador	17.89	7.59	76.7	3.247	6184	486.49	8.1	68.81	75·7
	Serbia	8.7	19.06	75.8	8.33	7359	641.03	8.5	51.62	81·9
	India	1393.41	6.57	69.3	6.497	2116	61.78	3.5	66.25	68·4
	Malaysia	32.78	7.18	75.9	8.165	11414	436.61	3.8	69.59	81·5
	China	1444.22	11.97	76.6	11.08	10004	535.13	5.4	66.38	75·4
	Iran	85.03	6.56	76.3	5.721	7282	470.43	8.7	52.25	77·5
	Turkey	85.04	8.98	77.3	4.906	9127	396.47	4.1	70.71	75·9
	Nigeria	211.4	2.74	54.2	8.868	2361	71.47	3.9	48.89	61·4
	Botswana	2.4	4.51	69.1	13.33	7961	481.53	5.8		73·9
	Ghana	31.73	3.14	63.7	12.402	2203	75.28	3.5		64·2
	Cameroon	27.22	2.72	58.8	11.887	1502	54	3.5		60·4
	Morocco	37.34	7.61	76.3	10.153	3282	174.22	5.3	45.81	63·0

GDP, Gross domestic product; HAQ, healthcare access and quality; HICs, high-income countries; LMICs, low- and middle-income countries.

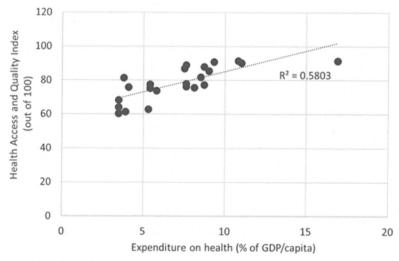

FIGURE 2 Preface description: This figure shows a plot. The y-axis is the health access and quality index out of 100, and the x-axis is the expenditure on health in the percentage of GDP per capita. The values are shown in blue circles, and a narrow dotted blue line indicates the linear regression of the plotted values. The dots and the line fitted show an ascending behavior. The R-squared value of this linear regression is shown to be 0.5803 to indicate how close data points are to the fitted line. *GDP*, Gross domestic product.

years of age in a country should correspond to higher death rates since stroke is associated with increased age. However, we see despite the fact that most LMICs have a smaller percentage of people over 65 years of age, the percentage of stroke deaths is typically higher suggesting that access issues and other factors may play a role in stroke care (Fig. 3).

The number of doctors and physical therapists per 1000 people in HICs versus LMICs is also concrete examples of healthcare disparities [7,8]. For example, in 2018, there were about 2.6 doctors for every 1000 persons in the United States, while for most LMICs in sub-Saharan Africa, there were between 0.05 and 0.8 doctors for every 1000 persons [7]. Similarly, there are about 0.66 physical therapists and 0.398 occupational therapists for every 1000 people in the United States, while there are about 0−0.0002 physical and occupational therapists for every 1000 people in most LMICs in sub-Saharan Africa [8]. Table 2 lists the numbers of healthcare personnel by featured country and population. Table 3 summarizes the average data from Tables 1 and 2 across HICs and LMICs.

If the number of trained medical and rehabilitation persons is limited in LMICs, then the need for technology-based solutions to bridge gaps in health care is even more urgent. So, if only 1 in 124,000 people are physical therapists in Africa where an estimate of 1.5 million stroke survivors live, then we can conclude that access to rehabilitation for stroke survivors is limited and will be infrequent, that is, not at the frequency recommended as standard

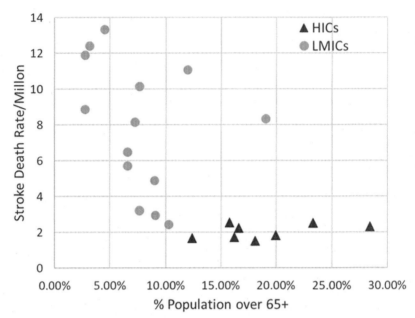

FIGURE 3 Preface description: This figure shows the stroke death rate in millions given the percentage of the population over 65 years of age in HICs and LMICs described in this book (Table 1). The data points for LMICs are shown in light gray circles, and those pertaining to the HICs are shown in dark gray triangles. The indicators for HICs show the percentage of population over 65 years of age approximately from 12% to 28%, and the death rate for all of them is around 2 cases per million of population. However, in the LMICs, the population over 65 years of age is approximately between 3% and 12%, and only in 1 case is around 19%, and the death rate for these countries is between 2 and 14 cases per million of population. *HICs*, High-income countries; *LMICs*, low- and middle-income countries.

of care defined by WHO [9,10]. The standard of care guidelines differs among countries making it even harder to clearly define the processes that lead to "good" quality of life (Chapter 2). Although the scientific literature offers 10 principles to elicit robust neuroplasticity (Chapters 1 and 5), these principles have not been consistently adopted in stroke guidelines for individual countries.

Faced with these disparities in stroke care, the use of therapy robots and assistive robots, appropriately designed for their environments, could bridge these care gaps and support the delivery of more frequent therapy and assistance to patients with stroke. In another scenario, since the use of mobile and cellular technologies is increasingly available in LMICs, new rehabilitation technology could also appropriately focus on capitalizing on these industry networks.

Robots are considered high-tech AT solutions that can leverage artificial intelligence, microcontrollers, microprocessors, and state-of-the-art mechanisms designed to create targeted therapy and assessment solutions. As a rehabilitation solution, robot-assisted therapy is assumed to be more pervasive in clinical

TABLE 2 Overview of healthcare and R&D personnel for countries in this book.

	Country	Population (million)	Number of physicians/ 1000	Number of PT/ 1000	Number of OT/ 1000
HICs	The United States	332.92	2.6	0.6634	0.398
	Canada	38.07	2.4	0.6650	0.379
	Italy	60.37	3.9	1.1721	0.033
	Spain	46.75	4.4	1.2790	0.167
	Japan	126.05	2.5	nr	0.593
	Australia	25.79	4.1	1.0820	0.667
	Republic of Korea	51.31	2.5	0.8524	0.196
	Israel	8.79	3.6	0.8152	0.523
LMICs	Mexico	130.26	2.4	nr	0.003
	Costa Rica	5.14	3.3	nr	nr
	Colombia	51.27	2.3	0.6657	0.082
	Ecuador	17.89	2.2	nr	nr
	Serbia	8.7	2.5	nr	nr
	India	1393.41	0.7	nr	0.0036
	Malaysia	32.78	2.3	nr	0.0549
	China	1444.22	2.2	nr	nr
	Iran	85.03	1.6	nr	0.0329
	Turkey	85.04	1.9	0.0775	0.0008
	Nigeria	211.4	0.4	nr	0.0002
	Botswana	2.4	0.4	nr	nr
	Ghana	31.73	0.2	nr	0.0001
	Cameroon	27.22	0.1	nr	nr
	Morocco	37.34	0.7	nr	nr

HICs, High-income countries; *LMICs*, low- and middle-income countries.

TABLE 3 Summary statistics in terms of averages across high-income countries (HICs) and low- and middle-income countries (LMICs) in this book.

Averages	HICs (in book)	LIMCs (in book)
Population (million)	86.26	237.59
People aged 60 + (%)	25%	12%
People aged 65 + (%)	19%	8%
Life expectancy at birth (years)	82.61	72.13
Stroke deaths (millions)	1.898651988	19.96115681
Stroke death/1,000,000	2.044	7.546333333
Number of physicians/1000	3.2500	1.5467
Number of physical therapists/1000	0.9327	0.3716
Number of occupational therapists/1000	0.3697	0.0222
GDP/capita (USD)	43491.88	6620.93
Health spending/capita (USD)	4681.87	389.45
Expenditure on health (% of GDP)	10.10	5.65
Expenditure on health (% of gov. exp.)	17.25	11.13
Health Care Index	74.90	61.87
Healthcare Access and Quality (HAQ) Index	89.30	72.75

settings in HICs and is not commonly being used in LMICs. By engaging colleagues worldwide, we set out to understand the pervasiveness of robot-assisted technologies globally and identify barriers and challenges to their use, and where possible identify opportunities to support easier uptake.

The book is organized into four main sections. The *first section* is a group of five chapters that covers fundamental concepts in stroke statistics and stroke care (Chapter 1), known guidelines used for stroke care worldwide in specific countries (Chapter 2), fundamentals in neurorehabilitation (Chapter 3), fundamentals of rehabilitation robotics (Chapter 4), and, finally, provides evidence for rehabilitation and assistive robotics (Chapter 5). After reading this section, readers will gain an appreciation of the complexity of the clinical and technological landscape in this field and the challenges and opportunities afforded by rehabilitation robotics.

The *second section* is a group of eight chapters each covering one of the following HICs: United States (Chapter 6), Canada (Chapter 7), Italy (Chapter 8), Spain (Chapter 9), Japan (Chapter 10), Australia (Chapter 11), South Korea (Chapter 12), and Israel (Chapter 13). The *third section* is a group of 15 chapters each covering one of the following LMICs: Mexico

(Chapter 14), Costa Rica (Chapter 15), Columbia (Chapter 16), Ecuador (Chapter 17), Serbia (Chapter 18), India (Chapter 19), Malaysia (Chapter 20), China (Chapter 21), Iran (Chapter 22), Turkey (Chapter 23), Nigeria (Chapter 24), Botswana (Chapter 25), Ghana (Chapter 26), Cameroon (Chapter 27), and Morocco (Chapter 28). Clinicians and engineers from each country came together to report on their country's healthcare system, their rehabilitation standards of care, stroke rehabilitation as practiced within the country, the processes used by those vested to develop and implement rehabilitation, and assistive robotics and challenges to the uptake of technologies within their healthcare infrastructure. By delving into sections two and three, readers will get to appreciate the similarities and differences within HICs and LMICs and across HICs and LMICs on subjects such as stroke incidence, stroke care, technology availability and use, and country-specific factors that impact disparities and access to services. Some of the results will be surprising. Chapters describe the health care for stroke survivors and the state of the art in robot-assisted therapy in selected HICs and offer contrasting insights into health care for stroke survivors in selected LMICs in terms of the use and/or barriers to the use of robot-assisted stroke rehabilitation, while authors highlight examples of the use of robot technology to bridge rehabilitation gaps in their countries.

The *fourth section* is a group of five chapters that together cover general barriers, best practices, and recommendations for the penetration of rehabilitation robotics in clinical practice. Chapter 29 discusses the need to consider psychosocial aspects of using robots to improve their uptake into clinical practice and their acceptability by patients. Chapter 30 focuses on providing human-centered guidelines to improve the usability and acceptability of robotic systems by both patients and clinicians and urges developers to implement metrics that quantify cost−benefit ratios and cost-effectiveness. Chapter 31 raises general concerns around global access to rehabilitation robot technologies and discusses this in the context of inclusivity, affordability, and other common concerns that may justify limiting or increasing access and use in low-resource settings. Finally, Chapter 32 looks forward into the future and discusses the potential growth of the field and key technical and medical advances that may unlock the doors currently barring worldwide use of robotics systems in stroke neurorehabilitation.

Michelle J. Johnson
Departments of Physical Medicine and Rehabilitation, Bioengineering, and Mechanical Engineering and Applied Mechanics, University of Pennsylvania, Philadelphia, PA, United States

Rochelle J. Mendonca
Department of Rehabilitation and Regenerative Medicine, Programs in Occupational Therapy, Columbia University, New York, NY, United States

References

[1] World Health Organization. Stroke, cerebrovascular accident [Internet], World Health Organization, <https://www.emro.who.int/health-topics/stroke-cerebrovascular-accident/index.html>; [cited 06.03.23].

[2] Feigin VL, Brainin M, Norrving B, Martins S, Sacco RL, Hacke W, et al. World Stroke Organization (WSO): Global Stroke Fact Sheet 2022. Int J Stroke 2022;17(1):18−29.

[3] Nichols-Larsen DS, Clark PC, Zeringue A, Greenspan A, Blanton S. Factors influencing stroke survivors' quality of life during subacute recovery. Stroke. 2005;36(7):1480−4.

[4] Barbay M, Diouf M, Roussel M, Godefroy O. Systematic review and meta-analysis of prevalence in post-stroke neurocognitive disorders in hospital-based studies. Dement Geriatric Cognit Disord 2018;46(5−6):322−34.

[5] Sexton E, McLoughlin A, Williams DJ, Merriman NA, Donnelly N, Rohde D, et al. Systematic review and meta-analysis of the prevalence of cognitive impairment no dementia in the first year post-stroke. Eur stroke J 2019;4(2):160−71.

[6] World Stroke Organization. Global Stroke Fact Sheet [Internet], World Stroke Organization, 2022, <https://www.world-stroke.org/assets/downloads/WSO_Global_Stroke_Fact_Sheet.pdf>; [cited 07.03.23].

[7] The World Bank. Physicians (per 1,000 people) | Data [Internet], <https://data.worldbank.org/indicator/SH.MED.PHYS.ZS?end = 2020&start = 1960>; [cited 08.03.23].

[8] The World Bank. Countries and Economics <https://data.worldbank.org/country>. [cited 07.03.23]

[9] World Health Organization. World report on disability [Internet], http://www.who.int, https://www.who.int/teams/noncommunicable-diseases/sensory-functions-disability-and-rehabilitation/world-report-on-disability>; [cited 07.03.23].

[10] Lindsay P, Furie KL, Davis SM, Donnan GA, Norrving B. World Stroke Organization global stroke services guidelines and action plan. Int J Stroke 2014;9(SA100):4−13.

Acknowledgments

First and foremost, we would like to thank all the stroke survivors and their families we have interacted with over the years, who inspired us to write this book. Additionally, we thank all the committed clinicians and innovative engineers across the world who, through their experiences and wisdom, led us to conceptualize this book.

We would like to thank all the authors who contributed to this book, without whose experience, knowledge, and patience this book would not have been possible. We appreciate their expertise in helping us understand neurorehabilitation and technology-assisted rehabilitation in high-, middle-, and low-income countries across the world. Each author brought a unique body of knowledge from both the clinical and engineering spheres to their chapters and has truly helped us develop a unique perspective.

We would like to thank Shafagh Keyvanian for her tireless and committed assistance throughout the process of completing this book. Your help is deeply appreciated.

We are also deeply thankful to our families for their relentless and never-ending support of us. Your encouragement and belief in us push us to want to do this work. Last, we are grateful for our continued and productive working relationship and friendship that allowed us to complete this book and continue our pursuit of equitable rehabilitation access across the world.

Part I

Fundamentals

Chapter 1

Stroke

Jakita Baldwin, Dixon Yang and Imama A. Naqvi
Columbia University Irving Medical Center, New York, NY, United States

Learning objectives

At the end of this chapter, the reader will be able to:
1. Understand the epidemiology and prevalence of stroke worldwide,
2. Understand healthcare systems for stroke and poststroke care, and
3. Identify disparities in care related to poststroke rehabilitation across countries with different socioeconomic status.

1.1 Introduction

The global incidence of stroke mortality declined by 36.2% between 1990 and 2016 according to the Global Burden of Disease Study [1]. Nevertheless, stroke remains a leading cause of death and disability-adjusted life years (DALYs). Some of this is attributable to overall increase in lifespans, the increase in aging populations, and uncontrolled secondary risk factors. Even so, the incidence of stroke in those under 65 years old has increased, especially in low- and middle-income countries (LMICs). Surprisingly, this is seen as people move from lower to higher socioeconomic status [1,2]. This is the opposite of what occurs in high-income countries (HICs) and may be due to an increase in stroke risk factors.

Stroke is a clinical diagnosis based on the sudden onset of focal neurological deficits that persist 24 hours or longer due to either ischemia, blockage of blood flow in a cerebral blood vessel, or hemorrhage into brain parenchyma, subarachnoid space, or ventricles. It is a noncommunicable disease associated with modifiable and nonmodifiable risk factors. Nonmodifiable risk factors include age, sex, race, ethnicity, family history of stroke, and cardiovascular disease. Stroke is traditionally considered a disease of the elderly. Conversely, a recent rise in stroke cases among those younger than 65 years has occurred with up to 25% increase in incidence worldwide [3]. This trend is most notable in LMICs such as India where

Rehabilitation Robots for Neurorehabilitation in High-, Low-, and Middle-Income Countries.
DOI: https://doi.org/10.1016/B978-0-323-91931-9.00026-8

12% of strokes affect those under 40 years old [3]. A similar trend is occurring in Russia and China where deaths attributable to stroke are higher than in Western countries [1]. Additionally, in 2016 the incidence of stroke in China exceeded 5.5 billion [1]. Women have a lower incidence of stroke than men. This difference may be due to protection conferred by estrogen at younger ages; however, above age 85 years, this trend reverses with more incidence among women [4]. Overall, the prevalence of stroke in women is higher, partially attributable to women generally living longer, having hypertension and atrial fibrillation more commonly than men. In the United States, African Americans and Latino Americans have between two and four times higher rates of stroke, stroke recurrence, and stroke-related deaths compared to non-Hispanic Whites (NHWs) [4]. Furthermore, after stroke, African Americans have less independence with ADLs at 1 year compared to their White American counterparts [4]. Hispanic Americans also have disparities in stroke compared to their White American counterparts. However, until recently, Hispanic Americans despite having higher rates of cerebrovascular risk factors and more limited access to care still maintained lower rates of mortality related to stroke [5]. This is known as the Hispanic Paradox. This survival advantage has lately been projected to lessen over time for unclear reasons. A recent study showed stagnation in the decline in stroke mortality specifically among Hispanic Americans [5]. From 2000 to 2010, lower mortality rates at 30 days and 1 year among Hispanic individuals disappeared. Based on this observation, it is also predicted that by 2028 there may even be a reversal in survival advantage [5]. Disparities in stroke occurrence and disability are seen globally with higher stroke risk, and stroke-related disability is seen in ethnic minorities in European countries [3]. Some of these differences may be attributable to differences in risk factor control.

Modifiable risk factors include hypertension, hyperlipidemia, diabetes mellitus, smoking, hypercoagulable state, obesity, sedentary lifestyle, and socioeconomic status. Hypertension is the most modifiable risk factor for stroke but remains uncontrolled in over half of stroke survivors [6]. In the United States, the prevalence of hypertension exceeds 20% in every state [4]. Furthermore, hypertension ranked in the top 5 risk factors that contributed to disease burden across the globe [1]. As of 2010, 31.1% of people worldwide had hypertension with prevalence of 28.5% in HICs compared to 31.5% in LMICs [7]. Obesity and metabolic syndrome linked to hyperlipidemia is also a significant risk factor for much of the US population with an expected rise in prevalence of 33% by 2030 [4]. Smoking is a behavioral risk factor, and in the National Health and Nutrition Examination Survey (NHANES), overall smoking prevalence among stroke survivors was 24% with secondary analysis showing no decrease in this prevalence over time [8]. Higher rates of tobacco use are associated with low socioeconomic status, sexual and gender minorities, and mental health disorders [4].

1.2 Healthcare structure and resources

Stroke care is divided into several phases of care: (1) primary prevention focusing on recognizing and reducing risk for stroke prior to first stroke occurrence, (2) acute phase of care involving emergent therapies, hospitalization, and secondary prevention which aims to reduce recurrent stroke risk by targeting modifiable risk factors, and (3) chronic phase with long-term post-stroke care involving tertiary prevention with shifted focus to rehabilitation strategies, maximizing functional status, and supporting patients and their caregivers.

There are clear guidelines for acute treatment of stroke after presentation to the emergency department. Acute treatments such as thrombolysis and thrombectomy for reperfusion are time-sensitive and mostly relevant for the first 24 hours after stroke symptom onset. Patients who could be eligible for these treatments often do not receive them due to delayed presentation to the hospital [9]. This occurs because of various barriers, one of which, relevant across all countries regardless of income, is not recognizing symptoms of stroke. This has sparked public awareness campaigns aimed at increasing public understanding. Even when stroke symptoms are identified early, patients often opt to contact or see their primary care physician first prior to having emergency medical attention, and this is true across high-, middle-, and low-income countries. Once symptoms are identified, having the infrastructure (i.e., ambulance, mobile stroke units, traffic organization) to quickly arrive at an acute care center as well as having access to specialists, appropriate diagnostic equipment, and therapeutic medications also prove to be barriers. In a study of stroke in African countries, among all countries surveyed, Tunisia had the highest amount at 43% of participants arriving at the hospital within 3 hours of onset of stroke symptoms [10]. Many LMICs do not have ambulance services or trained emergency medical professionals to operate an adequate number of these vehicles. Additionally, the lack of optimized traffic infrastructure and roads contributes to delayed presentations to the hospital. Once at the hospital, the percentage of patients receiving diagnostic imaging is low because of the lack of available technology. In Guatemala and Ecuador, for example, few places have comprehensive stroke centers [11]. In Ghana, one-third of hospitals do not have functional CT scanners during weekdays [11]. A study in India found only 12% of patients underwent any type of imaging when presenting for acute stroke [11].

The acute care model in HICs is hub and spoke, with comprehensive stroke centers and primary stroke centers (Fig. 1.1). This model serves as a central location for the most complicated stroke cases, with access to emergency therapies, stroke-trained clinicians, interventionalists, surgeons, researchers, and multidisciplinary teams for rehabilitation and social support in a stroke unit setting within a larger hospital. These hospitals often provide care for a certain geographic region and frequently accept transfers from

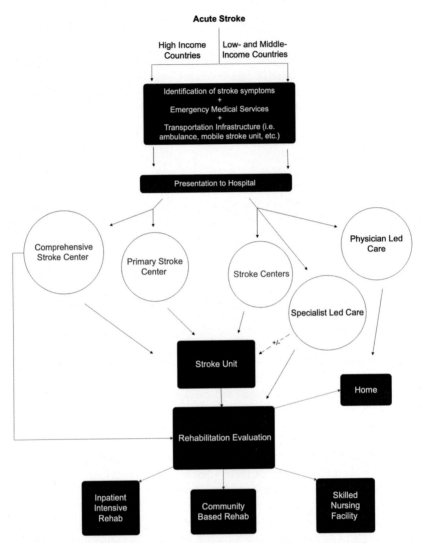

FIGURE 1.1   Systems of care: high-income versus low- and middle-income countries.

smaller hospitals unable to provide such advanced care. Primary stroke centers are similarly structured and have stroke specialists providing care within a stroke unit in the hospital; however, these centers are smaller, providing more local care, and do not have capability to care for more complex patients requiring specialized interventions [9]. More recently, endovascular-capable primary stroke centers provide acute therapies to reduce delay in treatment and relieve the workload on comprehensive centers. In high-income regions like North America and Europe, there recently has been the advent of mobile stroke units to assist with patient arrival for care [9]. These specialty ambulances are equipped with stroke specialists, CT scanners, and point-of-care laboratory testing that expedite time to diagnosis, appropriate therapies, and triaging to hospitals with specific capabilities. Tele-stroke often serves as an adjunct to these units. Once admitted to the hospital, placing stroke patients in a stroke unit has become a fundamental part of care in high-income countries and has been shown to improve stroke outcomes [9]. A stroke unit is a floor or wing within the hospital that provides stroke-specific care with a multidisciplinary team of stroke-trained physicians, nurses, rehabilitation therapists, and staff.

Differences in care are noted once patients leave acute stroke care. In the United Kingdom and Canada, there is an evidence-based system used for post-acute stroke care called early supported discharge (ESD) which has been shown to decrease hospital length of stay, enhance functional recovery, and improve patient and caregiver satisfaction [9]. This system is multidisciplinary involving all members of the healthcare team, social workers, and therapists. The COMPASS study was a recent trial to assess whether a standardized, multidisciplinary, post-acute care model like that observed in the United Kingdom and Canada is feasible and beneficial in the United States. However, it failed to show a significant effect on functional status at 90 days post-discharge, with less than half in the intervention arm receiving the full treatment dose [12,13]. Other studies outside of the United States have shown an impact of poststroke care interventions on risk factor reduction. In the Nurse-Based Age-Independent Intervention to Limit Evolution of Disease (NAILED) after stroke trial from Sweden, patients were placed either with nurse follow-up or usual care for blood pressure management. Results showed a systolic blood pressure reduction of 6.1 mmHg compared to 3.4 mmHg reduction in usual care [14]. If the technological capability is available and accessible, telemedicine for home-based post-acute stroke rehabilitation has also previously shown benefit [15].

Low- and middle-income countries utilize various models of care for poststroke patients (Fig. 1.1). Services include multidisciplinary teams at stroke centers, specialist-led care, and physician-led care [11]. Multidisciplinary care involves a team of specialty (i.e., stroke)-trained nurses and therapists and is led by a physician (neurologist or internist) with stroke training. It occurs in the setting of a stroke unit within the hospital. This model most closely resembles those utilized in high-income countries. These stroke centers often act as hubs for spoke hospitals like those found in HICs. Many smaller affiliated hospitals in rural areas can rely on the hub hospital for specialist services, including telemedicine for acute stroke assessments. Ideally, these spoke hospitals are located within a reasonable distance from the hub to facilitate transferring patients for escalation of care if needed. Specialist-led care is similar except that the physician (neurologist), nurses, and therapist are not required to have specialty stroke training.

This type of care may or may not take place at a stroke unit or center. Physician-led care is one where the leading physician is not formally trained in stroke and is also not a neurologist. Patients may not have access to therapies during the hospital stay, and it may not be set in a stroke unit. The primary goal of this physician-led model is to support the patient through the acute stroke period until they can discharge home for outpatient care. This model is used especially in resource-limited settings where access to stroke specialists is extremely limited.

Subacute and chronic stroke care transitions focus more on rehabilitation and functional independence within patients' communities. Ideally, functional assessments for rehabilitation needs are performed in the hospital after the patient is stable for discharge. These assessments then help with the transition to the appropriate disposition. Some patients recover with enough functional capacity to be discharged home with supportive outpatient community-based rehabilitation therapies either within their own homes or during the day at various community facilities while others may require more intensive therapies in an inpatient rehabilitation setting or long-term care at advanced care facilities. Others, either due to their pre-stroke functional status or due to current deficits related to stroke, may be discharged instead to a subacute rehabilitation facility often within a skilled nursing facility [9,11].

This figure depicts the systems of care for acute stroke starting with identifications of symptoms and activation of emergency medical services to presentation to the hospital. There is a comparison of different systems available in HICs versus LMICs. Typically, HICs have more access to comprehensive and primary stroke centers whereas LMICs often have physician- and specialist-led care and less stroke centers. Both can have stroke units and opportunities for rehabilitation evaluation. Disposition after the hospital is similarly disparate with more patients having access to facilities for rehabilitation such as inpatient- and community-based rehabilitation in HICs as well as skilled nursing facilities while patients in LMICs are more likely to be discharged home after stroke.

1.3   Healthcare disparities and challenges

Despite major advancements in acute stroke care systems, the burden of stroke increases worldwide [16]. Less attention has been paid to post-acute stroke care even though a substantial proportion of stroke survivors will have lasting physical, cognitive, and psychosocial deficits [17–19]. Barriers to adequate poststroke rehabilitation can exist throughout the care continuum in all types of healthcare systems. In HICs, a fragmented rehabilitation landscape and patient-level barriers to poststroke care can present significant obstacles to well-recovered stroke survivorship. While in LMICs, a severely limited rehabilitation workforce presents significant challenges [20].

During acute stroke hospitalization, the early introduction of therapy services in medically stable patients is generally safe, and those who initiate rehabilitation treatment sooner may have better outcomes [21]. In the United States, over a quarter of hospitalized stroke patients may not receive consistent physical and occupational therapies, partly due to a lack of system-level standardized practice guidelines [22]. At the patient level, socioeconomic disparities may contribute to delayed high-quality rehabilitation, even in high-resource settings with a universal healthcare system [23]. In a recent qualitative analysis, providers from a tertiary care center in the United States frequently cited access to care, health insurance barriers, and lack of standardized outcomes as obstacles to early stroke rehabilitation [24]. Access to inpatient rehabilitation after acute stroke hospitalization can vary widely in HICs, ranging from 10% to 60% of stroke survivors [25]. Published guidelines in HICs on eligibility for inpatient rehabilitation poststroke are not completely uniform, with some advising caution or excluding those with severe stroke or poor functional prognosis [26]. However, accurately identifying stroke survivors' functional prognosis is not entirely clear [27]. Furthermore, the choice and use of inpatient rehabilitation facility versus skilled nursing facility can be subjective, often incorporating demographic, clinical, and nonclinical factors, as there are no well-validated clinical prediction tools [28]. In the United States, socioeconomic factors may be associated with disparities in access to rehabilitation facilities [29]. Patient-level financial barriers which delay or limit choices, lack of cultural understanding and poorer communication from providers, and ineffective patient education may contribute to these disparities [30]. The transition to home for a stroke survivor can equally be challenging for both patients and caretakers. In the United States, given the highly heterogeneous landscape of rehabilitation services, patients often experience a high number of care transitions. This complexity adds to a fragmented transfer to outpatient care. Additionally, patient-specific gaps in social support, resources, and knowledge are contributing factors. Stroke survivors may not receive necessary support at home, adequate outpatient rehabilitation, or follow-up with neurologists [31].

LMICs may experience an even greater burden of stroke and poststroke disability than HICs yet face critically fewer poststroke care resources [20]. The WHO estimates that there can be fewer than ten skilled therapists per 1 million population in some LMICs [32]. Like HICs, the reported proportion of patients receiving adequate poststroke rehabilitation in LMICs is heterogeneous from 40% to 90% [33]. Providers also convey a lack of clear stroke rehabilitation guidelines [26]. International society guidelines may contain language or structures that are not easily implemented in certain LMICs, emphasizing that it is not necessarily one-size-fits-all [33]. Moreover, LMIC stroke centers may need greater interdisciplinary care, as discharge planning can often be without clear poststroke rehabilitation support [26]. This is largely due to the scarcity of therapists and rehabilitation centers in these healthcare systems [34]. Transitioning to outpatient care in LMICs, stroke survivors may face fewer community support systems, have more difficulty affording medications and follow-up therapy, and must navigate inaccessible environments for those with poststroke disability [33]. Stroke survivors are often not followed up due to access issues related to patient finances, lack of transport to clinics attached to distant hospitals, and physical barriers in rural settings [34]. Some stroke survivors may face cultural stigma related to dependency that may lead to social isolation and restricted societal roles poststroke [35]. Considering cost limitations in developing inpatient rehabilitation centers in these resource-limited settings, greater focus has been shifted toward community-based approaches [33,34].

1.4   Opportunities

Given the increasing burden of stroke and the fragmented landscape of post-acute stroke care, there is a need for a more patient-centered approach to transitioning care in stroke survivors. The 2019 American Stroke Association (ASA) policy statement made explicit evidence-based recommendations to improve systems for post-acute stroke care, which include recommendations for stroke systems to establish transitional support systems to ensure appropriate follow-up, education and training for the patient and caregivers, and comprehensive communication across the poststroke continuum of care. Further, the ASA recommended standardization of post-acute stroke care, while working in an interdisciplinary fashion to individualize care plans that may include referral to community services, reinforcing secondary prevention and self-management of stroke risk factors, and supporting lifestyle changes [36]. Specifically, the development of community-centered care networks has been the focus for many, in both HICs and LMICs. The COMPASS study, based in North Carolina, United States, evaluated a post-stroke transitional care model that was designed to integrate medical and community resources to support stroke survivors and caregivers in the community [37]. The intervention included follow-up telephone calls at 2, 30, and 60 days, a comprehensive 2-week follow-up visit, and an individualized care plan. Patients receiving intervention had clinically meaningful improvements in functional status, less disability, lower mortality, and were more satisfied with their poststroke care than those who did not. However, substantial difficulty in staff maintenance and interrupted interventions in over half of the participants were significant barriers to implementation, and the intention-to-treat analysis ultimately found no benefit. The Systemic Use of Stroke Averting Interventions (SUSTAIN) trial assessed a chronic care model in California, United States [38]. They demonstrated a reduction in systolic blood pressure in both arms, (17 mmHg vs 14 mmHg in usual care), but the difference was not significant. The follow-up trial was a multimodal coordinated care intervention, and against the systolic blood pressure reduction was not significant with only 14.5% of participants receiving the intended full dose of the intervention. Participation was better once transportation and financial barriers were addressed, suggesting an opportunity for improved accessibility [39]. To overcome these barriers, some have advocated for an expansion of quality metrics for poststroke care and stroke systems to incentivize resource allocation and commitment to improve poststroke care [31].

In LMICs, where resource constraints are greater, efforts to find low-cost improvements to post-acute stroke care are increasing. While each healthcare system's needs may vary, poststroke education may be an especially attractive first step. Some suggest the development of free-access stroke rehabilitation training standards to educate interdisciplinary providers with support from international professional organizations or universities [11]; however, this would require substantial investment and collaboration. Education should also extend to stroke survivors and caregivers. Like HICs, greater focus has been placed on community-based poststroke care given the limited inpatient rehabilitation services and skilled therapists. For some health systems, task shifting, or delegation of tasks to less specialized healthcare workforce such as community volunteers, may in part alleviate a lack of therapy services [11]. This involves the incorporation of community health workers to deliver supervision and therapy under the guidance of trained therapists. Community health workers additionally take on a greater role in health surveillance, social support, program referrals, and networking. Task shifting, however, can be taxing on caregivers and does not address physical environmental barriers or access to medication. Moreover, it may only be effective and safe in less complex patients [33,40]. Ultimately, collaboration with HICs is needed to develop effective well-tailored post-acute stroke care systems in LMICs.

1.5 Conclusion

Aging populations, modifiable risk factor differences, and changing regional socioeconomic climates contribute to stroke, remaining a leading cause of disability despite overall decline in stroke mortality. Healthcare systems have traditionally targeted increasing awareness of stroke symptoms to decrease time to activating medical services with faster presentation to hospital and receipt of emergency medical and interventional therapies. In HICs, healthcare systems focus on caring for stroke patients within stroke units at comprehensive stroke centers or primary stroke centers which have specialty-trained clinicians, interventionalists, rehabilitation services, and social workers. LMICs may not have the resources or specialists to replicate these models exactly and, however, do have multidisciplinary team care systems which most closely resemble those in HICs. In LMICs, there are specialist-led care teams which may take place in stroke units and physician-led care teams which have the capacity to medically support fewer complex patients usually without access to inpatient rehabilitation services. Uniform guidelines for and implementation of post-acute stroke care are often lacking worldwide and are magnified in low- and middle-income countries due to limited resources. In sum, this leads to a large proportion of stroke patients with compounded disability from lack of consistent access to rehabilitation services. Opportunities exist to address these disparities namely in developing multidisciplinary patient-centered approaches with more community-based interventions. This will require large investment and collaboration among both high-income and low- and middle-income countries.

Conflict of interest

No conflict of interest for all authors.

References

[1] Johnson CO, Nguyen M, Roth GA, Nichols E, Alam T, Abate D, et al. Global, regional, and national burden of stroke, 1990−2016: A systematic analysis for the global burden of disease study 2016. The Lancet Neurology 2019;18:439−58.

[2] Morgenstern LB, Kissela BM. Stroke disparities: Large global problem that must be addressed. Stroke 2015;46:3560−3.

[3] Katan M, Luft A. Global burden of stroke. Semin Neurol 2018;38:208−11.

[4] Benjamin EJ, Muntner P, Alonso A, Bittencourt MS, Callaway CW, Carson AP, et al. Heart disease and stroke statistics-2019 update: A report from the american heart association. Circulation 2019;139:e56−e528.

[5] Morgenstern LB, Brown DL, Smith MA, Sanchez BN, Zahuranec DB, Garcia N, et al. Loss of the mexican american survival advantage after ischemic stroke. Stroke 2014;45:2588−91.

[6] O'Donnell MJ, Chin SL, Rangarajan S, Xavier D, Liu L, Zhang H, et al. Global and regional effects of potentially modifiable risk factors associated with acute stroke in 32 countries (interstroke): A case-control study. Lancet 2016;388:761−75.

[7] Virani SS, Alonso A, Aparicio HJ, Benjamin EJ, Bittencourt MS, Callaway CW, et al. Heart disease and stroke statistics-2021 update: A report from the american heart association. Circulation 2021;143:e254−743.

[8] Parikh NS, Chatterjee A, Diaz I, Merkler AE, Murthy SB, Iadecola C, et al. Trends in active cigarette smoking among stroke survivors in the united states, 1999 to 2018. Stroke 2020;51:1656−61.

[9] Langhorne P, Audebert HJ, Cadilhac DA, Kim J, Lindsay P. Stroke systems of care in high-income countries: What is optimal? The Lancet 2020;396:1433−42.

[10] Urimubenshi G, Cadilhac DA, Kagwiza JN, Wu O, Langhorne P. Stroke care in africa: A systematic review of the literature. Int J Stroke 2018;13:797−805.

[11] Pandian JD, Kalkonde Y, Sebastian IA, Felix C, Urimubenshi G, Bosch J. Stroke systems of care in low-income and middle-income countries: Challenges and opportunities. The Lancet 2020;396:1443−51.

[12]   Duncan PW, Bushnell CD, Rosamond WD, Jones Berkeley SB, Gesell SB, D'Agostino Jr. RB, et al. The comprehensive post-acute stroke services (compass) study: Design and methods for a cluster-randomized pragmatic trial. BMC Neurol 2017;17:133.

[13]   Duncan PW, Bushnell CD, Jones SB, Psioda MA, Gesell SB, D'Agostino Jr. RB, et al. Randomized pragmatic trial of stroke transitional care: The compass study. Circ Cardiovasc Qual Outcomes 2020;13:e006285.

[14]   Ögren J, Irewall AL, Söderström L, Mooe T. Long-term, telephone-based follow-up after stroke and tia improves risk factors: 36-month results from the randomized controlled nailed stroke risk factor trial. BMC Neurol 2018;18:153.

[15]   Cramer SC, Dodakian L, Le V, See J, Augsburger R, McKenzie A, et al. Efficacy of home-based telerehabilitation vs in-clinic therapy for adults after stroke: A randomized clinical trial. JAMA Neurol 2019;76:1079–87.

[16]   Singh RJ, Chen S, Ganesh A, Hill MD. Long-term neurological, vascular, and mortality outcomes after stroke. Int J Stroke 2018;13:787–96.

[17]   Kwakkel G, Kollen BJ, van der Grond J, Prevo AJ. Probability of regaining dexterity in the flaccid upper limb: Impact of severity of paresis and time since onset in acute stroke. Stroke 2003;34:2181–6.

[18]   Ding MY, Xu Y, Wang YZ, Li PX, Mao YT, Yu JT, et al. Predictors of cognitive impairment after stroke: A prospective stroke cohort study. J Alzheimers Dis 2019;71: 1139–51.

[19]   Mitchell AJ, Sheth B, Gill J, Yadegarfar M, Stubbs B, Yadegarfar M, et al. Prevalence and predictors of post-stroke mood disorders: A meta-analysis and meta-regression of depression, anxiety and adjustment disorder. Gen Hosp Psychiatry 2017;47:48–60.

[20]   Langhorne P, O'Donnell MJ, Chin SL, Zhang H, Xavier D, Avezum A, et al. Practice patterns and outcomes after stroke across countries at different economic levels (interstroke): An international observational study. Lancet 2018;391:2019–27.

[21]   Paolucci S, Antonucci G, Grasso MG, Morelli D, Troisi E, Coiro P, et al. Early versus delayed inpatient stroke rehabilitation: A matched comparison conducted in italy. Arch Phys Med Rehabil 2000;81:695–700.

[22]   Kumar A, Adhikari D, Karmarkar A, Freburger J, Gozalo P, Mor V, et al. Variation in hospital-based rehabilitation services among patients with ischemic stroke in the united states. Phys Ther 2019;99:494–506.

[23]   Langagergaard V, Palnum KH, Mehnert F, Ingeman A, Krogh BR, Bartels P, et al. Socioeconomic differences in quality of care and clinical outcome after stroke: A nationwide population-based study. Stroke 2011;42:2896–902.

[24]   DiCarlo JA, Gheihman G, Lin DJ. Reimagining stroke rehabilitation and recovery across the care continuum: Results from a design-thinking workshop to identify challenges and propose solutions. Arch Phys Med Rehabil 2021;102:1645–57.

[25]   Zucker I, Laxer I, Rasooli I, Han S, Cohen A, Shohat T. Regional gaps in the provision of inpatient rehabilitation services for the elderly in israel: Results of a national survey. Isr J Health Policy Res 2013;227.

[26]   Eng JJ, Bird ML, Godecke E, Hoffmann TC, Laurin C, Olaoye OA, et al. Moving stroke rehabilitation research evidence into clinical practice: Consensus-based core recommendations from the stroke recovery and rehabilitation roundtable. Int J Stroke 2019;14: 766–73.

[27]   Guidelines for management of ischaemic stroke and transient ischaemic attack 2008. *Cerebrovasc Dis.* 2008;25:457-507

[28]   Simmonds KP, Burke J, Kozlowski AJ, Andary M, Luo Z, Reeves MJ. Rationale for a clinical trial that compares acute stroke rehabilitation at inpatient rehabilitation facilities to skilled nursing facilities: Challenges and opportunities. Arch Phys Med Rehabil 2021;.

[29]   Cao Y, Nie J, Sisto SA, Niewczyk P, Noyes K. Assessment of differences in inpatient rehabilitation services for length of stay and health outcomes between us medicare advantage and traditional medicare beneficiaries. JAMA Netw Open 2020;3:e201204.

[30]   Magwood GS, Ellis C, Nichols M, Burns SP, Jenkins C, Woodbury M, et al. Barriers and facilitators of stroke recovery: Perspectives from african americans with stroke, caregivers and healthcare professionals. J Stroke Cerebrovasc Dis 2019;28:2506–16.

[31]   Duncan PW, Bushnell C, Sissine M, Coleman S, Lutz BJ, Johnson AM, et al. Comprehensive stroke care and outcomes: Time for a paradigm shift. Stroke 2021;52: 385–93.

[32]   WHO Rehabilitation 2030: a call for action. The need to scale up rehabilitation. 2017; Available from: https://www.who.int/disabilities/care/Need-to-scale-up-rehab-July2018. pdf?ua=1 (cited 06.12.21).

[33]   Bernhardt J, Urimubenshi G, Gandhi DBC, Eng JJ. Stroke rehabilitation in low-income and middle-income countries: A call to action. Lancet 2020;396:1452–62.

[34]   Rhoda A, Cunningham N, Azaria S, Urimubenshi G. Provision of inpatient rehabilitation and challenges experienced with participation post discharge: Quantitative and qualitative inquiry of african stroke patients. BMC Health Serv Res 2015;15423.

[35]   Jenkins C, Ovbiagele B, Arulogun O, Singh A, Calys-Tagoe B, Akinyemi R, et al. Knowledge, attitudes and practices related to stroke in ghana and nigeria: A siren call to action. PLoS One 2018;13:e0206548.

[36]   Adeoye O, Nyström KV, Yavagal DR, Luciano J, Nogueira RG, Zorowitz RD, et al. Recommendations for the establishment of stroke systems of care: A 2019 update. Stroke 2019;50:e187–210.

[37]   Gesell SB, Bushnell CD, Jones SB, Coleman SW, Levy SM, Xenakis JG, et al. Implementation of a billable transitional care model for stroke patients: The compass study. BMC Health Serv Res 2019;19:978.

[38]   Cheng E, Cunningham W, Towfighi A, Sanossian N, Bryg R, Anderson T, et al. Randomized, controlled trial of an intervention to enable stroke survivors throughout the los angeles county safety net to "stay with the guidelines. ". Circulation. Cardiovascular quality and outcomes 2011;4:229–34.

[39]   Towfighi A, Cheng EM, Ayala-Rivera M, Barry F, McCreath H, Ganz DA, et al. Effect of a coordinated community and chronic care model team intervention vs usual care on systolic blood pressure in patients with stroke or transient ischemic attack: The succeed randomized clinical trial. JAMA Netw Open 2021;4:e2036227.

[40]   Gupta S, Rai N, Bhattacharrya O, Cheng AYY, Connelly KA, Boulet LP, et al. Optimizing the language and format of guidelines to improve guideline uptake. CMAJ 2016;188(14):e362–8.

Chapter 2

Rehabilitation guidelines for stroke care: a worldwide perspective

Cori Barger[1], Rochelle J. Mendonca[2], Michelle J. Johnson[3] and Beatriz Coto-Solano[4]

[1]Department of Rehabilitation Medicine, NewYork-Presbyterian Brooklyn Methodist Hospital, Brooklyn, NY, United States, [2]Department of Rehabilitation and Regenerative Medicine, Programs in Occupational Therapy, Columbia University, New York, NY, United States, [3]Departments of Physical Medicine and Rehabilitation, Bioengineering, and Mechanical Engineering and Applied Mechanics, University of Pennsylvania, Philadelphia, PA, United States, [4]Rehabilitation Service, Hospital R. A. Calderón Guardia, Caja Costarricense de Seguro Social, San José, Costa Rica

Learning objectives

By the end of this chapter, readers will be able to:

1. Understand how clinical practice guidelines inform stroke neurorehabilitation in high-, middle-, and low-income countries and
2. Understand and articulate the need for standardized practice guidelines to inform stroke rehabilitation worldwide.

2.1 Introduction

The United Nations Convention on the Rights of Persons with Disabilities (UNCRPD) supports the right to rehabilitation; however, access to and quality of rehabilitation services around the globe vary [1]. Rehabilitation guidelines can help to bridge this gap by providing evidence-based recommendations on both a broader systems level, to guide governments and policymakers, and on an individual intervention level, to guide clinicians, patients, caregivers, and family members.

Clinical practice guidelines help direct decision-making and establish best practices in clinical settings. The authors of these guidelines, typically teams of experts associated with organizations or departments of health,

Rehabilitation Robots for Neurorehabilitation in High-, Low-, and Middle-Income Countries.
DOI: https://doi.org/10.1016/B978-0-323-91931-9.00001-3

rigorously review and synthesize the literature to recommend evidence-based best practice as practice guidelines. However, the availability, quality, and implementation of practice guidelines vary from country to country across the globe [2]. The unique structure and capacity of each country's health system, however, determine the extent to which stroke guidelines can be successfully implemented. This chapter describes stroke rehabilitation guidelines from different countries and highlights potential areas to expand the implementation of stroke rehabilitation guidelines in low- and middle-income countries.

2.2 Guidelines

2.2.1 The World Health Organization, the World Stroke Organization, and the European Stroke Organization

Organizations such as the World Health Organization (WHO), the World Stroke Organization (WSO), and the European Stroke Organization (ESO) provide recommendations and action plans to improve rehabilitation for stroke patients in many countries (Table 2.1).

Rehabilitation services in many countries are often under-resourced and underdeveloped. The lack of access to rehabilitation services can lead to long-term complications for individuals who need those services the most. In recognition of this, WHO launched the Rehabilitation 2030 initiative in 2017 with member states, development partners, and civil society. The initiative declares a need to improve rehabilitation services around the world despite limited practical guidance available on how to do so, and it recognizes rehabilitation as an essential health service for all populations that should be integrated into all levels of healthcare. WHO subsequently published the *Rehabilitation in Health Systems* recommendations to provide information to government leaders and policymakers to extend and strengthen rehabilitation service delivery. The recommendations focus on integrating rehabilitation services into health systems, improving the quality of care via multidisciplinary rehabilitation teams and specialized rehabilitation units, and promoting

TABLE 2.1 Supranational and international guidelines and action plans.

Organization	Guideline
World Health Organization	Rehabilitation in Health Systems Document [3]
World Stroke Organization	Global Stroke Services Guideline and Action Plan [4]
European Stroke Organization	European Stroke Action Plan 2018–2030 [5]

equitable and affordable access to assistive products and rehabilitation services. Despite the quality of these recommendations, many countries face barriers such as limited finances, lack of knowledge and understanding of the benefits of rehabilitation by policymakers, lack of administrative structure, or an insufficient rehabilitation workforce which prohibit recommendations from being implemented successfully [6].

International nongovernmental organizations such as the World Stroke Organization (WSO), an international network of stroke specialists, and the European Stroke Organization (ESO), with individual and organizational members across Europe, work to promote institutional changes and improve stroke management and treatment across many countries. WSO published the first-ever global stroke guideline in 2014, targeted at low- to middle-income countries, to recommend evidence-based practice across the care continuum [4]. Key recommendations include treatment in specialized stroke rehabilitation units, initial assessments to develop individualized rehabilitation plans, repetitive and intensive task training, adaptive training, education for healthcare workers and families, fall prevention, and treatment for spasticity, contractures, and aphasia. The European Stroke Action Plan (ESAP 2018−2030) sets target goals for 2030 across Europe. Overarching targets include access to dedicated stroke units and established national plans to address all elements along the chain of care. Domain-specific targets include guaranteeing access to early rehabilitation within stroke units, provision of early supported discharge, community-based physical fitness programs, plan for community-based rehabilitation for stroke patients upon discharge, and 3-month, 6-month, and annual review of the rehabilitation for patients and caregivers [5].

2.2.2 National guidelines

2.2.2.1 High-income countries

The United States, Canada, Australia, South Korea, Japan, Spain, Italy, and Israel are classified as high-income countries, those with a gross national income per capita greater than $12,696 as of 2021, by the World Bank [7]. Many of these countries, aside from Italy and Israel, published country-specific guidelines for stroke neurorehabilitation. In Italy, guidelines focus on intravenous thrombolysis and intra-arterial interventions in acute ischemic stroke; in Israel, there is a national stroke registry, but stroke guidelines are lacking. The guidelines from the other countries surveyed within this income category lack uniformity, but there are notable similarities across recommendations (Table 2.2).

Country-specific stroke guidelines are commonly published by health ministries or national stroke-related organizations. Consequently, they may be more tailored to the health systems within the countries in which they are published. Stroke guidelines from high-income countries commonly span the

TABLE 2.2 High-income countries' stroke rehabilitation guidelines.

Country	Organization	Guideline
United States	American Heart Association	Guidelines for Adult Stroke Rehabilitation and Recovery [8]
Canada	Heart and Stroke Foundation, Canada	Canadian Stroke Best Practice Recommendations [9]
Australia	Stroke Foundation	Australian and New Zealand Clinical Guidelines for Stroke Management [10]
South Korea	Korean Society for Neurorehabilitation	Clinical Practice Guidelines for Stroke Rehabilitation in Korea [11]
Japan	Committee for Guidelines for Management of Aneurysmal Subarachnoid Hemorrhage, Japanese Society on Surgery for Cerebral Stroke	Evidence-based Guidelines for the Management of Aneurysmal Subarachnoid Hemorrhage [12]
Spain	Ministry of Health, Social Policy, and Equality	Clinical Practice Guidelines for the Management of Stroke in Primary Health Care [13]
Italy	Italian Stroke Organization	SPREAD Italian Guidelines for stroke. Indications for carotid endarterectomy and stenting [14]
Israel	Ministry of Health	The National Stroke Registry [15]

continuum of care, from acute care to home- or community-based care after hospital discharge, with a consensus for stroke patients to be treated by an interdisciplinary team in a specialized stroke unit in the hospital. An assessment of the severity of stroke and rehabilitation needs is typically recommended within 48 hours. The standard amount of interdisciplinary rehabilitation recommended by many high-income countries is 3 hours/day, 5 days/week; early mobilization is recommended after the first 24 hours. Task-specific training is strongly recommended to improve upper-extremity function, whereas repetitive, task-oriented mobility training and treadmill-based gait training are recommended for lower-limb motor rehabilitation. Additionally, robotic-assisted devices and virtual reality are recommended to rehabilitate stroke survivors with difficulty walking.

Screenings for cognitive deficits and poststroke depression are generally recommended, along with cognitive rehabilitation to improve attention and memory. For spatial neglect, it is commonly recommended that patients are assessed using validated tools, and education is provided on treatment recommendations. Spatial neglect interventions with consensus between two or more high-income countries include visual scanning, mirror therapy, eye patching, mental practice, and virtual reality; conflicting recommendations exist for prism adaptations, limb activation, and repetitive transcranial magnetic stimulation. Intensive speech and language therapy, computer-based treatment, and group treatment are recommended for aphasia as early as tolerated. Individually tailored interventions for patients with dysarthria and apraxia of speech are recommended.

Many guidelines include recommendations on managing complications from stroke, including dysphagia, pain, edema, bowel and bladder incontinence, falls, and deep vein thrombosis (DVT). Recommendations for driving, sexuality, return to work, recreation, and leisure activity are commonly included in high-income countries' guidelines as well. Additionally, guidelines frequently address early supported discharge, where patients are provided interdisciplinary rehabilitation in the home in lieu of a longer hospital stay. It is less common for guidelines to have recommendations on preventing skin breakdown, contractures, seizures, and poststroke osteoporosis.

2.2.2.2 Low- and middle-income countries

In a systematic review, Yaria and colleagues (2021) identified country-specific stroke guidelines in low- and middle-income countries published between 2010 and 2020. Low- and middle-income countries did not produce guidelines that met the standards for guideline development, and few demonstrated the transparency of high-income countries; they also neglected to reach a broad target audience and did not cover the full range of stroke services, including surveillance, prevention, acute care, and rehabilitation [2].

Upper-middle-income countries' gross national income per capita ranges from $4,096 to $12,695, and lower-middle-income countries' gross national income per capita range from $1,046 to $4,095 (4a). The upper- and lower-middle-income countries surveyed for this chapter include Mexico, Costa Rica, Colombia, Ecuador, Serbia, Bosnia and Herzegovina, India, Malaysia, China, Iran, Turkey, Nigeria, Botswana, Ghana, Cameroon, and Morocco. Less than one-third of these countries, including three upper-middle-income (Mexico, Colombia, and China) and two lower-middle-income (Cameroon and India) countries, produced stroke neurorehabilitation guidelines, whereas one country (Turkey) published only recommendations for patients with poststroke dysphagia (Table 2.3). No low-income countries were surveyed.

Guidelines from upper- and lower-middle-income countries lack uniformity; however, there is significant overlap in the recommendations from these countries. For instance, it is recommended that patients are treated in designated stroke units by multidisciplinary teams, although the makeup of

TABLE 2.3 Upper- and lower-middle-income countries' stroke rehabilitation guidelines.

Country	Organization/authors	Guideline
Mexico	Mexican Social Security Institute Directorate of Medical Benefits	Diagnóstico y Tratamiento Temprano de la Enfermedad Vascular Cerebral Isquémica en el segundo y tercer nivel de atención [16]
Colombia	Ministerio de Salud y la Protección Social in partnership with Universidad Nacional de Colombia	Clinical Practice Guidelines (CPGs) for the diagnosis, treatment, and rehabilitation of acute arterial ischemic stroke in adults over age 18 [17]
China	Chinese Stroke Association	Guidelines for the clinical management of cerebrovascular disorders [18]
Cameroon	North West Region (NWR) Best Practices in Stroke Rehabilitation Group	Best Practice Guidelines for the Management and Rehabilitation of Stroke in the North West Region of Cameroon [19]
India	Government of India's Directorate General of Health Services Ministry of Health and Family Welfare	Guidelines for Prevention and Management of Stroke [20]
Turkey	Umay et al. (2021)	Best Practice Recommendations for Stroke Patients with Dysphagia: A Delphi-Based Consensus Study of Experts in Turkey-Part II: Rehabilitation [21]

these teams varies between countries. Starting rehabilitation as early as possible once the patient is medically stable is generally recommended; some countries specify beginning within 24–48 hours, either bedside or away from bedside, while others recommend within 72 hours. The intensity of therapy depends on the patients' tolerance level; recommendations vary between 40- and 60-minute sessions anywhere from two times per week to two times per day. Early mobilization with functional use of the affected side is generally recommended. Interventions to treat spatial neglect are not as commonly recommended, apart from in China and India, where they recommend specific interventions ranging from pen-and-paper tasks and eye patching to prism adaptations and virtual reality. Screenings for cognitive impairment, poststroke depression, and communication disorders such as aphasia are recommended, but there is limited guidance and no consensus on the treatment for aphasia. The prevention of skin breakdown is addressed through patient education, positioning, and the use of a special air bed. Recommendations to prevent contractures include positioning, active exercise training, functional electrical stimulation, drugs, and Botox. A screening for swallowing function with dysphagia treatment is commonly included in the guidelines. Additionally, shoulder pain, bladder and bowel management, DVT, and fall prevention are commonly included in recommendations within this income level; recommendations on poststroke driving, sexuality, return to work, poststroke osteoporosis, and recreation and leisure are not.

Guidelines for stroke neurorehabilitation are lacking in many upper- and lower-middle-income countries. For instance, Costa Rica's National Guide for Stroke Management and Creation of Unified Stroke Units, published in 2010, does not include guidance on poststroke early rehabilitation. Rather than referencing clinical practice guidelines, hospitals establish their own methods to rehabilitate stroke patients and define the timeframe in which to begin therapy after stroke, including the duration and frequency of sessions [22,23]. In Serbia, existing guidelines focus on the prevention of stroke in patients with atrial fibrillation, and poststroke rehabilitation recommendations are incomplete [24]. In Malaysia, stroke guidelines [25] do not cover stroke neurorehabilitation, but evidence published on stroke rehabilitation between 2000 and 2014 in Malaysia suggests the effectiveness of stroke units and the potential for innovative rehabilitation programs [26]. Additionally, there are currently no published stroke rehabilitation guidelines in Morocco, Nigeria, Ghana, Iran, Botswana, Bosnia and Herzegovina, or Ecuador.

2.3 Future directions

Without guidelines or strategies to implement evidence-based practice, countries may rely on health provider discretion and potentially compromise the quality and effectiveness of care [27]. Expanding access to current best practice recommendations can help establish standards in stroke care in low- and middle-income countries, where the burden of stroke is high [28]. We can look to Australia's "living" clinical guidelines to illustrate the effectiveness of an online, publicly accessible platform that is updated as new evidence becomes available. On a broader scale, a "living" worldwide central stroke rehabilitation database could expand the availability of rehabilitation evidence in several languages while remaining flexible enough to be applied within country-specific contexts, where resources allow. It must be noted, however, that barriers to improving stroke rehabilitation services lie beyond a country's ability to access rehabilitation guidelines. Countries may need guidance and resources to strengthen the leadership and governance of their health systems, develop strategic planning, and integrate changes across systems, among other things. WHO is addressing this need on a global scale, but there is more work to be done. A concerted effort to prioritize the standards of stroke care on a national and subnational level within low- and middle-income countries is still needed. Given these challenges, it is understandable why many countries surveyed still do not have guidelines for the use of technologies within stroke rehabilitation. We recommend that as experts consider guidelines they also work to incorporate best practices for the use of technologies, robotics or otherwise.

Acknowledgments

We thank the authors of the country-based chapters for providing content for this chapter.

Conflict of interest

None.

References

[1] United Nations. Convention on the Rights of Persons with Disabilities (CRPD) [Internet]. New York: United Nations; 2006 [updated 2022 May 6; cited 2022 September 29]. Available from: https://www.un.org/development/desa/disabilities/convention-on-the-rights-of-persons-with-disabilities.html

[2] Yaria J, Gil A, Makanjuola A, Oguntoye R, Miranda J, Lazo-Porras M, et al. Quality of stroke guidelines in low- and middle-income countries: a systematic review. Bull World Health Org 2021;99(09):640–652E.

[3] World Health Organization. Rehabilitation in Health Systems [Internet]. Geneva: World Health Organization; 2017 [cited 2022 Sep 28]. 92 p. Available from: https://apps.who.int/iris/handle/10665/254506

[4] Lindsay P, Furie KL, Davis SM, Donnan GA, Norrving B. World stroke organization global stroke services guidelines and action plan. Int J Stroke 2014;9(SA100):4–13.

[5] Norrving B, Barrick J, Davalos A, Dichgans M, Cordonnier C, Guekht A, et al. Action plan for stroke in Europe 2018–2030. Eur Stroke J 2018;3(4):309–36.

[6] World Health Organization. Rehabilitation in Health Systems [Internet]. Geneva: World Health Organization; 2017. Available from: https://www.who.int/publications/i/item/9789241549974

[7] Hamadeh, N., Van Rompaey, C., Metreau, E. New World Bank country classifications by income level: 2021–2022 [Internet]. Worldbank.org; 2021 July 01. Available from: https://blogs.worldbank.org/opendata/new-world-bank-country-classifications-income-level-2021-2022

[8] Winstein CJ, Stein J, Arena R, Bates B, Cherney LR, Cramer SC, et al. Guidelines for adult stroke rehabilitation and recovery: a guideline for healthcare professionals from the American Heart Association/American Stroke Association. Stroke 2016;47(6):e98. Available from: https://www.ahajournals.org/doi/10.1161/STR.0000000000000098.

[9] Teasell R., Salbach N.M. Part One: Rehabilitation and Recovery following Stroke; 2019;127.

[10]   Stroke Foundation. Living Clinical Guidelines for Stroke Management [Internet]. Australia: Stroke Foundation; 2022 [cited 2022 September 29]. Available from: https://informme.org.au/guidelines/living-clinical-guidelines-for-stroke-management

[11]   Kim DY, Kim YH, Lee J, Chang WH, Kim MW, Pyun SB, et al. Clinical practice guideline for stroke rehabilitation in Korea 2016. Brain Neurorehabil 2017;10(Suppl 1):e11.

[12]   Committee for Guidelines for Management of Aneurysmal Subarachnoid Hemorrhage, Japanese Society on Surgery for Cerebral Stroke. Evidence-Based Guidelines for the Management of Aneurysmal Subarachnoid Hemorrhage English Edition. Neurol Med Chir (Tokyo). 2012;52(6):355−429.

[13]   Ministry of Science and Innovation. Clinical Practice Guidelines for the Management of Stroke in Primary Health Care. Madrid: Estilo Estugraf Impresores, S.L.; 2009. 228 p.

[14]   Setacci C, Lanza G, Ricci S, Cao PG, Castelli P, Cremonesi A, et al. SPREAD Italian Guidelines for stroke. Indications for carotid endarterectomy and stenting. J Cardiovasc Surg (Torino) 2009;50(2):171−82.

[15]   State of Israel Ministry of Health. National Stroke Registry [Internet]. Israel: Israel Center for Disease Control; 2022 [cited 2022 September 29]. Available from: https://www.health.gov.il/English/MinistryUnits/ICDC/disease_Registries/Pages/stroke.aspx

[16]   Instituto Mexicano del Seguro Social. Diagnóstico y Tratamiento Temprano de la Enfermedad Vascular Cerebral Isquémica en el segundo y tercer nivel de atención. Mexico: 2017. 84 p.

[17]   Colombian General Social Security System for Health. Clinical Practice Guideline for diagnosis, treatment and rehabilitation of Acute Ischemic Stroke in population over the age of 18 [Internet]. Colombia: Colombian General Social Security System for Health; 2025 [cited 2022 September 29]. Available from: https://extranet.who.int/ncdccs/Data/COL_D1_stroke_CLINICAL_PRACTICE_GUIDELINE.pdf

[18]   Zhang T, Zhao J, Li X, Bai Y, Wang B, Qu Y, et al. Chinese Stroke Association guidelines for clinical management of cerebrovascular disorders: executive summary and 2019 update of clinical management of stroke rehabilitation. Stroke Vasc Neurol 2020;5 (3):250−9.

[19]   The NWR Best Practices in Stroke Rehabilitation Group. Best Practice Guidelines for the Management and Rehabilitation of Stroke in the North West Region of Cameroon. Cameroon; 2013. 44 p.

[20]   Directorate General of Health Services Ministry of Health and Family Wellfare. Guidelines for Prevention and Management of Stroke. India: Government of India; 2019. 90 p.

[21]   Umay E, Eyigor S, Ertekin C, Unlu Z, Selcuk B, Bahat G, et al. Best practice recommendations for stroke patients with dysphagia: a Delphi-based consensus study of experts in Turkey-Part II: rehabilitation. Dysphagia. 2021;36(5):800−20.

[22]   Ouriques Martins SC, Sacks C, Hacke W, Brainin M, de Assis Figueiredo F, Marques, et al. Priorities to reduce the burden of stroke in Latin American countries. Lancet Neurol 2019;18(7):674−83.

[23]   Fernández H., Carazo K., Henríquez F., Montero M., Valverde A. Guía Nacional de Manejo del Evento Cerebrovascular y Creación de Unidades de Ictus Unificadas. San José, Costa Rica; 2010.

[24]   Potpara TS, Lip GYH, Blomström-Lundqvist C, Chiang CE, Camm AJ. Viewpoint: stroke prevention in recent guidelines for the management of patients with atrial fibrillation: an appraisal. Am J Med 2017;130(7):773−9.

[25] The Stroke Council of the Malaysian Society of Neurosciences. Clinical Practice Guidelines Management of Ischaemic Stroke. 3rd edn Malaysia: Medical Development division of the Ministry of Health; 2020. p. 156. February.

[26] Cheah WK. A review of stroke research in Malaysia from 2000 − 2014. Med J Malaysia. 2016;71:58−69.

[27] Woolf SH, Grol R, Hutchinson A, Eccles M, Grimshaw J. Clinical guidelines: potential benefits, limitations, and harms of clinical guidelines. BMJ. 1999;318(7182):527−30.

[28] Mukherjee D, Patil G. Epidemiology and the Global Burden of Stroke. World Neurosurg 2011;76(6S):S85−90.

Chapter 3

Fundamentals of neurorehabilitation

Dawn M. Nilsen[1,2], Lauren Winterbottom[1,3] and Carly Goldberg[1,2]
[1]Programs in Occupational Therapy, Columbia University, New York, NY, United States,
[2]Department of Rehabilitation, New York Presbyterain, New York, NY, United States, [3]Columbia
Department of Rehabilitation and Regenerative Medicine, New York, NY, United States

Learning objectives

At the end of this chapter, the reader will be able to:

1. Describe the common diagnoses treated by practitioners specializing in neurorehabilitation,
2. Describe the clinical sequelae associated with stroke,
3. Describe the continuum of stroke care in the United States,
4. Identify common stroke assessments and evidence-based interventions, and
5. Identify the key principles of experience-dependent neuroplasticity and motor learning that serve as the cornerstones of neurorehabilitation.

3.1 Introduction

Neurodegenerative diseases and traumatic injuries to the nervous system cause a myriad of neurological impairments that negatively impact engagement in daily life activities. These impairments and functional limitations are most frequently addressed by an interdisciplinary team of healthcare practitioners specializing in neurorehabilitation. Of the diagnoses treated by neurorehabilitation specialists, stroke is by far the most common.

According to the World Health Organization, 15 million people per year will suffer a stroke. About 10 million of these individuals will survive, with 5 million experiencing permanent disability [1]. Consistent with these findings, stroke is considered the number one cause of long-term disability in the United States [2]. Stroke survivors typically require extensive rehabilitation services in order to reduce dysfunction and maximize independence in

Rehabilitation Robots for Neurorehabilitation in High-, Low-, and Middle-Income Countries.
DOI: https://doi.org/10.1016/B978-0-323-91931-9.00013-X

ADLs. These services are typically provided across a continuum of care that begins soon after the onset of stroke and continues several months or years poststroke onset [3]. The purpose of this chapter is to provide the reader with an overview of the fundamentals of neurorehabilitation, with a specific focus on stroke rehabilitation.

3.2 Overview of common diagnoses

The following section provides a brief overview of some of the most common diagnoses treated by practitioners specializing in the field of neurorehabilitation. While the specific clinical signs and symptoms differ among the varying diagnoses, commonalities include impairments in sensorimotor, cognitive, and psychosocial functions that limit engagement in ADLs.

3.2.1 Progressive neurodegenerative diseases

Neurodegenerative disease is an umbrella term for a variety of diseases that attack the neurons within the central nervous system. Most of these diseases have no cure, and treatment is aimed to either slow the progression and/or relieve the symptoms caused by the disease. The prognosis for this variety of diseases is varied based on specific diagnosis as well as the age of onset [4]. Rehabilitative services are aimed at preventing decline and maximizing independence in ADLs.

Dementias, such as Alzheimer's disease and Lewy body dementia, are one class of neurodegenerative diseases that primarily affect mental functioning. Although the dementias initially and primarily affect cognition, the progression of these diseases ultimately affects one's ability to participate in ADLs, including functional mobility [5,6].

Parkinson's disease (PD) and Huntington's disease (HD) are movement disorders that result from the degeneration of select areas of the basal ganglia. In addition to non-motor signs, those with PD display resting tremor, rigidity, and bradykinesia [7], while those with HD present with chorea (involuntary and irregular, writhing-like movement) [8]. Postural instability and loss of balance are common in both disease processes impacting functional mobility [7,8].

Amyotrophic lateral sclerosis is a motor neuron disease. Degeneration of motor neurons in the brain and spinal cord produces a combination of upper and lower motor neuron signs. It is a very rapidly progressing disease, with a typical life expectancy of 3–5 years [9].

Demyelinating diseases, such as multiple sclerosis and Guillain–Barre syndrome, are not always progressive in nature. Both are characterized by the demyelination of nerve fibers resulting in impaired nerve conduction. This most commonly leads to pain, vision loss, muscle weakness, muscle stiffness, muscle spasms, and/or decreased bladder/bowel control [10].

3.2.2 Traumatic injuries

Traumatic injuries commonly requiring treatment by neurorehabilitation specialists include brain injury, spinal cord injury, and stroke. Neurological impairments caused by these injuries can affect both physical and cognitive abilities and impair one's ability to function in ADL [11].

Traumatic brain injury (TBI) is caused by an external force on the head, disrupting typical brain functioning. While falls are the most common source of TBI in adults, motor vehicle crashes and assaults are other common ways TBIs are caused [12]. In addition to the initial insult to the brain, secondary complex pathophysiological mechanisms may continue long after the injury resulting in impairments in both cognitive and motor functions [13]. Patients with TBI typically require extensive rehabilitation services to maximize independence in ADLs.

Spinal cord injuries can result in a variety of clinical presentations based on the level and completeness of the cord injury. Cognition is typically spared in these patients; however, physical impairments can involve any or all limbs in addition to the trunk. Impairments can include respiratory failure or neurogenic bowel and/or bladder. As care following a spinal cord injury has significantly improved, treatment is now concerned with optimizing QOL rather than extending survival [14]. Rehabilitation following spinal cord injury involves not only the physical rehabilitation of limbs and bodily functions but also the psychosocial impact of a life-changing injury.

As indicated earlier, stroke is by far the most common diagnosis treated by neurorehabilitation specialists. Depending on the location of the stroke, stroke survivors can experience impairments in sensory processing, motor functions, speech and language functions, cognition and perception, and emotional expression and regulation [3]. Because of these deficits, many stroke survivors are unable to independently carry out basic self-care activities or manage their home and community environments necessitating extensive rehabilitation services. The following section will present typical clinical signs and symptoms associated with stroke.

3.3 Clinical sequelae of stroke

A stroke in one hemisphere of the brain often results in contralateral weakness (hemiparesis) or paralysis (hemiplegia) of the body. For example, a stroke involving the right side of the brain often results in left hemiplegia, while a stroke on the left side of the brain results in right hemiplegia. In addition to weakness, there may be additional upper motor neuron signs, such as hyperreflexia, hypertonicity, and spasticity [3]. In addition to these motor deficits, a variety of other impairments may result including the following [3]:

- sensory impairments (e.g., hemianesthesia; pain syndromes)
- cognitive and perceptual impairment (e.g., executive functions, memory, praxis)

- visual disturbances (e.g., visual field deficits, inattention)
- behavioral changes (e.g., emotional lability, depression)
- difficulty swallowing
- speech and language function impairment (e.g., aphasia, dysarthria)

Impairments from stroke can create significant challenges for stroke survivors during participation in ADLs and meaningful life roles. Stroke survivors may experience challenges when completing ADLs such as getting dressed, bathing, or moving from one place to another. Participation in life situations such as home management, employment, hobbies, social engagement, or leisure activities is also commonly affected. Independence in ADLs is influenced by the severity of the stroke, cognitive impairments, and age. Stroke survivors who require inpatient rehabilitation may have more difficulty regaining independence in daily activities such as bathing and stairclimbing within the first 3 months and may continue to have difficulty with ADLs, home management, and leisure activities after 1 year [15,16]. These limitations may continue to persist for stroke survivors long term [17]. For working-age adults, returning to employment may be challenging and can be predicted by cognitive ability, neurological impairments, and independence in daily activities [18]. Stroke survivors may also have difficulty participating in other important life roles, including parenting and social activities [19,20]. Over two-thirds of stroke survivors require rehabilitative services after discharge from the hospital [21].

3.4 Key components of stroke rehabilitation

Stroke rehabilitation is complex and multifaceted. Its primary focus is to help the patient recover lost functions and to maximize independence in the performance of valued activities. Published rehabilitative guidelines exist across various countries [21−23], and the reader is referred to these for review. The following sections will provide an overview of poststroke rehabilitation in the United States [21].

3.4.1 Interdisciplinary team and continuum of care

Rehabilitative services are typically provided by an interdisciplinary healthcare team that consists of an MD, RN, therapy services (OT, PT, SLP), and SW. Typically, services are provided in various settings across a continuum of care as follows [21]:

- Acute care hospital setting: the focus of care is on stabilizing the patient and preventing further damage; OT, PT, and SLP are started as tolerated.
- Post-acute care: care provided after an initial acute hospital stay; intensity and goals of therapy vary based on the setting.

- Inpatient rehabilitation facility: hospital-level care with most intense level of therapy (minimum of 3 hours of OT, PT, and SLP per day for at least 5 days per week); the goal is to maximize recovery and independence in ADLs.
- Skilled nursing facility (sub-acute rehabilitation): less intense therapy; admission to this level of care may be due to the patient's inability to tolerate intensive services or to maintain/prevent decline in function.
- Long-term care (nursing home): long-term residential care for individuals who are unable to live in the community.
- Long-term care hospital: provides extended medical and rehabilitative care to patients with complex medical needs.
- Home care: rehabilitative care provided to stroke patients who have been discharged home after an acute hospital stay and are homebound.
- Outpatient clinic: rehabilitative care provided to stroke patients who have been discharged home and are able to access the community (not homebound).

Evidence suggests that organized interdisciplinary care reduces the likelihood of long-term disability, enhances recovery, and increases independence in ADLs [21].

3.4.2   Assessment

Assessment is an important part of the rehabilitative process and involves all members of the healthcare team. Assessment allows the team to (1) gather information about the patient's current functional status; (2) identify factors that support or hinder independence; (3) generate targeted outcomes; and (4) evaluate the effectiveness of the treatment program [24]. There is no single assessment that meets the needs of all stroke survivors. In fact, there are a variety of standardized and valid measures that evaluate body structure/function, activity limitations, participation restrictions, and QOL [21]. The patient's stage of recovery and rehabilitative setting are factors that are considered when selecting appropriate assessments. Examples of common standardized measures used to assess stroke survivors are as follows [1,21,24]:

- Global measure of impairment: National Institutes of Stroke Scale, World Health Organization Disability Assessment Schedule.
- Cognitive screening: Neurobehavioral Cognitive Status Examination; Mini-Mental Test; Montreal Cognitive Assessment Test.
- Speech/language impairments: Western Aphasia Battery; Boston Naming Test
- Visual neglect: Behavioral Inattention Test; Line Cancelation Test; Catherine Bergego Scale.
- Depression: Beck Depression Inventory II; Hamilton Depression Scale.

- Sensorimotor impairments: Fugl-Meyer Assessment; Chedoke McMaster Stroke Assessment.
- UE function: Action Research Arm Test; Wolf Motor Function Test; Box and Blocks.
- Balance: Berg Balance Scale; Functional Reach Test; Postural Assessment Scale for Stroke.
- Mobility: Timed Up and Go; 6-Minute Walk Test.
- ADL/IADL; Functional Independence Measure, Barthel Index; Frenchay Activity Index; ADL-CAT.
- Participation: Stroke Impact Scale; Activity Specific Balance Confidence Scale.
- QOL: Stroke Adapted Sickness Impact Profile; Stroke Specific QOL.

A proper assessment of the patient allows the treatment team to establish an appropriate plan of care, and it guides the intervention process.

3.4.3 Intervention

Intervention planning is a collaborative process that involves the patient, care partners, and the healthcare team. The focus of intervention varies depending on the patient, stage of recovery, and rehabilitative setting. When determining treatment plans, practitioners will often combine aspects of *preventative, restorative, compensatory,* and *adaptive* approaches [24]. Preventative approaches focus on preventing secondary complications (e.g., contractures, pain, edema, rotator cuff tear, falls, etc.) from developing. An example of a preventative intervention is splinting of the hand to prevent contractures or passive range of motion exercise to maintain joint mobility [24].

The restorative approach focuses on remediation of impairments and restoration of function to previous levels. This approach is typically used with stroke survivors early in the recovery process. Examples of restorative approaches will be presented later in this section. Compensation is aimed at education and instruction of alternative techniques used to maximize function in the absence of typical movement. An example of compensation is the use of the nondominant limb to feed oneself when the dominant limb is too weak to support self-feeding. The compensatory approach is more appropriate for clients who are later poststroke and/or for those who do not regain sufficient recovery of function with remediation approaches. Finally, adaptation involves changing the activity or the environment in order to make task performance easier. The goal of adaptation is to make the activity achievable for the client and often involves the use of adaptive equipment and various technologies to allow successful task performance [24]. An example of the application of the adaptive approach is the use of a rocker knife that allows a patient to cut food with one hand.

Restorative, compensatory, and adaptive approaches involve the patient relearning skills that were lost or learning new skills to compensate for lost function. This skill learning requires practice. Setting up optimal practice conditions that promote learning is an important part of the neurorehabilitative process. Thus neurorehabilitation intervention approaches are guided by principles derived from neuroscience literature and studies investigating motor skill acquisition. The subsequent section provides an overview of these key principles.

3.4.3.1   Key principles of neuroplasticity and motor learning

Our ability to learn is dependent on the adaptive capacity of the brain, otherwise known as neuroplasticity. According to Kleim and Jones [25], the ability of neurons and other brain cells to change their structure and function in response to behavioral training is termed "experience-dependent neural plasticity." Based on an extensive review of neuroscience literature, they identified ten key principles of "experience-dependent neural plasticity" that are used to guide rehabilitation approaches and optimize functional outcomes. These principles are as follows [25]:

- Use it or lose it: failure to drive specific brain functions can lead to functional degradation.
- Use it and improve it: training that drives a specific brain function can lead to an enhancement of that function.
- Specificity: the nature of the training experience dictates the nature of the plasticity.
- Repetition matters: induction of neural plasticity requires repetition.
- Intensity matters: induction of plasticity requires sufficient training intensity.
- Time matters: different forms of plasticity occur at different times during training.
- Salience matters: the training experience must be sufficiently salient to induce plasticity.
- Age matters: training-induced plasticity occurs more readily in younger brains.
- Transference: plasticity in response to one training experience can enhance the acquisition of similar behaviors.
- Interference: plasticity in response to one experience can interfere with the acquisition of other behaviors.

Building on this previous work, Maier and colleagues [26] identified 15 principles that should be considered when establishing an effective neurorehabilitative approach [26]:

- Massed/repetitive practice: work episodes with very brief or no rest periods.

- Spaced (distributed) practice: rest periods between repetitions or training sessions.
- Dosage: ill-defined; the amount of training that induces learning; exact dose-response for different therapies at different stages poststroke needs to be determined.
- Task-specific practice: practice of the task; conditions of practice shape the internal sensorimotor representation of the skill learned.
- Goal-oriented practice: emphasis is placed on achieving the goal rather than the movement patterns used to achieve the goal.
- Variable practice: providing variability within a training sequence (variability of practice) or by randomizing the presentation of different training sequences (contextual interference); leading to better skill retention and generalization.
- Increasing difficulty: difficulty inherent in the task itself (nominal task difficulty) and difficulty of the training relative to the skill of the performer (functional task difficulty); just-right challenge; shaping or graded practice.
- Multisensory stimulation: integration of various modalities requires probabilistic estimations to enhance perception; multimodal stimulation training may help patients recover from unimodal deficits.
- Rhythmic cueing: various sensory modalities may aid motor skill learning via neuroentrainment.
- Explicit feedback/knowledge of results: verbal, terminal, and augmented feedback about goal achievement.
- Implicit feedback/knowledge of performance: feedback given about movement execution via verbal descriptions, demonstrations, or replays of recordings.
- Modulate effector selection: use of the impaired limbs as opposed to unimpaired limbs.
- Action observation/embodied practice: observation of another performing a task action might facilitate movement execution and motor learning by facilitating the excitability of the motor system.
- Mental practice/motor imagery: simulation of actions without overt behavior primes the motor system and may be beneficial for motor learning.
- Social interaction: enriched environments with social interaction may enhance activity.

As illustrated in the next section, many evidence-based treatment interventions for stroke survivors incorporate these principles.

3.4.3.2   Examples of evidence-based treatment strategies to reduce disability and improve motor function after stroke

Evidence suggests that interventions grounded upon the previously presented principles reduce impairment and improve function. Examples of these evidence-based interventions appear below.

3.4.3.2.1   Task-oriented training

Treatment strategies for stroke must be targeted to focus on the specific functional deficits and goals of the individual. Task-oriented training (TOT) involves the practice of the specific task that the individual needs to perform. The difficulty of the task is graded to provide an appropriate challenge, and practice must occur with enough intensity, frequency, and duration so that motor skills can be relearned by the individual [21]. TOT has been shown to improve lower extremity function, gait, and balance for stroke survivors through repetitive practice and circuit training [27]. Additionally, repetitive practice of sit-to-stand interventions may help stroke survivors improve their time and symmetry in performing the task [28]. TOT has also been applied to upper limb interventions. Constraint-induced movement therapy (CIMT) involves massed repetitive practice with the weaker upper limb while the stronger limb is intentionally restrained, and the intervention has been shown to improve arm motor function [29]. Bilateral upper-limb training may also produce similar improvements in the weaker limb as compared to CIMT [21].

3.4.3.2.2   Strategies to augment task-oriented training

Cognitive strategies, such as motor imagery training, action observation, mirror therapy, and problem-solving techniques, have been developed for stroke survivors to help enhance the effects of TOT. As mentioned previously, motor imagery training involves mentally visualizing oneself performing a specific task without actually completing it physically and may help improve balance, mobility, and upper-limb function for stroke survivors [30]. Action observation and mirror therapy are adjunct interventions that activate the mirror neuron system of the brain to enhance motor recovery. In action observation, the stroke survivor watches a video of an able-bodied individual performing a motor task and may then be asked to practice the action they observed. This technique has been shown to improve arm and hand function, walking, and performance of daily activities [31,32]. Mirror therapy involves using the stronger limb to perform movements while watching its reflection in a mirror that is positioned in front of the weaker side, creating an illusion that the weaker side is moving more normally. Mirror therapy has been shown to improve upper and lower limb motor function, mobility, and activities of daily living for stroke survivors [33,34]. Finally, emerging evidence indicates that the use of problem-solving techniques in addition to task-oriented training may help stroke survivors generalize learning to untrained functional tasks [35].

3.4.3.2.3   Technology interventions

Interventions that use technology, such as robotics, virtual reality, and electrical stimulation, have also been investigated for use with stroke survivors.

Low-tech adaptive equipment and various one-handed devices can help stroke survivors with hemiparesis to perform daily activities with limited use of one hand. Reachers, button hooks, one-handed nail clippers, and adaptive kitchen tools are just a few examples of available options. Mobility devices (e.g., walker, cane, wheelchair) and durable medical equipment (e.g., tub transfer bench, shower chair, commode) are also commonly prescribed to assist stroke survivors when performing functional mobility and ADLs. Robotic interventions include exoskeletons devices that directly control specific joints of the human body and end-effector devices that apply forces to a distal point on the targeted limb. These interventions may improve motor control and muscle strength for the upper limb but have not been shown to be more effective than conventional therapy for upper-limb capacity, walking ability, or ADLs performance [36,37]. Virtual reality and interactive gaming interventions in addition to usual care may also provide stroke survivors more opportunities to practice simulated activities and can improve ADLs performance [38]. Physical modalities such as neuromuscular electrical stimulation may be a reasonable adjunct therapy for stroke survivors who have limited volitional movement, foot drop, or shoulder subluxation [21]. Generally, these interventions are adjunctive and should be considered in addition to function-based rehabilitation programs.

3.5   Conclusion

Neurorehabilitation is a dynamic field directed at reducing dysfunction and improving the health and well-being of those with impairments of the nervous system. Although practitioners work with patients with varying diagnoses, stroke is by far the most common. Stroke rehabilitation in the United States is carried out by an interdisciplinary team and is administered across a continuum of care. The goal of this care is to reduce impairments and maximize independence in ADLs. Specialists utilize a variety of assessments and interventions that are grounded in current best evidence. As the fields of neuroscience and motor learning advance, so too will the field of neurorehabilitation. The remaining chapters in this book provide a more detailed look at some or more of these interventions from the perspectives of high-, middle-, and low-income countries.

References

[1] World Health Organization. Stroke, Cerebrovascular Accident. http://www.emro.who.int/health-topics/stroke-cerebrovascular-accident/index.html. [accessed 08.07.21.]

[2] Benjamin EJ, Virani SS, Callaway CW, et al. Heart disease and stroke statistics-2018 update: a report from the American Heart Association. Circulation 2018;137(12): e67—e492.

[3]   Nilsen DM, Gillen G. Cerebrovascular accident . Philadelphia: Wolters Kluwer In: Powers Dirette D, Gutman SA, editors. Occupational therapy for physical dysfunction. Lippincott: Williams, & Wilkins; 2020. p. 735−64.

[4]   Bertram L, Tanzi RE. The genetic epidemiology of neurodegenerative disease. J Clin Invest 2005;115(6):1449−57.

[5]   Lane CA, Hardy J, Schott JM. Alzheimer's disease. Eur J Neurol 2018;25(1):59−70.

[6]   Taylor JP, McKeith IG, Burn DJ, Boeve BF, Weintraub D, Bamford C, et al. New evidence on the management of Lewy body dementia. Lancet Neurol 2020;19(2):157−69.

[7]   Beitz JM. Parkinson's disease: a review. Front Biosci (Sch Ed) 2014;6(6):65−74.

[8]   Novak MJ, Tabrizi SJ. Huntington's disease. BMJ. 2010;3:40 Jun 30.

[9]   Morris J. Amyotrophic lateral sclerosis (ALS) and related motor neuron diseases: an overview. Neurodiagn J. 2015;55(3):180−94.

[10] Rosenthal JF, Hoffman BM, Tyor WR. CNS inflammatory demyelinating disorders: MS, NMOSD and MOG antibody associated disease. J Investig Med 2020;68(2):321−30.

[11] World Health Organization. Neurological Disorders: Public Health Challenges. World Health Organization; 2006. https://www.who.int/publications/i/item/9789241563369 [accessed 12.14.21]

[12] Centers for Disease Control and Prevention. Report to congress on traumatic brain injury in the United States: epidemiology and rehabilitation. National Center for Injury Prevention and Control; 2015. 72 pages.

[13] Masel BE, DeWitt DS. Traumatic brain injury: a disease process, not an event. J Neurotrauma 2010;27(8):1529−40.

[14] Simpson LA, Eng JJ, Hsieh JT. Wolfe and the spinal cord injury rehabilitation evidence (SCIRE) research team DL. The health and life priorities of individuals with spinal cord injury: a systematic review. J Neurotrauma 2012;29(8):1548−55.

[15] Morone G, Paolucci S, Iosa M. In what daily activities do patients achieve independence after stroke? J Stroke Cerebrovasc Dis [Internet] 2015;24(8):1931−7. Available from: https://doi.org/10.1016/j.jstrokecerebrovasdis.2015.05.006.

[16] Hartman-Maeir A, Soroker N, Ring H, Avni N, Katz N. Activities, participation and satisfaction one-year post stroke. Disabil Rehabil 2007;29(7):559−66.

[17] Gadidi V, Katz-Leurer M, Carmeli E, Bornstein NM. Long-term outcome poststroke: predictors of activity limitation and participation restriction. Arch Phys Med Rehabil [Internet] 2011;92(11):1802−8. Available from: https://doi.org/10.1016/j.apmr.2011.06.014.

[18] Edwards JD, Kapoor A, Linkewich E, Swartz RH. Return to work after young stroke: a systematic review. Int J Stroke 2018;13(3):243−56.

[19] Harris GM, Prvu Bettger J. Parenting after stroke: a systematic review. Top Stroke Rehabil [Internet] 2018;25(5):384−92. Available from: http://www.ncbi.nlm.nih.gov/pubmed/29607739.

[20] Foley EL, Nicholas ML, Baum CM, Connor LT. Influence of environmental factors on social participation post-stroke. Behav Neurol [Internet] 2019;2019:2606039. Available from: http://www.ncbi.nlm.nih.gov/pubmed/30800187.

[21] Winstein CJ, Stein J, Arena R, Bates B, Cherney LR, Cramer SC, et al. Guidelines for adult stroke rehabilitation and recovery: a guideline for healthcare professionals from the American Heart Association/American Stroke Association. Stroke. 2016;1−169.

[22] Teasell R, Salbach NM, Foley N, Mountain A, Cameron JI, Jong AD, et al. Canadian stroke best practice recommendations: rehabilitation, recovery, and community

participation following stroke. Part one: rehabilitation and recovery following stroke; Update 2019. Int J Stroke 2020;7:763–88 Oct;15.

[23] Stroke Foundation of Australia Clinical Guidelines for Stroke Management 2017. https:// informme.org.au/en/Guidelines/Clinical-Guidelines-for-Stroke Management (accessed 07.03.21).

[24] Wolf T, Nilsen DM. Occupational therapy practice guidelines for adults with stroke. Maryland: AOTA Press; 2015. p. 11–8.

[25] Kleim JA, Jones TA. Principles of experience-dependent neuroplasticity: implications for rehabilitation after brain damage. J Speech Lang Hear Res 2008;51:S225–39 Feb.

[26] Maier M, Ballester BR, Verschure PFMJ. Principles of neurorehabilitation after stroke based on motor learning and brain plasticity mechanisms. Front Syst Neurosci 2019;13:74 Dec.

[27] Jeon B-J, Kim W-H, Park E-Y. Effect of task-oriented training for people with stroke: a meta-analysis focused on repetitive or circuit training. Top Stroke Rehabil [Internet] 2015;22(1):34–43. Available from: http://www.ncbi.nlm.nih.gov/pubmed/25776119.

[28] Pollock A, Gray C, Culham E, Durward BR, Langhorne P. Interventions for improving sit-to-stand ability following stroke. Cochrane Database Syst Rev [Internet] 2014;5: CD007232. Available from: http://www.ncbi.nlm.nih.gov/pubmed/24859467.

[29] Corbetta D, Sirtori V, Castellini G, Moja L, Gatti R. Constraint-induced movement therapy for upper extremities in people with stroke. Cochrane Database Syst Rev 2015;2017(9).

[30] Guerra ZF, Lucchetti ALG, Lucchetti G. Motor imagery training after stroke: a systematic review and meta-analysis of randomized controlled trials. J Neurol Phys Ther [Internet] 2017;41(4):205–14. Available from: http://www.ncbi.nlm.nih.gov/pubmed/28922311.

[31] Borges LR, Fernandes AB, Melo LP, Guerra RO, Campos TF. Action observation for upper limb rehabilitation after stroke. Cochrane Database Syst Rev [Internet] 2018;10: CD011887. Available from: http://www.ncbi.nlm.nih.gov/pubmed/30380586.

[32] Peng T-H, Zhu J-D, Chen C-C, Tai R-Y, Lee C-Y, Hsieh Y-W. Action observation therapy for improving arm function, walking ability, and daily activity performance after stroke: a systematic review and meta-analysis. Clin Rehabil [Internet] 2019;33 (8):1277–85. Available from: http://www.ncbi.nlm.nih.gov/pubmed/30977387.

[33] Louie DR, Lim SB, Eng JJ. The efficacy of lower extremity mirror therapy for improving balance, gait, and motor function poststroke: a systematic review and meta-Analysis. J Stroke Cerebrovasc Dis [Internet] 2019;28(1):107–20. Available from: http://www.ncbi. nlm.nih.gov/pubmed/30314760.

[34] Yang Y, Zhao Q, Zhang Y, Wu Q, Jiang X, Cheng G. Effect of mirror therapy on recovery of stroke survivors: a systematic review and network meta-analysis. Neuroscience [Internet] 2018;390:318–36. Available from: http://www.ncbi.nlm.nih.gov/pubmed/29981364.

[35] McEwen S, Polatajko H, Baum C, Rios J, Cirone D, Doherty M, et al. Combined cognitive strategy and task-specific training improve transfer to untrained activities in subacute stroke: an exploratory randomized controlled trial. Neurorehabil Neural Repair [Internet] 2015;29(6):526–36. Available from: http://www.ncbi.nlm.nih.gov/pubmed/25416738.

[36] Veerbeek JM, Langbroek-Amersfoort AC, van Wegen EEH, Meskers CGM, Kwakkel G. Effects of robot-assisted therapy for the upper limb after stroke. Neurorehabil Neural Repair [Internet] 2017;31(2):107–21. Available from: http://www.ncbi.nlm.nih.gov/pubmed/7597165.

[37] Lo K, Stephenson M, Lockwood C. Effectiveness of robotic assisted rehabilitation for mobility and functional ability in adult stroke patients: a systematic review. JBI Database Syst Rev Implement Rep [Internet] 2017;15(12):3049−91. Available from: http://www.ncbi.nlm.nih.gov/pubmed/29219877.

[38] Laver KE, Lange B, George S, Deutsch JE, Saposnik G, Crotty M. Virtual reality for stroke rehabilitation. Cochrane Database Syst Rev [Internet] 2017;11:CD008349. Available from: http://www.ncbi.nlm.nih.gov/pubmed/29156493.

Chapter 4

Fundamentals of neurorehabilitation robotics (engineering perspective)

Vincent Crocher and Justin Fong
Department of Mechanical Engineering, University of Melbourne, Parkville, VIC, Australia

Learning objectives (minimum 2)

At the end of this chapter, the reader will be able to:

1. Understand the different robot structures and functionalities available for neurorehabilitation;
2. Understand the necessary engineering trade-off in the design of such devices;
3. Understand the current challenges in designing and controlling such robots for effective neurorehabilitation.

4.1 Introduction

4.1.1 What's in a rehabilitation robot?

Robots can be defined as actuated mechanical structures (i.e., devices capable of exerting force or producing movement) with programmable actuation which can be controlled based on sensory inputs from its environment.

Even if no formal robot taxonomy exists, we can still locate rehabilitation robots within the subfamily of robotic manipulators, as opposed to other distinct types such as drones or mobile robots. Robotic manipulators aim to manipulate an object and/or a tool and are often seen in industrial plants and assembly lines. Rehabilitation robots can be further considered part of the family of devices geared toward physical human—robot interaction (pHRI), which have specific safety constraints and interaction abilities required due to direct physical interaction with the human body.

The need for a smooth physical interaction requires that these systems provide a good trade-off between compliance (the ability to follow smoothly

Rehabilitation Robots for Neurorehabilitation in High-, Low-, and Middle-Income Countries.
DOI: https://doi.org/10.1016/B978-0-323-91931-9.00027-X

39

user movements) and rigidity (being able to enforce movements when the user is passive). This desired behavior is specific to pHRI and has consequences on both the mechanical design and the control of the system.

4.1.2   Origins and rationale

The adoption and development of rehabilitation robots have primarily been driven by the need to increase therapy intensity and quantity—two key factors which are demonstrably linked with recovery. The potential for robots to tirelessly physically assist and correct patients' movements to achieve this goal has been pursued since the early 1990s. Their controllability and reactivity differentiate them from their passive counterparts—which can provide passive weight support or act as simple joysticks—and continuous passive motion (CPM) machines, also encountered in rehabilitation. This allows the adjustment of the physical interaction to patient needs but also to the specific rehabilitation task while preserving the patient's intention.

This chapter thus explores the building blocks of achieving this goal. Our discussion starts with the mechanical design of rehabilitation robots and continues with their controllers, attempting to translate the principles of neurorehabilitation into defined control laws and finally discuss their user interfaces, critical to patients' engagement and therapists' interaction.

4.2   Mechanical design: what does the robot look like and what can it do?

The physical manifestation of a rehabilitation robot can come in all sorts of shapes and sizes, designs and structures. While many options exist, there are a number of important characteristics, which increase the ease or effectiveness of their abilities to perform its function as a robot in a clinical environment. Of particular interest are the force control capabilities of the device, the complexity of the device, and the portability and footprint of the device. We discuss these in more detail below:

- Force control: The ideal behavior of a rehabilitation device is to possibly render every physical interaction behavior—from very rigid, high-force interactions to fully compliant ones (i.e., the ability to feel and react to the user's physical input and ultimately make the feel of the robot disappear). This constitutes the main specificity and difficulty in the design of pHRI systems and, indeed, rehabilitation robots.
- Device complexity: While a somewhat subjective concept, the complexity of a device's mechanical structure may be evaluated through metrics such as the number and shape of (moving or not moving) parts. While a more complex device may be able to affect a larger quantity of different movements, it also has clear implications for manufacturability and cost

as well as usability—for example, increasing setup time. Typically, exoskeleton-type systems, which have more components running closer to the limb, provide a higher level of control than their manipulator counterparts, but at the cost of more adjustments and a higher number of joints.

- Portability and footprint: The portability and footprint of any clinical device are increasingly important concepts. A smaller, more portable device may (1) decrease the amount of space required in the clinic when in use, (2) decrease the space requirements in a clinic when *not* in use, and/or (3) eliminate the requirement for any clinical space through use outside clinical settings, such as the home.

With the vast majority of devices either targeting the upper extremities, that is, the arms, or the lower extremities, that is, the legs, but not both, this section introduces different structures per their targeted body segment. Given that several hundred robotic systems dedicated to rehabilitation have been developed and presented in the scientific literature over the last 30 years (refer to [1] for UE systems and [2] for LE ones), this section does not intend to provide an exhaustive list of the different design approaches, but instead describes the main categories of devices, selecting the most notable examples.

4.2.1   Systems for the upper extremity

Robots targeting the upper limb have to deal with the versatility of use and functions, as well as the high redundancy of the UE. As such, these devices generally do not attempt to address this problem as a whole but are dedicated to subparts of the anatomy: hand and fingers, wrist or gross upper-limb movements (shoulder and/or elbow) and can be restricted to some functions by limiting the practice workspace. No single type of UE device clearly demonstrates a clinical advantage compared to others [3] but are rather complementary in their approach. This thus constitutes a relevant classification between the set of available robots and their selection.

4.2.1.1   Planar manipulanda

Due to their relative design simplicity, end-effector devices, also referred to as manipulanda, are the largest category and oldest type of robots used for UE rehabilitation. Most of these devices consist of a handle and/or forearm support attached to the patient's hand and/or wrist and allow movement and force interaction at the hand level. They allow practice and control of gross UE movements in two or three dimensions. Most common are two-dimensional (or planar) systems, which allow movements of the hand in a transverse (horizontal) plane while providing direct physical support of the arm against gravity.

The MIT-Manus is the most notable example of a planar manipulandum, developed in 1992 at MIT [4] and commercialized as the InMotion Arm.

It introduced a state-of-the-art design for pHRI and proposed exercises coupled to game activities displayed on a computer screen (Fig. 4.1). Planar manipulanda can be of smaller size and made transportable to ease their use in hospitals and clinics (where available floor space can be limited) or allow home-based use, such as the H-MAN [5], commercialized by Articares.

Addressing gross UE movements only, these systems are particularly appropriate for early stage of recovery and rehabilitation of patients with more severe movement deficits. They are currently the go-to systems due to their relatively simple design and lower cost.

4.2.1.2 Three-dimensional manipulanda

To extend the capabilities of planar manipulandum to a larger workspace and set of possible movements, manipulanda with a 3D workspace have also been developed and evaluated. They add an extra degree-of-freedom (DoF), allowing movement against gravity in a 3D workspace in front of the patient.

The first of this kind, the MIME [6], was naturally derived from an existing industrial robotic manipulator (a PUMA 560) equipped with a force sensor (Fig. 4.2A). This was followed by the ADLER project, leveraging a robot developed for haptic interaction: the HapticMaster (Motek Medical, Netherlands). This device shows that such systems can allow real-world task practice—by freeing the patient hand—while providing various levels of assistance [7]. While these two systems took an admittance approach, the impedance one, with higher mechanical complexity, is also possible and shows similar interaction behaviors (Section 4.3.1). This is the case of the

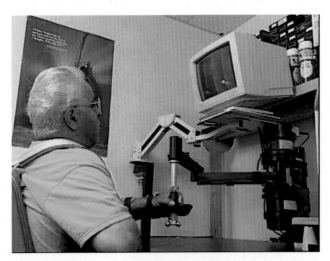

FIGURE 4.1 The MIT Manus [4], a planar manipulandum. Description: A subject seated with his back strapped to a chair, holding the extremity of a robotic device on a table, looking at a screen.

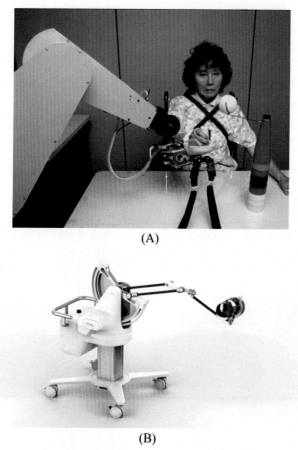

(A)

(B)

FIGURE 4.2   Two 3D manipulandum examples: (A) the MIME [6], using an industrial PUMA manipulator, and (B) the ArmMotusEMU (Fourier Intelligence, Shanghai, China). Description: On the left, an industrial robotic arm holding a woman forearm while she is reaching for a colored foam ball. On the right, a view of a white robotic system, mounted on wheels, with an arm which is terminated by a cuff with straps.

EMU [8] recently commercialized by Fourier Intelligence (Fig. 4.2B), or the BURT, derived from the haptic technology company Barrett Medical, United States.

These systems, compared to their exoskeleton counterparts discussed in the following sections, have the advantage of a relatively simple setup (one point of attachment, no adjustment required) while still allowing a large variety of movements and the patients to directly and physically interact with objects. They come with the drawback of not controlling every individual joint of the arm, and while they provide arm weight support, it is not distributed along the arm but only from the forearm, wrist, or hand.

The versatility of these systems may allow their use at various stages of recovery, and they constitute an interesting compromise between design cost and complexity and usability. Their open kinematic chain could also allow bedside use for early rehabilitation, an avenue not yet well explored.

4.2.1.3 Proximal upper-limb multi-contact systems

Some hybrid systems allowing three-dimensional motion of the hand with several points of attachment have also been developed. Research systems of this type such as the REHAROB [9] (Fig. 4.3), which uses two 6-DoF industrial manipulators, pursued the use of existing robotics systems, but their safety and convenience remain questionable for clinical translation.

Nevertheless, the Tyromotion Diego (Tyromotion GmbH, Austria) is an example of a commercial rehabilitation robot adapting this concept. It provides active assistance against gravity, with two points of support (one on the forearm and one on the upper arm) through hanging cables and allows bilateral practice in three dimensions.

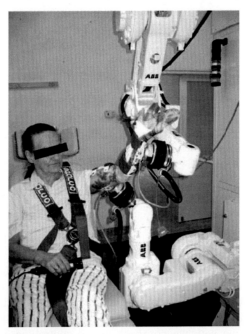

FIGURE 4.3 An example of multi-contact systems: the REHAROB [9], using two industrial manipulators. Description: Two industrial robotic arms which are attached to a man arm. The man is seated and strapped in a chair.

4.2.1.4   Proximal upper-limb exoskeletons

As opposed to manipulanda, exoskeleton systems aim to fit along the limb of the patient, with several attachment points (at least one per segment to be controlled)—similar to some insects' external structures from which they borrow their name. For the support of gross motor movements of the UE, they commonly target the control of the shoulder joint (often restricted to the glenohumeral, leaving out the scapula motion), the elbow joint, and the wrist joint [10]. They aim to provide full support and control of the limb in an intrinsically safe fashion (their allowable movements closely matching the anatomical ones). This configuration has the potential to independently control all the UE joints and enforce or correct movement patterns, although this is not always fully exploited. The main notable exoskeleton example is the ArmeoPower by Hocoma, with six actuated DoF (Fig. 4.4). It is derived from the ARMin series developed at ETH Zurich [11].

The main drawback of these systems is indeed their cost, due to a more complex mechanical structure and often a higher number of actuated joints. Additionally, their use generally requires some manual adjustment of the link lengths to fit every user's morphology, increasing the setup time while never allowing a perfect alignment [12].

By providing full support and control of the UE, they are more suitable for whole limb practice specifically at early stage of recovery in hospital and clinics where cost and space are not limiting.

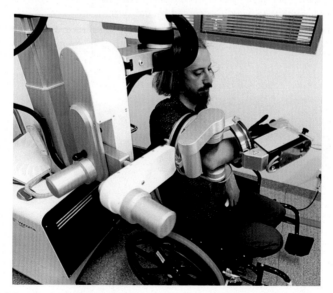

FIGURE 4.4   The ArmeoPower (Hocoma AG, Volketswil, Switzerland) UE exoskeleton with the three attachment points: on the upper arm, forearm, and hand. Description: A man has his arm strapped at both the upper arm and the forearm in a gray and white robotic exoskeleton.

4.2.1.5 Systems for the hand and fingers

Due to the complexity of the hand kinematics, exoskeleton structures are often used to address fine motor movements of the fingers, often in the form of a glove, on which actuation is provided to the different joints through a transmission system (cables or cams) to deport the actuation to the wrist or on a separate unit. Glove-type systems have the benefit of being portable and thus allow some assistive practice or rehabilitation practice in a variety of environments. The Tyromotion Amadeo (Tyromotion GmbH, Austria), a grounded system, is a different example taking an end-effector approach: each finger is attached to a single separate actuator moving linearly, allowing individual finger control.

A less complex approach is to place the actuation within the patient's hand and control only its gross movements (opening—closing actions). The mechanism then takes the shape of two plates or half-handles, one secured to the thumb and one to the other four fingers. This is the case of the InMotion HAND (Fig. 4.5), which can be used in coordination with the InMotion ARM to train full reach-and-grasp movements.

The practicability of the use of devices for the hand is an important factor to balance with the function to be trained: exoskeleton-type systems with independent finger control and multiple points of attachment allow fine motor skills practice but at the cost of a more complex setup and potentially time-consuming length adjustment. This system setup time should not be neglected—particularly in the frequently encountered case of patients with stiff hands. On the other hand, in-hand grasping systems do not allow the patient to physically interact with objects since the space is occupied by the system.

The reader can refer to Balasubramian et al. [13] for a detailed review of existing hand rehabilitation robots and their clinical evaluation.

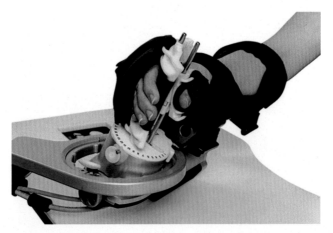

FIGURE 4.5 One example of a hand rehabilitation robot: the InMotion HAND (BIONIK, Toronto, Canada). Description: A hand is grasping a white handle, made of multiple parts and mounted on a rotational mechanism over a tabletop.

4.2.2   Systems for the lower limb

Unlike the upper limbs, which have a large variety in their movement goals and ranges, lower limbs are mostly used for mobility, the recovery of which is often the primary goal of neurologically impaired patients.

This has some significant implications on the robotic devices used for the rehabilitation of the lower limb. First—their movement range can be explicitly designed around the same movement—most lower-limb systems target effective gait rehabilitation rather than joint-specific movements. Second, the devices generally require some means of supporting the weight of the patient. Third, as the goal of the movement being rehabilitated involves the traversal of ground, some mechanism to account for or simulate this is required.

There are some exceptions to this, with some robotic devices having been developed for conditioning. Typically, these act on a single joint (e.g., the ankle/knee/hip) and work toward increasing strength. These are relatively simple devices and are often used while seated (refer to examples in Fig. 4.6 or in [14]). However, the majority of lower-limb devices are geared toward

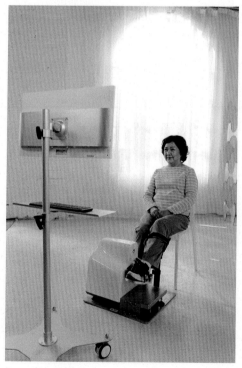

FIGURE 4.6   AnkleMotus (Fourier Intelligence, Shanghai, China) ankle rehabilitation device. Description: A woman seated on a chair, in front of a screen, with her foot strapped on the footplate of a small white device placed on the ground.

gait rehabilitation, and the remainder of the discussion in this chapter focuses on these (Fig. 4.7).

Therapist-facilitated gait rehabilitation generally requires a significant amount of therapist involvement, often requiring individuals to assist with the support of the bodyweight of the patient, along with mobilizing the limbs. This is an often-strenuous activity for multiple therapists and plays into one of the main advantages of robotics that they can be used to support and move the patient without fatigue. Different robotic devices address this through different means and degrees; however, they can be broadly categorized into grounded and overground devices.

4.2.2.1 Grounded devices

Grounded devices attempt to rehabilitate gait, without requiring traversal of ground. This has the significant advantage of reducing the amount of space required for rehabilitation, at the disadvantage of having to have a space dedicated for the device.

The most basic grounded device "robot" is a treadmill, often combined with an active (i.e., controllable) bodyweight support. Such devices do not require therapists to physically support the patient, allowing them to focus on the movements themselves.

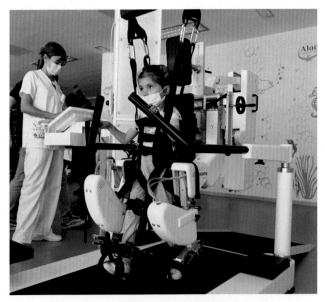

FIGURE 4.7 An example of an exoskeleton-based grounded device (Lokomat, Hocoma AG, Switzerland). Picture: José Ramón Márquez//JCCM – CC BY-ND 2.0). Description: A system comprising a treadmill, a screen, and a white exoskeleton structure in which a girl is strapped while supported by slings.

End-effector devices, such as the Gait Trainer GT II—BWS (Reha-Stim Medtech AG, Switzerland) and Lokohelp (Woodway, USA), provide additional functionality—specifically moving the end effectors (i.e., the feet) in a manner consistent with gait. Such devices take the lifting of the limb away from the therapist, who may then be able to focus on joint-level kinematics and dynamics. However, they do not constrain the movement to "natural" gait patterns—this is generally left to the therapist to ensure.

Finally, but perhaps most significantly, grounded exoskeleton devices may be considered the most advanced of this category of robots. These devices typically have links aligned with the thigh and tibia/fibula, and a footplate, combined with actuated DoF for hip and knee flexion/extension, with ankle dorsi/plantarflexion also sometimes actuated (e.g., the Lokomat, Hocoma, Switzerland), along with a bodyweight support system. Additional DoF in hip adduction/abduction have also been developed (e.g., the LOPES), which encourage more natural weight shifting, outside that allowed by the natural compliance in the strapping systems and the patient's softer tissue.

As previously discussed, all grounded devices require relatively high force/torque capabilities, as they must support the weight and move the mass of the human. Their grounded nature means that they are obviously not portable, and thus require a lot of space within the clinical rehabilitation environment. Furthermore, devices to date have generally only focused on one task, overground walking, with some ability to allow balance control. Other types of movements (e.g., sit to stand) are usually not possible.

4.2.2.2   Overground devices

Contrary to grounded devices, overground devices allow patient traversal during rehabilitation. Such devices are much more portable than their grounded counterparts and do not generally require a dedicated space, but do require more space during use. However, this space is typically already available in clinical environments dedicated to gait rehabilitation. A second major difference is that overground devices typically require more personnel to allow clinical use due to the fact that arresting falls are not typically within their capabilities.

These robots are almost exclusively exoskeletons, with their development likely evolving from orthoses in more typical therapy, which gives rise to another common label—powered orthoses. Such devices have been developed since the 1960s [15]; however, advances in battery and actuator technologies over the past few decades have enabled the development of many exoskeleton devices, including the EksoNR (Ekso Bionics, USA) and the ReWalk ReStore (Rewalk Robotics, USA). Structurally, these devices are often similar to their grounded counterparts, with links aligned with the thigh, tibia/fibula and a footplate, and actuated hip and knee flexion/extension. Typically, these devices are also used with arm crutches to enable the patient some degree of self-balancing capabilities. In addition, soft devices of this sort are also emerging, such as the Rewalk ReStore.

However, in terms of capability, many of these devices focus only on overground gait rehabilitation, with limited focus given to sit-to-stand training—however, this is in theory possible if motors can be responsive enough and provide enough power.

4.3 Control: what should the robot do?

If there are a large variety of devices and design approaches, there is an even larger variety when it comes to robot control, which aims to define what the robot should do at any instant. In our descriptions here, we divide the robot controller into three levels, which work together in a hierarchical manner:

- The low-level control manages the quality of the physical interaction—something which is more relevant to engineers with no direct clinical relevance other than facilitating the implementation of the interaction strategy level.
- The interaction strategy level defines the physical action of the robot over time (over the course of the movement/action); that is, it defines the force that is applied to the patient.
- The supervisory level aims to achieve a higher degree of automation by adjusting the task(s) to practice and/or alter the parameters of the interaction strategies at an iteration level (for each action, movement, or each rehabilitation session).

While the two first levels of control are necessary and always present, the third level is optional and not encountered in every system. Instead, the setting of control parameters and tasks can be completely left to the therapists' accord, or simply left constant.

4.3.1 Control for pHRI: low-level control

The low-level control layer is essential to allow the robot to offer a range of possible interactions of interest. Its main objective is to abstract the kinematics and imperfection of the mechatronics aspects of the device (friction, weight, inertia, etc.) to offer an interface to the higher levels of control to enable a physical interaction between the human part and the robot faithful to that desired, that is, the actual interaction force applied onto the limb and the actual displacement of the system (and so the attached limb).

As the desirable characteristics of this control layer are to be real time with a high bandwidth (usually with a frequency of more than 1 kHz), they are usually implemented separately on a dedicated controller or at least clearly separated from the higher-level control layers. This layer is also generally in charge of critical safety features of the system, imposing classical force, speed, and position limits which cannot be overridden and are constantly active with a short reaction time.

This layer is commonly formulated by an approach introduced by Hogan in 1984 [16], which consists of the definition of a virtual desired mechanical interaction between the robotic system and its environment, here the human limb. The higher-level controller has then the choice to prescribe a selected interaction (e.g., rigid for passive mobilization of the limb or an adjustable assistance or resistance).

Depending on the device design approach, this layer is either formulated as an impedance, measuring position/velocity and controlling force or as its dual, the admittance, measuring the force interaction (by means of a force sensor) and controlling the position/velocity. Both approaches can provide similar behaviors, but, as depicted in Fig. 4.8, the impedance approach typically has a better performance when the environment in contact with the robot (here, the human limb) is relatively stiff and admittance when the environment offers low stiffness.

While some research is still ongoing to extend and ease the implementation of the range of possible interactions, this low-level control is no longer a challenge but relies on a proper combination of the selected design and the control implementation.

4.3.2   Interaction strategies and physiological sensing

Leveraging the low-level controller interaction capabilities, the objective of the interaction strategies of a rehabilitation robot is to translate motor learning and rehabilitation principles into defined control strategies applicable to

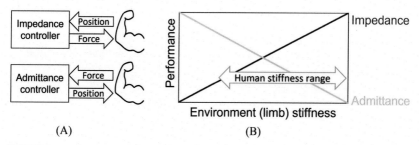

(A)   (B)

FIGURE 4.8   (A) Admittance and impedance principles and (B) typical performance graph of both approaches. Description: On the left, a schematic depicting on the top an impedance controller box to which an incoming arrow "Position" and an outcoming arrow "Force" are connected and in the bottom an admittance controller with an incoming "Force" arrow and an outgoing "Position" arrow. On the right, a graph showing performance on the vertical axis, environment (limb) stiffness on the horizontal axis, and two lines: a blue line going from bottom left to upper right noted "Impedance" and a yellow line, going from top left to bottom right noted "Admittance." An area on the bottom three-quarter right of the graph is denoted "Human stiffness range."

the robot. This control is performed by using physiological measures as an input and applying a force or position as an output at every time instant (ideally thousands, but at least hundreds of times per second).

Control approaches are commonly classified as belonging to one of the four main categories, as depicted in Fig. 4.9:

- Assistive defines strategies where the robot acts toward the completion of the task with or instead of the patient. They are the most commonly encountered. The most classic examples include passive mobilization— fully driving the patient limb along a trajectory, positive viscosity— amplifying patients' movement velocity, de-weighting—compensating fully or partially for the patient's limb weight—or simply the application of a constant positive force in the direction of the movement.
- Resistive defines the opposite strategies, acting against the completion of the task, aiming to promote additional patient effort in some direction. They are typically for strength training, for example, increased artificial weight (inverse of de-weighting) or damping.
- Corrective, where the robot acts in an orthogonal direction to the task completion, neither directly assisting nor preventing its completion, but rather correcting how the task is achieved. The most common example is the "tunneling" approach, creating a virtual wall around a path to enforce movement in straight-line, for example, but strategies correcting joint coordination (synergies) also exist [17].
- Error augmentation is an approach that aims to emphasize (amplify) the motor control error, by pushing the patient's limb in the direction of the error. Based on the idea of "error-driven learning," it aims to improve learning by emphasizing the error and thus encourages a correction from the patient. While research examples exist [18], this is not to date used in any commercially available system.

Obviously, the exact implementation that can take each of these approaches also depends on the design of the device. Manipulanda will have

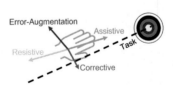

FIGURE 4.9 Schematic representation of the different interaction strategies approaches: assistive, resistive, corrective, and error augmentation (EA) for a simplified reaching action. Description: A schematic depicting a black line denoted task leading to a target. A hand is depicted next to the black line. Four colored arrows are drawn from the center of the hand: a green "Assistive" arrow pointing toward the target, a yellow "Resistive" arrow in the opposite direction of the green one, a red "EA" arrow pointing away from the task line, and a blue "Corrective" arrow pointing toward the "Task" line.

to define this interaction in the space of the task (the hand) whereas exoskeletons can—and have to—interact at the joint level. For the UE, some control strategies have been developed specifically to influence joint coordination or synergies, and in most cases for LE, the control strategy is directly expressed at the joint level.

Most of these interaction approaches rely on a knowledge of the task to be performed to provide a suitable interaction. While the task may be predefined, it is desirable to leave some movement intention (where to move, when to move) and execution freedom (how fast to move, how to move) to the patient. In addition, even if some interactions can be task-agnostic and do not require this knowledge (such as a de-weighting, a viscous resistance, or some movement amplification assistance), it is still desirable for these interactions to evaluate and react to the patient's immediate movement contribution.

To perform this intention detection and capability estimation, different physiological sensing modalities are used, sometimes combined together, in order to define controllers which are reactive to the user.

- Force and/or movement is the most commonly encountered, present in every robot, and used in many interaction strategies. It can be implemented without any additional sensors and provide useful information on the user action intention and capability but requires the patient to have some volitional limb control.
- Electromyography (EMG) measuring muscle activity allows a finer sensing of movement intention by not requiring any volitional movement capability and is thus potentially usable at earlier stages of recovery. It requires a slightly more complex setup to place electrodes on the dedicated muscles and advanced signal processing capabilities. To date, it has been used successfully as a movement trigger to detect movement initiation from the patient which can then be "amplified" by the robotic assistance.
- Gaze information, sensed through the mean of glass-mounted or screen-mounted eye trackers, offers the ability to estimate where the user intends to move, so as to leave some goal decision to the patient [19,20]. Existing systems have to date mostly been used in research settings only due to their relatively low accuracy and practicality.

Electroencephalography (EEG) is a noninvasive brain—computer interface (BCI) which could be used to incorporate brain activity within the interaction control. Its significant interindividual variability and processing complexity has led its use to remain anecdotal in practice [21].

The interaction strategies will thus leverage one or more of these modalities to achieve the desired behavior (the reader can refer to the systematic review by Basteris et al. [22] which proposes an extensive list of available strategies and use of some of these modalities).

It is noted that all these different control strategies aim to promote faster and more efficient motor relearning, but to date there is no strong clinical evidence favoring any specific approach. This is not surprising, as it is likely that this efficiency may be highly dependent on the patient lesion, symptoms, and recovery stage as well as on the specific therapy goal and the evaluation of the outcome, either in terms of body functions or activities. Most robotic systems thus offer a set of possible interaction strategies [22] to cater for the specific patient's presentation and specific goal of the rehabilitation session(s) and sometimes with high-level automatic adjustment policies for their parameters.

The development of appropriate interaction strategies is still today an active field of research but is still lacking structured clinical effectiveness evaluations.

4.3.3 High-level supervision and adaptation

For most of the interaction strategies introduced above, some key parameters are to be adjusted. This includes specific movement(s) to be practiced, but also a number of tuning parameters for a given strategy. Typically, this is the level of assistance or resistance, the percentage of de-weighting, or the speed and length of the gait cycle.

These parameters can be left to be tuned by the therapist; however, in order to provide the robot with a higher level of autonomy, it is desirable to automatically adjust these parameters based on the measured patient evolution over time and the variability of capacity during sessions. This adaptation is particularly important for semi-supervised or unsupervised settings, such as home-based therapy where there is no therapist intervention for long periods.

The main objective in automatically adjusting the assistance or difficulty over movement iterations is to promote, and even optimize, the motor learning. The most common approach attempts to "assist-(only-)as-needed," which aims to assist the patient as little as possible while ensuring the completion of the task. This can be done through an optimization problem formulation whose goal is to minimize concurrently a task error and the robot contribution to the task [23]. This approach raises the problem of selecting a suitable performance measure and associated cost formulation. A more pragmatic approach is to assume that the patient contribution will improve over time and to adjust the assistance—or task difficulty—based on the previous iteration performance and an arbitrary reduction of this assistance—or increase of difficulty [24].

While these strategies aim to adjust assistance parameters, it is also possible to automatically adjust the task to be practiced. Rosenthal et al. recently proposed a method of automatically selecting reaching tasks based on measurements of the performance of those tasks [25]. The algorithm selected tasks where the difference between best and worst performances are located (i.e., the "steepest gradient" areas).

One of the major difficulties for these automatic adjustments lies in the antagonism between the task difficulty and the robotic assistance: for a given patient's capability, the assistance provided may be increased, or conversely the task difficulty can be reduced, with both approaches leading to the task completion. Defining what is the right level of difficulty for optimal motor learning without impeding patient motivation remains an important challenge complexified by the difficulty in estimating the true capacity of the patient which may be "hidden" behind the robotic assistance, or altered by a slacking effect, difficult to evaluate [26].

Formalizing rules on these questions to allow a higher level of autonomy of the robots is thus still very much a work in progress, and the expertise from the therapist to prescribe and adjust the robotic action often remains the preferred option.

4.4   Non-robotic interactions: why physical interaction is not enough

To this point, the focus of this chapter has been on what may immediately come to mind with the suggestion of a "robot"—the physical interaction. However, equally, if not more important to the success of a rehabilitation device, are the nonphysical interactions, which instruct, motivate, and provide nonphysical feedback to the users of such devices.

These nonphysical interactions are covered here in two broad categories: activity-level interactions, defined as those which occur during the rehabilitation exercise or movement itself, and supervisory-level interactions, which involve the configuration of, or assessment provided by, the rehabilitation devices.

4.4.1   Activity-level interaction

There are two key objectives of activity-level interfaces: instruction and motivation.

Providing instruction involves some indication of some aspect of the movement to be performed, such as the timing of the movement, direction or position of the movement, or even some indication of the quality of the movement. This nonphysical interaction may also complement the physical interaction, through appropriate interaction strategy in the control of the robot. It is interesting to note that the adoption of robotics removes an avenue of feedback for the therapist—often they no longer have access to the physical feedback which they use to evaluate the patient's voluntary (or involuntary) movements and impairments. As such, the activity-level interface may also provide some indication of what the patient is doing, through the robot's own sensor readings.

On the other hand, it has become clear that motivating patients to complete their rehabilitation exercises is important and challenging. Gaming (and other) technologies have clearly demonstrated their ability to engage the human mind, and as such, the integration or adoption of such technologies into a robotic rehabilitation system is obvious.

4.4.1.1 Feedback and motivational content

Core to the activity-based interface is the content of the material presented on this interface. The myriad of sensors on a robotic device can be interpreted and then back to the patient—for example, the position of the hand or the pose of the leg may be shown to the patient, or the amount of force measured by a force sensor can be displayed on the screen, such that the patient and/or the therapist can understand how much the patient is "contributing" to the movement. However, the accuracy of such content should be scrutinized, to ensure that the presented information is indicative of the reported phenomena. This is easier to achieve with measurements such as position, than those associated with force or acceleration, and can be highly reliant on the mechanism behind the sensor arrangement.

Although basic feedback can be displayed to the patient, in many cases they do not motivate the patient, potentially affecting the patient's engagement and performance. As such, this feedback is often combined with motivational content, which has several potential forms. Many developers have taken advantage of digital expertise to present games on an activity-based interface. These games typically map the DoF of the robotic device to the DoF in the game. Thus "classic" games such as Pong can be used to motivate the patient to move their arms, increasing movement repetitions and promoting neuroplasticity. Such interfaces are prevalent in devices ranging from the InMotion Arm to the Hocoma Armeo series of devices, to the Tyromotion Diego. These interfaces have the advantage of being easy to implement, but care should be taken to ensure that the goals of the game align with the goals of the therapy, such that the patient cannot "cheat" to get a better score by performing movements which have reduced therapeutic benefit.

In more recent years, devices such as the ADLER [27] and EMU have explored the idea of using activities related to daily activities as part of their interface (see Fig. 4.10) —such as the use of their upper limbs to drink or comb hair. Such interfaces are designed to provide context and relevance to the patients, although no evidence has yet emerged suggesting that such interfaces are more effective than their counterparts. Interfaces designed to invoke such relationships are also common in lower-limb rehabilitation, with interfaces commonly involving 2D or 3D avatars traversing an environment.

FIGURE 4.10 The Haptic Master used for ADLs assisted training ([29] CC BY 2.0). Description: A white stroke tagged "HapticMaster" is connected to a man wrist through a cuff. The man is seated in front of a table and trying to grasp a cylindrical object.

Engagement can also be promoted through interpersonal interactions. While devices may typically be placed together in a room, simply promoting a sense of community, interfaces can also be designed explicitly for multi-person play. Such interfaces engage two or more individuals in either cooperative or competitive play [28].

4.4.1.2 Modalities: from screens to social robots

By far, the most common type of interface is a typical screen—be it television screen or computer monitor. Such interfaces are ubiquitous in our lives, and indeed among rehabilitation robots, and again take advantage of existing and accessible graphical development capabilities, as well as existing code bases—particularly for gaming content. However, it can be difficult to accommodate patients with neurological symptoms that affect the visual cortex. Such interfaces are also typically indirect—meaning that the patient must map the movement of their limb to a movement on the screen as depicted in Fig. 4.11. This results in the proprioceptive feedback from the limb not necessarily aligning with the visual feedback on the screen, requiring an additional cognitive load on the patient, which may be prohibitive in some cases [30]. For upper-limb systems, this can be removed by using a touch screen as an interface. With appropriately designed content, the touch screen can act as a target for the reaching action, resulting in direct mapping.

Virtual or augmented reality can also be used to provide a direct interface, reducing this cognitive load [31]. However, the head-mounted devices often used to provide this modality can be cumbersome to set up, which has significant implications for the adoption of this as an interface.

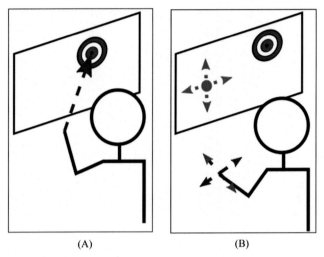

        (A)   (B)

FIGURE 4.11   Schematic representation of (A) a direct visual feedback and (B) indirect one where while the blue arrows represent the actual limb movements, the green arrows represent the mapped cursor movements on screen. Description: Two schematics, each showing a screen with a target on it and a character in front of the screen. On the left one, a direct blue arrow is connecting the character hand to the target on-screen. On the right one, two directional blue short arrows are connected to the hand, and separately, two small green arrows are drawn on the screen.

Additionally, such interfaces come at a significantly higher expense, although the development of VR and AR technologies for the wider community is resulting in more affordable devices becoming more and more available.

Some systems have explored the use of real objects with their devices. Such objects may include those that already exist within a clinical environment—such as those used for box and block tests. Such objects have obvious advantages in context and relevance; however, coordination with the robotic control can be a challenge. This has been somewhat addressed with instrumented objects [32], which can provide additional information about how the location of the object and how the patient is handling it (e.g., whether they are grasping it or not) to inform more advanced control strategies or to trigger certain control events.

These interfaces can also be combined with audio interfaces. Audio cues indicating success or failure are commonly used in gaming implementations, but they can also be used to provide instruction. However, as robotic rehabilitation systems are often used in busy clinical environments, audio may be undesirable.

Finally, social assistive robots (SARs), as shown in Fig. 4.12, are relatively new in rehabilitation. Key to these devices is their ability to measure

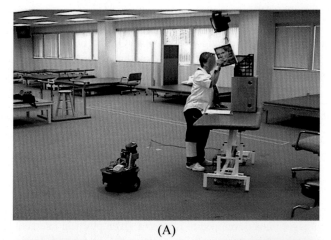

(A)

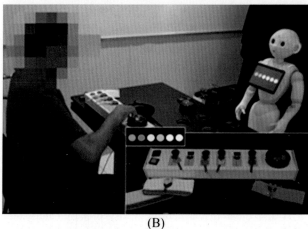

(B)

FIGURE 4.12 Socially assistive robots for rehabilitation—physical (non-humanoid) robot [33] and a humanoid [34]. Description: On the left, in a large room, a person is placing an object within a folder placed at sight height while a mobile robot equipped with a camera is "watching" her movements. On the right, a person is seated in front of a keyboard equipped with color buttons, and a humanoid robot with a screen on its chest is placed in front. A closer view of the keyboard with colored button is depicted on the bottom right.

and react to the actions of the patient, in a way which motivates and engages the patients beyond simply presenting the same response on a computer screen. Social robots promote an interaction with which the patient can develop trust, which may echo a patient−therapist relationship. Results suggest that the physicality of the robot presence has an impact on the level of engagement with the patient—compared with videos of a robot, or a virtual (computer-generated) robot [33]. These systems can be used either in conjunction with pHRI systems or on their own. However, in the absence of pHRI, this typically requires an additional method of monitoring patient capabilities, such as that provided by vision. On the other hand, this lack of physical interaction means that social robots have significantly simpler regulatory approval processes.

4.4.2 Supervisory

Supervisory interfaces allow the device's users—the patient and the therapist—to configure the device, monitor the patient's progress, and make suitable changes.

4.4.2.1 Device configuration—who and what

The control algorithms developed to impose a given physical interaction are typically parameterized by some values—such as the amount of de-weighting, or strength of the assistive force field. As discussed in Section 4.3.3, in many cases to date, these parameters are exposed to therapists, independent of the activity being performed—that is, independent of the activity-level interface. While this may be preferable in many cases, it remains a challenge to label and expose these parameters in a way which is intuitive to the end users of the device.

4.4.2.2 Assessment, analysis, and reporting

The use of rehabilitation robotics (and technologies in general) has clearly disrupted the approaches which are and can be taken in understanding the progress and capabilities of a given patient. These devices are capable of precise measurement of the interaction forces between the patient and the device, which can in theory lead to consistent, objective measurements. However, it is also clear that a neurologically impaired individual is complex, and clinical assessments must be made considering the entire patient—and not just observable physical interactions. Thus it is not as simple as "translating" the clinical assessments to the robotic device.

Most devices perform some sort of high-level feedback or assessment, for example, some indication of a "score" (most simply, the number of movements or steps performed with the robot, but also assessment such as range of motion, or speed of movements), which can then be used to motivate the patient or track their progress in subsequent sessions. However, while such scores can be somewhat reflective of the patients' capabilities, they are also influenced by the robotic interaction active during the movement, and even psychological factors, such as the level of motivation on the day.

However, interactions specifically designed to be assessed are growing in popularity and play to the strengths of robotic devices. Such devices create a structured task for the patient to complete while interacting with the device, and report on how the patient completes it. Basic examples include range of

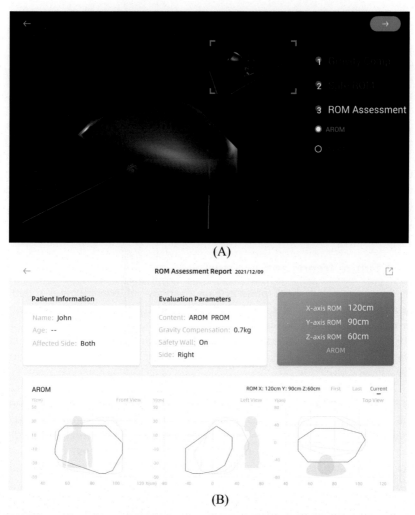

FIGURE 4.13 ArmMotus EMU: (A) range of motion assessment and (B) associated report. Description: On the left, a screenshot of a computer interface displaying a blue 3D blob on black background with the interface indicating "ROM Assessment." On the right, a screenshot of a computer interface displaying a report with three schematics of human shapes with a drawing of a ROM and information about the patient (name, weight, etc.) and size of the ROM.

motion assessments for the upper limb (as shown in Fig. 4.13), which asks the patient to reach as far as they can in all directions while the device runs in a transparent mode, or isometric force measurements, which lock the patients into specific postures and measure their ability to generate force.

While the above simple examples allow measurement of certain parameters, the development of "new" scales which can only be measured with the accuracy of a robotic device is generating momentum [35]. For example, the use of robotic devices to measure spasticity—often described as a velocity-dependent increase in muscle tone—has been reported in numerous cases [36]. The ability of robotic devices to produce accurate velocity movements and accurate force measurements allows a level of quantification not possible with manual (human) handling of the patient. As a second example, cleverly designed mechanisms, control-, and activity-level interfaces can be used to meaningfully separate the effects of perception and action [37]—a challenging task for clinical practitioners.

Such developments have the potential to improve the quality of rehabilitation and have the potential to be a clear differentiator between robotic rehabilitation devices and traditional therapy—that is, it is through these mechanisms that robotic devices may provide previously unattainable information which may be used to improve rehabilitation outcomes.

4.5 Future trends and conclusions

In this chapter, we have presented an overview of rehabilitation robotic devices, for both the upper and lower limbs. Clearly, there are a wide variety of mechanisms, control strategies, and auxiliary systems that cater to a variety of different impairments and situations, which need to be considered when designing or even choosing a system, either for a clinic or an individual patient.

There remain emerging technologies not discussed in depth here, including soft robotics (particularly exoskeletons which can conform to the body); the use of adjunct technologies such as functional electrical stimulation; and remote or connected devices. Additionally, there is significant interest in the use of machine learning algorithms to leverage data collected by robotics devices for both assessment and control. However, these approaches still often remain very abstract and without clear clinical meaning.

Despite not having completely reached their potential, rehabilitation robots have proven to be effective at reducing some of the physical load on the treating therapists and increasing treatment intensity, which will be discussed in detail in the next chapter.

Conflict of interest

The authors are coinventors of the EMU rehabilitation device mentioned in this chapter and have financial interests in the commercialized version of the device.

As part of their research activities, the authors are partially funded by industry grants (ARC-LP scheme) co-awarded to the University of Melbourne and Fourier Intelligence (Shanghai, China).

References

[1]   Maciejasz P, Eschweiler J, Gerlach-Hahn K, Jansen-Troy A, Leonhardt S. A survey on robotic devices for upper limb rehabilitation. J Neuroeng Rehabil 2014;11:3 January.

[2]   Hobbs B, Artemiadis P. A review of robot-assisted lower-limb stroke therapy: unexplored paths and future directions in gait rehabilitation. Front Neurorobot 2020;14:19.

[3]   Mehrholz J, Pollock A, Pohl M, Kugler J, Elsner B. Systematic review with network meta-analysis of randomized controlled trials of robotic-assisted arm training for improving activities of daily living and upper limb function after stroke. J Neuroeng Rehabil 2020;17:83 June.

[4]   Hogan N., Krebs H.I., CharnnarongJ., Srikrishna P., SharonA. MIT-MANUS: a workstation for manual therapy and training. I," In: 1992 IEEE International Workshop on Robot and Human Communication, pp. 161−165; 1992.

[5]   TommasinoP., MasiaL., Gamage W.G.K.C., MuhammadA., Hughes C.M.L., Campolo D. H-Man: Characterization of a novel, portable, inexpensive planar robot for arm rehabilitation. In *5th IEEE RAS/EMBS International Conference on Biomedical Robotics and Biomechatronics*; 2014.

[6]   Lum PS, Burgar CG, Van Der Loos M, Shor PC, Majmundar M, Yap R. MIME robotic device for upper-limb neurorehabilitation in subacute stroke subjects: a follow-up study. J Rehabil Res Dev 2006;43:631.

[7]   Timmermans AAA, Lemmens RJM, Monfrance M, Geers RPJ, Bakx W, Smeets RJEM, et al. Effects of task-oriented robot training on arm function, activity, and quality of life in chronic stroke patients: a randomized controlled trial. J Neuroeng Rehabil 2014;11:45 March.

[8]   Fong J, Crocher V, Klaic M, Davies K, Rowse A, Sutton E, et al. Promoting clinical best practice in a user-centred design study of an upper limb rehabilitation robot. Disabil Rehabil: Assist Technol 2020;0:1−8 July.

[9]   Toth A, Fazekas G, Arz G, Jurak M, Horvath M. "Passive robotic movement therapy of the spastic hemiparetic arm with REHAROB: report of the first clinical test and the follow-up system improvement," 2005.

[10] Proietti T, Crocher V, Roby-Brami A, Jarrasse N. Upper-limb robotic exoskeletons for neurorehabilitation: a review on control strategies. IEEE Rev Biomed Eng 2016;9:4−14.

[11] Klamroth-Marganska V, Blanco J, Campen K, Curt A, Dietz V, Ettlin T, et al. Three-dimensional, task-specific robot therapy of the arm after stroke: a multicentre, parallel-group randomised trial. Lancet Neurol 2014;13:159−66 February.

[12] Jarrasse N, Morel G. Connecting a Human Limb to an Exoskeleton. IEEE Trans Robot 2012;28:697−709 June.

[13] Balasubramanian S, Klein J, Burdet E. Robot-assisted rehabilitation of hand function. Curr Op Neurol 2010;23:661−70 December.

[14] Saglia JA, Tsagarakis NG, Dai JS, Caldwell DG. A high-performance redundantly actuated parallel mechanism for ankle rehabilitation. Int J Robot Res 2009;28:1216−27.

[15] Seireg A, Grundmann JG. Design of a multitask exoskeletal walking device for paraplegics. Biomechanics of medical devices. New York: Marcel Dekker, Inc; 1981. p. 569−644.

[16] Hogan N. "Impedance Control: An Approach to Manipulation," 1984.

[17] Jarrassé N, Proietti T, Crocher V, Robertson J, Sahbani A, Morel G, et al. Robotic exoskeletons: a perspective for the rehabilitation of arm coordination in stroke patients. Front Hum Neurosci 2014;8:1−13.

[18] Abdollahi F, Case Lazarro ED, Listenberger M, Kenyon RV, Kovic M, Bogey RA, et al. Error augmentation enhancing arm recovery in individuals with chronic stroke: a randomized crossover design. Neurorehabil Neural Repair 2014;28:120−8 February.

[19] Novak D, Omlin X, Leins-Hess R, Riener R. Predicting targets of human reaching motions using different sensing technologies. IEEE Trans Biomed Eng 2013;60:2645−54 September.

[20] Frisoli A, Loconsole C, Leonardis D, Banno F, Barsotti M, Chisari C, et al. A New Gaze-BCI-Driven Control of an Upper Limb Exoskeleton for Rehabilitation in Real-World Tasks. In: *IEEE Transactions on Systems, Man, and Cybernetics, Part C (Applications and Reviews)*, vol. 42, p. 1169−1179, November 2012.

[21] Baniqued PDE, Stanyer EC, Awais M, Alazmani A, Jackson AE, Mon-Williams MA, et al. Brain−computer interface robotics for hand rehabilitation after stroke: a systematic review. J Neuroeng Rehabil 2021;18:15 January.

[22] Basteris A, Nijenhuis SM, Stienen AHA, Buurke JH, Prange GB, Amirabdollahian F. Training modalities in robot-mediated upper limb rehabilitation in stroke: a framework for classification based on a systematic review. J Neuroeng Rehabil 2014;11:111 July.

[23] Emken JL, Bobrow JE, Reinkensmeyer DJ. Robotic movement training as an optimization problem: designing a controller that assists only as needed. In: *2005 9th International Conference on Rehabilitation Robotic*s, pp. 307−312, 2005.

[24] Marchal-Crespo L, Reinkensmeyer D. Review of control strategies for robotic movement training after neurologic injury. J Neuroeng Rehabil 2009;6:20.

[25] Rosenthal O, Wing AM, Wyatt JL, Punt D, Brownless B, Ko-Ko C, et al. Boosting robot-assisted rehabilitation of stroke hemiparesis by individualized selection of upper limb movements − a pilot study. J Neuroeng Rehabil 2019;16:42 March.

[26] Casadio M, Sanguineti V. Learning, retention, and slacking: a model of the dynamics of recovery in robot therapy. IEEE Trans Neural Syst RehabilitatiEng 2012;20:286−96 May.

[27] Johnson MJ, Wisneski KJ, Anderson J, Nathan D, Smith RO. Development of ADLER: The Activities of Daily Living Exercise Robot. In: 2006 *First IEEE/RAS-EMBS International Conference on Biomedical Robotics and Biomechatronics*, pp. 881−886, 2006.

[28] Mace M, Kinany N, Rinne P, Rayner A, Bentley P, Burdet E. Balancing the playing field: collaborative gaming for physical training. J Neuroeng Rehabil 2017;14:1−18.

[29] Timmermans AA, Lemmens RJ, Monfrance M, Geers RP, Bakx W, Smeets RJ, et al. Effects of task-oriented robot training on arm function, activity, and quality of life in chronic stroke patients: a randomized controlled trial. J Neuroeng Rehabil 2014;11 (no 1):1−12.

[30] Crocher V, Fong J, Klaic M, Tan Y, Oetomo D. Direct versus indirect visual feedback: the effect of technology in neurorehabilitation. In: *2019 9th International IEEE/EMBS Conference on Neural Engineering (NER)*, 2019.

[31] Wenk N, Penalver-Andres J, Palma R, Buetler KA, Müri R, Nef T, et al. Reaching in Several Realities: Motor and Cognitive Benefits of Different Visualization Technologies. In: *2019 IEEE 16th International Conference on Rehabilitation Robotics (ICORR)*, 2019.

[32] Hussain A, Balasubramanian S, Roach N, Klein J, Jarrassé N, Mace M, et al. SITAR: a system for independent task-oriented assessment and rehabilitation. J Rehabil Assist Technol Eng 2017;4 2055668317729637, January.

[33] Matarić MJ, Eriksson J, Feil-Seifer DJ, Winstein CJ. Socially assistive robotics for post-stroke rehabilitation. J Neuroeng Rehabil 2007;4:1−9.

[34] Feingold-Polak R, Barzel O, Levy-Tzedek S. A robot goes to rehab: a novel gamified system for long-term stroke rehabilitation using a socially assistive robot—methodology and usability testing. J Neuroeng Rehabil 2021;18:1−18.

[35] Shirota C, Balasubramanian S, Melendez-Calderon A. Technology-aided assessments of sensorimotor function: current use, barriers and future directions in the view of different stakeholders. J Neuroeng Rehabil 2019;16:1−17.

[36] de-la-Torre R, Ona ED, Balaguer C, Jardon A. Robot-aided systems for improving the assessment of upper limb spasticity: a systematic review. Sens (Basel) 2020;20.

[37] Zbytniewska M, Kanzler CM, Jordan L, Salzmann C, Liepert J, Lambercy O, et al. Reliable and valid robot-assisted assessments of hand proprioceptive, motor and sensorimotor impairments after stroke. J Neuroeng Rehabil 2021;18:115 July.

Chapter 5

Evidence for rehabilitation and socially assistive robotics

Hermano I. Krebs[1,2,3,4,5] and Taya Hamilton[6,7]

[1]*Department of Mechanical Engineering, Massachusetts Institute of Technology, Cambridge, MA, United States,* [2]*School of Medicine, University of Maryland, Baltimore, MD, United States,* [3]*School of Medicine, Fujita Health University, Nagoya, Aichi Prefecture, Japan,* [4]*Department of Mechanical Science and Bioengineering, Osaka University, Osaka, Osaka Prefecture, Japan,* [5]*The Wolfson School of Engineering, Loughborough University, Loughborough, Leicestershire, United Kingdom,* [6]*Massachusetts General Hospital, Boston, MA, United States,* [7]*Perron Institute of Neurological and Translational Science, University of Western Australia, Perth, WA, Australia*

Learning objectives

At the end of this chapter, the reader will be able to
1. Explain the principles, applications, evidence, and challenges with the adoption of upper- and lower-extremity rehabilitation robotics.
2. List and describe the applications and evidence for some of the more common socially assistive robotic solutions used in older adult care and stroke rehabilitation.

5.1   Introduction

The ebb-and-flow of progress and evolution in any field requires the convergence of several different factors, technologies, and stakeholders. This view on the potential of robotics in general, and rehabilitation robotics in particular, was eloquently summarized by Bill Gates over a decade ago:

> *Imagine being present at the birth of a new industry. It is an industry based on groundbreaking new technologies, wherein a handful of well-established corporations sell highly specialized devices for business use and a fast-growing number of start-up companies produce innovative toys, gadgets for hobbyists and other interesting niche products...... (like the computer industry) ...trends are now starting to converge and I can envision a future in which robotic devices will become a nearly ubiquitous part of our day-to-day lives.*

Rehabilitation Robots for Neurorehabilitation in High-, Low-, and Middle-Income Countries.
DOI: https://doi.org/10.1016/B978-0-323-91931-9.00023-2

Technologies such as distributed computing, voice and visual recognition, and wireless broadband connectively will open the door to a new generation of autonomous devices that enable computers to perform tasks in the physical world on our behalf. We may be on the verge of a new era, when the PC will get up off the desktop and allow us to see, hear, touch, and manipulate objects in places where we are not physically present.

This vision is highly ambitious, necessitating a new industrial revolution strongly anchored on the creation of robotic partners that will cooperate with humans in close contact. It includes widespread automation such as smart homes with a multitude of interconnected devices embedded in the homes to aid people with everyday activities. It includes self-driving cars, trucks, drones, and the associated changes in urban development. It also includes service robotics that will assist in the rehabilitation and support of a person with disabilities, either through direct physical manipulation of a paralyzed limb or indirectly via social interaction.

One cannot overstate the need for this technology globally to assist all stakeholders. Aging of the world population is increasing the demand for caregivers, rehabilitation services, and payors. By 2050, the contingent of older adults in the United States alone is expected to grow from 50 to 80 million. With this growth comes an increased incidence of age-related maladies, including stroke. There are already over 50 million stroke survivors worldwide, with this number projected to expand rapidly (Fig. 5.1). The most recent survey from World Physiotherapy (2021) indicates there are insufficient therapy resources to support and service the aging and stroke survivor population. On average, there are 5–10 practicing physical therapists per 10,000 people in both Europe and the United States (https://world.physio/ and https://world.physio/membership/profession-profile). This situation creates a vital need for new approaches to improve the effectiveness and efficiency of rehabilitation and an unprecedented opportunity to deploy technologies such as robotics.

The industry interest has been preceded by an exponential growth in related academic research. Fig. 5.2 shows the number of citations for academic papers in journals searched with typical search engines using the keywords: "rehabilitation robotics, social robotics, service robotics, and arm prosthesis." In 1980, there were 26 papers on arm prosthesis, with none employing the remaining keywords. By the end of the decade, there was a trickle of papers using the aforementioned terms. The initial development of the MIT-Manus, which was the first robot deployed in clinical trials delivering rehabilitation therapy, occurred in 1989. By the end of the 1990s, the numbers were 138, 20, 168, and 19 papers, respectively, for rehabilitation robotics, social robotics, service robotics, and arm prosthesis. By 2015, the number of papers on rehabilitation robotics and service robotics shows a strong and sustained growth of activity with a small but noticeable increase

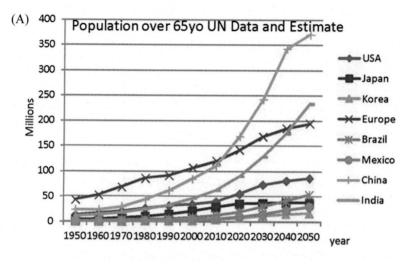

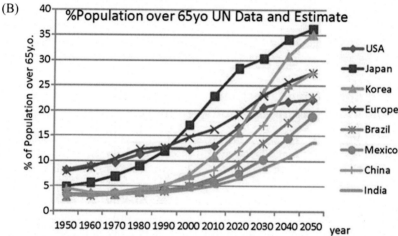

FIGURE 5.1   Size (A) and percentage (B) of the population above 65 years old (y.o) (United Nations Data and Estimates).

in prosthesis work (1900, 1480, 1410, and 68, respectively). The magnitude of this change goes beyond the usual ebb-and-flow of activity in technology-related fields, and more change is predicted as society ages.

Given the rapid growth, development, and interest in the rehabilitation robotic industry, it would be a disservice to the field to attempt to summarize in a single chapter the whole spectrum of robotic solutions aimed at minimizing impairment and disability. Here we will focus on reviewing the applications of rehabilitation robotics, including social interactions, compared to conventional rehabilitation. In particular, we will highlight selected pieces of evidence in stroke rehabilitation that explore the use of robotic technology

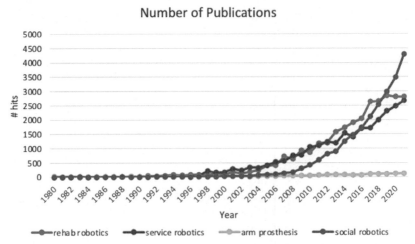

FIGURE 5.2 Number of hits of academic papers and keywords.

for both upper- and lower-extremity recovery either through direct manipulation or coaching aimed at behavior modification (social robotics).

5.2 Principles and mechanisms of rehabilitation robots: Hebbian conjecture and motor learning

Prior to the 1970s and 1980s, the predominant perception was that the adult brain was hardwired, and restitution of motor control after brain injury was seemingly futile. One of the most well-known promoters of such view was the famous Spanish Nobel Laureate Santiago Ramón y Cajal. According to Ramón, beyond edema resolution and optimization of natural recovery, rehabilitation of patients following brain injury should involve training compensatory techniques (how to accomplish activities of daily living [ADLs] often using the unaffected upper extremity or assistive devices), and not motor training of the paretic limb or addressing the underlying impairment. Out of the late 1980s and 1990s, an era of discovery and technological development was born for the neuroscience field, whereby animal models showed the potential for training-induced change in sensorimotor function following central nervous system injury [1–3]. An increased understanding and promotion for brain plasticity and the development of the Hebbian theory, colloquially synthesized by the saying that "neurons that fire together, wire together," led to the opportunity to investigate how to harness and optimize brain plasticity beyond natural recovery [4]. In addition to the therapeutic potential of robotics, the 1980s also marked the development of robotics for the objective quantification and standardized assessment of motor control of the upper extremity both in healthy populations and for those with injury or disease.

As conventional therapy practices of the time did not focus on the affected limb post-stroke, there was no clear design target for the robotic technology nor any reliable "gold standard" against which to gauge its effectiveness. The motor learning framework served as a basis for modeling motor recovery and consequently to design rehabilitation robots and their associated interventions. Kleim and Jones elegantly summarized basic principles of experience-dependent neural plasticity in an easy-to-understand format as shown in Table 5.1 [5].

Compared to manual therapy, the design of rehabilitation robots provides the potential for greater intensity and dosage of movement repetitions (addressing principles 4 and 5), exposure to intensive or more challenging therapy even in the acute or early subacute stages (principles 1, 2, and 6) in addition to other motor control strategies not possible with conventional methods (such as error augmentation or the application of force fields to emphasize error in reaching trajectory, addressing principles 7, 9, and 10

TABLE 5.1 Ten motor learning concepts (discussed in more detail in Chapter 3).

Principle	Description
1. Use it or lose it	Failure to drive specific brain functions can lead to functional degradation
2. Use it and improve it	Training that drives a specific brain function can lead to an enhancement of that function
3. Specificity	The nature of the training experience dictates the nature of the plasticity
4. Repetition matters	Induction of plasticity requires sufficient repetition
5. Intensity matters	Induction of plasticity requires sufficient training intensity
6. Time matters	Different forms of plasticity occur at different times during training
7. Salience matters	The training experience must be sufficiently salient to induce plasticity
8. Age matters	Training-induced plasticity occurs more readily in younger brains
9. Transference	Plasticity in response to one training experience can enhance the acquisition of similar behaviors
10. Interference	Plasticity in response to one experience can interfere with the acquisition of other behaviors

[6−8]). Nonetheless, while movement repetition has been shown to induce plasticity, it does not necessarily induce learning of voluntary behavior. In fact, the intended or expected mechanisms for improved motor learning following therapeutic assist-as-needed robotic devices are not fully understood. Some benefits may be explained by training the desired sensory and proprioceptive components of the task. Nonetheless, successful motor learning requires an individual with intact learning mechanisms to select an action guided by instruction, error, or reward feedback and subsequent dosage-appropriate practice of the task to improve the desired task outcome. Although robots can facilitate the intensity of the task practice and provide reward and error feedback, the intent and nature of the robotic assistance are often heterogeneous between studies and across devices which may dilute the perceived benefits of the technology when reviewing the literature in this field.

5.3 Rehabilitation robotics evidence: big picture in stroke care

Clinical guidelines from the American Heart Association (AHA) and the literature review from the Cochrane report summarize and analyze the available evidence for the use of upper- and lower-extremity rehabilitation robotics in post-stroke. In 2010 and 2016, the AHA endorsed the use of robotic technology for the upper extremity post-stroke but not for the lower extremity [9,10]. For the upper extremity, the AHA guidelines state that "Procedure/Treatment SHOULD be performed" or "IT IS REASONABLE to perform procedure/administer treatment," depending on the phase of recovery (acute/subacute or chronic phase). For the lower extremity, AHA guidelines state that "Procedure/Treatment MAY BE CONSIDERED." The Cochrane Reviews also endorsed the use of robotic technology for both the upper and lower extremities [11−13].

In addition to the heterogeneity of robotic technology, the current state of usual care for poststroke rehabilitation across the globe also presents a challenge in determining the benefit of the interventions. Despite the influx in research to inform clinical guidelines, the Cochrane team in their letter published in Lancet caution how significantly usual care differs across the world, which constitutes a hindrance to progress in evidence-based rehabilitation medicine (Usual Care: the big but unmanaged problem in rehabilitation medicine) [14]. If usual care represents the standard by which the effectiveness of rehabilitation interventions is evaluated in most studies, then the results of clinical guidelines are likely to differ depending on the region. For example, the AHA includes primary studies conducted in the United States which differ from the results reported in the Cochrane Review, which takes a global perspective [11,12].

5.3.1   Upper-extremity applications and evidence

As there are many studies to consider which use different devices, across a wide variety of settings and stages of recovery, we selected three comparable studies that employed the same MIT-Manus robotic technology, lasted the same 12 weeks, 3 times per week, for a total of 36 sessions to highlight the potential of upper-extremity robotics.

The first study demonstrated the potential of rehabilitation robotics to promote motor recovery. We enrolled 82 volunteers with a single chronic ischemic stroke and right hemiparesis who received anodal transcranial direct current stimulation (tDCS) or sham stimulation, prior to robotic therapy [15−17]. Both tDCS and sham stimulation were administered at rest for 20 minutes immediately prior to each robotic session (anodal tDCS, 2 mA, affected hemisphere, M1/SO montage). Robotic therapy involved 1024 repetitions, alternating shoulder, elbow, and wrist robots, for a total of 36 sessions. Shoulder, elbow, and wrist kinematic and kinetic metrics were collected in addition to clinical scores at intervention completion and 6 months since completion. Clinical and robotic biomarkers demonstrated significant improvements beyond the minimal clinically important difference (MCID). For example, the Fugl-Meyer Assessment of Upper Extremity motor function (FMA-UE) showed an average group improvement of 7.36 at the end of the intervention and 7.65 at 6-months follow-up (FMA MCID = 5). No between-groups difference was detected in either clinical or robotic metrics, demonstrating that intensive robotic therapy is beneficial, but no additional gain from anodal tDCS. Perhaps because the effect of robotic therapy is so dominant, additional benefits of anodal tDCS were not observed. Figs. 5.3 and 5.4 summarizes these results. Further information on the observed lack of additional benefit of anodal tDCS, delivered prior to intensive robotic therapy, can be found in [16,17]. In addition, these studies highlight the high correlation of robot metrics with clinical outcomes and larger effect size. One should, however, take these results with the appropriate caveat as we have not tested whether anodal tDCS has an effect on stroke recovery, just that it does not have an additional effect beyond intensive robotic therapy.

While the Edwards et al. study did not include a comparison of groups that received robotic therapy to usual care, the Veterans Affairs' (VA) ROBOTICS randomized clinical trial and the National Health Service's RATULS trial did compare robotic training with a positive control group and usual care among participants with moderate-to-severe stroke [18−21]. ROBOTICS enrolled 127 participants and employed an adaptive trial design and three primary outcome scales at 12 weeks to assess impairment reduction (FMA-UE), functional improvement (Wolf Motor Function), and quality of life (Stroke Impact Scale). Furthermore, it offered an incentive for participants randomized to usual care to complete the study: at completion, these participants could select to receive the positive control or robotic therapy. RATULS enrolled

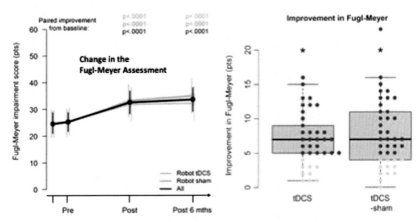

FIGURE 5.3   The impact of robotic therapy and anodal tDCS on chronic stroke. The left plot shows changes in the combined group Pre-to-Post of 7.36 points in the Fugl-Meyer Assessment for the Upper Extremity (max 66 points) and of 7.65 points Pre-to-Post 6 months. Note that anodal tDCS did not confer any additional benefit over and beyond robotic therapy. The right plot shows individual impairment changes with most participants improving more than the MCID for chronic stroke (5 points).

770 participants and chose a more traditional trial design and outcome measures recommended by the Stroke Recovery and Rehabilitation Roundtable [22]. The primary outcome was the Action Research Arm Test (ARAT) with the Fugl-Meyer Assessment as the secondary outcome. RATULS allocated equal number of participants to the three arms of the study. Both studies require careful reading to augment our understanding and to improve upon future trial designs. For example, contrary to the VA study, RATULS did not offer any incentive to minimize attrition in the usual care group or change in participant's behavior. Table 5.1 shows the attrition rate for each group in the two studies. Note that the attrition rate for the usual care group in RATULS is almost double the attrition rate for the other groups at 25.2% or 1 in 4 participants in the usual care group withdrawing from the study. Furthermore, the improvement in the usual care group was very good and should be better controlled for, as highlighted by the Cochrane team in Lancet [14]. Table 5.2 summarizes the changes from admission to 12 and 36 weeks in the ROBOTICS trial [23]. Note that adjusting for comorbidities, the active intervention groups are significantly better than the usual care group and the improvement is at the MCID level. Fig. 5.5 summarizes the changes in the RATULS trial for the FMA_UE. RATULS primary outcome, the Action Research Arm Test (ARAT), did show to be a poor choice for moderate-to-severe stroke participants (the target group of RATULS study). The ARAT is heavily dependent on hand function, and we speculate that it might be a good scale for assessing recovery after mild stroke, which was not the focus of the study, but not for severe to moderate stroke, the focus of the

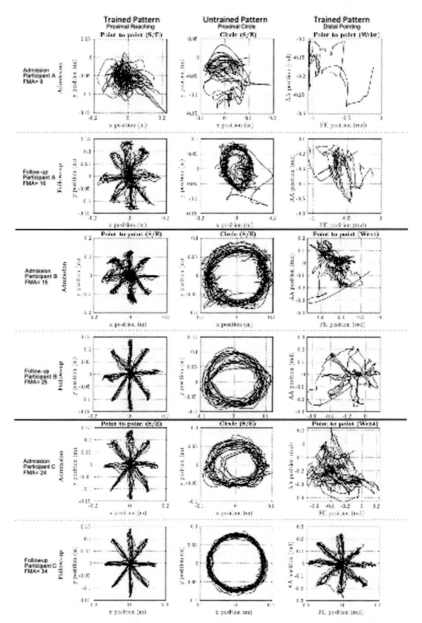

FIGURE 5.4 Unassisted proximal (S/E = shoulder-elbow) and distal (wrist) movement attempts of three representative stroke participants at study baseline and follow-up. The left column shows participants attempts to draw a star (trained movement) demonstrating participants at different impairment levels and improvement from baseline to follow-up. The middle column shows the participants attempts to draw a circle (untrained movement) demonstrating generalization. The right column shows the attempts to point with the wrist in a star pattern (trained movement). It demonstrates improvement and that the recovery of distal movements appears to be slower than proximal but possibly with the same traits (for more details, refer to Moretti et al. part 1, 2021). Note: FMA = Fugl-Meyer assessment, FE = flexion−extension, AA = abduction−adduction.

TABLE 5.2 Attrition rates at two comparable large rehabilitation robotics randomized controlled trials (RCTs).

Group	Robot therapy	Active comparison	Usual care
6 Months	Attrition rate %	Attrition rate %	Attrition rate %
VA-ROBOTICS	10.2	16	10.7
NHS - RATULS	13.2	14.3	25.2

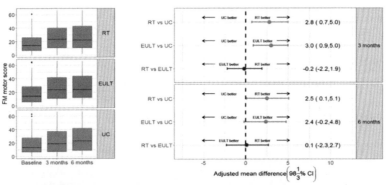

Mean (SD)	Baseline	3 months	6 months	Attrition
Robot Therapy	18.0 (13.1)	26.2 (17.7)	27.2 (18.4)	13.2
Enhanced Upper Limb Therapy	18.2 (14.1)	27.1 (18.3)	28.2 (19.3)	14.3
Usual Care	18.2 (13.9)	24.2 (18.4)	26.3 (18.9)	25.2

FIGURE 5.5   RATULS impairment outcome. The figure shows on the top left panel and on the bottom table that all groups had a significant reduction in impairment during the intervention, well beyond the minimal clinically important difference (MCID). The top right panel shows the group difference adjusted for comorbidities and attrition. To be significant as defined in the pre-trial statistical analysis plan, the adjusted odds ratio should not cross 0 (determining statistical significance) and include a difference of 5 points in the Fugl-Meyer assessment (FM motor score, determining MCID level). Both active interventions were superior to usual care with the robotic therapy group maintained its significant advantage over the usual care at the 6-months follow-up. Note: UC = usual care, EULT = enhanced upper-limb therapy, RT = robotic therapy.

RATULS study (for more details see [24]: "Analyzing the Action Research Arm Test (ARAT): a cautionary tale from RATULS Trial"). All groups improved significantly and above the MCID from admission to the completion of treatment with the active groups improving more than the usual care at 12 weeks. Note that only the robot group maintaining this statistically significant advantage over the usual care at 6 months (as defined in the statistical

TABLE 5.3 VA-ROBOTICS: Final results adjusted for comorbidities [23]. Note that in terms of impairment reduction, the Robotic Group advantage over the Usual Care in the VA Study maintained its statistical significance at 6-month follow-up.

Outcome	FMA-UE	Wolf motor function	Stroke impact scale
Group comparison	Mean difference (95% CI), P value	Mean difference (95% CI), P value	Mean difference (95% CI), P value
12-weeks therapy			
Robot vs usual care	4.2 (1.3–7.1), 0.005	−7.6(−15.1 to −0.1), 0.046	8.3 (2.7–14.0), 0.005
Intensive comparison vs usual care	4.2 (1.2–7.3), 0.007	−7.8 (−17.5 to 1.8), 0.1	6.7 (1.8–11.6), 0.008
36-weeks follow-up			
Robot vs. usual care	4.6 (0.6–8.6), 0.026	−7.8 (−15.7 to 0.05), 0.051	3.2 (−2.9 to 9.4), 0.3
Intensive comparison vs. usual care	2.6 (−1.0 to 6.2), 0.2	−7.8 (−14.1 to −1.5), 0.014	2.5 (−2.7 to 7.8), 0.3

analysis plan the adjusted odds ratio should not cross 0 and include the MCID of 5 points) [21] (Table 5.3).

5.3.2   Lower-extremity applications and evidence

There is considerably more debate on the impact of the younger and arguably more complex lower-extremity (LE) robotic counterpart devices when compared to usual care. For example, the 2010 Veterans Administration/ Department of Defense (VA/DOD) guidelines recommended against the use of robotics for the LE. The VA/DOD guidelines state that "there is no sufficient evidence supporting the use of robotic devices during gait training in patients post-stroke" and the "recommendation is made against routinely providing the intervention to asymptomatic patients." They go on to also say that "At least fair evidence was found that the intervention is ineffective ..." [25]. The AHA recommendation is more neutral when comparing robotic outcomes to usual LE stroke rehabilitation care (as practiced in the United States) [9,10]. Results are marginally better when considering a mixture of usual care practices from around the world, with the Cochrane report stating that while robotics alone were not superior to a mixture of usual care, robotic

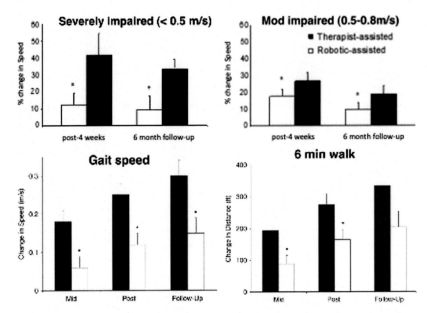

FIGURE 5.6 Clinical results of robotic therapy in stroke using Lokomat. The top row shows the results with chronic stroke (enrollment >6 months post-stroke), and the bottom row shows the results of subacute stroke trials (enrollment between 3 and 6 months post-stroke).

walking training plus usual care leads to better outcomes than equal time in usual care alone [13].

This negative perception of LE robotic rehabilitation is not without merit. Initial studies employing the Lokomat (Hocoma, Zurich, Switzerland) showed statistically significant inferior results when compared to those produced by usual care for both chronic as well as subacute stroke patients in North America [26,27]. Fig. 5.6 shows the results of two studies comparing LE rehabilitation robotics with usual care. The top row shows results with chronic stroke patients (stroke onset > 6 months), who trained 3 times per week for 30 minutes for 4 weeks, demonstrating improvements for the Lokomat-trained (white bars) and the usual care group (black bars). The usual care group (as practiced in North America) improved significantly more than the Lokomat-trained group and retained that advantage at 6 months follow-up. This was true for both patients with severe as well as moderate LE impairments [26]. For subacute stroke patients (stroke onset <6 months) who trained for 8 weeks, a qualitatively similar result was observed. Both groups improved from admission to mid-point, to completion, and to 3 months follow-up, but patients in the usual care group improved more, and the differences between groups were statistically significant [27].

There are many plausible reasons for these results, all of which indicate the initial immaturity of LE robotic therapy. First, the technologists developing the robots assumed that bodyweight-supported treadmill training (BWSTT) delivered by two or three therapists was an effective form of therapy. Their devices are elegant engineering solutions aiming to automate this labor-intensive and demanding form of therapy, which is based on the conjecture that by "strengthening" spinal cord central pattern generators (CPGs), gait in stroke patients might be enhanced [28]. Lacquanity and colleagues demonstrated that load receptor input is essential for physiological leg muscle activation during stance and gait [29]. However, an NIH-sponsored randomized controlled study demonstrated that contrary to the hypothesis of its clinical proponents, bodyweight-supported treadmill training administered by 2 or 3 therapists for 20−30 minutes followed by 20−30 minutes of overground carry-over training did not lead to superior results when compared to a home program of strength training and balance (LEAPS Study) [30]. This is a landmark result that must be seriously acknowledged by roboticists: the goal of rehabilitation robotics is to optimize care and augment the potential of individual recovery. It is not simply automating current rehabilitation practices, which for the most part lack a scientific basis, primarily due to the lack of tools to properly assess the practices themselves. In fact, the lead authors of the LEAPS study have scathingly urged a wake-up call in their paper: "Should Body Weight−Supported Treadmill Training and Robotic-Assistive Steppers for Locomotor Training Trot Back to the Starting Gate?" [31].

In what follows, we will review some of our evidence that has shown the role of supraspinal circuitry in gait, indicating that the traditional CPG model must be revised for the development and optimization of stroke rehabilitation.

5.3.3   Beyond central pattern generators

To move LE robotics beyond its infancy, we must determine what constitutes best practice and how to assess it. Alternatives must be carefully examined and the mechanisms understood. Gait rehabilitation after stroke often utilizes treadmill training delivered by either therapists or robotic devices. However, clinical results have shown a limited benefit of this modality when compared to usual care. On the contrary, results were inferior; perhaps because in its initial form it was not interactive and at least for stroke, central pattern generators at the spinal level do not appear to be the key to promote recovery. This result can be attributed to several factors [32]. The symmetric CPG models used to justify these rehabilitation approaches may not be capturing critical physiological mechanisms needed to initiate and maintain gait. An asymmetric CPG model that sets different weights to the peripheral and supraspinal inputs might do better. This model underscores the need for supraspinal input to initiate gait (mainly activating leg flexors). Gait initiation depends on commands originating from the mesencephalic (or midbrain) locomotor region (MLR), and this influence cannot be obtained from peripheral mechanical stimulation, as provided by robotic devices. While it is commonly considered that the lower leg motion is less precise than upper-limb motion, locomotion tasks must take into account not only the dynamics of walking, but also biped stability constraints. The maintenance of biped stability requires the integration of vestibular, visual, and proprioceptive information, thus supraspinal control. In the following sections, we explore the work done on perturbing walking, investigate basic psychophysical aspects of lower-limb control that have been hitherto unknown, and then review our working model for walking and describe some clinical results.

5.3.4 Perturbation response—haptic perturbation at the spinal level

Locomotion involves complex neural networks responsible for automatic and volitional actions. During locomotion, motor strategies can rapidly compensate for any obstruction or perturbation that could interfere with forward progression [32–34]. In a pilot study, Seiterle and colleagues examined the contribution of interlimb pathways for evoking muscle activation patterns in the contralateral limb when a unilateral perturbation was applied while bodyweight was externally supported [35]. Specifically, the latency of neuromuscular responses was measured, while the stimulus to afferent feedback was limited. The pilot experiment was conducted with six healthy young subjects. It employed the MIT-Skywalker [36]. Subjects were asked to walk on the split-belt treadmill, while a fast unilateral perturbation was applied mid-stance by unexpectedly lowering one side of the split-treadmill walking surfaces. The subject's weight was externally supported via the bodyweight support system consisting of a bicycle seat and a loosely fitted chest harness [37,38]. Both the weight support and the chest harness limited the afferent feedback. The unilateral perturbations evoked changes in the electromyographic activity of the non-perturbed contralateral leg. The latency of all muscle responses exceeded 100 ms, which precludes the conjecture that the spinal cord alone is responsible for the perturbation response (Table 5.4). It suggests the role of supraspinal or midbrain level pathways in the inter-leg coordination during gait.

TABLE 5.4 Muscle activity onset time for the four muscles: rectus femoris (RF), semitendinosus (ST), tibialis anterior (TA), and soleus (SOL). The values shown are mean ± one standard deviation over all subjects.

Muscle	RT	ST	TA	SOL
Onset time (s)	163 ± 22	129 ± 68	193 ± 80	207 ± 74

5.3.5 Perturbation response—visual perturbation at the supraspinal level

In addition, Seung-Jae Kim and Krebs tested healthy subjects to see whether an implicit subliminal "visual feedback distortion" influences gait spatial pattern [39,40]. Subjects were not aware of the visual distortion nor did they recognize changes in their gait pattern. The visual feedback of step length symmetry was distorted so that subjects perceived their step length as being asymmetric during treadmill training. We found that a gradual subliminal distortion of visual feedback, without explicit knowledge of the manipulation, systematically modulated gait step length away from symmetry and that the visual distortion effect was robust even in the presence of cognitive load (Fig. 5.7A). This indicates that, although the employed visual feedback display did not create a conscious and vivid sensation of self-motion, experimental modifications of subjects' visual information of movement were found to cause implicit gait modulation. Nevertheless, our results also indicate that modulation with visual distortion may require cognitive resource, because during the distraction task the amount of gait modulation was reduced (Fig. 5.7A). This suggests that gait patterns are not only explicitly but also implicitly influenced by visual feedback. We demonstrated that the implicit visual feedback distortion influenced all subjects and induced them to modulate their gait step symmetry. Moreover, the observed step symmetry changes systematically increased or decreased according to the distortion levels. This implies that subjects spontaneously modulated their spatial gait pattern to compensate for the asymmetric representations of visual feedback. In addition, it has previously been shown that changing the speed of virtual environments using virtual reality technology influences balance or modulates walking speed [41–44]. These results indicate there is a link between the supraspinal visual perception of movement and locomotion [45].

Interestingly, during post-experiment debriefing, all subjects confirmed the subliminal essence of the protocol: they neither noticed the visual distortions nor realized the changes in their gait symmetry. They reported that they were maintaining their natural gait pattern, remaining unaware of changes away from a symmetric gait pattern. While there were some suggestions that CPG residing at the spinal cord level is responsible for gait in humans, our results suggest that (at least for humans) visual perturbations can affect the other afferent signals coming from the LE [41,46].

5.3.6 Perturbation response—auditory perturbation at the supraspinal level

Different types of sound cues have also been used to adapt the human gait rhythm. We investigated whether young healthy volunteers followed subliminal

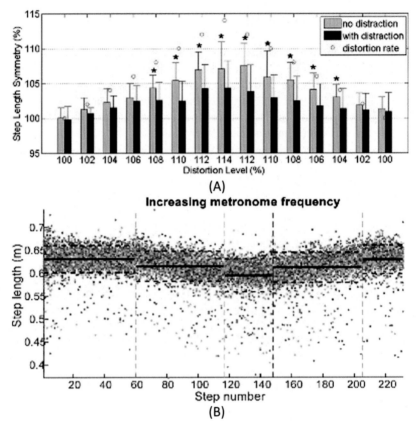

FIGURE 5.7 Supraspinal influence in gait. (A) Changes in step length symmetry (ratio) as a function of varying visual feedback distortion ± 1 standard deviation, averaged across all of the subjects during no-distraction and with-distraction conditions. The gray bars show changes for the no-distraction condition, and the black bars show them for the distraction condition. The horizontal axis shows changes in the distortion rate from the onset of the trial, and the vertical axis shows changes in mean step length symmetry ratio. The step length ratio was calculated by 100* (the measured left step length)/(the measured right step length). The small circles indicate distortion changes applied during trials. The asterisks (*) were marked at distortion levels where the induced step symmetry values were shown to be significantly different between the no-distraction and with-distraction conditions. (B) Step length (in m) for all participants under auditory increasing frequency condition. The vertical lines indicate the instants of the metronome frequency transitions. The horizontal lines represent the mean (solid) and standard deviation (dashed). The panel shows the subliminal decrease and increase step size with increased frequency which are closely followed by the subjects.

metronome rhythm changes during gait. The metronome rhythm was increased or decreased, without informing the subjects, at 1 msec increments or decrements to reach, respectively, a low- (596 msec) or a high-frequency (536 msec) plateau. After 30 steps at one of these isochronous conditions, the rhythm returned to the original period with decrements or increments of 1 msec. Motion data were recorded with an optical measurement system to determine the timing of heel strike during the gait cycle. The relative phase between sound cue (stimulus) and heel strike or foot contact (response) was compared. The results demonstrated that gait was entrained to the rhythmic auditory stimulus and subjects subconsciously adapted the step time and length to maintain treadmill speed, while following the rhythm changes (Fig. 5.7B). In most cases, there was a lead asynchrony: the foot contact occurred before the sound cue. These results showed that the gait period is strongly "entrained" with the first metronome rhythm, while subjects still followed metronome changes with larger error. Results from Forner-Cordero and colleagues' study suggest two processes: one slow-adapting, supraspinal oscillator with persistence that predicts the foot contact to occur ahead of the stimulus, and a second fast process linked to sensory inputs that adapts to the mismatch between peripheral sensory input (foot contact) and supraspinal sensory input (auditory rhythm) [47].

5.3.7 Perturbation response—sleep deprivation

Additional support for the role of supraspinal control of gait is provided by sleep deprivation studies [48]. Different levels of sleep restriction affect human performance in multiple aspects. However, it is unclear how sleep deprivation affects gait control. We applied a paced gait paradigm (described above) that included subliminal auditory rhythm changes to analyze the effects of different sleep restriction levels (acute, chronic, and control) on performance. The acute sleep deprivation (one night) group exhibited impaired performance in the sensorimotor synchronization gait protocol, such as a decrease in the period error between the heel strike and the auditory stimulus, as well as more frequently missing the auditory cues. The group with chronic sleep restriction also underperformed when compared to the control group, suggesting that partial or total sleep deprivation leads to a decrease in the performance in the sensorimotor control of gait. The results showed that subliminal supraspinal rhythmic compensation in gait is affected by sleep restriction.

In summary, human locomotion appears to be far more than CPGs, particularly in stroke rehabilitation. It appears to require the integration of different sensory information including visual, vestibular, and somatosensory systems [49,50]. With so much conflicting evidence in LE and gait rehabilitation after stroke, we opted to focus on incomplete spinal cord injury (iSCI) where the influence of supraspinal damage does not confound the recovery of gait. In particular, we will examine a recently published study employing an exoskeleton in chronic iSCI. This patient population is expected to benefit from robotic interventions that target load receptor input for physiological leg muscle activation during stance and gait [28,29]. Furthermore, the study somewhat mimics previously mentioned upper-extremity trials: the trial lasted for

FIGURE 5.8 One of the authors trialing the Ekso Bionics device at an event at Moss Rehabilitation Hospital (Philadelphia, USA).

12 weeks, 36 sessions, and afforded the comparison of the active intervention with usual care in addition to no training—but no active control [51].

Edwards and colleagues ran a multisite randomized controlled trial in the United States comparing exoskeleton gait training (Ekso Bionics, CA, USA— see Fig. 5.8), consisting of a minimum of 300 steps in 45 minutes (excluding donning–doffing), with standard gait training or no training (2:2:1 randomization). The trial lasted 12 weeks, 3 times per week, each session lasting 1 hour, and included a "phase-in" period that involved 1−3 participants per site to eliminate any procedural issues. Participants had chronic iSCI (> 1 year post-injury, American Spinal Injury Association Impairment Scale C and D) with residual stepping ability. The primary outcome measure was change in robot-independent gait speed (10-m walk test, 10MWT) post-12-week intervention. Secondary outcomes included: Timed-Up-and-Go (TUG), 6-min walk test (6MWT), Walking Index for Spinal Cord Injury (WISCI-II) (assistance and devices), and treating therapist NASA-Task Load Index. The study included 25 participants in the analysis (excluding phase-in participants). The mean change in gait speed at the primary outcome was not significant. Improvements in secondary outcomes were not significant. While there was no group difference, it is important to notice that 1/2 of the participants in the exoskeleton group and 1/3 of the usual care group changed clinical ambulatory category (household ambulation defined as self-selected walking speed ≤ 0.44 m/s to limited or full community ambulation >0.44 m/s). As anticipated, no participant in the no-training group changed category.

One must take these results with the appropriate caveats, particularly because the number of participants is small. While none of the outcomes demonstrated the superiority of any approach, the fact that both groups receiving therapy showed changes in the ambulatory category ($> 1/3$ of the participants) strongly suggest that there is some merit to this kind of gait intervention in iSCI. Perhaps, the choice of intervention might be dictated by additional considerations, which future studies should aim to explore and understand.

5.4 Socially assistive robots: the potential to bridge the gap in human care

Socially assistive robots (SARs) are distinct from their rehabilitation robot counterparts by providing largely social rather than physical assistance and interaction. The SARs field is a subset of service robotics that aims to empower and improve the quality of life of humans across all ages and various conditions by providing nonphysical assistance and behavioral change through social motivation and reward [52]. The importance of socially acceptable and meaningful human−robot interactions has become more apparent as service robots have expanded beyond the industrial and manufacturing arenas into various everyday environments including hospital, education, care, and rehabilitation settings (as in the previous UE and LE robotic sections of this chapter) and even the home.

The coronavirus (COVID-19) pandemic has led to an unparalleled global crisis for alternative means of social interaction and companionship. Never in modern history have humans been so isolated. SARs have the potential to bridge the gap in human care, but not replace, by providing purposes such as monitoring, coaching, motivating, and companionship [53,54]. The technology employs recognition and analysis software to process verbal tones and cues, facial expressions, and gestures to tailor assistance and guidance. Like your favorite mentor, teacher, or coach, SARs can be designed to provide individualized social, emotional, and cognitive support and prompts to facilitate learning, development, and rehabilitation. In addition, an appropriate and effective SAR must do all this in a safe and ethical way. The primary domain in which SARs are currently being employed and researched is in the healthcare field. In this brief introduction, we will highlight the evidence and application of SARs specifically in stroke rehabilitation and the care of older adults.

5.4.1   Socially assistive robots caring for older adults

The world's aging population means the demand for care to assist with daily tasks will increase, and there will be an unprecedented need for aging-in-place services to allow older adults to remain safely in their homes and live independently. Although older adults will undoubtedly require physical assistive tools (for mobility and carrying out activities of daily living [ADLs] such as groceries, washing, and cleaning), there will also be a high demand for nonphysical assistance such as cognitive strategies (medication or appointment reminders), social supports (to reduce the risk of isolation and deterioration in mental health), and lifestyle tools (encouragement to engage in physical activity and make healthy food choices to reduce the risk of obesity and chronic disease).

There is both a growing body of literature and clinical adoption of SARs in the care of older adults around the world. One of the most renowned SARs for the care of older adults, PARO a seal robot (Fig. 5.9A), has been used to explore the effectiveness of SARs in improving social interaction, mental health, and engagement in older adults with dementia. PARO is a nonverbal SAR with the ability to move, blink, and make sounds and has temperature, light, sound, posture, and tactile sensors which respond to being stroked and can learn to respond by name [55]. Overall studies reported improved depression, anxiety, mood, stress (using urinary samples) [56], physical activity [57], and quality of life when PARO was used in supervised or unsupervised groups compared to a range of control conditions (reading groups, music, physical activity, toy, or dog interactions) [56,58−60].

The Sony AIBO therapy (Fig. 5.9B) and PARO have also been investigated as a solution for companionship. The AIBO is also a nonverbal dog-like robot which has visual sensors, the ability to walk and interpret commands as well as learn human interaction, and express emotional responses [55]. Banks et al. and Kanamori et al. showed SARs can be effective at reducing loneliness, improving quality of life, amount of speech generated, and satisfaction in an older adult population [61,62].

Studies using SoftBank Robotics' humanoid SAR technologies, NAO and Pepper (Fig. 5.9C and D, respectively), have evaluated the effectiveness of monitoring, coaching, and providing feedback during health monitoring and exercise regimes for older adults [63]. Although usability issues impacted the

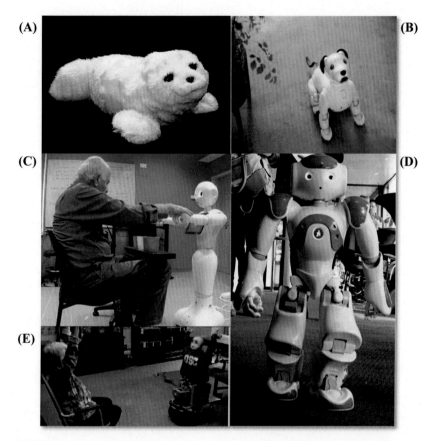

FIGURE 5.9 The PARO robot (A) and Sony AIBO Dog (B) have largely been studied for their effectiveness in providing companionship, engagement, and social interaction in older adult care settings. The Pepper (C) and NAO robot (D) have been used to investigate the use of SARs for providing exercise coaching in the care or rehabilitation environments whereas (E) shows Maja Matarić's work out of the University of Southern California, performing seated exercise coaching within the home environment.

participant's experience in the study by Olde Keizer et al., the NAO was still rated positively regarding its personal approach, usefulness, and enjoyment. One of the many Pepper robot experiments with older adults involved a gamified cognitive motor task (Fig. 5.4) where the robot provided instructions and feedback about the performance or outcome, with the robotic feedback being both verbal and gestural. Similar to the NAO, the results showed a high level of interaction, engagement, and enjoyment with the Pepper robot [64].

As highlighted in the brief summary of the evidence available to date, there is great promise for SARs in optimizing older adult care through cognitive training, social interaction, health monitoring, physical training, companionship, and affective therapy. Nonetheless, the scale of the studies, the usability of the robots, and the quality of the outcome measures selected leave many more unaddressed issues and questions which should be tackled before the practical and widespread translation of these robots into care or home environments.

5.4.2 SARs for hands-off stroke rehabilitation

The high incidence of stroke in the growing aging population means there will be a shortage of resources to provide the intensity and frequency of therapy as well as the motivating, engaging, and meaningful rehabilitation required to improve functional outcomes post-stroke. There is mounting evidence for the value of noncontact therapy within the toolbox of rehabilitation care as a way of supplementing, enhancing, and optimizing physical therapy for certain patient populations [52]. The potential for noncontact therapy provision in stroke rehabilitation has been highlighted by success in both constraint-induced movement therapy [65] and telerehabilitation studies. In the work leading up to the telerehabilitation intervention design, the authors emphasize the importance of the quality of the patient's rehabilitation experience; brain plasticity is greater when a task is challenging and varied, accompanied by meaningful feedback, motivating, interesting, and relevant [66]. In addition, poor adherence to therapy is often linked to lack of motivation [67]. It seems that motivation, coaching, and appropriate feedback are key elements of successful rehabilitation programs for stroke survivors. Yet human therapists are not only in increasingly short supply but are also vulnerable to the flaws of human nature, fatigue, inconsistent feedback, impatience, bias, and unrealistic expectations to name a few. This is where SAR technology may add significant value to the patient experience in rehabilitation by providing high-quality, individualized, meaningful encouragement, coaching, and motivation. As eloquently stated by Maja Matarić, SARs "are infinitely patient...they have fewer biases, and they have no expectations" [68].

In addition, SARs have the potential to provide detailed and objective assessment data on patient progress through motion capture, camera, and wearable sensor technology, like the robotic outcome measures generated by physically assisted rehabilitation robots. In one study, the authors used inertial measurement unit (IMU) sensors to detect and record real-time data on the functional use of the stroke survivor's UE during their exercise program [52]. The data were also used by the SAR to provide feedback and coach the patient by encouraging the use of their UE or correcting UE movement patterns as appropriate (Fig. 5.9E). The findings of the study reported the robot was well accepted by patients and therapists, and there was improved exercise adherence with the SAR compared to the control condition.

A noteworthy trial using the autonomous Bandit SAR (out of the University of Southern California, Fig. 5.9E) for a seating UE exercise program with stroke survivors investigated the most effective form of feedback and motivation for improving self-efficacy [69]. Self-efficacy, or one's belief in their ability to complete a task or achieve a goal, has close ties to motivation and has been proposed to mediate motor performance and rehabilitation outcomes after stroke. Swift-Spong et al. [69] hypothesized that SAR coaching and motivating strategies may facilitate greater self-efficacy and, in turn, improve rehabilitation outcomes. The results of the study showed performance of the UE reaching task improved with the use of the SAR and that self-comparative or no-comparative feedback was most effective (as opposed to other or external comparative feedback) for optimizing timed motor performance. Self-efficacy responses were however mixed, suggesting that optimizing self-efficacy using SARs will likely require a personalized approach to system design.

These study results and others prompted investigation into the individualization of the SAR to optimize human−robot interaction and patient experience for stroke survivors. The previously mentioned study by Mataric et al. [52] and subsequent follow-up studies explored the patient's personality and the design of adaptive robot-assisted protocols [70−72]. The methods compared a computer agent to a physically present colocated SAR. The content of the program was identical, but the medium for the delivery of feedback, motivation, and encouragement varied between the computer and robot groups. Participants rated the colocated SAR condition as significantly more enjoyable and entertaining, more valuable/useful, and a significantly better companion and exercise coach.

Like any technology, there are limitations and considerations that prevent the adoption, integration, and translation of research findings into the clinical environment. Given the nature of SARs, the ethical considerations for their design and implementation are vast and extensive [73]. There are many concerns, with the main themes revolving around safety, security, and privacy of personal data, deception and violation of human dignity, loss of autonomy, loss of human contact, and responsibility of care to name just a few. Given SARs immense potential to benefit populations who are vulnerable and at risk, it is important that these limitations and concerns are addressed so the full potential of SARs can be realized in our society.

Another logical next step in this research field is to combine the best attributes of physically and socially assisted rehabilitation robotic technologies to provide a holistic patient-centered therapy platform. If both physically and socially assistive technologies can be incorporated into one device, we may be able to optimize and enhance the functional outcomes of patients beyond what has been achieved to date in either field alone. This future direction highlights the importance of ongoing collaborations among clinicians, engineers, researchers, consumers, and administrators to optimize the quality, value, and adoption of robotics across rehabilitation settings around the world.

5.5 Conclusions

While robots may play an important role in the advancement and optimization of rehabilitation of neurological disease or care of older adults, they are not standalone tools. There is further room for improvements in the technology, combination with other therapies to augment patient outcomes, and the prescription of individualized therapy based on an in-depth understanding of the neurophysiological mechanisms underlying both the patient's condition and the therapy provided. In addition, there is an increasing focus on rehabilitation technology being able to promote functional independence and restore the performance of ADLs. With this, it is possible we may see a shift toward more wearable UE, LE, and social robotics systems that can integrate into daily life and have both therapeutic and assistive applications.

A longstanding challenge for the rehabilitation robotic field has been the lack of widespread adoption of the technology into clinical settings or acceptance into the home environment. This has been partly driven by high costs, inconsistencies in international regulatory requirements, and the absence of a clear reimbursement pathway to offset the investment in the technology and the associated training. Rehabilitation robotic commercial solutions must also consider many factors and overcome many barriers, including but not limited to the size and space required for the installation of the device, upfront and ongoing costs, training requirements and the time commitment for staff, patient setup time, clinician or carer supervision, required cleaning, infection control procedures, and maintenance, patient comfort, clinician and patient ease of use, reporting capabilities for outcomes and utilization of the device, and efficiency and effectiveness of patient care compared to conventional techniques.

The likely solution to addressing these barriers and challenges and the parting message for this chapter is to emphasize the importance of collaboration: collaboration from the major stakeholders within the industry, collaboration between industry leaders and government bodies, and collaboration among the engineers, researchers, clinicians, and patients to have a chance in realizing the potential of rehabilitation robotics and optimizing the outcomes of our patients across the globe.

References

[1] Nudo RJ, Wise BM, SiFuentes F, Milliken GW. Neural substrates for the effects of rehabilitative training on motor recovery after ischemic infarct (in eng) Science 1996; 272(5269):1791−4.

[2] Nudo RJ, Plautz EJ, Frost SB. Role of adaptive plasticity in recovery of function after damage to motor cortex. Muscle Nerve 2001;24(8):1000−19.

[3] Edgerton VR, Tillakaratne NJ, Bigbee AJ, de Leon RD, Roy RR. Plasticity of the spinal neural circuitry after injury. Annu Rev Neurosci 2004;27:145−67.

[4] Hebb DO. The organization of behavior: a neuropsychological theory. New York: Wiley and Sons; 1949.

[5]   Kleim JA, Jones TA. Principles of experience-dependent neural plasticity: implications for rehabilitation after brain damage. J Speech Lang Hear Res 2008;51(1):S225—39.

[6]   Krebs HI, Hogan N, Hening W, Adamovich SV, Poizner H. Procedural motor learning in Parkinson's disease (in eng) Exp Brain Res 2001;141(4):425—37.

[7]   Patton JL, Mussa-Ivaldi FA. Robot-assisted adaptive training: custom force fields for teaching movement patterns. IEEE Trans Biomed Eng 2004;51(4):636—46.

[8]   Patton JL, Stoykov ME, Kovic M, Mussa-Ivaldi FA. Evaluation of robotic training forces that either enhance or reduce error in chronic hemiparetic stroke survivors. Exp Brain Res 2006;168(3):368—83.

[9]   Miller EL, et al. Comprehensive overview of nursing and interdisciplinary rehabilitation care of the stroke patient: a scientific statement from the American Heart Association (in eng) Stroke 2010;41(10):2402—48.

[10] Winstein CJ, et al. Guidelines for adult stroke rehabilitation and recovery: a guideline for healthcare professionals from the American Heart Association/American Stroke Association. Stroke 2016;47(6):e98—e169.

[11] Mehrholz J, Pohl M, Platz T, Kugler J, Elsner B. Electromechanical and robot-assisted arm training for improving activities of daily living, arm function, and arm muscle strength after stroke. Cochrane Database Syst Rev 2018;9:CD006876.

[12] Mehrholz J, Pollock A, Pohl M, Kugler J, Elsner B. Systematic review with network meta-analysis of randomized controlled trials of robotic-assisted arm training for improving activities of daily living and upper limb function after stroke. J Neuroeng Rehabil 2020;17(1):83.

[13] Mehrholz J, Thomas S, Kugler J, Pohl M, Elsner B. Electromechanical-assisted training for walking after stroke. Cochrane Database Syst Rev 2020;10:CD006185.

[14] Negrini S, Arienti C, Kiekens C. Usual care: the big but unmanaged problem of rehabilitation evidence. Lancet 2020;395(10221):337.

[15] Edwards DJ, et al. Clinical improvement with intensive robot-assisted arm training in chronic stroke is unchanged by supplementary tDCS. Restor Neurol Neurosci 2019;37 (2):167—80.

[16] Moretti CB, et al. Robotic Kinematic measures of the arm in chronic stroke: part 1 - Motor Recovery patterns from tDCS preceding intensive training. Bioelectron Med 2021;7(1):20.

[17] Moretti CB, et al. Robotic Kinematic measures of the arm in chronic stroke: part 2 - strong correlation with clinical outcome measures. Bioelectron Med 2021;7(1):21.

[18] Lo AC, et al. Multicenter randomized trial of robot-assisted rehabilitation for chronic stroke: methods and entry characteristics for VA ROBOTICS (in eng) Neurorehabil Neural Repair 2009;23(8):775—83.

[19] Lo AC, et al. Robot-assisted therapy for long-term upper-limb impairment after stroke (in eng) N Engl J Med 2010;362(19):1772—83.

[20] Rodgers H, et al. Robot assisted training for the upper limb after stroke (RATULS): study protocol for a randomised controlled trial. Trials 2017;18(1):340.

[21] Rodgers H, et al. Robot assisted training for the upper limb after stroke (RATULS): a multicentre randomised controlled trial. Lancet 2019;394(10192):51—62.

[22] Kwakkel G, et al. Standardized measurement of sensorimotor recovery in stroke trials: consensus-based core recommendations from the Stroke Recovery and Rehabilitation Roundtable. Int J Stroke 2017;12(5):451—61.

[23] Wu X, Guarino P, Lo AC, Peduzzi P, Wininger M. Long-term effectiveness of intensive therapy in chronic stroke. Neurorehabil Neural Repair 2016;30(6):583—90.

[24] Wilson N, Howel D, Bosomworth H, Shaw L, Rodgers H. Analysing the Action Research Arm Test (ARAT): a cautionary tale from the RATULS trial. Int J Rehabil Res 2021; 44(2):166−9.

[25] Management of Stroke Rehabilitation Working Group. VA/DOD Clinical practice guideline for the management of stroke rehabilitation (in eng) J Rehabil Res Dev 2010; 47(9):1−43.

[26] Hornby TG, Campbell DD, Kahn JH, Demott T, Moore JL, Roth HR. Enhanced gait-related improvements after therapist- versus robotic-assisted locomotor training in subjects with chronic stroke: a randomized controlled study (in eng) Stroke 2008;39(6):1786−92.

[27] Hidler J, et al. Multicenter randomized clinical trial evaluating the effectiveness of the Lokomat in subacute stroke (in eng) Neurorehabil Neural Repair 2009;23(1):5−13.

[28] Dietz V. Spinal cord pattern generators for locomotion (in eng) Clin Neurophysiol 2003;114(8):1379−89.

[29] Ivanenko YP, Grasso R, Macellari V, Lacquaniti F. Control of foot trajectory in human locomotion: role of ground contact forces in simulated reduced gravity (in eng) J Neurophysiol 2002;87(6):3070−89.

[30] Duncan PW, et al. Body-weight-supported treadmill rehabilitation after stroke (in eng) N Engl J Med 2011;364(21):2026−36.

[31] Dobkin BH, Duncan PW. Should body weight-supported treadmill training and robotic-assistive steppers for locomotor training trot back to the starting gate? (in eng) Neurorehabilit Neural Repair 2012;26(4):308−17.

[32] Duysens J, Forner-Cordero A. Walking with perturbations: a guide for biped humans and robots. Bioinspir Biomim 2018;13(6):061001.

[33] Schillings AM, Van Wezel BM, Mulder T, Duysens J. Widespread short-latency stretch reflexes and their modulation during stumbling over obstacles. Brain Res 1999; 816(2):480−6.

[34] Schillings AM, van Wezel BM, Mulder T, Duysens J. Muscular responses and movement strategies during stumbling over obstacles. J Neurophysiol 2000;83(4):2093−102.

[35] Seiterle S, Susko T, Artemiadis PK, Riener R, Igo Krebs H. Interlimb coordination in body-weight supported locomotion: a pilot study (in eng) J Biomech 2015; 48(11):2837−43.

[36] Susko T, Swaminathan K, Krebs HI. MIT-skywalker: a novel gait neurorehabilitation robot for stroke and cerebral palsy. IEEE Trans Neural Syst Rehabil Eng 2016; 24(10):1089−99.

[37] Goncalves RS, Krebs HI. MIT-Skywalker: considerations on the design of a body weight support system. J Neuroeng Rehabil 2017;14(1):88.

[38] Goncalves RS, Hamilton T, Daher AR, Hirai H, Krebs HI. MIT-Skywalker: evaluating comfort of bicycle/saddle seat. IEEE Int Conf Rehabil Robot 2017;2017:516−20.

[39] Kim SJ, Krebs HI. Effects of implicit visual feedback distortion on human gait (in eng) Exp Brain Res 2012;218(3):495−502.

[40] Kim SJ, Krebs HI. Implicit visual distortion modulates human gait," (in eng) Conf Proc IEEE Eng Med Biol Soc 2011;2011:3079−82.

[41] Varraine E, Bonnard M, Pailhous J. Interaction between different sensory cues in the control of human gait (in eng) Exp Brain Res 2002;142(3):374−84.

[42] Sheik-Nainar MA, Kaber DB. The utility of a virtual reality locomotion interface for studying gait behavior (in eng) Hum Factors 2007;49(4):696−709.

[43] Prokop T, Schubert M, Berger W. Visual influence on human locomotion. Modulation to changes in optic flow (in eng) Exp Brain Res 1997;114(1):63−70.

[44] Lamontagne A, Fung J, McFadyen BJ, Faubert J. Modulation of walking speed by changing optic flow in persons with stroke (in eng) J Neuroeng Rehabil 2007;4:22.

[45] Patla AE, Niechwiej E, Racco V, Goodale MA. Understanding the contribution of binocular vision to the control of adaptive locomotion (in eng) Exp brain Res 2002; 142(4):551−61.

[46] Pailhous J, Ferrandez AM, Fluckiger M, Baumberger B. Unintentional modulations of human gait by optical flow (in eng) Behav Brain Res 1990;38(3):275−81.

[47] Forner-Cordero A, et al. Effects of supraspinal feedback on human gait: rhythmic auditory distortion. J Neuroeng Rehabil 2019;16(1):159.

[48] Umemura GS, Pinho JP, Duysens J, Krebs HI, Forner-Cordero A. Sleep deprivation affects gait control. Sci Rep 2021;11(1):21104.

[49] Armstrong DM. Supraspinal contributions to the initiation and control of locomotion in the cat (in eng) Prog Neurobiol 1986;26(4):273−361.

[50] Rossignol S. Visuomotor regulation of locomotion (in eng) Can J Physiol Pharmacol 1996;74(4):418−25.

[51] Edwards DJ, et al. Walking improvement in chronic incomplete spinal cord injury with exoskeleton robotic training (WISE): a randomized controlled trial. Spinal Cord 2022; 60(6):522−32.

[52] Mataric MJ, Eriksson J, Feil-Seifer DJ, Winstein CJ. Socially assistive robotics for post-stroke rehabilitation. J Neuroeng Rehabil 2007;4:5.

[53] Getson C, Nejat G. The adoption of socially assistive robots for long-term care: during COVID-19 and in a post-pandemic society. Healthc Manage Forum 2022;35(5):301−9.

[54] Van Assche M, Moreels T, Petrovic M, Cambier D, Calders P, Van de Velde D. The role of a socially assistive robot in enabling older adults with mild cognitive impairment to cope with the measures of the COVID-19 lockdown: a qualitative study. Scand J Occup Ther 2021;1−11.

[55] Abdi J, Al-Hindawi A, Ng T, Vizcaychipi MP. Scoping review on the use of socially assistive robot technology in elderly care. BMJ Open 2018;8(2):e018815.

[56] WadaK., ShibataT., SaitoT., TanieK. Psychological, physiological and social effects to elderly people by robot assisted activity at a health service facility for the aged. In: Presented at the IEEE/ASME International Conference on Advanced Intelligent Mechatronics, AIM, 1., 2003.

[57] Sabanovic S, Bennett CC, Chang WL, Huber L. PARO robot affects diverse interaction modalities in group sensory therapy for older adults with dementia. IEEE Int Conf Rehabil Robot 2013;2013:6650427.

[58] Moyle W, et al. Exploring the effect of companion robots on emotional expression in older adults with dementia: a pilot randomized controlled trial. J Gerontol Nurs 2013; 39(5):46−53.

[59] Petersen S, Houston S, Qin H, Tague C, Studley J. The utilization of robotic pets in dementia care. J Alzheimers Dis 2017;55(2):569−74.

[60] Thodberg K, et al. Therapeutic effects of dog visits in nursing homes for the elderly. Psychogeriatrics 2016;16(5):289−97.

[61] Banks J, Koban K. Framing effects on judgments of social robots' (Im)Moral Behaviors. Front Robot AI 2021;8:627233.

[62] Kanamori S, et al. Social participation and the prevention of functional disability in older Japanese: the JAGES cohort study. PLoS One 2014;9(6):e99638.

[63] Olde Keizer CM, et al. Using socially assistive robot monit preventing frailty older adults: a study usability user experience challenges. Health Technol 2019;9(36).

[64] Kellmeyer P, Mueller O, Feingold-Polak R, Levy-Tzedek S. Social robots in rehabilitation: a question of trust. Sci Robot 2018;3(21).

[65] Wolf SL, et al. Effect of constraint-induced movement therapy on upper extremity function 3 to 9 months after stroke: the EXCITE randomized clinical trial (in eng) JAMA 2006;296(17):2095−104.

[66] Cramer SC, et al. Harnessing neuroplasticity for clinical applications. Brain 2011; 134(Pt 6):1591−609.

[67] Oyake K, Suzuki M, Otaka Y, Tanaka S. Motivational strategies for stroke rehabilitation: a descriptive cross-sectional study. Front Neurol 2020;11:553.

[68] CarusoC. Grandma's Little Robot. Scientific American. Available from: https://www.scientificamerican.com/article/grandma-rsquo-s-little-robot/; 2017.

[69] Swift-Spong K, Short E, Wade E, Mataric MJ. Effects of comparative feedback from a Socially Assistive Robot on self-efficacy in post-stroke rehabilitation. IEEE Int Conf Rehabil Robot 2015.

[70] Fasola J, Mataric M. Comparing physical and virtual embodiment in a socially assistive robot exercise coach for the elderly. Technical Rep Cres-11-003 2011.

[71] Fasola J, Mataric M. A socially assistive robot exercise coach for the elderly. J Human-Robot Interact 2013;2(2).

[72] Matarić M, Tapus A, Winstein C, Eriksson J. Socially assistive robotics for stroke and mild TBI rehabilitation. Stud Health Technol Inform 2009;145.

[73] Boada JP, Maestre BR, Genís CT. The ethical issues of social assistive robotics: a critical literature review. Technol Soc 2021;67:101726.

Part II

Background on country healthcare systems, rehabilitation standards of care, stroke rehabilitation, and rehabilitation and assistive robotics in selected HICs

Chapter 6

North America and Caribbean region: USA

Rochelle J. Mendonca[1], Carol A. Wamsley[2], Chung-Ying Owen Tsai[3], Hao Su[4] and Michelle J. Johnson[5]

[1]*Department of Rehabilitation and Regenerative Medicine, Programs in Occupational Therapy, Columbia University, New York, NY, United States,* [2]*Therapy Division, Good Shepherd Penn Partners, Penn Institute for Rehabilitation Medicine, Philadelphia, PA, United States,* [3]*Department of Rehabilitation and Human Performance, Icahn School of Medicine at Mount Sinai, New York, NY, United States,* [4]*Mechanical and Aerospace Engineering Department, Lab of Biomechatronics and Intelligent Robotics, North Carolina State University, Raleigh, NC, United States,* [5]*Departments of Physical Medicine and Rehabilitation, Bioengineering, and Mechanical Engineering and Applied Mechanics, University of Pennsylvania, Philadelphia, PA, United States*

Learning objectives

By the end of this chapter, readers will be able to:
1. Describe the healthcare and rehabilitation care for stroke survivors in the United States.
2. Describe the advances in research and implementation of rehabilitation robotics in the United States.
3. Articulate the challenges and needs related to rehabilitation robotics in the United States.

6.1 Introduction

6.1.1 Overview of the United States of America and health facts

The United States is located primarily in North America and includes 50 states, the District of Columbia, and unincorporated territories and minor possessions. As per the World Bank categorization, the United States is considered a high-income country (HIC) based on gross national income per capita [1]. It is the third-largest country in the world based on landmass, has a population of approximately 332 million [2], and is geographically diverse.

Rehabilitation Robots for Neurorehabilitation in High-, Low-, and Middle-Income Countries.
DOI: https://doi.org/10.1016/B978-0-323-91931-9.00035-9

Cardiovascular diseases, including heart disease and stroke, are among the top 5 causes of mortality, and stroke is the leading cause of significant long-term disability in the United States [3]. Currently, 7 million persons, aged 20 years and older, report living with stroke; projections indicate that by 2030 another 3.4 million adults will experience a stroke and 10.8 million people will be living with residual disability due to stroke [4,5]. With the country's geographical diversity, aging population, and projected shortage of rehabilitation service providers, there are inequities in providing health and rehabilitation care to large segments of the US population [6−8]. This highlights the need to explore new and innovative techniques, including technology, to provide rehabilitation care.

6.1.2   Stroke care and rehabilitation in the United States

The United States has a multitiered system of care for patients with stroke [9]. The average costs for stroke care in the first year range from $44,929 to $61,354, and the lifetime cost is over $140,000 including inpatient care, rehabilitation, and long-term care. It is estimated that the average yearly cost of rehabilitation services is $11,689 and increases as the level of disability increases [10−12]. In the United States, healthcare is provided via private insurance and government national insurance (Medicare or Medicaid) [13] and includes the cost of rehabilitation.

Medical advances are leading to more stroke survivors living longer with disabilities, including physical, cognitive, sensory, and emotional impairments. About 65%−75% of individuals with stroke in the United States have residual motor impairment in the upper and lower limbs, and about 39%−61% have cognitive impairment even 10 years poststroke [14−17]. Medical care for stroke usually begins in intensive care units within acute care hospitals. Depending on the complexity of medical care required, the patient may stay in the hospital on average for 4−7 days [18].

It is usually recommended that rehabilitation in the acute stage begins as soon as the patient can tolerate it [19]. Research has shown that early rehabilitation after the incidence of stroke prevents complications and secondary weakness and restores function [20,21]. After receiving rehabilitation in the acute care unit of the hospital, the patient and their family (with the help of a social worker or case manager) choose the best next care setting. Studies approximate that 43.1% of stroke survivors are discharged to home, 34.2%−74% to inpatient rehabilitation facilities, 13%−33.4% to skilled nursing facilities, 0.7% to long-term acute care hospitals, and 13% to home care programs [18]. It is estimated that 45% of hospitalized Medicare recipients are discharged to home directly, with 4 out of 10 not receiving post-acute care. Studies suggest that patients who go to IRF have better outcomes, fewer readmissions, and lower mortality than those who go to SNF, though at a greater cost.

The duration of therapy and length of stay will vary depending on the setting. Patients in inpatient rehabilitation facilities are expected to undergo 3 hours of therapy for at least 5 days a week, and the typical length of stay in IRFs is 8−23 days [22]. In SNFs, also known as subacute rehabilitation, patients receive about 1−2 hours of therapy 5 days a week from physical, occupational, and speech therapy, in addition to skilled nursing care. The average length of stay for poststroke rehabilitation in a SNF is about 32 days [23]. Outpatient therapy can be provided through comprehensive outpatient rehabilitation facilities (CORFs) or outpatient departments of acute care and rehabilitation hospitals, in rehabilitation clinics, and doctor's offices where diagnostic, therapeutic, and restorative services are provided about 1−3 times a week for about 60 minutes per service [24]. After being discharged from acute care hospitals, homebound patients typically receive home-based rehabilitation programs provided by home health agencies, inpatient rehabilitation facilities, or skilled nursing facilities approximately 1−3 times a week [19]. Long-term acute care hospitals are special hospital units that care for patients with major medical problems, where comprehensive rehabilitation services (OT/PT/ST) are provided based on the patient's tolerance level [13,19]. Stroke care can also be provided through community-based rehabilitation programs [25].

Rehabilitation across all these settings usually includes an assessment by occupational therapy (OT), physical therapy (PT), and speech therapy (ST) as needed and focuses on early mobilization, emotional support, and education prior to discharge to different settings. Therapy interventions across these care systems may include early mobilization, therapeutic exercises and activities, mobility and gait training, activities of daily living (ADL) training, cognitive and linguistic services, swallowing treatment, driving evaluation and rehabilitation, pool therapy, orthotic services, adaptive and assistive technology services, wheelchair clinics, vocational rehabilitation, emotional support, education, and robotic and virtual reality services [26,27].

Post-acute stroke care is effective; however, costs are relatively high. Health insurance coverage and high costs lead to increasing disparity in access to quality care and contribute to poor health outcomes for individuals who require rehabilitation. The forecasted shortage of healthcare providers adds to the need for innovative solutions to augment health service delivery [28−30]. Advances in state-of-the-art technology-assisted rehabilitation, especially robot-assisted ones, show promise to help assess impairment, augment care, leverage the limited supply of rehabilitation practitioners, and improve overall patient outcomes. Section 6.2 of this chapter reviews robot-assisted neurorehabilitation in the United States, and Section 6.3 discusses perceived barriers to widespread use and possible future directions including implications of COVID-19 for rehabilitation care delivery.

6.2   Robot-assisted neurorehabilitation

Technology can help bridge the gap in healthcare, especially rehabilitation. In the United States, several types of technology qualify as assistive technology (AT). Cooper and colleagues [31] define therapeutic AT as "tools for remediation or rehabilitation that are not a part of a person's daily life and functional activities," primarily to rehabilitate limitations occurring due to some disabling condition. Advanced technologies such as rehabilitation robots are ATs that can function as therapy robots and/or assistive robots that focus on supporting the motor recovery of the upper or lower limbs. More recently, there has been a focus on providing cognitive rehabilitation, including for persons with dementia or Alzheimer's. The benefits of rehabilitation robots are that they can provide customized, task-oriented, prolonged, intensive, standardized, and reproducible training to patients as well as objective assessment of therapy effects.

6.2.1   Upper-limb robot-assisted neurorehabilitation

Long-term impairments in stroke survivors can impact the ability to perform ADL and, consequently, the ability to live independently. Upper-extremity impairments may present as weakness, limited range of motion, limited fine and gross hand dexterity for grasping and manipulation, spasticity, or abnormal flexor synergies. The practice of common ADLs, such as drinking, is a crucial goal of occupational therapy and often involves the practice of reaching, grasping, and manipulation skills, in and out of the plane.

In the United States, the development efforts around rehabilitation robots began in the 1980s. In the early 1990s, several leading institutions began collaborations to advance the field, including the Massachusetts Institute of Technology (MIT) with Burke Rehabilitation Hospital and Spaulding Hospital, Stanford University with the Department of Veterans Affairs (VA) Palo Alto, the University of Delaware with A.I. Dupont, and Northwestern University with the Rehabilitation Institute of Chicago (now called the Shirley Ryan Ability Labs). In addition, this decade saw the development of "helper robots" for upper-limb therapy, designed to provide force assistance to patients with impaired motor control due to stroke.

Research at Newman Laboratory for Biomechanics and Human Rehabilitation at MIT developed the MIT-MANUS that later became the commercial InMotion robots (Bionik Labs) [32]. MIT-MANUS is a 2-degrees-of-freedom (DOF) planar robot focused on training the elbow and shoulder for reaching tasks. The impedance-based control of the robot allowed severe to mild stroke patients to perform an exercise. In this clinical trial, the efficacy of high-intensity therapy with an upgraded version of MIT-MANUS with horizontal, vertical, wrist, and hand modules was compared to non-robot therapy consisting of standard therapy and intensive matched standard therapy. The study included stroke survivors having moderate to severe upper-limb impairment for at least 6 months and lesions due to single and multiple strokes. The results indicated that robot therapy did not lead to significantly more improvements than the non-robot control groups of usual care or intensive therapy but had modest improvements over 36 weeks. In addition, the cost of robot therapy was comparable to that of non-robot therapies [33].

Research at Stanford University began in 1978 where the Palo Alto Rehabilitation Engineering Research and Development Center was established under the direction of Professor Larry Leifer to replace lost functional abilities using robotic technologies [34]. One of the group's key therapy robot systems is the Mirror Image Movement Enabler (MIME) that later saw some Phase 1 commercial development as ARC-MIME. The initial version of MIME allowed for a novel bilateral training mode where the less impaired arm of stroke survivors guided movements of the impaired arm. This version limited arm movement to the horizontal plane and used a 6-DOF robot arm (PUMA-260) that applied forces and torques to the impaired forearm through one of the mobile arm supports. Clinical trials with this system demonstrated the utility of unilateral and bilateral training with rehabilitation robots and their therapeutic benefits in reducing motor impairment and increasing muscle strength [35].

Researchers in the United States have explored other configurations of upper-limb therapy robots such as ARM Guide [36] to practice out-of-the-plane reaching, REOgo (Motorika, LLC) to practice exergaming with a joystick-style robot, and T-WREX now marketed as Armeo Spring (Hocoma Ltd) for the practice of 3D reaching using an exoskeleton [37]. In addition, Activities of Daily Living Exercise Robot (ADLER) (Fig. 6.1) focused on the practice of ADL tasks such as drinking, requiring, reaching, and grasping and other robots targeted to the improvement of hand and finger movements [38]. In response to some criticism that therapy using these robots did not transfer to real ADLs, many of these systems began to combine rehabilitation robots with functional electrical stimulation systems such as Ness H200 (Bioness, Inc., USA) and wearable gloves to enable the training of hand movements.

Early efforts to develop robots that allow patients to be able to exercise their upper limb in more community settings began with the wearable robotic-assisted upper-extremity repetitive therapy (RUPERT) at Arizona State University [39] and the RICEWrist [40]. RUPERT via 4-DOF of support at shoulder, elbow, and wrist provided assistive forces to move the arm during ADL. Uncomplicated prototypes are considered driving only the elbow [41]. These early ideas have led to successful commercial efforts such as the Myomo (Myomo Inc.) to assist in upper-limb function for everyday reaching and grasping ADLs.

FIGURE 6.1 ADLER—Activities of Daily Living Exercise Robot.

Other efforts include the development of demonstrator and observer robots to support upper-limb rehabilitation beyond hospitals. Experiments with socially assistive robots (SARs) such as Bandit out of the University of Southern California (USC) [42] illustrate that noncontact assistive robots can be programmed to support exercise activities, monitor patient safety, and provide social support activities for people with disabilities including stroke.

Recently, considerable efforts have been made in the United States to move rehabilitation robots out of the hospital and into the community and homes. These efforts revolved around two directions: (1) decrease cost by using simplified robots/mechatronic systems and (2) make robots wearable for usability outside of the hospitals. The Hand Mentor (Motus, Inc.), an example of 1-DOF wrist robot, and HEXORR, an example of a 2-DOF hand robot, are designed to deliver home rehabilitation and exemplify these two strategies [43,44]. Developing robots/mechatronic systems that are more affordable than many of the current commercial systems may help the uptake of rehabilitation robots into more low-resource and community-based rehabilitation settings. Some early low-cost (<5000 USD) examples of these robots/mechatronic systems are Driver's SEAT [45] and TheraJoy [46]. More recent developments in the United States have revolved around the use of gaming [47], mobile health platforms [48], 3D printing, haptics, and soft materials [49,50] to develop robots that can move into the community.

6.2.2 Lower-limb robot-assisted neurorehabilitation

More than 80% of stroke survivors have gait deficits [14] which increase fall risk and greatly influence the performance of ADLs, social participation, and quality of life. Highly repetitive and task-specific gait training is recommended as an essential part of stroke rehabilitation [51]. However, conventional overground gait training, including those in which body weight is supported while clinicians manually facilitate stepping movements, requires a considerable amount of assistance from therapists or aides to provide trunk stability, facilitate correct gait patterns, and prevent buckling and falling. Therefore, often patients can only practice a few steps in the regular conventional overground gait training session due to the difficulty of facilitating correct gait patterns and setting up a body-weight support system to prevent falling. This type of training also increases the risk of injury to therapists and aides [52].

Lower-limb-powered exoskeletons are one type of rehabilitation robot that can offer upright, weight-bearing overground locomotion for people with stroke. These devices have external framing for support and computer-controlled motors and sensors on the hip, knee, or ankle joints. The coordinated motion of robots can assist in sit-to-stand, stand-to-sit, and walking activities.

Previously, activities in robot-assisted gait training revolved around the use of end-effector use of imported end-effector robots such as the Lokomat and Gait Trainer which were developed overseas. However, more recently lower-limb rehabilitation robots developed in the United States such as Gait-Assist KineAssist [53] (Woodway, Inc.) and Alter G antigravity treadmill [54] have been introduced. These lower-limb rehabilitation robots combined with a treadmill and a body-weight support system focus on stationary gait training. They can greatly increase the safety and efficiency of gait training by providing a higher intensity and more repetitive step practice in the same amount of time compared to conventional gait training. However, these devices usually occupy a lot of space and are limited to providing only stationary gait training.

Recently in the United States, a wide variety of research and translational work has been conducted for designing new lower-limb rehabilitation robots, called powered robotic exoskeletons. These exoskeletons do not rely on a treadmill and body-weight support system and incorporate new robot design approaches on actuators, sensors, and intelligent controllers. Several FDA-approved lower-limb wearable exoskeletons, including EksoTM (Ekso Bionics, Inc. $\sim$ 140k), IndegoTM (Parker Hannifin Corp. $\sim$ 140k), ReStoreTM (ReWalk Robotics, Inc.), were originally developed at UC Berkeley, Vanderbilt University, and Harvard University, respectively. Studies show that individuals with stroke can achieve more steps during exoskeletal-assisted walking (EAW) training compared to conventional walking training in the same amount of training time [55] because the exoskeleton device (Ekso Bionics, Inc.) can provide appropriate assistance to move users' limbs automatically. Furthermore, EAW is an overground and more task-specific training with less effort from therapists.

However, the lower-limb exoskeleton fitting process sometimes may take up to 30 minutes due to the exoskeleton's rigid and bulky structures [56,57]. The fitting process may take a lot of time away from the actual walking training time. Studies have shown that misalignment and nonoptimal fitting can increase metabolic cost and pain and cause skin abrasion and even fractures [56]. In addition, users may not be able to don and doff the exoskeletons by themselves. Because of the bulky structures of exoskeletons, users often have difficulties transferring in and out of the exoskeletons, which may lead to falling accidents. Thus independently using an exoskeleton by users with stroke is very challenging [58].

Due to the limitations mentioned previously, researchers and engineers are seeking new design strategies for lower-limb exoskeletons. Current challenges in the design of lower-limb rehabilitation robots revolve around meeting the multifaceted design requirements for physical interaction with humans. The robots need to be lightweight and have high force/torque output. They also need to be highly compliant and have high bandwidth. Researchers in the United States have been trailblazing new design approaches for lower-limb rehabilitation robots, including soft exosuits and universal actuation platforms. The wearable interface of soft exosuits [59,60] is made of fabrics and cables; hence, the exosuit is unobtrusive and does not have joint misalignment issues. To accelerate the robot design process, instead of relying on a specialized robot prototype that is time-consuming to develop, Stanford University [61] developed a cable-driven tethered robotic emulator that can provide 175 Nm torque to evaluate the biomechanical impact of control strategies (Fig. 6.2).

The robustness and stability of the lower-limb exoskeletons are imperative to ensure the safety of the users. Most lower-limb exoskeletons are limited by the user's weight being lower than 100 kg, which may limit people with stroke from using them due to obesity [62]. Most exoskeletons do not have a self-balance function and require users to operate exoskeletons with a walker, cane, or crutches. Using these walking aids also limits the user's ability to interact with the environment, such as carrying bags and reaching, and at times hinders their movement. Furthermore, there is a big learning curve for both patients and therapists to use exoskeletons [57]. Clinicians may have concerns about putting persons with stroke who do not have good postural control into lower-limb exoskeletons due to the risks of injury and falling [57]. Lower-limb exoskeletons such as Rex ($\sim$150k) [63] and Atalante [64] have been designed to be self-balanced without the need for using walking aids during EAW. However, these exoskeletons are expensive, heavier, and limited by a slower walking speed. Thus there is a need for the development of controllers that allow for robust control of the exoskeletons during a wide range of activities and with external perturbations without the need for walking aids for stability.

To facilitate the clinical adoption of wearable robots, control algorithms need to be adaptive to human movements and an individual user's physiology. Low assistance (20%$\sim$40% of biological torques) delivered with an exosuit is shown to facilitate more normal walking in ambulatory individuals after stroke due to increased dorsiflexion of the paretic ankle during the swing phase and increased paretic limb's generation of forward propulsion.

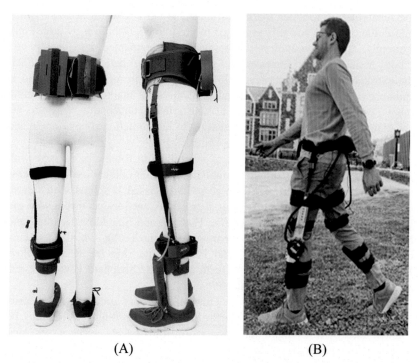

(A)   (B)

FIGURE 6.2   (A) An untethered soft exosuit to alleviate foot drop in patients with stroke; (B) highly compliant and lightweight knee exoskeleton to assist gait during activities of daily living.

To overcome the limitation of mainstream robot controllers that are based on discrete gait phase detection and require complicated switching rules, continuous controllers [65] are proposed to ensure smooth and synergistic assistive torque for different overground walking patterns. Since robot controllers typically require time-intensive manual tuning of control policies, human-in-the-loop adaptive controllers were investigated to automatically optimize multiple parameters to minimize the energy expenditure of walking with ankle exoskeletons using an evolution strategy [66]. Furthermore, to ensure the robustness and stability of the controller without relying on walking aids, novel artificial intelligence-based controllers are proposed that can adapt to external perturbation [67].

6.3   Barriers to widespread clinical use of robot-assisted therapy

The past four decades have seen tremendous growth in the use of robots, artificial intelligence, and other technologies within healthcare. Reports on the rehabilitation robotics market estimate the current worth at $789.9 million dollars (USD), and it is projected to grow to $3.17 billion by 2028 [68]. Currently, the rehabilitation robot market is segmented into therapeutic robots, assistive robots, exoskeleton robots, and prosthetic robots. In 2021, the exoskeleton robots segment held the largest share of the market and is anticipated to continue exponential growth. The market for lower-limb robots, especially exoskeletons, is driving growth. Given this, why are we not seeing the widespread implementation of rehabilitation robots inside and outside hospital environments in the United States?

When reflecting on the success of rehabilitation robots for therapy with stroke survivors, it is important to highlight that their clinical use and success have been primarily attached to large research-based hospitals with ties to large universities. Some of the largest programs are located throughout the United States in Massachusetts (Burke Hospital and Spaulding Hospital and Mass General), Texas (TIRR), Georgia (Georgia Tech), Ohio (Cleveland Clinic), Delaware (AI Dupont), and Pennsylvania (MaGee Rehab), Moss Rehab Hospital, Good Shepherd (Allentown), Good Shepherd Penn Partners, and UPMC Rehabilitation Institute (Pittsburgh).

The reasons for the lack of penetration may vary greatly with culprits being cost and lack of affordability for low-resource settings, which may be characterized by low patient−clinician ratios, low budgets, space issues, lack of access to technologies, lack of adequate standards, and lack of training of clinician users. Two recent systematic reviews indicate that cost is the major barrier to penetration into low-resource source settings—most of the rehabilitation robots for therapy in the United States cost over $20,000 USD with some systems going for >$150,000 USD [69]. Many rural and community centers in the United States are not able to afford such funds. We discuss two key reasons below: (1) high cost and (2) lack of clear clinical evidence and guidelines.

6.3.1 Cost

The current high implementation cost of commercial systems remains beyond the reach of many rehabilitation hospitals, skilled nursing facilities, nursing homes, and community rehab clinics in the United States. The costly components of robotic systems include electric motors, transmission mechanisms (e.g., harmonic drive gearbox), motor driver electronics, various sensors (e.g., torque sensors for feedback control), metallic structures, and control computers. In addition, it typically takes several years to develop a reliable rehabilitation robot even for experts. Reduction in mechanical complexity and the use of novel low-cost materials and fluid-driven soft materials are some potential cost-saving solutions [69]. The following case examples illustrate the promotion of cost-effective rehabilitation.

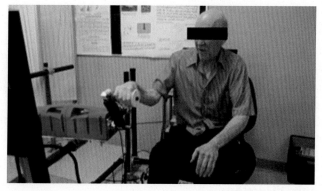

FIGURE 6.3 Haptic theradrive mounted on the rehab cares platform and can be oriented in both the horizontal and vertical position.

Upper-limb case example: Three robot systems [70,71]—Haptic Theradrive (Fig. 6.3), Rehab CARES, and Lil Flo—sponsored by the University of Pennsylvania are being explored to address issues of affordability. The technologies reduce mechanical complexity and use 3D printing as substitution for typical metal parts. The development of Haptic Theradrive focused on the use of multiple 1-DOF robots together (in different configurations and group settings) to address the diverse upper-limb training needs of stroke survivors. Other efforts focused on augmented telerehabilitation using 3D printing and off-the-shelf components. Although not yet commercial, the goal is to develop these systems at price points <$5000 USD.

Lower-limb case example: Powered exoskeletons are formidably expensive, and their costs are typically above $100 K. Their rigid structures, actuators, and batteries often make them costly, heavy, and bulky [56]. Researchers at the University of Michigan [67] and North Carolina State University [72] developed high torque density electrical motors with small ratio transmission mechanisms that are able to significantly reduce mass, make them compact, and improve compliance to human movements and at the same time lower the cost. Fluid-driven soft robots also have the potential to significantly reduce cost. These robots can provide satisfactory physical assistance with lower force control accuracy since rehabilitation robots do not require very high accuracy like robots for automotive manufacturing. Pneumatic soft exosuits can reduce cost by several factors, including low-cost pneumatic actuation (instead of high-precision electric motors and harmonic drive gear of rigid robots), fabric-based wearable frames (instead of metallic structures), and simple assembly of the robotic systems [73].

6.3.2 Clinical evidence and guidelines

Many clinicians in the United States support the use of rehabilitation robots in the care of stroke survivors and other clinical populations [74]. However, the lack of clinical practice and training guidelines on the use of these robots creates another barrier for widespread use. Specifically, with respect to robotic treatment after stroke, there is a lack of guidelines about the applicability of rehabilitation robots and clarity on patient subgroups that would benefit [27].

The lack of clear clinical evidence is also problematic. There are inconsistencies in the delivery of therapy with or without technology in terms of timing, intensity, and modality. Furthermore, the exact characteristics of patients who would benefit from rehabilitation, the timing, and the dose are still being debated [75,76].

6.4 Future directions

The widespread use of rehabilitation robots in healthcare will depend on how successful we are in increasing access to the robots outside of hospitals and within low-resource settings. Technological innovations will help advance a paradigm shift of rehab robots from heavy and expensive equipment in clinical settings to lightweight and affordable personal devices in community settings.

The face of rehabilitation care in the United States was already changing, but the COVID-19 pandemic accelerated the need for safe technology-driven solutions that can be deployed in the face of a widespread shortage of healthcare workers. Some COVID-19-driven changes included increased need for inpatient healthcare facilities to operate safely and effectively by reducing the need for in-person training, restricting visitors to selected areas, and using telehealth video and audio calls [77]. Patients were often restricted to receiving rehabilitation in their rooms in inpatient facilities and receiving telehealth visits for outpatient care. The pandemic impacted the emotional care provided by in-person rehabilitation and the experience of caregiver training due to social distancing [78]. COVID-19 sped up the digital transformation of healthcare with the use of computer and phone applications for communicating, tablets on wheels for demonstrating, real-time interpreter electronic language applications, efficiency and ease of delivery of information, and decreased the need for in-person appointments when information could be relayed by phone or video [79].

There is also a need for laws and regulations that support the use of technology and reduce the barriers for testing, reimbursement, and the use of rehabilitation robots. We saw how the intervention of the US government with the Coronavirus Aid, Relief, and Economic Security (CARES) Act legislation during COVID-19 impacted the use of telehealth [80]. The CARES Act along with other governmental healthcare agency policy changes removed barriers to telehealth by allowing clinicians to easily get reimbursed for telehealth, expanding the type of patients eligible for healthcare, waiving geographical barriers, and providing guidelines for use; these all allowed for increased use of digital media and Zoom in healthcare with reported increases of 30% with 3 months across United States [81,82]. Likewise, without the recognition of the clear utility of rehabilitation robot systems in the healthcare infrastructure, through legislation and policy changes, clinicians' ability to use them will be stymied.

Healthcare crises and shortages in healthcare workers will also help drive the need to quickly incorporate more technology into rehabilitation. One study suggests the escalating demand for better and quicker healthcare services drives the growth of the overall rehabilitation robotics market. Rehab clinicians also came face to face with limitations imposed when care was limited solely to phone or digital medium. These realizations are opportunities for therapy and assistive robots to be seen and adopted as potential solutions to augmenting care at a distance by supporting motor and cognitive evaluations and rehabilitation treatment at a distance [27,83,84].

The future of rehabilitation robotics in the United States must drive systems to be more inclusive and affordable. A commitment to affordability and inclusivity will allow the United States to see the penetration of rehabilitation robots in more low-resource settings such as into community-based rehabilitation areas, rural rehabilitation settings, and homes. Clearly, making rehabilitation more affordable and inclusive will not be easy to address and will require paradigm shifts in the use of materials, manufacturing methods, and business practices.

Acknowledgments

Dr. Michelle Johnson's work is supported by the University of Pennsylvania's Center for AIDS Research (P30AI 045008); the University of Pennsylvania's Center for Biomedical Image Computing and Analytics; and the University of Pennsylvania's Departments of Bioengineering and Physical Medicine and Rehabilitation.

Dr. Hao Su's work is supported in part by the National Science Foundation (NSF) Future of Work under grant 2026622, National Institute on Disability, Independent Living, and Rehabilitation Research (NIDILRR) under grant 90DPGE0011, and National Institutes of Health (NIH) under grant R01EB029765.

Conflict of interest

Dr. Michelle Johnson and Dr. Rochelle Mendonca are cofounders of Recupero Robotics, LLC, a company that licenses the Rehab CARES from the University of Pennsylvania. Dr. Johnson is funded by the National Institute of Health to conduct several studies using haptic theradrive for stroke and HIV.

References

[1] The World Bank. Data for high income, United States. [Internet]. 2021. Available from: https://data.worldbank.org/?locations = XD-US.

[2] United States Census Bureau. U.S. and World Population Clock [Internet]. 2021 (cited March 25, 2022). Available from: https://www.census.gov/popclock/.

[3] Ahmad FB, Anderson RN. The leading causes of death in the US for 2020. JAMA. 2021;325(18):1829 May 11.

[4] Virani SS, Alonso A, Benjamin EJ, Bittencourt MS, Callaway CW, Carson AP, et al. Heart disease and stroke statistics—2020 update: a report from the American Heart Association Circulation [Internet] 2020;141(9)Mar 3 [cited 2022 Mar 25]. Available from: https://www.ahajournals.org/doi/10.1161/CIR0.0000000000000757.

[5] Virani SS, Alonso A, Aparicio HJ, Benjamin EJ, Bittencourt MS, Callaway CW, et al. Heart disease and stroke statistics—2021 update: a report from the American Heart Association Circulation [Internet] 2021;143(8)Feb 23 [cited 2022 Mar 25]. Available from: https://www.ahajournals.org/doi/10.1161/CIR.0000000000000950.

[6] Fischer G, Keehn M. Workforce Needs and Issues in Occupational and Physical Therapy. [Internet]. University of Illinois at Chicago; 2007. Available from: http://www.westga.edu.

[7] Levanon G, Cheng B, Paterra M. The risk of future labor shortages in different occupations and industries in the United States. Bus Econ 2014;49(4):227–43 Oct.

[8] Bureau U.C. The graying of America: More older adults than kids by 2035. [Internet]. Census.gov. Available from: https://www.census.gov/library/stories/2018/03/graying-america.html. (cited 25 March 2022).

[9] Higashida R, Alberts MJ, Alexander DN, Crocco TJ, Demaerschalk BM, Derdeyn CP, et al. Interactions within stroke systems of care: a policy statement from the American Heart Association/American Stroke Association. Stroke. 2013;44(10):2961–84.

[10] Ma VY, Chan L, Carruthers KJ. Incidence, prevalence, costs, and impact on disability of common conditions requiring rehabilitation in the United States: stroke, spinal cord injury, traumatic brain injury, multiple sclerosis, osteoarthritis, rheumatoid arthritis, limb loss, and back pain. Arch Phys Med Rehabil 2014;95(5):986–95 e1.

[11] Godwin KM, Wasserman J, Ostwald SK. Cost associated with stroke: outpatient rehabilitative services and medication. Top Stroke Rehabil 2011;18(sup1):676–84.

[12] Johnson B, Bonafede M, Watson C. Short- and longer-term health-care resource utilization and costs associated with acute ischemic stroke. CEOR 2016;53 Feb.

[13] MedPac.gov. Chapter 11: Long Term Care Hospital Services. [Internet]. 2020. Available from: https://www.medpac.gov/wp-content/uploads/import_data/scrape_files/docs/default-source/reports/mar20_medpac_ch11_sec.pdf.

[14] Duncan PW, Zorowitz R, Bates B, Choi JY, Glasberg JJ, Graham GD, et al. Management of adult stroke rehabilitation care: A Clinical Practice Guideline Stroke [Internet] 2005;36 (9)Sep [cited 2022 Mar 25]. Available from: https://www.ahajournals.org/doi/10.1161/01. STR0.0000180861.54180.FF.

[15] Patel M, Coshall C, Rudd AG, Wolfe CD. Natural history of cognitive impairment after stroke and factors associated with its recovery. Clin Rehabil 2003;17(2):158–66.

[16] Mayo NE, Wood-Dauphinee S, Côté R, Durcan L, Carlton J. Activity, participation, and quality of life 6 months poststroke. Arch Phys Med Rehabil 2002;83(8):1035–42.

[17] Delavaran H, Jönsson A-C, Lövkvist H, Iwarsson S, Elmståhl S, Norrving B, et al. Cognitive function in stroke survivors: a 10-year follow-up study. Acta Neurol Scand 2017;136(3):187–94.

[18] Hong I, Karmarkar A, Chan W, Kuo Y-F, Mallinson T, Ottenbacher KJ, et al. Discharge patterns for ischemic and hemorrhagic stroke patients going from Acute Care Hospitals to inpatient and skilled nursing rehabilitation. Am J Phys Med Rehabil 2018;97(9):636−45.

[19] Winstein CJ, Stein J, Arena R, Bates B, Cherney LR, Cramer SC, et al. Guidelines for adult stroke rehabilitation and recovery: a guideline for healthcare professionals from the American Heart Association/American Stroke Association Stroke [Internet] 2016;47(6)Jun [cited 2022 Mar 25]. Available from: https://www.ahajournals.org/doi/10.1161/STR.0000000000000098.

[20] Bernhardt J, Dewey H, Thrift A, Collier J, Donnan G. A very early rehabilitation trial for stroke (AVERT): phase II safety and feasibility. Stroke. 2008;39(2):390−6.

[21] Cumming TB, Thrift AG, Collier JM, Churilov L, Dewey HM, Donnan GA, et al. Very early mobilization after stroke fast-tracks return to walking: further results from the phase II AVERT randomized controlled trial. Stroke. 2011;42(1):153−8.

[22] Camicia M, Wang H, DiVita M, Mix J, Niewczyk P. Length of stay at inpatient rehabilitation facility and stroke patient outcomes. Rehabil Nurs 2016 Mar;41(2):78−90.

[23] DaVanzo J, El-Gamil A, Li J, Shimer M, Manolov N, Dobson A. Assessment of Patient Outcomes of Rehabilitative Care Provided in Inpatient Rehabilitation Facilities (IRFs) and After Discharge [Internet]. 2014. Available from: https://amrpa.org/portals/0/dobson%20davanzo%20final%20report%20-%20patient%20outcomes%20of%20irf%20v_%20snf%20-%207_10_14%20redated.pdf.

[24] MedicareInteractive.org. Comprehensive Outpatient Rehabilitation Facilities. [Internet]. 2020. Available from: https://www.medicareinteractive.org/get-answers/medicare-covered-services/rehabilitation-therapy-services/comprehensive-outpatient-rehabilitation-facilities-corfs.

[25] Taylor RR, Jones CJ. Community-based rehabilitation [Internet]. 2014. Available from: https://www.britannica.com/topic/community-based-rehabilitation.

[26] Sunnerhagen KS, Olver J, Francisco GE. Assessing and treating functional impairment in poststroke spasticity. Neurology 2013;15;80(Issue 3):S35−44 Supplement 2.

[27] Morone G, Palomba A, Martino Cinnera A, Agostini M, Aprile I, et al. on behalf of "CICERONE" Italian Consensus Conference on Robotic in Neurorehabilitation. Systematic review of guidelines to identify recommendations for upper limb robotic rehabilitation after stroke. Eur J Phys Rehabil Med [Internet]. 2021; 57(2). Available from: https://www.minervamedica.it/index2.php?show = R33Y2021N02A0238 (cited 2022 Mar 25).

[28] Zimbelman JL, Juraschek SP, Zhang X, Lin VW-H. Physical therapy workforce in the United States: forecasting nationwide shortages. PM R. 2010;2(11):1021−9.

[29] Lin V, Zhang X, Dixon P. Occupational therapy workforce in the United States: forecasting nationwide shortages. PM R. 2015;7(9):946−54.

[30] Colwill JM, Cultice JM. The future supply of family physicians: implications for rural America. Health Aff 2003;22(1):190−8.

[31] Cooper RA, Ohnabe H, Hobson DA, editors. An introduction to Rehabilitation Engineering. Boca Raton: Taylor & Francis; 2007. p. 444.

[32] Hogan N, Krebs HI, Charnnarong J, Srikrishna P, Sharon A. MIT-MANUS: a workstation for manual therapy and training. I. In: Proceedings IEEE International Workshop on Robot and Human Communication [Internet]. Tokyo, Japan: IEEE; 1992, p. 161−5. Available from: http://ieeexplore.ieee.org/document/253895/.(cited 25 Mar 2022).

[33] Lo AC, Guarino PD, Richards LG, Haselkorn JK, Wittenberg GF, Federman DG, et al. Robot-assisted therapy for long-term upper-limb impairment after stroke. N Engl J Med 2010;362(19):1772−83.

[34] Burgar CG, Lum PS, Shor PC, Machiel Van der Loos HF. Development of robots for rehabilitation therapy: the Palo Alto VA/Stanford experience. J Rehabil Res Dev 2000;37 (6):663−73.

[35] Lum PS, Burgar CG, Shor PC, Majmundar M, Van der Loos M. Robot-assisted movement training compared with conventional therapy techniques for the rehabilitation of upper-limb motor function after stroke. Arch Phys Med Rehabil 2002;83(7):952−9.

[36] Kahn LE, Lum PS, Rymer WZ, Reinkensmeyer DJ. Robot-assisted movement training for the stroke-impaired arm: does it matter what the robot does? JRRD 2006;43(5):619.

[37] Housman SJ, Le V, Rahman T, Sanchez RJ, Reinkensmeyer DJ. Arm-Training with T-WREX After Chronic Stroke: Preliminary Results of a Randomized Controlled Trial. In: 2007 IEEE 10th International Conference on Rehabilitation Robotics [Internet]. Noordwijk, Netherlands: IEEE; 2007. p. 562−8. Available from: http://ieeexplore.ieee. org/document/4428481/.(cited 25 Mar 2022).

[38] Johnson MJ, Wisneski KJ, Anderson J, Nathan D, Smith RO. Development of ADLER: The Activities of Daily Living Exercise Robot. In: The First IEEE/RAS-EMBS International Conference on Biomedical Robotics and Biomechatronics, 2006 BioRob 2006 [Internet]. Pisa, Italy: IEEE; 2006. p. 881−6. Available from: http://ieeexplore.ieee. org/document/1639202/.(cited 25 Mar 2022).

[39] Huang J, Tu X, He J. Design and evaluation of the RUPERT wearable upper extremity exoskeleton robot for clinical and in-home therapies. IEEE Trans Syst Man Cybern Syst 2016;46(7):926−35.

[40] Pehlivan AU, Lee S, O'Malley MK. Mechanical design of RiceWrist-S: A forearm-wrist exoskeleton for stroke and spinal cord injury rehabilitation. In: 4th IEEE RAS & EMBS International Conference on Biomedical Robotics and Biomechatronics (BioRob) [Internet]. Rome, Italy: IEEE; 2012. p. 1573−8. Available from: http://ieeexplore.ieee. org/document/6290912/.(cited 25 Mar 2022).

[41] Stein J, Narendran K, McBean J, Krebs K, Hughes R. Electromyography-controlled exo-skeletal upper-limb−powered orthosis for exercise training after stroke. Am J Phys Med Rehabil 2007;86(4):255−61.

[42] Matarić MJ, Eriksson J, Feil-Seifer DJ, Winstein CJ. Socially assistive robotics for post-stroke rehabilitation. J Neuroeng Rehabil 2007;4(1):5.

[43] Butler AJ, Bay C, Wu D, Richards K, Buchanan S, Yepes M. Expanding tele-rehabilitation of stroke through in-home robot-assisted therapy. Int J Phys Med Rehabil 2014;02(184):1−11.

[44] Schabowsky CN, Godfrey SB, Holley RJ, Lum PS. Development and pilot testing of HEXORR: hand EXOskeleton rehabilitation robot. J Neuroeng Rehabil 2010;7(1):36.

[45] Johnson MJ, Van der Loos HFM, Burgar CG, Shor P, Leifer LJ. Design and evaluation of Driver's SEAT: a car steering simulation environment for upper limb stroke therapy. Robotica 2003;21(1):13−23.

[46] Johnson MJ, Feng X, Johnson LM, Winters JM. Potential of a suite of robot/computer-assisted motivating systems for personalized, home-based, stroke rehabilitation. J Neuroeng Rehabil 2007;4(1):6.

[47] Gorsic M, Novak D. Design and pilot evaluation of competitive and cooperative exercise games for arm rehabilitation at home. In: 38th Annual International Conference of the IEEE Engineering in Medicine and Biology Society (EMBC) [Internet]. Orlando, FL, USA: IEEE; 2016. p. 4690−4. Available from: http://ieeexplore.ieee.org/document/ 7591774/.(cited 25 Mar 2022).

[48] Bhattacharjya S, Stafford MC, Cavuoto LA, Yang Z, Song C, Subryan H, et al. Harnessing smartphone technology and three dimensional printing to create a mobile rehabilitation system, mRehab: assessment of usability and consistency in measurement. J Neuroeng Rehabil 2019;16(1):127.

[49] Polygerinos P, Wang Z, Galloway KC, Wood RJ, Walsh CJ. Soft robotic glove for combined assistance and at-home rehabilitation. Robot Auton Syst 2015;73:135−43.

[50] Rose CG, O'Malley MK. Hybrid rigid-soft hand exoskeleton to assist functional dexterity. IEEE Robot Autom Lett 2019;4(1):73−80.

[51] Cooke EV, Mares K, Clark A, Tallis RC, Pomeroy VM. The effects of increased dose of exercise-based therapies to enhance motor recovery after stroke: a systematic review and meta-analysis. BMC Med 2010;8(1):60.

[52] Darragh AR, Campo M, King P. Work-related activities associated with injury in occupational and physical therapists. Work. 2012;42(3):373−84.

[53] KineAssist - Enabling Advanced Recovery [Internet]. Woodway. Available from: https://www.woodway.com/products/kineassist/.(cited 4 April 2022).

[54] AlterG | Anti-Gravity TreadmillTM for Rehab & Training [Internet]. AlterG. Available from: https://www.alterg.com/.(cited 4 April 2022).

[55] Nolan KJ, Karunakaran KK, Roberts P, Tefertiller C, Walter AM, Zhang J, et al. Utilization of robotic exoskeleton for overground walking in acute and chronic stroke. Front Neurorobot 2021;15:689363 Sep 1.

[56] Rodríguez-Fernández A, Lobo-Prat J, Font-Llagunes JM. Systematic review on wearable lower-limb exoskeletons for gait training in neuromuscular impairments. J Neuroeng Rehabil 2021;18(1):22.

[57] Vaughan-Graham J, Brooks D, Rose L, Nejat G, Pons J, Patterson K. Exoskeleton use in post-stroke gait rehabilitation: a qualitative study of the perspectives of persons post-stroke and physiotherapists. J Neuroeng Rehabil 2020;17(1):123.

[58] Tefertiller C, Hays K, Jones J, Jayaraman A, Hartigan C, Bushnik T, et al. Initial Outcomes from a Multicenter Study Utilizing the Indego Powered Exoskeleton in Spinal Cord Injury. Top Spinal Cord Inj Rehabi. 2018;24(1):78−85.

[59] Awad LN, Bae J, O'Donnell K, De Rossi SMM, Hendron K, Sloot LH, et al. A soft robotic exosuit improves walking in patients after stroke. Sci Transl Med 2017;9(400): eaai9084 Jul 26.

[60] Salmeron LJ, Juca GV, Mahadeo SM, Ma J, Yu S, Su H. An Untethered Electro-Pneumatic Exosuit for Gait Assistance of People With Foot Drop. In: Proceedings of the 2020 Design of Medical Devices Conference. 2020 Design of Medical Devices Conference. Minneapolis, Minnesota, USA. April 6−9, 2020. V001T09A009. ASME.

[61] Caputo JM, Collins SH. A universal ankle−foot prosthesis emulator for human locomotion experiments. J Biomech Eng 2014;136(3):035002.

[62] Liou T-H, Pi-Sunyer FX, Laferrere B. Physical disability and obesity. Nutr Rev 2005;63 (10):321−31.

[63] Rex Bionics US - Reimagining Rehabilitation [Internet]. Rex Bionics US. Available from: https://www.rexbionics.com/us/.(cited 26 March 2022).

[64] Wandercraft | Ordinary Life for Extraordinary People [Internet]. Available from: https://www.wandercraft.eu/.(cited 26 March 2022).

[65] Li M, Wen Y, Gao X, Si J, Huang H. Toward expedited impedance tuning of a robotic prosthesis for personalized gait assistance by reinforcement learning control. IEEE Trans Robot 2022;38(1):407−20.

[66] Zhang J, Fiers P, Witte KA, Jackson RW, Poggensee KL, Atkeson CG, et al. Human-in-the-loop optimization of exoskeleton assistance during walking. Science. 2017;356 (6344):1280−4.

[67] Zhu H, Nesler C, Divekar N, Peddinti V, Gregg RD. Design principles for compact, back-drivable actuation in partial-assist powered knee orthoses. IEEE/ASME Trans Mechatron 2021;26(6):3104−15.

[68] Partners TI. Rehabilitation Robots Market $3.17Bn by 2028 Growth Forecast at 21.8% CAGR During 2021 to 2028 COVID Impact and Global Analysis by TheInsightPartners. com [Internet]. GlobeNewswire News Room. 2021. Available from: https://www.globe-newswire.com/news-release/2021/09/24/2303049/0/en/Rehabilitation-Robots-Market-3-17Bn-by-2028-Growth-Forecast-at-21-8-CAGR-During-2021-to-2028-COVID-Impact-and-Global-Analysis-by-TheInsightPartners-com.html. (cited 4 April 2022).

[69] Demofonti A, Carpino G, Zollo L, Johnson MJ. Affordable robotics for upper limb stroke rehabilitation in developing countries: a systematic review. IEEE Trans Med Robot Bionics 2021;3(1):11−20.

[70] Johnson MJ, Rai R, Barathi S, Mendonca R, Bustamante-Valles K. Affordable stroke therapy in high-, low- and middle-income countries: from theradrive to rehab CARES, a compact robot gym. J Rehabil Assist Technol Eng 2017;4 205566831770873.

[71] Sobrepera MJ, Lee VG, Garg S, Mendonca R, Johnson MJ. Perceived usefulness of a social robot augmented telehealth platform by therapists in the United States. IEEE Robot Autom Lett 2021;6(2):2946−53.

[72] Yu S, Huang T-H, Yang X, Jiao C, Yang J, Chen Y, et al. Quasi-direct drive actuation for a lightweight hip exoskeleton with high backdrivability and high bandwidth. IEEE/ASME Trans Mechatron 2020;25(4):1794−802.

[73] Salmeron LJ, Juca GV, Mahadeo SM, Ma J, Yu S, Su H. An Untethered Electro-Pneumatic Exosuit for Gait Assistance of People With Foot Drop. In: Design of Medical Devices Conference [Internet]. Minneapolis, Minnesota, USA: American Society of Mechanical Engineers; 2020. p. V001T09A009. Available from: https://asmedigitalcollec-tion.asme.org/BIOMED/proceedings/DMD2020/83549/Minneapolis, %20 Minnesota, %20 USA/1085743. (cited 26 March 2022).

[74] Wolff J, Parker C, Borisoff J, Mortenson WB, Mattie J. A survey of stakeholder perspectives on exoskeleton technology. J Neuroeng Rehabil 2014;11(1):169.

[75] Veerbeek JM, Langbroek-Amersfoort AC, van Wegen EEH, Meskers CGM, Kwakkel G. Effects of robot-assisted therapy for the upper limb after stroke: a systematic review and meta-analysis. Neurorehabilit Neural Repair 2017;31(2):107−21.

[76] Mehrholz J, Thomas S, Werner C, Kugler J, Pohl M, Elsner B. Electromechanical-assisted training for walking after stroke. Cochrane Stroke Group, editor. Cochrane Database of Systematic Reviews [Internet]. 2017 May 10; Available from: https://doi.wiley.com/10.1002/14651858.CD006185.pub4. (cited 25 March 2022).

[77] Centers for Disease Control and Prevention (CDC) National Center for Immunization and Respiratory Diseases (NCIRD), Division of Viral Diseases. Healthcare Facilities: Managing Operations During the COVID-19 Pandemic. [Internet]. 2020. Available from: https://www.cdc.gov/coronavirus/2019-ncov/hcp/guidance-hcf.html#print.

[78] Sutter-Leve R, Passint E, Ness D, Rindflesch A. The caregiver experience after stroke in a COVID-19 environment: a qualitative study in inpatient rehabilitation. J Neurol Phys Ther 2021;45(1):14−20.

[79]   EXPRESSweekly. How COVID-19 Has Sped the Digital Transformation of Health Care. [Internet]. 2021. Available from: https://www.pennmedicine.org/news/internal-newsletters/system-news/2021/march/how-covid19-has-sped-the-digital-transformation.

[80]   Demeke HB, Pao LZ, Clark H, Romero L, Neri A, Shah R, et al. Telehealth practice among health centers during the COVID-19 Pandemic—United States, July 11–17, 2020. MMWR Morb Mortal Wkly Rep 2020;69(50):1902–5.

[81]   Byrne MD. Telehealth and the COVID-19 Pandemic. J Perianesth Nurs 2020;35 (5):548–51.

[82]   American Medical Association. American Medical Association CARES Act: AMA COVID-19 pandemic telehealth fact sheet. [Internet]. Available from: https://www.ama-assn.org/delivering-care/public-health/cares-act-ama-covid-19-pandemic-telehealth-fact-sheet. (cited 13 February 2022).

[83]   Wamsley CA, Rai R, Johnson MJ. High-force haptic rehabilitation robot and motor outcomes in chronic stroke. Int J Clin Case Stud 2017;3.

[84]   Bui KD, Wamsley CA, Shofer FS, Kolson DL, Johnson MJ. Robot-based assessment of HIV-related motor and cognitive impairment for neurorehabilitation. IEEE Trans Neural Syst Rehabil Eng 2021;29:576–86.

Chapter 7

North America and Caribbean region: Canada

Amanda McIntyre[1,2], Javad K. Mehr[3,4], Marcus Saikaley[2],
Mahdi Tavakoli[3], Dalton L. Wolfe[2] and Ana Luisa Trejos[5,6]
[1]*Arthur Labatt Family School of Nursing, Western University, London, ON, Canada,* [2]*Parkwood
Institute Research, Lawson Health Research Institute, London, ON, Canada,* [3]*Department of
Electrical and Computer Engineering, University of Alberta, Edmonton, AB, Canada,*
[4]*Department of Medicine, University of Alberta, Edmonton, Edmonton, AB, Canada,*
[5]*Department of Electrical and Computer Engineering, Western University, London, ON,
Canada,* [6]*School of Biomedical Engineering, Western University, London, ON, Canada*

Learning objectives

At the end of this chapter, the reader will be able to:

1. Describe the unique characteristics of the Canadian healthcare system that affect stroke rehabilitation.
2. Identify the efforts by the research community to advance the field of rehabilitation robotics.

7.1 Overview

This chapter provides insight into how stroke is treated in Canada, what research is being done to develop novel robotic technologies, and what can be done to improve the use of rehabilitation robotics in clinical care. All patient data relating to stroke in Canada are collected and organized by the Canadian Chronic Disease Surveillance System (CCDSS). Led by the Public Health Agency of Canada (PHAC), the CCDSS identifies chronic disease cases from provincial and territorial chronic disease administrative health databases and links them to health insurance registry records using personal identifiers. Data on more than 97% of the Canadian population are captured; thus it is possible to easily estimate the burden of stroke in Canada using this national network.

Rehabilitation Robots for Neurorehabilitation in High-, Low-, and Middle-Income Countries.
DOI: https://doi.org/10.1016/B978-0-323-91931-9.00029-3
Copyright © 2024 Elsevier Inc. All rights reserved.

7.1.1   Stroke statistics

Stroke is the third leading cause of death in Canada [1] and, among all conditions, one of the largest contributors to disability-adjusted life years [2]. According to CCDSS [1], the number of individuals who experienced a stroke rose significantly between 2003−2004 and 2012−2013. During this time, more than 215,000 individuals experienced a stroke and survived. After accounting for age, stroke incidence increased from 2.3% in 2003−2004 to 2.6% in 2012−2013; this presents a yearly increase in total cases of 1.0%. However, the incidence rate of first stroke decreased by 2.7% per year on average during the same period, from 383.0 per 100,000 (2003−2004) to 297.3 per 100,000 (2012−2013) [1]. The rate of the first stroke is higher among women (reflective of their longer life expectancy) and older adults in Canada. Approximately, 10% of Canadians who are 65 years of age or older have had a stroke. Interestingly, men have a greater occurrence of stroke over 60 years of age and greater survivability between 50 and 79 years. However, among those over 80 years of age, women have a greater survivability poststroke. After adjusting for age, CCDSS data show that all-cause mortality among those with stroke has decreased over a decade from 38.0 per 1000 (2003−2004) to 28.1 per 1000 (2012−2013) [1].

7.1.2   Canadian healthcare system

Canada has a publicly funded healthcare system that is dynamic and responsive to the changes that occur in medicine and general society [3]. Significant reform has occurred over the last four decades; however, it is established on the principle of universal coverage. That is, medically necessary healthcare services are provided to individuals based on their need, rather than their ability to pay. Under the Canada Health Act, the Canadian healthcare system is legislated to be publicly administered, comprehensive, universal, accessible, and portable. It is the role and responsibility of the provincial and territorial governments (as opposed to the federal government) to administer health insurance plans; to plan and fund hospitals, healthcare facilities, and services by health professionals; to plan and implement health promotion and public health initiatives; and to negotiate fee schedules with health professionals [3]. The healthcare system is financed by general revenue from federal, provincial, and territorial taxation. In 2018, the total health expenditure was $254.6 billion [4].

7.1.3   Healthcare continuum after stroke in Canada

Typically, an individual who first experiences a stroke will receive hyperacute care management in the emergency department at a hospital. Some Canadian provinces and territories have dedicated regional stroke centers where patients are transferred to receive specialized stroke services. For example, in Ontario, Canada, CorHealth Ontario [5] (formerly Ontario Stroke Network) is made up of 11 regional centers dedicated to ensuring quality stroke care. These centers have uniquely skilled staff who specialize in hyperacute and acute stroke care, as well as the diagnostic equipment and medications available to diagnose and treat those with stroke. In Ontario, those who are suspected of having a stroke are transported "code stroke bypass" via an ambulance to the closest regional stroke center for care, not necessarily the closest hospital. Walk-in patients suspected of having a stroke are typically stabilized and then receive transfer to the regional stroke center. Hyperacute management involves assessment with diagnostic investigations (i.e., computed tomography scan), medical stabilization, and treatment with thrombolytic medication (e.g., tissue plasminogen activator) or other procedures (e.g., endovascular treatment), if applicable. Depending on stroke severity and other factors, patients may be admitted to a hospital neurology unit, or a specialized stroke neurology observation unit in a regional stroke center. After acute care, patients are discharged to hospital-based inpatient rehabilitation or back to the community (e.g., home, long-term care). In the community, patients can be followed by a physician in an outpatient rehabilitation setting or attend private rehabilitation programs provided in both medical and nonmedical settings [6].

7.1.4 Rehabilitation standard of care for people with stroke

Rehabilitation is critical for recovery after stroke; it is essential for patients to help regain skills and independence. The Canadian Stroke Best Practice Recommendations: Rehabilitation, Recovery, and Reintegration following Stroke [7] is a comprehensive set of evidence-based guidelines for rehabilitation of sequelae after a stroke, including impairments, activity limitations, and participatory restrictions. It addresses common impairments related to hemiparesis, pain, balance, swallowing, vision and neglect, mobility limitations, and activities of daily living, among others. Further, the recommendations are useful for healthcare providers and system planners across multiple rehabilitation settings. Up until the last decade, organized stroke rehabilitation in Canada did not adhere to best practice recommendations, and for many Canadians, stroke care was not available [8]. However, since then, Canada has made substantial progress in shifting toward standardization, which has dramatically impacted care. Standardization focuses on specialized, interdisciplinary rehabilitation units with specific aspects of care, particularly therapy intensity, a focus on the timing of rehabilitation, and task-specific therapy. In some provinces (e.g., Ontario), in large urban centers with highly specialized stroke rehabilitation units in particular, healthcare funding is tied to quality-based measures (e.g., rehabilitation length of stays, rehabilitation intensity of core therapists). For example, each year

CorHealth publishes an Ontario Annual Stroke Report, which provides an overview of stroke system performance in the province, stratified by the 11 provincial stroke centers. This detailed report allows for monitoring, system planning, decision-making, and evaluation of patient outcomes [9].

Interdisciplinary rehabilitation teams tend to consist of physiotherapists (PTs), occupational therapists (OTs), speech-language pathologists (SLPs), rehabilitation nurses, social workers, rehabilitation therapists, recreational therapists, dieticians, and physiatrists (rehabilitation physicians). Depending on the setting and point in the recovery, rehabilitation is provided in person (i.e., home, hospital, or community) or virtually (i.e., telerehabilitation) using individual or group-therapy methods. The team assesses, monitors, guides, and facilitates therapy activities that are important to the patient in achieving their goals and maximizing their recovery.

7.1.5 Current attitudes and/or barriers to effective rehabilitation

The core principles now underpinning stroke rehabilitation in Canada are well known at a systems level. However, at the individual level, the vast majority of therapies provided to patients are functionally based, repetitive therapies, implemented clinically using one-on-one interactions with a "one-size-fits-all" approach (i.e., conventional). However, conventional therapies are less responsive to the high degree of heterogeneity in stroke-related deficits among patients. While conventional therapies are necessary, the application of adjunct therapies could deliver more nuanced and personalized care. Adjunct therapies are those that aim to modify the effect of primary behavioral intervention (i.e., conventional therapy); they are well studied in the literature and are categorized as either "priming" the brain (e.g., action observation, mirror therapy, repetitive transcranial magnetic stimulation) or "facilitating" peripheral motor deficits (e.g., robotics, cycle ergometry, truck training, wheelchair use). Of the more than 3000 stroke rehabilitation randomized controlled trials (RCTs) published to date, 87% evaluate adjunct, as opposed to conventional, therapies, and the evidence is strong for further improving recovery after stroke [10]. Recently, McIntyre et al. [11] surveyed PTs, OTs, and SLPs in English and French across Canadian provinces to understand current clinical practice better with respect to knowledge and use of adjunct interventions; nearly 900 therapists responded. Overall, therapists across all professional roles were highly aware of the breadth of adjunct interventions studied in the literature; however, preliminary findings confirmed the notion that knowledge and use of adjuncts across these disciplines do not correlate, and that uptake into practice was very low. For example, 64.3% of physiotherapists (N = 175) were aware of robotics and electromechanical devices for use as a stroke rehabilitation intervention, and 93.1% did not use them at all in clinical practice citing access (81.7%), cost (31.0%), and/or time (14.1%) as demotivating factors. In fact, just 12 PTs indicated that they used them, primarily on a monthly basis. The national survey also showed that for a significant number of adjunct interventions, therapists largely believed that the application of the intervention was outside of their professional role/scope of practice and that they did not know where to learn about their use. Few therapists cited (lack of) evidence as a driver of nonuse.

Patients face barriers to rehabilitation not only on an individual level but on a systems level as well. When an individual in Canada experiences a stroke, the quality of their assessment, treatment, and management depends on where they live. Not unlike other large, heterogeneous countries, the geospatial distribution of the Canadian population impacts the nature and delivery of healthcare services [12], perhaps more so than other factors (e.g., funding structures, population sociodemographic compositions, varying models of care); it is a keen area of interest and study in the Canadian research literature. In a recent geospatial analysis of Canada, Eswaradass et al. [12] reported that 53.3%−96.8% of Canadians live within 6 hours of a hyperacute stroke care center. Kapral et al. [13] reported similar findings with respect to geographic access to rehabilitation centers including rural communities. Importantly, rural settings in Southwestern Ontario are distinctly different than in Northern Ontario and the other Canadian provinces and territories. As such, the findings cannot be generalized to these other communities, and further study is required for these jurisdictions.

7.2 Technological solutions for stroke therapy

The previous section provided a general picture of rehabilitation care in Canada, highlighted the need for adjunct therapies that can be tailored to individual needs, and can address the vast geospatial distribution of the population. Integrating robotic rehabilitation systems into stroke care would allow technologies to track progress and provide individualized motion assistance. They can also provide a means for improving remote care for those who are unable to attend treatment centers. Toward these goals, rehabilitation robotics research in Canada is focused on three main areas: teleoperated robots, autonomous collaborative robots, and wearable mechatronic devices. This section outlines examples of the most significant Canadian advances in these three areas.

7.2.1 Teleoperated robotic systems

A bilateral teleoperated or telerobotic system involves two synchronized systems called the leader and follower connected by a communication channel. Multilateral telerobotic systems are an extension of telerobotic systems, consisting of several robots communicating through a multiport network. The main advantage of such a technology is to transfer the agency and motor control of the human operator(s) over barriers such as distance, danger, and scale [14]. Remote operations can be performed while providing sensory awareness feedback from the remote environment(s) to the leading operator(s). These teleoperated systems can be categorized into three main areas considering their applications: rehabilitation, assistance, and assessment.

Rehabilitation: Telerobotic rehabilitation systems can be used to train the movements of impaired patients [15]. They can facilitate a physical collaboration between patients and therapists even when physically distanced, allowing patients to complete rehabilitation training exercises under the guidance of therapists. For example, a therapist can provide online force assistance to the patient, while the robotic system scales the therapist's force value and removes the patient's involuntary movements such as tremors. In addition to patient rehabilitation goals, multilateral teleoperation can be used to train new clinicians [16].

Assistance: One of the principal advantages of assistive telerobotic systems is the ability to assist people with severe movement disorders and conditions such as cerebral palsy (CP) [14,17]. With these assistive systems, the main objective is to enhance the ability of individuals to interact with objects in a physical environment by using telerobotic methods. This, for instance, enables users to perform tasks on a larger scale than what their own range of motion allows, and in a smoother and more coordinated manner than would have been possible without assistance.

Assessment: Telerobotic systems have also been used to assess users' performance in a number of applications [18]. The visual attention of participants with hand–eye discoordination has been studied and analyzed where one robot interacts with the environment and the other robot is held by the user's hand [18].

7.2.2 Collaborative robotics systems

Collaborative robots are autonomous machines designed to perform a task with a high level of autonomy and situational awareness while collaborating with humans. The major advantage of using collaborative robots is that they can take over repetitive and precise tasks that humans would need to perform [14]. Overall, collaborative robots can be divided into fixed-base and mobile-base robots. Fixed-base articulated arm robots are an example of the first category and are beginning to be more widely used in rehabilitation in Canada. In some applications, manipulator robots are mounted on a mobile base (often called mobile manipulators) to perform tasks in a bigger workspace. Collaborative robots can be used in the following three application domains.

Rehabilitation: The majority of collaborative robotic systems have been developed for the rehabilitation of people after stroke [19–22]. One of the key requirements in poststroke rehabilitation is constant and long-term access to rehabilitation exercises, which is mostly restricted due to limited access to clinics and also a shortage of therapists [20]. One of the reasons for the success of collaborative robotic systems in poststroke rehabilitation is their ability to replicate repetitive tasks and minimize the required engagement of therapists [23]. Collaborative robots in the form of exoskeletons are being used to study how to achieve rehabilitation goals in Canada. ETS-MARSE [24] and Kinarm [25] are two exoskeletons developed in Canada and have been designed to support shoulder, elbow, forearm, and wrist joint motion in both unimanual [24] and bimanual [25] tasks.

Assistance: In addition, robots have been developed to assist in the rehabilitation of patients. People with CP and severe motion impairments are another important target population of assistive collaborative robots [26,27]. These robotic systems have been developed to enhance the hand movements of people with disabilities when performing functional tasks, such as sorting objects and coloring. The development of remotely controlled social robots to assist people with Alzheimer's disease has also demonstrated benefits [22].

Assessment: Although the majority of collaborative robotic systems have been used for rehabilitation and assistance purposes, a few studies have examined their application in assessment activities [28].

7.2.3 Wearable mechatronic devices and robotics systems

The availability of fully worn devices that people can use in their homes would also be of great benefit. There is little research in Canada aimed at developing lower-limb wearable robots. A lower-limb exoskeleton called the Keeogo + is manufactured in Canada by B-Teemia, and an initial study [29] was performed identifying that moderately impaired individuals have the most to gain from the use of this device. Nevertheless, Canada is one of the leaders in wearable rehabilitation technologies for the upper limb.

A wearable vibrotactile device was developed to provide real-time feedback and support rehabilitation when linked to a computer game, demonstrating improvements in performance [30]. Using motors, a series of devices have been developed for stroke rehabilitation, including the preliminary design of a device for wrist and hand to support motion [31], and a device for wrist rehabilitation [32] that consists of a wearable wrist robot controlled using force myography with passive, active-assistive, and active-resistive modes. Also, a bimanual wearable robotic device for elbow motion was configured as a leader−follower system, where the leader is the nonparetic limb, and the follower is the paretic limb. The paretic limb can then follow the motion of the nonparetic limb, while force feedback is provided to the patient for awareness of the level of impairment. Preliminary testing was performed with three participants demonstrating feasibility [33].

One of the major contributions to the field is called the Hero glove [34], which has been used extensively in clinical trials demonstrating significant improvements in hand performance after stroke. Other efforts have been dedicated toward the design of the different components of wearable mechatronic systems, including sensing and data processing [35], actuation [36], control system design [37], and their integration into a fully wearable stroke rehabilitation glove [38]. A preliminary prototype of a hand exoskeleton device was developed using coiled nylon actuators [39]. Finally, an integrated axiomatic approach to design was implemented in [40] to design a hand rehabilitation device resulting in an interesting concept that would require further research and development prior to implementation.

These studies show that most of the devices are still under development and have been tested only in laboratory settings. Further work is needed to fully evaluate the value of rehabilitation robotics in practice.

7.3 Trends in neurorehabilitation care in Canada

According to the Evidence-Based Review of Stroke Rehabilitation [10], a total of 181 RCTs have been published up to the year 2020 examining the use of robotics as an intervention for upper-limb rehabilitation and 70 RCTs for lower-limb rehabilitation. Among upper-limb and lower-limb RCTs, 49.2% and 42.9% were published in the last 5 years (2016−2020), respectively. Among all RCTs evaluating the use of robotics for stroke rehabilitation, just four studies [41−44] have originated in Canada, and they were for use on the upper limb [10]. A recent study by [45] presents a national, multisite RCT that evaluated patients during early stroke rehabilitation and found no difference between traditional PT and exoskeleton-based therapy using the EksoGT-powered exoskeleton.

Actual use of robotics or other advanced technologies as a method for assessment or treatment remains infrequent in Canada in routine stroke rehabilitative care, and this is similar to other neurological conditions such as spinal cord or acquired brain injury. There is little literature characterizing actual patterns of use in clinical practice other than the survey results of McIntyre et al. [11] described previously. The current state of affairs is that there are a small number of leading Canadian tertiary rehabilitation centers that are involved in research of various mechatronic devices, and some of these have implemented strategies for clinical adoption. Moreover, over the past decade, Canadian researchers and clinicians have increasingly sought intentional strategies to overcome the divide between research and clinical practice.

Of note, condition-specific Canadian clinical networks operating as "communities of practice" have been developed with a focus on enhancing care using approaches of benchmarking and building capacity in implementation science (e.g., Canadian Stroke Consortium and the Heart & Stroke Canadian Partnership for Stroke Recovery [10], SCI Implementation and Evaluation Quality Care Consortium [46], Activity-Based Therapy Community of Practice [47]). Canadians have a strong history in knowledge translation and implementation science [48]. The application of these approaches to advanced technologies such as robotics is in relatively early days. However, what is occurring organically within these Canadian networks is that researchers and clinicians are intentionally exploring different models integrating clinical practice and research efforts with the ultimate aim of improving patient outcomes. There are several funding proposals currently under consideration to move this work forward across various collaborative networks, and it is anticipated that this work will unfold over the next few years.

7.4   Challenges caused by the pandemic

The COVID-19 pandemic caught societies, healthcare systems, and governments off-guard. The remarkable speed and severity of the coronavirus pandemic were unparalleled in the modern age. Social isolation and quarantine policies for containing COVID-19 significantly impacted the most at-risk and vulnerable populations, including seniors and adults with chronic conditions and special needs, particularly when they are in long-term, group, or home care. For long periods, nonlife-threatening healthcare services as well as visits to senior/retirement homes and long-term care facilities were suspended, despite the fact that a staggering 57% of people in long-term care centers in Alberta had neurological conditions [49], and 22% of long-term care residents in Ontario had a stroke [50]. Stroke is the third most common diagnosis in long-term care [50]. These statistics highlight the urgent and critical need for the delivery of healthcare for both physical and mental wellness which includes assistive and assessment services for isolated populations.

When a health crisis is further complicated by extreme socioeconomic stress, such as during the COVID-19 pandemic, wearable, robotic, and autonomous systems can be part of the solution [51]. For instance, robotic systems can help to provide safe, intelligent, and effective assistance, therapies, and assessments [52]. These advanced technologies can significantly help by increasing the level of independence and quality of life of seniors and adults with chronic conditions and special needs. They allow caregivers to ensure that the person they are caring for can maintain their independence, even when there is a need to physically distance. Finally, these technologies enable teleassessment and telerehabilitation, thereby providing equitable access to healthcare for those living in isolated communities.

7.5   Future of neurorehabilitation and opportunities

Looking toward the future, it is possible to identify issues with the practical implementation of rehabilitation robots, limiting their acceptability and effectiveness. These limitations include discomfort, lack of sufficient force application, motion that does not match natural joint movements, durability, and a lack of responsiveness [53]. The low uptake of the technologies is exacerbated by studies with insufficient participants, or participants that do not complete the studies, leading to a lack of evidence that the technologies work. However, patients and clinicians continue to be optimistic and excited about the prospect of emerging technologies, and stakeholders recognize that the research directions being pursued are in line with the clinical needs. Robotic rehabilitation technologies will continue to advance through the development of unobtrusive sensors and actuators, better data processing strategies, and more intelligent control systems.

Conflict of interest

The authors have no conflicts of interest to disclose.

References

[1] Statistics Canada. Leading causes of death, total population, by age group and sex, Canada. CANSIM (death database). Available from: <http://www150.statcan.gc.ca/can-sim/a05?lang = eng&id = 1020561>; 2022 (Accessed 1 March 2022).

[2] GBD DALYs and HALE Collaborators. Global, regional, and national disability-adjusted life-years (DALYs) for 315 diseases and injuries and healthy life expectancy (HALE), 1990−2015: a systematic analysis for the Global Burden of Disease Study. Lancet 2015;388 (10053):1603−58.

[3] Government of Canada. Canada's health care system. Available from: <https://www.canada.ca/en/health-canada/services/health-care-system/reports-publications/health-care-system/canada.html>; 2019 (Accessed 1 March 2022).

[4] (CIHI) CIfHI. National health care expenditure trends, 2020. Ottawa, ON, 2021.

[5] CorHealth. Who we are. Available from: <https://www.corhealthontario.ca/who-we-are>; 2021 (Accessed 1 March 2022).

[6] Allen L, Richardson M, McIntyre A, Janzen S, Meyer M, Ure D, et al. Community stroke rehabilitation teams: providing home-based stroke rehabilitation in Ontario, Canada. Can J Neurol Sci 2014;41(6):697−703.

[7] Teasell R, Salbach NM, Foley N, Mountain A, Cameron JI, Jong A, et al. Canadian stroke best practice recommendations: rehabilitation, recovery, and community participation following stroke. Part one: rehabilitation and recovery following stroke. Int J Stroke 2020;15 (7):763−88.

[8] Meyer M, Foley N, Pereira S, Salter K, Teasell R. Organized stroke rehabilitation in Canada: redefining our objectives. Top Stroke Rehabil 2012;19(2):149−57.

[9] CorHealth. Ontario Stroke Report 2019/20 Overview. CorHealth. Available from: <https://www.corhealthontario.ca/data-&-reporting/ontario-stroke-reports>; 2021 (Accessed 1 March 2022).

[10] Teasell R, Iruthayarajah J, Saikaley M, Longval M. Evidence based review of stroke rehabilitation. Heart and Stroke Foundation, Can Partnersh Stroke Recovery. Available from: <http://www.ebrsr.com>; 2021 (Accessed 1 March 2022).

[11] McIntyre A, Saikaley M, Janzen J, Cao P, Lala D, Teasell R, et al. Awareness and use of stroke rehabilitation interventions in clinical practice among physiotherapists. Physiotherapy Canada 2023;75:169−76. Available from: https://doi.org/10.3138/ptc-2022-0095.

[12] Eswaradass PV, Swartz RH, Rosen J, Hill MD, Lindsay MP. Access to hyperacute stroke services across Canadian provinces: a geospatial analysis. CMAJ Open 2017;5(2):E454–9.

[13] Kapral MK, Hall R, Gozdyra P, Yu AYX, Jin AY, Martin C, et al. Geographic access to stroke care services in rural communities in Ontario, Canada. Can J Neurol Sci 2020;47 (3):301–8.

[14] Atashzar SF, Jafari N, Shahbazi M, Janz H, Tavakoli M, Patel RV, et al. Telerobotics-assisted platform for enhancing interaction with physical environments for people living with Cerebral Palsy. J Med Robot Res 2017;2(2):1740001.

[15] Sharifi M, Salarieh H, Behzadipour S, Tavakoli M. Patient-robot-therapist collaboration using resistive impedance controlled tele-robotic systems subjected to time delays. J Mech Robot 2018;10(6):061003.

[16] Sharifi I, Talebi HA, Patel RV, Tavakoli M. Multi-lateral teleoperation based on multi-agent framework: application to simultaneous training and therapy in telerehabilitation. Front Robot AI 2020;7:538347.

[17] Najafi M, Sharifi M, Adams K, Tavakoli M. Robotic assistance for children with cerebral palsy based on learning from tele-cooperative demonstration. Int J Intell Robot Appl 2017;1(1):43–54.

[18] Castellanos-Cruz JL, Gómez-Medina MF, Tavakoli M, Pilarski PM, Adams K. Preliminary testing of a telerobotic haptic system and analysis of visual attention during a playful activity. In: IEEE International Conference on Biomedical Robotics and Biomechatronics, Enschede, The Netherlands, August 26–29, 2018, pp. 1280–1285.

[19] Nicholson-Smith C, Mehrabi V, Atashzar SF, Patel RV. A multi-functional lower- and upper-limb stroke rehabilitation robot. IEEE Trans Med Robot Bionics 2020;2(4):549–52.

[20] Khoshroo P, Muñoz JE, Boger J, McPhee J. Games For robot-assisted upper-limb post-stroke rehabilitation: a participatory co-design with rehabilitation therapists. Assist Technol 2021;33(3):165.

[21] Ghannadi B, Razavian RS, McPhee J. Configuration-dependent optimal impedance control of an upper extremity stroke rehabilitation manipulandum. Front Robot AI 2018;5:124.

[22] Wang RH, Sudhama A, Begum M, Huq R, Mihailidis A. Robots to assist daily activities: views of older adults with Alzheimer's disease and their caregivers. Int Psychogeriatr 2017;29(1):67–79.

[23] Fong J, Rouhani H, Tavakoli M. A therapist-taught robotic system for assistance during gait therapy targeting foot drop. IEEE Robot Autom Lett 2019;4(2):407–13.

[24] Rahman MH, Rahman MJ, Cristobal OL, Saad M, Kenné JP, Archambault PS. Development of a whole arm wearable robotic exoskeleton for rehabilitation and to assist upper limb movements. Robotica 2015;33(1):19–39.

[25] Mochizuki G, Centen A, Resnick M, Lowrey C, Dukelow SP, Scott SH. Movement kinematics and proprioception in post-stroke spasticity: assessment using the Kinarm robotic exoskeleton. J Neuroeng Rehabil 2019;16(1):1–13.

[26] Jafari N, Adams KD, Tavakoli M. Development of an assistive robotic system with virtual assistance to enhance play for children with disabilities: a preliminary study. J Med Biol Eng 2018;38(1):33–45.

[27] Sakamaki I, Adams K, Gomez MF, Castanellos JL, Jafari N, Tavakoli M, et al. Preliminary testing by adults of a haptics-assisted robot platform designed for children with physical impairments to access play. Assist Technol 2018;30(5):242–50.

[28] Atashzar SF, Shahbazi M, Tavakoli M, Patel RV. A grasp-based passivity signature for haptics-enabled human-robot interaction: application to design of a new safety mechanism for robotic rehabilitation. Int J Rob Res 2017;36(5–7):778–99.

[29] Mcleod JC, Ward SJM, Hicks AL. Evaluation of the Keeogo TM Dermoskeleton. Disabil Rehabil: Assist Technol 2019;14(5):503−12.

[30] Hung CT, Croft EA, Van Der Loos HFM. A wearable vibrotactile device for upper-limb bilateral motion training in stroke rehabilitation: A case study. In: Proceedings of the Annual International Conference of the IEEE Engineering in Medicine and Biology Society, EMBS, November 2015, 3480−3483.

[31] Henrey M, Sheridan C, Khokhar ZO, Menon C. Towards the development of a wearable rehabilitation device for stroke survivors. In: IEEE Toronto International Conference - Science and Technology for Humanity, Toronto, ON, Canada, September 26−27, 2009, pp. 12−17.

[32] Sangha S, Elnady AM, Menon C. A compact robotic orthosis for wrist assistance. In: IEEE International Conference on Biomedical Robotics and Biomechatronics (BioRob), June 26−29, 2016, Singapore, pp. 1080−1085.

[33] Herrnstadt G, Alavi N, Neva J, Boyd LA, Menon C. Preliminary results for a force feedback bimanual rehabilitation system. In: IEEE International Conference on Biomedical Robotics and Biomechatronics (BioRob), June 26−29, 2016, Singapore, pp. 768−773.

[34] Yurkewich A, Kozak IJ, Ivanovic A, Rossos D, Wang RH, Hebert D, et al. Myoelectric untethered robotic glove enhances hand function and performance on daily living tasks after stroke. J Rehabil Assist Technol Eng 2020;7(1−14):205566832096405.

[35] Farago E, Chinchalkar S, Lizotte DJ, Trejos AL. Development of an EMG-based muscle health model for elbow trauma patients. Sens (Switz) 2019;19(15):3309.

[36] Edmonds BPR, Degroot CT, Trejos AL. Thermal modeling and characterization of twisted coiled actuators for upper limb wearable devices. IEEE/ASME Trans Mechatron 2021;26 (2):966−77.

[37] Desplenter T, Trejos AL. A control software framework for wearable mechatronic devices. J Intell Robot Syst 2020;99:757−71.

[38] Zhou Y, Desplenter T, Chinchalkar S, Trejos AL. A wearable mechatronic glove for resistive hand therapy exercises. In: IEEE International Conference on Rehabilitation Robotics, Toronto, ON, Canada, June 24−28, 2019, pp. 1097−1102.

[39] Bahrami S, Dumond P. Testing of coiled nylon actuators for use in spastic hand exoskeletons. In: International Conference of the IEEE Engineering in Medicine and Biology Society, EMBS, 2018-July, 1853−1856.

[40] Yang J, Peng Q, Zhang J, Gu P. Design of a hand rehabilitation device using integrated axiomatic and benchmarking methods. Procedia CIRP 2018;78:295−300.

[41] Valdés BA, Schneider AN, Van der Loos HFM. Reducing trunk compensation in stroke survivors: a randomized crossover trial comparing visual and force feedback modalities. Arch Phys Med Rehabil 2017;98(10):1932−40.

[42] Valdés BA, Van der Loos HFM. Biofeedback vs. game scores for reducing trunk compensation after stroke: a randomized crossover trial. Top Stroke Rehabil 2018;25(2):96−113.

[43] Abdullah HA, Tarry C, Lambert C, Barreca S, Allen BO. Results of clinicians using a therapeutic robotic system in an inpatient stroke rehabilitation unit. J Neuroeng Rehabil 2011;8:50.

[44] Bouchard AE, Corriveau H, Milot MH. A single robotic session that guides or increases movement error in survivors post-chronic stroke: which intervention is best to boost the learning of a timing task? Disabil Rehabil 2017;39(16):1607−14.

[45] Louie DR, Mortenson WB, Durocher M, Teasell R, Yao J, Eng JJ. Exoskeleton for poststroke recovery of ambulation (ExStRA): study protocol for a mixed-methods study investigating the efficacy and acceptance of an exoskeleton-based physical therapy program during stroke inpatient rehabilitation. BMC Neurol 2020;20:35.

[46] Bateman EA, Sreenivasan VA, Farahani F, Casemore S, Chase AD, Duley J, et al. Improving practice through collaboration: early experiences from the multi-site Spinal Cord Injury Implementation and Evaluation Quality Care Consortium. J Spinal Cord Med 2021;44(sup1):S147–58.

[47] Musselman KE, Walden K, Noonan VK, Jervis-Rademeyer H, Thorogood N, Bouyer L, et al. Development of priorities for a Canadian strategy to advance activity-based therapies after spinal cord injury. Spinal Cord 2021;59(8):874–84.

[48] Wolfe DL, Walia S, Burns AS, Flett H, Guy S, Knox J, et al. Development of an implementation-focused network to improve healthcare delivery as informed by the experiences of the SCI knowledge mobilization network. J Spinal Cord Med 2019;42 (sup1):34–42.

[49] Continuing care statistics. COVID19 Taskforce for Albertans with neuro/physical disabilities. Alberta Health Services Neurosciences, Rehabilitation & Vision Strategic Clinical Networks. May 2020.

[50] Ontario Stroke Network. The integration of stroke best practice into long term care resident care planning. Available from: <http://www.strokenetworkseo.ca/sites/strokenetworkseo.ca/files/background_overview_july_2013.pdf>; 2013 (Accessed 1 March 2022).

[51] Tavakoli M, Carriere J, Torabi A. Robotics, smart wearable technologies, and autonomous intelligent systems for healthcare during the COVID-19 pandemic: an analysis of the state of the art and future vision. Adv Intell Syst 2020;2:2000071.

[52] Atashzar SF, Carriere J, Tavakoli M. Review: how can intelligent robots and smart mechatronic modules facilitate remote assessment, assistance, and rehabilitation for isolated adults with neuro-musculoskeletal conditions? Front Robot AI 2021;8:610529.

[53] Desplenter T, Zhou Y, Edmonds BPR, Goldman A, Lidka M, Trejos AL. Rehabilitative and assistive wearable mechatronic upper-limb devices: a review. J Rehabil Assist Technol Eng 2020;7:1–26.

Chapter 8

Europe region: Italy

Paolo Boldrini[1], Donatella Bonaiuti[1], Stefano Mazzoleni[2], Federico Posteraro[3] and Loredana Zollo[4]

[1]*Italian Society of Physical and Rehabilitation Medicine (SIMFER), Rome, Italy,* [2]*Department of Electrical and Information Engineering, Politecnico di Bari, Bari, Italy and The BioRobotics Institute, Scuola Superiore Sant'Anna, Pisa, Italy,* [3]*Rehabilitation Department, Versilia Hospital, AUSL Toscana Nord Ovest, Lucca, Italy,* [4]*Advanced Robotics and Human-Centred Technologies Laboratory, University Campus Bio-Medico of Rome, Rome, Italy*

Learning objectives

At the end of this chapter, the reader will be able to:
1. understand the healthcare system, rehabilitation system, and rehabilitation robotics in Italy,
2. describe the results of the National CC on robotics for neurorehabilitation, and
3. describe the social and ethical implications, and future trends of scientific and technological research.

8.1 Healthcare and stroke rehabilitation

The population in Italy is 59,030,133 as of 2021, which fell by 0.3% compared to 2020. The Italian population is rapidly aging: people aged between 0 and 14 are 13% of the whole population, 15 and 64 years are 68%, and people aged over 65 are 23.2%, with an average population age of 46.2 years. The mean life expectancy at birth is 82.4 years, and the number of children for each woman is 1.25 [1].

Italy is the sixth largest country in Europe and has the second-highest average life expectancy, reaching 80.1 years for men and 84.7 years for women in 2021. There are regional differences in most health indicators, reflecting the economic and social imbalance between the north and south of the country. The main diseases affecting the population are circulatory diseases, malignant tumors, and respiratory diseases.

Rehabilitation Robots for Neurorehabilitation in High-, Low-, and Middle-Income Countries.
DOI: https://doi.org/10.1016/B978-0-323-91931-9.00015-3

Italy's healthcare system is a regionally based national health service (Servizio Sanitario Nazionale, SSN) providing universal coverage largely free of charge. The central government sets the fundamental principles and goals of the health system and determines which core health services (Livelli Essenziali di Assistenza, LEA) must be provided to all citizens. The regions are responsible for organizing and delivering healthcare services. At a peripherical level, geographically based local health authorities (Aziende Sanitarie Locali, ASL) deliver public health, community health services, and primary care directly, and secondary and specialist care directly or through public hospitals or accredited private providers.

The Italian National Healthcare Service (Sistema Sanitario Nazionale, SSN) was created in 1978 to replace a previous system based on insurance. In 2001, the role of the regional authorities in ensuring basic healthcare services was increased. The National Health Service is largely funded through national and regional taxes, supplemented by co-payments for pharmaceuticals and outpatient care. Hospitals are reimbursed by the SSN according to a national diagnosis-related group (DRG)-like system.

In 2012, Italy had a total of 202,676 hospital beds (including general hospital, mental health, and other specialized hospital beds) of which 80% were dedicated to acute care and the remaining to rehabilitation and long-term care activities. On average, 68.4% of hospital beds are public while the remaining are in private facilities accredited by the SSN. Rehabilitation (inpatient, outpatient, and home-based treatments) is included in the health services provided by the SSN. In 1998, the Guidelines for Rehabilitation Care were issued by the Ministry of Health, establishing general rules for the organization and delivery of services.

Rehabilitation medicine specialists (i.e., physiatrists) and other healthcare professionals provide services to people affected by various disabling conditions through the implementation of an individual rehabilitation program, based on a comprehensive multidisciplinary assessment.

In 2011, the Italian Ministry of Health issued a strategic plan for rehabilitation ("Piano di Indirizzo per la Riabilitazione") in which rehabilitation is seen as a core component of the healthcare system and proposes the establishment of rehabilitation departments within ASLs to uniformly coordinate the provision of care.

In 2017, rehabilitation treatments supported by robotic equipment have been included in the core health services (LEA) [2].

Cause of death, by noncommunicable diseases (NCDs) (% of total), in Italy was reported at 90.64% in 2019, according to the World Bank collection of development indicators [3]. Specifically, 36% of deaths are due to cardiovascular diseases, 27% to cancers, 5% to communicable, maternal, perinatal, and nutritional conditions, 4% to injuries, 6% to chronic respiratory diseases, 3% to diabetes, and 18% to other NCDs [4]. Currently in Italy, there are about 90,000 hospitalizations per year due to stroke (20% are relapses), and the prevalence in the elderly population is 6.5% [5]. In the last two decades, the incidence of stroke has decreased from 293 to 143 cases per 100,000 inhabitants, being slightly higher in women (147 cases per 100,000 per year) than in men (139 cases per 100,000 per year), with an increase from 35.7% to 47.8% in persons over 80 years. In Italy, stroke is the third cause of death. It causes 10%−12% of all deaths every year and represents the main cause of disability. While the incidence of stroke remained stable over time, mortality and disability showed a decreasing trend. Mortality in the first month is 30%. Among survivors, 40% present with significant residual disability.

The progressive reduction in the incidence and mortality of cerebrovascular events is correlated to primary and secondary prevention, acute and post-acute care, and home and community care for patients with stroke provided by the SSN. National management guidelines for stroke have been issued including support for the development of regional and subregional care pathways. Unfortunately, they were launched and implemented in an uneven way, leading to major geographical inequalities [6].

8.2   Rehabilitation robotics in Italy

The introduction of the first robotic systems for rehabilitation in Italy started around 2005, mainly for research led by some pioneering universities and hospitals. In the subsequent years, a growing number of robots for rehabilitation were installed in different public and private rehabilitation centers, especially concentrated in the central-north areas. Moreover, some university spinoff companies have started to manufacture robotic devices for rehabilitation (mainly for the upper limb), capitalizing on the clinical evidence gathered in the previous years during research trials. Currently, many different robotic systems for rehabilitation are being used in a growing number of hospitals and rehabilitation centers, especially for stroke, spinal cord injuries, multiple sclerosis, Parkinson's disease, and cerebral palsy, in experimental clinical trials and clinical practice.

The incorporation of neuroscientific findings on the central nervous system in the design of robots for rehabilitation (such as those on upper-limb kinematics and those on locomotion assisted by a body weight system and a treadmill) represents a key factor for the effectiveness of rehabilitation treatments based on these devices. An in-depth understanding of the neural substrates that underlie motor recovery after neurological impairments has led to the development of innovative rehabilitation strategies and tools that incorporate key elements of motor skill relearning, such as intensive motor training involving goal-oriented repeated movements.

The use of robotic technologies in the rehabilitation field has experienced a constant increase in recent years, and a further significant expansion is expected in the coming years. However, there are discrepancies in the criteria and practical methodologies for the clinical use of these technologies, in the organizational contexts where they are used, and in the evaluation of their outcomes. A shared reference framework is currently missing, which can clarify the many different aspects to be considered for a scientifically grounded introduction of these technologies into rehabilitation services in an effective, reliable, and safe way. The Italian Society of Physical and Rehabilitative Medicine (SIMFER) and the Italian Society of Neurological Rehabilitation (SIRN) have tried to address this issue by organization of a National Consensus Conference (CC) on robotics for neurorehabilitation (2018−21) [7].

8.3 Clinical aspects of robot-assisted rehabilitation

The CC Working Groups reviewed evidence-based studies on robot-assisted rehabilitation in neurological diseases in the following databases: Cochrane Library, PEDro, PubMed, and Google Scholar. The study quality was assessed by AMSTAR 2 and PEDro Scale, and for the quality assessment of the guidelines AGREEII.

8.3.1 Recommendations for pediatric neurological disabilities

Rehabilitation robots represent a recent therapeutic opportunity for motor disorders in children with neurological disabilities. Although studies on pediatric patients are promising, they are limited and do not identify guidelines or provide specific recommendations. However, the literature highlights some common benefits of the use of robotics in pediatric rehabilitation such as a playful approach to robotic training that motivates and stimulates the children to improve their performance and the absence of side effects. The literature (16 studies) presents generally promising results for robotics for the upper limb, even without the need for a high number of sessions. The potential feedback (sensorimotor, motor learning) from robotic devices can be customized to increase the effects on motor learning for the child, specifically smoothness and speed of movements.

For robot-assisted gait training (RAGT), the studies used mostly exoskeleton robots on treadmills in children with cerebral palsy classified with the Gross Motor Function Classification System from II to IV. The proposed treatment protocols were heterogeneous: the sessions lasted 30−60 minutes, two to five times a week, for 2−6 weeks, up to a maximum of 10 weeks [8]. In some studies, the robot was associated with conventional treatment [9,10], characterized by a heterogeneity of interventions. Robot-assisted gait training in children seems promising, demonstrating improvements in gait speed, endurance, and balance. As regards the rehabilitation for the lower limb, it is difficult to generalize the positive results reported from the studies examined because in most cases they used experimental settings. The overall outcomes report an improvement in biomechanical, clinical, spatiotemporal, kinetic, kinematic, and electromyographic parameters in gait analysis [11−13].

8.3.2   Recommendations for neurological disabilities in adults

8.3.2.1   Dysfunction of the upper limb and recovery of the reaching abilities and manipulation

Literature on robot-assisted training for the upper limb, primarily in post-stroke rehabilitation, is widely available. The evaluation of the effectiveness of the robot-assisted rehabilitation in this case is made difficult by the complexity of upper-limb function. The widespread diffusion of robotic devices in rehabilitation does not correspond to an agreement on effectiveness [14,15] and in particular on which patients can benefit most.

8.3.2.1.1   Stroke

The results of the literature search (70 articles included) concluded that robotic training seems to improve motor control and muscular strength of the paretic upper limb in stroke patients. There is insufficient evidence to conclude that robot training can reduce spasticity and pain, and improve function and dexterity. Robot-assisted treatment can be considered a therapeutic option to improve the overall autonomy of the poststroke patient, but there is no clear evidence of increased participation in activities of daily living (ADLs). Patients with greater motor impairment could use robotic devices to increase training intensity. It is not possible to hypothesize a greater effectiveness of one device compared to another, rather their different use based on the characteristics of the patient, the phase subacute or chronic, and the aim for which the rehabilitation is carried out for the upper limb. Although the intensity and repetitiveness of the functional tasks are recognized as "determinants" of the rehabilitation useful for motor recovery, the literature does not recommend any optimal quality and quantity of the robotic sessions [15−22].

8.3.2.1.2   Spinal cord injury

The studies included in the analysis (five studies) propose robotic training as a feasible and safe therapeutic opportunity for upper-limb treatment in patients with spinal cord injury. Patients with mild-to-moderate residual function seem to benefit more from robot-assisted training than conventional therapy. Robot-assisted treatment seems to be effective in improving kinematics and smoothness of movement. However, there is insufficient evidence to support its efficacy in improving muscular strength, function, and independence in ADLs. There are no indications on the duration and frequency of treatment due to the wide heterogeneity of the protocols reported and the different devices used. The combination of two different robots during the training, one with the aim of improving the motion of the shoulder and the elbow and one with the manipulation, is feasible and equally effective at a comparable dose of conventional treatment [23,24].

8.3.2.1.3 Multiple sclerosis

The evaluation of the literature (eight studies) did not reveal any additional benefits of robot-assisted therapy compared to conventional rehabilitation. Robotic devices in upper-limb rehabilitation in multiple sclerosis (MS) patients still represent a therapeutic opportunity to improve manual dexterity, coordination, functional ability, and efficiency in upper-limb motor strategies. Patients with greater impairment could use robotic treatments to increase the intensity of the exercise during rehabilitation. The expected benefits of these treatments should affect not only narrow motor skills but in general the degree of autonomy and quality of life of MS patients. The beneficial effects obtained after intensive treatment tend to be exhausted over 6 months, so regular management is recommended for these patients [25−27].

8.3.2.1.4 Other pathologies

The scarcity of literature (one study for Parkinson's disease and none for acquired brain injury) does not allow us to draw any conclusions about efficacy. However, it is conceivable to use robot-assisted rehabilitation for the upper limb to increase rehabilitation practice and counteract the learned nonuse [28−30].

8.3.2.2 Lower limb dysfunction and locomotor recovery

The literature on RAGT is more extensive than that of the upper limb; therefore it is possible to deepen the discussion on its use in different disabilities.

8.3.2.2.1 Stroke

Besides 11 guidelines on stroke management, 58 studies were included. Several of these technologies are available on the market, the remaining are prototypes, and their characteristics are not always specified. As a result, reviews with meta-analysis should be cautiously interpreted. However, it is possible to propose the introduction of robotics in standard treatment protocols for stroke rehabilitation. First, RAGT seems to have different effects based on the distance from the acute event and the severity of functional impairment: nonambulatory patients starting early to robot-assisted gait rehabilitation have more benefits, the worse their disability is [31].

In patients with severe disability (Functional Ambulation Classification, FAC < 4), acute and subacute robot-assisted training is suggested. Merholz reports that on seven subjects treated within 3 months, one will switch from FAC < 4 to FAC ≥ 4, regardless of the device used [32]. The speed and temporospatial parameters of gait increase significantly mainly in subacute but not in chronic patients [33], while the improvement in the Six-Minute Walk Test (6MWT) and endurance is observable even in chronic patients [32,34]. Patients with FAC > 3 show improvements in the Time Up and Go test while those with FAC < 3 increased endurance in 6MWT [34]. Patients with FAC ≥ 2 [35] achieve an increase in muscular strength and balance with robotics as an "add-on."

We can also identify robotic devices more suitable for the severity of disability based on the implemented control strategy, but with regard to dosage of the exercise and protocols, there is a great variability in the studies that makes it difficult to identify standards of training.

8.3.2.2.2 Spinal cord injury

In total, 51 studies were included in the analysis of both with exoskeleton with body weight support on treadmill and dynamic overground.

a) Exoskeletons with body weight support on treadmill

In patients with incomplete lesions, these devices in addition to conventional therapy [35] are able to improve outcomes, such as Walking Index for Spinal Cord Injury, 10-Meter Walk Test (10MWT), 6MWT, lower extremity motor score (LEMS), and items related to locomotor Functional Independence Measure, compared to conventional treatment alone. There is heterogeneity in the protocols proposed: the sessions last on average 40−45 minutes, 3−5 times weekly (for the most part 3 times), for 3 weeks up to 3 months. The association of resistance applied to the lower limbs during training has been shown to improve walking speed on the ground in people who initially had greater gait ability [36]. Furthermore, robotic systems compared to other strategies for restoring gait (e.g., walking on the ground, body-weight supported treadmill training, functional electrical stimulation) are less effective in increasing the speed of walking, but strategies that take into account the most advanced technologies such as virtual reality, biofeedback besides the application of resistances, in addition to robotics, seem to have the most promising results.

b) Dynamic (or overground) robotic exoskeletons

Improvement, and in some cases recovery, of gait is shown after training with overground exoskeleton robots [37,38] which have also proven effective in increasing speed and endurance and the ability to walk in both people with incomplete and complete spinal cord injuries [38,39]. In some studies, beneficial effects on muscle spasticity and pain are observed [38,40]. These devices are relatively safe in both people with incomplete and complete spinal cord injury, with mild side effects, although episodes of shoulder pain secondary to the use of the necessary walker and/or crutches are described, and episodes of heel fracture secondary to osteoporosis [39].

8.3.2.2.3 Multiple sclerosis

From an analysis of the recent literature (17 studies analyzed), no benefits are highlighted in addition to those obtained from conventional treatment of equal intensity. However, RAGT is an advantage when used on patients with severe disability. On the other hand, when the patient's gait is less compromised, it is proposed to favor conventional training.

In MS patients with severe disability ($6 < EDSS < 7.5$), RAGT with body weight support on treadmill improves gait speed and endurance [41,42], increases muscle strength [42], and reduces energy consumption fatigue and spasticity, with a consequent increase in autonomy in ADLs [43]. Treatment with an exoskeleton overground device, on patients with maintained gait ability ($EDSS < 6.5$), has been shown to be effective in improving walking endurance and stair climbing, but does not improve overall physical activity measured as a number of steps/day or activity [44].

The results obtained after intensive treatment tend to be exhausted within 6 months, so the possibility of repeating the intensive treatment over time is to be evaluated.

8.3.2.2.4 Parkinson's disease

Ten studies were considered; of these, 60% highlight greater effectiveness of RAGT versus control in improving gait parameters, balance, and function of patients; the remaining studies highlight the efficacy, but not the superiority of this treatment over control (treadmill training and/or other conventional therapies). However, it is important to consider that the treatment described in the different studies is aimed at patients with different severity of disability and used heterogeneous endpoints [45].

8.3.2.2.5 Traumatic brain injury

The available literature (three studies) does not allow to give conclusive evidence on the use of robot-assisted treatment in traumatic brain injury. However, some observed improvements invite us not to discard the use of robotic devices by integrating them into daily clinical use for their motivational impact [46,47].

8.3.2.3 Balance dysfunctions

Scientific literature shows promising results with robotics in balance dysfunctions.

8.3.2.3.1 Stroke

Gait rehabilitation with robotic devices has positive effects on balance in the subacute and chronic phase (30 studies). The best outcome is observed in subjects who perform RAGT in addition to conventional physiotherapy. Training with high-intensity exoskeleton can bring additional benefits to balance in chronic stroke subjects. There are also significant positive effects with regard to the fear of falling and the gait parameters related to balance and postural stability [48,49].

8.3.2.3.2 Spinal cord injury

The scientific literature (two studies) underlines an activation of the trunk musculature with robot-assisted rehabilitation, which seems to be induced more by an overground exoskeleton than an exoskeleton with body weight support on treadmill [50].

8.3.2.3.3 Multiple sclerosis

The exoskeleton training can improve balance and balance-related step parameters but with a disappearance of the effect in long-term follow-up. There is significant evidence for combining robot-assisted training with conventional physiotherapy and virtual reality in improving static balance [Sitting Balance Scale; Berg Balance Scale (BBS)] compared to robot-assisted treatment alone [42,51].

8.3.2.3.4 Parkinson's disease

Literature shows a significant improvement in both the Unified Parkinson's Disease Rating Scale (UPDRS-III) score and the BBS, immediately after training and at one month of follow-up. Robot-assisted treatment can also have beneficial effects on the frequency and severity of the freezing of gait [52,53].

8.3.2.3.5 Acquired brain injury

The literature proposes to use both end-effector robots on tilt table and robotic exoskeletons on treadmill in combination with conventional physiotherapy for gait recovery. More specifically, a tilt table-based robotic device appears to improve static balance immediately at the end of treatment, not maintained at follow-up [54].

8.3.2.3.6 Other pathologies

Studies show the usefulness of RAGT combined with multidisciplinary intensive rehabilitation treatment for the improvement of postural control and the reduction of the number of falls in patients with progressive supranuclear paralysis.

8.4 Social and ethical implications

The rapid technological development in robotics and human—machine interactions, and their application to health care, leads to the need to look at their social implications, for example, the social acceptability of the technology, and their ethical and legal implications. The research on user perception of robot-based devices in rehabilitation, and their psychosocial effects, seems to suggest that such technologies have in general a positive impact, but the results are inconclusive because of methodological limitations and the lack of a common conceptual framework [55]. As for the ethical aspects, a reflection on the "classical" ethical principles (such as dignity and responsibility) is needed in relation to the interaction between humans and machines arising from the use of robotics, and the possibility of autonomous "decisions" of the devices.

The final report of an Italian National Consensus Conference on the use of robotics in the rehabilitation of persons with disability of neurological origin suggested that the principle of "centrality of the human being" must be respected in the use of technologies, with specific emphasis on maintaining human control and responsibility over the whole process of care, on preserving the value of the patient—carer relationship, and on protecting human dignity. From a practical viewpoint, specific attention should be given to: (1) research, with the evaluation of possible ethical implications; (2) legal responsibility of producers and users; (3) standardization and certification of devices, with the integration of ethical components into the design process; (4) informed consent to offer transparent, clear information on the possibilities and limits of the technologies; and (5) quality of the interactions between patients, professionals, and equipment.

8.5 Future trends of scientific and technological research

Rehabilitation robotics is a fast-growing discipline that has received significant attention over time due to increased acceptance of its validity by clinicians, cost reductions in sensors and actuators, and expansion through different applications that increase the range of impairments that can be targeted by robotic technologies. This chapter provides an overview of the outcomes from the Italian National Consensus Conference on robot-assisted neurological rehabilitation by presenting a picture of where the field is now. Going forward, we desire to transform Italy's healthcare response from reactive medicine to proactive medicine, the so-called *"P4 Medicine"*—a discipline that is predictive, personalized, preventive, and participatory (P4). P4 medicine will be fueled by systems approaches to diseases, emerging technologies, and analytical tools that can be used throughout the care process. This opens intriguing challenges that can be targeted through advanced robotic technologies, as briefly described below.

8.5.1   Ubiquitous and personalized rehabilitation treatment

Each patient can be stratified based on objective and multifaceted assessments enabled by current sensor technologies, and the most appropriate robot-aided interventions can be selected [56]. Then, these interventions can be administered in the clinic and/or at home, moving through sequences when the patient's motor recovery reaches a plateau. Current digital technologies—wearable sensors, cloud computing, connectivity and data exchange over the internet, and so on—allow establishing continuous communication with the therapist, real-time monitoring, and personalization of the treatment including telerehabilitation.

8.5.2   Closed-loop device and patient-in-the-loop approach

One of the main challenges of rehabilitation robotics is to design advanced devices that are able to promote mutual human—robot interaction and tailor the treatment for each specific user. One such enabling technology is the *closed-loop device* [57], which has the potential to reestablish bidirectional communication with the nervous system and rehabilitate as well as replace functions. Bio-cooperative robotic systems are a paradigmatic example of closed-loop devices that represent the new generation of robotic platforms that promote bidirectional interactions between the robot and the patient based on multimodal interfaces [58,59]. These platforms use multisensory monitoring systems to objectively assess patient performance during therapy and dynamically adapt robot behavior. Data coming from biomechanical, physiological, and psychological measurements, as well as data related to the user's intention and the environmental factors, are used to include the patient in the control loop, by providing continuous feedback on patients' global conditions [60] and personalizing therapy.

8.5.3   Brain—computer interface controlled robots for neurorehabilitation

Driven by advancements in sensor technology, wearable robotics, availability of computational capacities, and other recent innovations, brain—computer interfaces (BCIs) are rapidly becoming powerful tools to control robots. Noninvasive hybrid systems merging brain and neural signals, for example, eye movements, are key enabling technologies that coupled with robots increase accuracy and reliability during assistance. A recent study on healthy and poststroke subjects has demonstrated the feasibility and safety of shared electroencephalography/ electrooculography to intuitively and reliably control the upper-limb exoskeleton [61,62]. Neuroimaging studies have shown that BCI-controlled robot-aided upper-limb training in chronic stroke can lead to functional reorganization between ipsilesional motor regions (M1 and SMA) and contralesional areas

(SMA, PMd, SPL). Moreover, increased interhemispheric functional connectivity among the sensorimotor areas was observed [63].

8.5.4 Affordable rehabilitation robots

Despite the interesting outcomes and the intriguing recent advancements, one of the biggest limitations on the widespread access of the technology is its cost. Presently, rehabilitation robotic devices are priced in the range of $75,000−$350,000 USD prior to any additional hidden costs related to shipping, taxes, maintenance, and installation/training [64]. However, there is an expectation that this cost will reduce with the increased availability of clinic-based rehabilitation robots and the advancements in the field of 3D printing and low-cost technologies. Affordable robotics [65] aims to develop rehabilitation robots that can be used in low-resource clinical settings by: (1) developing end-effector and exoskeleton robots for the rehabilitation of proximal and distal joints, separately; (2) reducing the number of degrees of freedom of the robots at minimum (one or two), limiting the complexity of the machine, and related costs; (3) avoiding the use of conventional materials that could make the robot bulky, expensive, and difficult to transport.

8.6 Conclusions

Rehabilitation robotics is a fast-growing field that is inherently interdisciplinary. In Italy, a growing awareness of stroke prevention and the implementation of appropriate treatments, including rehabilitation, are leading to significant improvements in functional outcomes for patients with stroke. Our expectation is that robotic technology, suitably coupled with the digital transformation, may be implemented in Italy as standard rehabilitation treatment offered by the National Healthcare Service in order to address the challenges raised by P4 medicine.

The outcomes of the National CC offer a comprehensive review of the scientific literature on robotics for neurorehabilitation and provide significant insights into this field for the stakeholders, especially for rehabilitation professionals, patients, and healthcare policymakers. Future prospects should address the development of robotic devices in which the technical specifications of the devices themselves refer to existing theoretical models. The research should therefore be directed not only toward the measurement and quantification of motor recovery but also toward the evaluation of neurophysiopathological mechanisms induced by robot-assisted treatment. Particular attention should be paid to the distinction between adaptive motor recovery and recovery of functions.

Finally, the research guidelines should consider the following aspects: (1) the effect of the combination of different rehabilitation programs to distinguish functional profiles of patients; (2) the integration of different technologies (e.g., virtual reality, augmented reality, sensors, artificial intelligence) aimed at improving the usability, adaptability, operability, and versatility of the robot as a therapeutic aid; (3) quantification of robot-assisted therapy dosages, which are justified by clinically significant learning/recovery curves; and (4) integration between artificial intelligence methods (e.g., machine learning algorithms also applied to large volumes of data) and robotics for rehabilitation, in order to predict personalized recovery trends for each patient, which enables adaptation and modification.

References

[1] Available from: http://dati-censimentopopolazione.istat.it/Index.aspx?lang = en.

[2] Ferré F, de Belvis AG, Valerio L, Longhi S, Lazzari A, Fattore G, et al. Italy: health system review. Health Syst Transit 2014;16(4):1−168.

[3] Available from: https://data.worldbank.org/indicator/SH.DTH.NCOM.ZS?locations = IT.

[4] Available from: https://www.who.int/publications/m/item/noncommunicable-diseases-ita-country-profile-2018.

[5] Available from: https://www.salute.gov.it/imgs/C_17_pagineAree_5782_0_file.pdf.

[6] SPREAD. Stroke Prevention and Educational Awareness Diffusion. Ictus cerebrale: linee guida italiane di prevenzione e trattamento. VIII edizione, 2016.

[7] Boldrini P, Bonaiuti D, Mazzoleni S, Posteraro F. Rehabilitation assisted by robotic and electromechanical devices for people with neurological disabilities: contributions for the preparation of a national conference in Italy. Eur J Phys Rehabil Med 2021;57(3):458−9. Available from: https://doi.org/10.23736/S1973-9087.21.07084-2.

[8] Sarhan R, Chevidikunnan M, Gaowgzeh RAM. Locomotor treadmill training program using driven gait orthosis versus manual treadmill therapy on motor output in spastic diplegic cerebral palsy children. NUJHS 2014;4(4):10−17 ISSN 2249-7110.

[9] Druzbicki M, Rusek W, Szczpanik M, Dudek J, Snela S. Assessment of the impact of orthotic gait training on balance in children with cerebral palsy. Acta Bioeng Biomech 2010;12(3):53−8.

[10] Smania N, Bonetti P, Gandolfi M, Cosentino A, Waldner A, Hesse S, et al. Improved gait after repetitive locomotor training in children with cerebral palsy. Am J Phys Med Rehabil 2011;90:137−49.

[11] Chen YP, Howard AM. Effects of robotic therapy on upper-extremity function in children with cerebral palsy: a systematic review. Dev Neurorehabil 2016;19(1):64−71.

[12] Cheung EYY, Chau RMW, Chein GLY. Effects of robot-assisted body weight supported treadmill training for people with incomplete spinal cord injury. A pilot study. Physiotherapy 2015;101(S1):eS26−eS426.

[13] Wu M, Kim J, Gaebler-Spira DJ, Schmit BD, Arora P. Robotic resistance treadmill training improves locomotor function in children with cerebral palsy: a randomized controlled pilot study. Arch Phys Med Rehabilitation 2017;98:2126−33.

[14] Rodgers H, Bosomworth H, Krebs HI, van Wijck F, et al. Robot assisted training for the upper limb after stroke (RATULS): a multicentre randomised controlled trial. Lancet 2019;394(10192):51−62.

[15] Mehrholz J, Pohl M, Platz T, Kugler J, Elsner B. Electromechanical and robot-assisted arm training for improving activities of daily living, arm function, and arm muscle strength after stroke. Cochrane Database of Syst Rev 2015;2015(11):CD006876.

[16] Veerbeek JM, van Wegen EEH, van Peppen RPS. KNGF clinical practice guideline for physical therapy in patients with stroke: royal Dutch Society for physical therapy. The Netherlands. 2014.

[17] Zhang C, Li-Tsang CW, Au RK. Robotic approaches for the rehabilitation of upper limb recovery after stroke: a systematic review and meta- analysis. Int J Rehabil Res 2017;40 (1):19−28.

[18] Lo K, Stephenson M, Lockwood C. Effectiveness of robotic assisted rehabilitation for mobility and functional ability in adult stroke patients: a systematic review. JBI Database Syst Rev Implement Rep 2017;15(12):3049−91.

[19] Prange GB, Jannink MJ, Groothuis-Oudshoorn CGM, Hermens HJ, IJzerman MJ. Systematic review of the effect of robot-aided therapy on recovery of the hemiparetic arm after stroke. J Rehabil Res Dev 2006;43:171.

[20] Kwakkel G, Kollen BJ, Krebs HI. Effects of robot-assisted therapy on upper limb recovery after stroke: a systematic review. Neurorehabil Neural Repair 2008;22(2):111−21.

[21] Lin I-H, Tsai H-T, Wang C-Y, Hsu C-Y, Liou T-H, Lin Y-N. Effectiveness and superiority of rehabilitative treatments in enhancing motor recovery within 6 months poststroke: a systemic review. Arch Phys Med Rehabil 2019;100(2):366−78.

[22] Ferreira FMRM, Chaves MEA, Oliveira VC, Van Petten AMVN, Vimieiro CBS. Effectiveness of robot therapy on body function and structure in people with limited upper limb function: a systematic review and meta-analysis. PLoS One 2018;13(7):e0200330.

[23] Yozbatiran N, Francisco GE. Robot-assisted therapy for the upper limb after cervical spinal cord injury. Phys Med Rehabil Clin N Am 2019;30(2):367−84.

[24] Singh H, Unger J, Zariffa J, Pakosh M, Jaglal S, Craven BC, et al. Robot-assisted upper extremity rehabilitation for cervical spinal cord injuries: a systematic scoping review. Disabil Rehabil Assist Technol 2018;13(7):704−15.

[25] Carpinella I, Cattaneo D, Bertoni R, Ferrarin M. Robot training of upper limb in multiple sclerosis: comparing protocols with or without manipulative task components. IEEE Trans Neural Syst Rehabil Eng 2012;20(3):351−60.

[26] Feys P, Coninx K, Kerkhofs L, et al. Robot-supported upper limb training in a virtual learning environment: A pilot randomized controlled trial in persons with MS. J Neuroeng Rehabil 2015;12(1):1−12.

[27] Gandolfi M, Valè N, Dimitrova EK, et al. Effects of high-intensity Robot-assisted hand training on upper limb recovery and muscle activity in individuals with multiple sclerosis: a randomized, controlled, single-blinded trial. Front Neurol 2018;9:1−10.

[28] Picelli A, Tamburin S, Passuello M, Waldner A, Smania N. Robot-assisted arm training in patients with Parkinson's disease: a pilot study. J Neuroeng Rehabil 2014;11:28.

[29] Chew E, Straudi S, Fregni F, Zafonte RD, Bonato P. Transcranial direct current stimulation enhances the effect of upper limb functional task training in neurorehabilitation. In Abstract presented at 5th World Congress of ISPRM, 2009.

[30] Schmidt H, Hesse S, Werner C, Bardeleben A. Upper and lower extremity robotic devices to promote motor recovery after stroke -recent developments. The 26th Annual International Conference of the IEEE Engineering in Medicine and Biology Society, 2004, pp. 4825−4828.

[31] Cho JE, Yoo JS, Kim KE, Cho ST, Jang WS, Cho KH, et al. Systematic review of appropriate robotic intervention for gait function in subacute stroke patients. Biomed Res Int 2018;2018:4085298.

[32] Mehrholz J, Thomas S, Werner C, Kugler J, Pohl M, Elsner B. Electromechanical-assisted training for walking after stroke. Cochrane Database Syst Rev 2017;5(5):CD006185.

[33] Husemann B, Müller F, Krewer C, Heller S, Koenig E. Effects of locomotion training with assistance of a robot-driven gait orthosis in hemiparetic patients after stroke a randomized controlled pilot study. Stroke. 2007;38(2):349−54.

[34] Mazzoleni S, Focacci A, Franceschini M, Waldner A, Spagnuolo C, Battini E, et al. Robot-assisted end-effector-based gait; training in chronic stroke patients: a multicentric uncontrolled observational; retrospective clinical study. NeuroRehabilitation. 2017;40(4): 483−92.

[35] Alcobendas-Maestro M, Esclarín-Ruz A. Lokomat robotic-assisted versus overground training within 3 to 6 months of incomplete spinal cord lesion: randomized controlled trial. Neurorehabil Neural Repair 2012;26(9):1058−63.

[36] Wu M, Landry JM, Kim J, Schmit BD, Yen SC. Repeat exposure to leg swing perturbations during treadmill training induces long-term retention of increased step length in human SCI. Am J Phys Med Rehabil 2016;95(12):911.

[37] Atif SK, Donna CL, Caitlin LH, Jennifer D, John EM, et al. Retraining walking over ground in a powered exoskeleton after spinal cord injury: a prospective cohort study to examine functional gains and neuroplasticity. J Neuroeng Rehabil 2019;16:145.

[38] Shackleton C, Evans R, Shamley D, West S, Albertus Y. Effectiveness of over-ground robotic locomotor training in improving walking performance, cardiovascular demands, secondary complications and user-satisfaction in individuals with spinal cord injuries: a systematic review. J Rehabil Med 2019;51(10):723−33.

[39] Gagnon DH, Escalona MJ, Vermette M, Carvalho LP, Karelis AD, et al. Locomotor training using an overground robotic exoskeleton in long-term manual wheelchair users with a chronic spinal cord injury living in the community: lessons learned from a feasibility study in terms of recruitment, attendance, learnability, performance and safety. J Neuroeng Rehabil 2018;15(1):12.

[40] Platz T, Gillner A, Borgwaldt N, Kroll S, Roschka S. Device-training for individuals with thoracic and lumbar spinal cord injury using a powered exoskeleton for technically assisted mobility: achievements and user satisfaction. Biomed Res Int 2016;2016:8459018.

[41] Beer S, Aschbacher B, Manoglou D, Gamper E, Kool J, Kesslring J. Robot-assisted gait training in multiple sclerosis: a pilot randomized trial. J Mult Scler 2008;14:231−6.

[42] Straudi S, Manfredini F, Lamberti N, Martinuzzi C, Maietti E, Basaglia N. Robot-assisted gait training is not superior to intensive overground walking in multiple sclerosis with severe disability (the RAGTIME study): a randomized controlled trial. Mult Scler J 2019;26:716−24.

[43] Pompa A, Morone G, Iosa M, Pace L, Catani S, Casillo P, et al. Does robot-assisted gait training improve ambulation in highly disabled multiple sclerosis people? A pilot randomized control trial. Mult Scler 2017;23(5):696−703.

[44] McGibbon C, Sexton A, Jayaraman A, Deems-Dluhy S, Gryfe P, Novak A, et al. Evaluation of the Keeogo exoskeleton for assisting ambulatory activities in people with multiple sclerosis: an open-label, randomized, cross-over trial. J Neuroeng Rehabil 2018;15:117.

[45] Alwardat M, Etoom M, Al Dajah S, Schirinzi T, Di Lazzaro G, Sinibaldi Salimei P, et al. Effectiveness of robot-assisted gait training on motor impairments in people with Parkinson's disease: a systematic review and meta-analysis. Int J Rehabil Res 2018;41(4): 287−96.

[46] De Tanti A, Zampolini M, Pregno, CC3 Group S. Recommendations for clinical practice and research in severe brain injury in intensive rehabilitation: the Italian Consensus Conference. Eur J Phys Rehabil Med 2015;51(1):89−103.

[47] Lapitskaya N, Nielsen JF, Fuglsang-Frederiksen A. Robotic gait training in patients with impaired consciousness due to severe traumatic brain injury. Brain Inj 2011;25(11): 1070−9.

[48] Kim HY, Shin JH, Yang SP, Shin MA, Lee SH. Robot-assisted gait training for balance and lower extremity function in patients with infratentorial. J Neuroeng Rehabil 2019;16 (1):99.

[49] Bang D-H, Shin W-S. Effects of robot-assisted gait training on spatiotemporal gait parameters and balance in patients with chronic stroke: a randomized controlled pilot trial. Neurorehabilitation 2016;38(4):343−9.

[50] Alexander JH, Brooke M, Lombardo L, Audu ML, Triolo RJ. Reactive stepping with functional neuromuscular stimulation in response to forward-directed perturbations. J Neuroeng Rehabil 2017;14(1):54.

[51] Calabrò RS, Russo M, Naro A, De Luca R, Antonini L, Tomasello P, et al. Robotic gait training in multiple sclerosis rehabilitation: can virtual reality make the difference? Findings from a randomized controlled trial. J Neurol Sci 2017;377:25−30.

[52] Paker N, Bugdayci D, Goksenoglu G, Sen A, Kesiktas N. Effects of robotic treadmill training on functional mobility, walking capacity, motor symptoms and quality of life in ambulatory patients with Parkinson's disease: a preliminary prospective longitudinal study. NeuroRehabilitation. 2013;33(2):323−8.

[53] Picelli A, Melotti C, Origano F, Neri R, Verzè E, Gandolfi M, et al. Robot-assisted gait training is not superior to balance training for improving postural instability in patients with mild to moderate Parkinson's disease: a single-blind randomized controlled trial. Clin Rehabil 2015;29(4):339−47.

[54] Ancona E, Quarenghi A, Simonini M, Saggini R, Mazzoleni S, De Tanti A, et al. Effect of verticalization with Erigo® in the acute rehabilitation of severe acquired brain injury. Neurol Sci 2019;40(10):2073−80.

[55] Koumpouros Y. A systematic review on existing measures for the subjective assessment of rehabilitation and assistive robot devices. J Healthc Eng 2016;2016:1048964.

[56] Coscia M, Wessel MJ, Chaudary U, Millán JDR, Micera S, Guggisberg A, et al. Neurotechnology-aided interventions for upper limb motor rehabilitation in severe chronic stroke. Brain. 2019;142(8):2182−97.

[57] Jackson A, Zimmermann JB. Neural interfaces for the brain and spinal cord—restoring motor function. Nat Rev Neurol 2012;8:690−9.

[58] Simonetti D, Zollo L, Papaleo E, Carpino G, Guglielmelli E. Multi-modal adaptive interfaces for 3D robot-mediated upper limb neurorehabilitation: an overview of biocooperative systems. Robot Auton Syst 2016;85:62−72.

[59] Scotto di Luzio F, Simonetti D, Cordella F, Miccinilli S, Sterzi S, Draicchio F, et al. Biocooperative approach for the human-in-the-loop control of an end-effector rehabilitation robot. Front Neurorobot 2018;12:67.

[60] Riener R, Munih M. Guest editorial special section on rehabilitation via bio-cooperative control. IEEE Trans Neural Syst Rehabil Eng 2010;18:337−8.

[61] Nann M, Cordella F, Trigili E, Lauretti C, Bravi M, Miccinilli S, et al. Restoring activities of daily living using an EEG/EOG-controlled semiautonomous and mobile whole-arm exoskeleton in chronic stroke. IEEE Syst J 2020;15:3214−21. Available from: https://doi. org/10.1109/JSYST.2020.3021485.

[62] Crea S, Nann M, Trigili E, Cordella F, Baldoni A, Badesa FJ, et al. Feasibility and safety of shared EEG/EOG and vision-guided autonomous whole-arm exoskeleton control to

perform activities of daily living. Sci Rep 2018;8. Available from: https://doi.org/10.1038/s41598-018-29091-5.

[63] Yuan K, Wang X, Chen C, Lau CC, Chu WC, Tong RK. Interhemispheric functional reorganization and its structural base after BCI-guided upper-limb training in chronic stroke. IEEE Trans Neural Syst Rehabil Eng 2020;28(11):2525−36.

[64] Duret C, Grosmaire AG, Krebs HI. Robot-assisted therapy in upper extremity hemiparesis: overview of an evidence-based approach. Front Neurol 2019;10:412.

[65] Demofonti A, Carpino G, Zollo L, Johnson MJ. Affordable robotics for upper limb stroke rehabilitation in developing countries: a systematic review. IEEE Trans Med Robot Bionics 2021;3(1):11−20.

Further reading

Boldrini P, Bonaiuti D, Mazzoleni S, Posteraro F. Rehabilitation assisted by robotic and electromechanical devices for persons with neurological disabilities: an Italian consensus conference. Funct Neurol 2019;34(2):123−4. Available from: https://doi.org/10.23736/S1973-9087.21.07084-2.

Gandolfi M, Valè N, Posteraro F, Morone G, Dell'Orco A, Botticelli A, et al. State of the art and challenges for the classification of studies on electromechanical and robotic devices in neurorehabilitation: a scoping review Italian Consensus Conference on Robotics in Neurorehabilitation (CICERONE)Eur J Phys Rehabil Med 2021;57:831−40. Available from: https://doi.org/10.23736/S1973-9087.21.06922-7.

Chapter 9

Europe region: Spain

Nicolás García-Aracil[1,2], Jose López Sánchez[3,4], José María Catalán Orts[2], Andrea Blanco Ivorra[2] and Javier Sánchez Aguilar[3]

[1]Department of Control and Systems Engineering, Miguel Hernandez University, Alicante, Spain, [2]Robotics and Artificial Intelligence Unit, Bioengineering Institute, Spain, [3]European Neurosciences Center, Madrid, Spain, [4]Spanish Association for Intensive Therapy in Neurorehabilitation, Spain

Learning objectives

By the end of this chapter, the reader will learn about:
1. The structure and resources of healthcare in Spain, focusing on the structures of classical stroke rehabilitation care and the use of neurorehabilitation technologies.
2. The different Spanish companies that develop technologies for neurorehabilitation therapies.

9.1 Introduction

9.1.1 Country history and demographics

Spain is located in Southern Europe, also known as Mediterranean Europe, as it is surrounded by the Mediterranean Sea. Spain is a country with profound historical roots in Europe. Its identity and unique idiosyncrasies have been forged by a variety of phenomena, such as the discovery of the Americas and its neutral position during the two world wars.

In 2019, the population of Spain exceeded 47 million, reaching 47,332,614 million during the first half of 2020. Since that time, the population has grown slower, remaining below 47 million according to the latest data for 2021 [1]. The population density grew to reach 94.8 inhabitants per square kilometer in 2020, one more inhabitant compared to 2018 [2]. In 2019, the total fertility rate was 1.24 children per woman (360.617 births). In the last decade, the fertility rate has decreased; in 2010, it was 1.37 children per woman.

Currently, 19.6% of the population of Spain is older than 65, with an average life expectancy of 83 years, one of the highest in Europe. The population

Rehabilitation Robots for Neurorehabilitation in High-, Low-, and Middle-Income Countries.
DOI: https://doi.org/10.1016/B978-0-323-91931-9.00005-0

decreased by 72,007 people during the first half of the year. According to the Instituto Nacional de Estadística, this decrease was due to a negative natural balance of 70,736 people and a practically zero migratory balance [3].

Spain's gross domestic product (GDP) in 2021 reached 1,205,063 million euros. Spanish economy ranks 15th in the world and 5th in Europe, after Germany, the United Kingdom, France, and Italy. The weakness of the Spanish economy can be summarized in three main key points: (1) a high public debt: 1,205,063 million euros, with a debt of 120% of GDP in 2021; (2) a high structural unemployment rate, mainly among young people; and (3) high dependency rate on energy imports. Moreover, Spain spent only 1.45% of GDP on research and development (R&D) in 2020, but the new "Strategy for Science, Technology and Innovation 2021−2027" aims to increase Spain's public and private R&D investment to 2.1% of GDP by 2027. These data show that Spanish expenditure on R&D is low compared with 2.35% in France and 3.14% in Germany, according to OECD statistics.

9.1.2 Communicable and noncommunicable disease prevalence

In March 2022, a total of 418,703 deaths were registered in Spain [4]. Cerebrovascular disease, breast cancer, coronary heart disease, and lung or colon cancer are the major causes of death in Spain. On average, the Spanish population spends more time living in good health compared to other EU countries, but risk factors for disease and disability such as alcohol consumption have increased. Prevalence and mortality rates for stroke and heart attack are one of the lowest in Europe, but in 2016, both were still among the leading causes of death [4]. However, there are more than 552,000 people who have survived a stroke and more than 1.5 million people live with coronary heart disease, and these numbers are on the rise [5]. In 2020, life expectancy at birth in Spain experienced the largest reduction among EU countries caused by the COVID-19 pandemic, a temporary fall of 1.6 years [4].

9.1.3 Stroke incidence and prevalence

In Spain, according to available data, the annual incidence of stroke is 187.4 cases per 100,000 inhabitants, which represents a total of 71,780 new cases considering the Spanish population aged 18 or over as of January 1, 2018 [6].

9.2 Healthcare structure and resources

The 1978 Spanish Constitution established a region-based organization of the national territory that allowed the devolution of central healthcare powers to the 17 autonomous communities and two autonomous cities (Ceuta and Melilla). Therefore the healthcare devolution process began in 1981 and ended in 2002, with the central government retaining the responsibility for healthcare management in the autonomous cities of Ceuta and Melilla, through the National Health Management Institute (INGESA) [7]. From 2022, the Spanish health system has been decentralized, which has given the regional healthcare authorities the power and duties in health planning, public health, and healthcare.

In summary, the current health system consists of three hierarchical levels: (1) central: the Ministry of Health is in charge of issuing health proposals, planning and implementing government health guidelines, and coordinating activities; (2) autonomous community: each of Spain's 17 autonomous communities is responsible for offering integrated health services to the regional population through the centers, services, and establishments of that community; and (3) local: the health areas are responsible for the unitary management of the health services offered at the level of the autonomous community and are defined by taking into account factors of demography, geography, climate, socioeconomics, employment, epidemiology, and culture.

From an economic point of view, Spain's health spending has increased in recent years but remains below the EU average (115,400 billion euros per year, 9.3% GDP, 2451 euros per inhabitant) [4]. The national health system is based on universal coverage and is mainly funded by taxes. Health spending is 81.6 billion euros per year in public hospitals (6.6% GDP, 1732 euros per inhabitant) and 33.8 billion euros per year in private hospitals (2.7% GDP, 719 euros per inhabitant). In both cases, most of the costs are for curative care and rehabilitation.

According to data from the Ministry of Health, Spain has around 3000 healthcare centers and a total number of 831 hospitals, with a capacity of 158,648 beds. The public health system has 418 hospitals, equipped with 111,214 beds, and two Ministry of Defense hospitals with 700 beds. The remainder 411 hospitals are private hospitals equipped with 46,734 beds. Public hospitals are generally much larger than private hospitals and deal with a much higher number of patients [8].

9.2.1   Rehabilitation structure

9.2.1.1   Rehabilitation access

Rehabilitation access is heterogeneous in the different regions of Spain. It is also different depending on the type of service provided, differentiating between inpatient and outpatient rehabilitation.

Based on official data from 2015 [9], intensive rehabilitation in the public network is restricted mostly to the acute and early subacute phases, and there is a marked deficiency of some specialized resources and professionals in the late subacute and chronic phases. In these phases, the lack of public resources is supplemented and complemented, in most cases, with private resources.

All different regions of Spain offer physical rehabilitation in the acute phase. However, the provision of the whole package of basic intervention areas (neuropsychology, occupational therapy, physical therapy, and speech therapy) is rare and only happens in 33.3% of the cases [10].

In addition, there are protocols regulating the transition between acute and subacute phases only in 33.3% of the cases. Once rehabilitation in the hospital is finished, the rehabilitation resources vary enormously between the different regions of Spain. For example, all regions have outpatient rehabilitation, but only half of them have a specialized hospital unit. The duration of the rehabilitation process is, on average, less than 3 months for hospitalization and less than 6 months for outpatient rehab. Physical therapy is always provided, unlike other specialists like occupational therapy and speech therapy (94% and 88%, respectively) and neuropsychology (58.8%) [10].

In the chronic phase, only one region has regulations for the continuity of services, once the subacute phase is finished. In the rest of the regions, the patients and their families need to make arrangements to continue receiving social support, and rehabilitation normally ends abruptly. Some regions (38.9%) have alternatives like day centers or residences; however, for the rest of them, the resources are not specific for stroke patients or are dependent on associations founded by patients and families.

9.2.1.2 Rehabilitation capacity

In Spain, stroke is the second leading cause of death, the leading cause of disability in adulthood, and the second highest cause of dementia. Because of this, the national health system and the autonomous communities have implemented a stroke strategy that specifically establishes organizational systems based on the recommendations of scientific societies. These plans include the implementation of stroke units as the most effective and efficient healthcare resource in the treatment of stroke patients, as it is the one that benefits the greatest number of patients, reduces mortality, dependency, and the need for institutionalization, and creates stroke teams in hospitals with a lower level of complexity [11].

According to the latest report of the Ministry of Health, Spain has 133 sterilization units and 71 intensive care units [12]. Most stroke units (80%) have an average of five beds with continuous noninvasive monitoring. On the other hand, 36% of stroke teams reported having monitored beds for patient care (two to five beds). The mean number of non-monitored beds assigned to stroke units was 14 ± 8 and 12 ± 7 in the case of stroke teams.

Fig. 9.1 shows the average duration of rehabilitation treatment in each region of Spain [10]. Rehabilitation protocols are not normally intensive, and patients receive conventional therapy sessions (between 1 and 2 hours per day combining all different specialties). Table 9.1 summarizes the kind of specific and nonspecific rehabilitation resources and the percentage of regions in Spain where those resources are available [10].

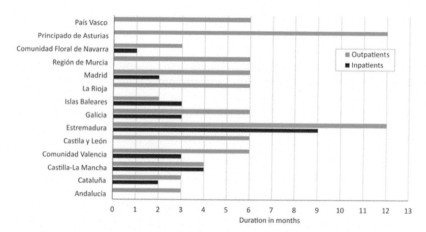

FIGURE 9.1   Average duration of rehabilitation treatment in each region of Spain, differentiated between inpatients and outpatients.

TABLE 9.1 Summary of the type of specific and nonspecific rehabilitation resources and the percentage of regions in Spain where they are available.

Kind of specific resources available	Percentage of regions having the resource
Day rehab center	85.7%
Residential center	42.3%
Associations	42.3%
Nonspecific resources available	**Percentage of regions having the resource**
Rehabilitation center for physically handicapped	27.8%
Personal independence promotion center	11.1%

9.2.1.3   Stroke rehabilitation standard of care

The predominant approach in physical rehabilitation in Spain focuses on task parts and overall exercise rather than practicing the entire task [13]. Thus it emphasizes normal kinematics, compensation avoidance, and the idea of avoiding challenging tasks or conditions until they are consistent. This approach is circumscribed by a general lack of practical application of the principles of experience-dependent neuroplasticity [14] and motor learning

[15]. This entails general rehabilitation, which even using technology does not rise from the extraction of the mechanisms of action, active ingredients, and targets of the treatment [16]. This neurorehabilitation paradigm based on the emphasis on normal movement, tone inhibition, and manual facilitation, added to the traditional paradigm of compensatory training, could be a barrier to the promotion of physical activity in stroke patients. The beginning of a sedentary lifestyle takes place in the hospital environment [17] and, therefore, affects the acute and subacute phases of recovery when spontaneous recovery takes place and greater improvements can be observed. Descriptive studies show that people with stroke spend more than 75% of their waking hours sitting [18,19] and that at least 30% of all therapy sessions are not aimed at active therapy [20].

Motor learning studies in healthy animals suggest that hundreds of goal-directed repetitions are required to induce changes in cortical synaptic density [21,22]. This is less than what is typically seen in rehabilitation centers (approximately 32 repetitions per task) [23]. However, Birkenmeier and Waddell have shown that patients can achieve and tolerate around 300 repetitions per task in a 1-hour session [24,25]. Therefore, and as stated by Barrett et al., on many occasions there is no relationship between the level of activity and personal factors (such as the severity of the stroke) [26]. This leads the authors to think that the causes of the low levels of activity are mainly of an institutional nature and the structure of the rehabilitation programs. Therefore low levels of therapy, together with low intensity of therapy and an excessively sedentary lifestyle, are part of a paradigm that conspires to reduce the potential for recovery after a stroke [26].

Until relatively recently, stroke patients were thought to be neurologically and physiologically fragile, and not able to tolerate intensive exercise programs [27]. In current neurorehabilitation, concerns about strength and resistance training still exist, although the latest clinical practice guidelines recommend such forms of training for stroke recovery [28,29]. John Krakauer has speculated that low intensity of therapy may be responsible for rehabilitation not influencing spontaneous recovery [30], supporting the idea that there is an intensity threshold that allows improvements beyond biological recovery. Faced with this reality, we may find ourselves with rehabilitation systems that may not be influencing the recovery of stroke patients.

9.2.1.4 Implementation of neurorehabilitation technologies

The use of neurorehabilitation technologies in Spain is not as widespread as in other European countries. Still, they are a rapidly expanding field that many institutions are starting to invest in as a complement or tool for conventional neurorehabilitation therapies. However, it is still not easy to find clinics or hospitals that have this type of technology. In general, public hospitals in Spain do not have them. However, some use them because they collaborate in their development with universities, companies, or research centers. Some of these examples are the National Hospital for Paraplegics in Toledo and the Hospital La Pedrera located in Denia, Alicante.

Another example is the Fundació Institut Guttmann [31]. They offer different initiatives dedicated to treating people with disabilities of neurological origin based on technologies such as robotic rehabilitation platforms, exoskeletons, and virtual reality. In addition, they use techniques such as noninvasive brain stimulation.

Another institution that focuses on the use of new neurorehabilitation technology is the European Center of Neurosciences (CEN) [32]. Among the different technologies they have, they use the Lokomat and the Armeo Spring exoskeletons, the Rysen system, several motion sensors, and functional electrostimulation devices, among others. CEN is the first clinic in Spain to introduce intensive therapy protocols, combining evidence-based approaches with advanced technologies in neurorehabilitation.

The Congregation of Sisters Hospitallers of the Sacred Heart of Jesus has numerous centers throughout Spain [33]. They not only use robotics and virtual reality in neurorehabilitation therapies but also invest in research, development, training, and teaching activities in this field.

In Spain, there is also the Vithas group [34]. It consists of 19 hospitals and 31 medical and healthcare centers, distributed in 13 provinces of Spain. Among the different specialties, they have the Vithas Hospitals Neurorehabilitation Unit. This unit has adult neurorehabilitation and pediatric neurorehabilitation programs that include activities carried out with neurorehabilitation technologies such as robotic rehabilitation platforms, exoskeletons, and virtual reality. In addition, this unit has a dedicated team that conducts research into the application of new technologies. They use the Lokomat and the Armeo Spring exoskeletons and some systems based on virtual reality environments.

One of the first rehabilitation clinics for acquired brain damage and spinal cord injury in Spain was Neuron [35]. Its neurological rehabilitation centers have different intensive rehabilitation programs in which they combine conventional therapies with the use of technology. At Neuron, they have a wide range of robotic devices and virtual reality environments that make them an international reference center for treatment, training, and research in dealing with acquired brain damage and neurological pathologies. Neuron is the only rehabilitation clinic in Spain that has the complete neurological rehabilitation circuit of the Tyrosolution by Tyromotion [36].

Movex Clinics is a neurorehabilitation clinic specializing in the use of robotics [37]. They have different exoskeletons among which are the Hank and Belk lower-limb exoskeletons and the Hand of Hope exoskeleton. They also have different virtual reality systems.

Los Madroños Hospital has an advanced neurorehabilitation unit that is committed to comprehensive and multidisciplinary rehabilitation, incorporating robotics and immersive virtual reality [38]. They have a Walkbot lower-limb exoskeleton and the Armeo Power upper-limb exoskeleton.

9.3 Technology-assisted rehabilitation

Robotic-assisted rehabilitation services are an emerging and growing market in Spain. However, there are a reduced number of Spanish companies producing technology devices to deliver rehabilitation therapies. In this section, we are going to list some of these companies based on the type of technologies they develop.

9.3.1 Robotics

As it is well known, rehabilitation robotic devices can be categorized into end-effector and exoskeleton devices according to their mechanical structures. Most Spanish companies producing robotic rehabilitation devices are focused on exoskeleton devices and more specifically on lower-limb exoskeleton devices.

One example is Marsi Bionics [39]. This company is a small and medium enterprise (SME) founded in 2013 as a spin-off of the Centro de Automática y Robótica (CAR), a joint center of the Universidad Politécnica de Madrid (UPM) and the Consejo Superior de Investigaciones Científicas (CSIC). Its activities are focused on the development of lower-limb exoskeletons. One of its products is the MAK Active Knee, which consists of a knee exoskeleton designed for use in gait rehabilitation in patients affected by hemiplegia or knee weakness. But its flagship product is the Atlas Pediatric Exo (Fig. 9.2), which is the first pediatric gait exoskeleton developed in the world. It provides assistance to the patient, collects data to generate their walking pattern, and monitors the information of each session.

Another one is ABLE Human Motion [40]. This company is a start-up that designs, develops, and commercializes exoskeleton technology for people with disabilities. They produce and commercialize the ABLE lower-limb exoskeleton (Fig. 9.3).

Gogoa is a start-up that designs and develops exoskeletons to improve people's quality of life [41]. They are the first European company to obtain the CE mark for an exoskeleton for the rehabilitation of lower limbs, the HANK exoskeleton (Fig. 9.4). They also have another lower-limb exoskeleton called BELK for knee rehabilitation.

Technaid specializes in developing robotic assistive systems [42]. They develop two lower-limb exoskeletons called Exo-H2 and Exo-H3, and an ankle exoskeleton called Robotic Ankle H3. The Exo-H3 is the most technologically advanced exoskeleton of the company (Fig. 9.5).

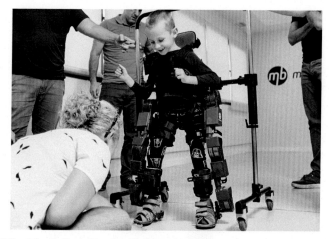

FIGURE 9.2   Atlas Pediatric Exo developed by Marsi Bionics.

FIGURE 9.3   ABLE lower-limb exoskeleton developed by ABLE Human Motion.

Regarding end-effector robotic devices for delivering rehabilitation therapies, Innovative Devices for Rehabilitation & Assistance (iDRhA) [43] is an SME founded in 2019 as a spin-off of the Robotics and Artificial Intelligence (RAI) research group [44] of the Universidad Miguel Hernández (UMH). iDRhA produces Helium, which is an active exoskeleton for the rehabilitation and assistance of the hand (Fig. 9.6). Helium can be used in both hands and is the only hand exoskeleton in the world that incorporates

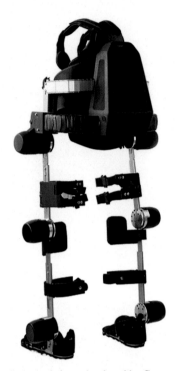

FIGURE 9.4 HANK lower-limb exoskeleton developed by Gogoa.

optical force sensors in each finger. These sensors measure the user's movement intention in real time and therefore have assistance modes that allow adaptation of the level of assistance provided to the patient.

iDRhA is also the only company in Spain that markets an end-effector-based rehabilitation robot called Rubidium (Fig. 9.7). It is a portable rehabilitation robotic platform that can be used for rehabilitation activities both in the clinic and at home.

Both devices work with a software platform that has different therapy games specially designed for rehabilitation activities. In addition, Helium and Rubidium can be integrated into the same system for simultaneous use (Fig. 9.8).

9.3.2 Non-actuator devices

Cyber Human Systems is a spin-off of Gogoa Mobility Robots. They are the only Spanish company that designs and markets exoskeletons for the industry [41]. Two products are available—the ALDAK lumbar exoskeleton and the BESK G arm exoskeleton (Fig. 9.9).

© Technaid S.L.

FIGURE 9.5 Exo-H3 lower-limb exoskeleton developed by Technaid.

FIGURE 9.6 Helium, the hand exoskeleton developed by iDRhA.

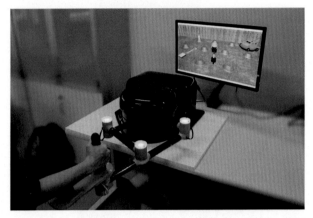

FIGURE 9.7 Rubidium, end-effector-based rehabilitation robot developed by iDRhA.

FIGURE 9.8 Robotic rehabilitation platform developed by iDRhA from the union of the Rubidium robotic rehabilitation platform, together with the Helium hand exoskeleton.

9.3.3 Sensor technology

Eodyne specializes in real-time interactive systems and technologies for virtual and augmented reality [45]. The main product of Eodyne is the Rehabilitation Gaming System (RGS). This system is based on the use of an RGB-D camera to detect the movements of the patient.

Evolvis a certified medical device manufacturer specializing in software and hardware development of technological solutions for rehabilitation [46]. Their solutions called EvolvRehab are mainly based on the use of motion

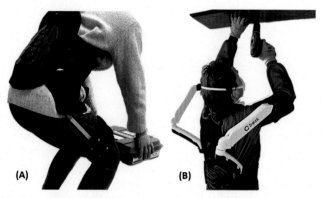

(A)   **(B)**

FIGURE 9.9   The ALDAK lumbar exoskeleton (A) and the BESK G arm exoskeleton (B) developed by Cyber Human Systems.

capture sensors to play video games especially designed to perform rehabilitation exercises.

On the other hand, DyCare [47] has a series of systems specially designed to perform rehabilitation exercises at home similar to those mentioned above. DyCare has two main products. ReHub is a personalized digital musculoskeletal solution for home rehabilitation. It can also incorporate inertial sensors for motion tracking. They also have a product called Lynx. Lynx is a software platform that incorporates inertial magnetic units to measure the mobility of each joint in the human body. This system allows for evaluating the patient's range of motion before, during, and after therapy.

Along the same line, the company BJ Adaptaciones [48] also has an online platform for rehabilitation and cognitive stimulation called Guttmann NeuroPersonalTrainer (GNPT) developed together with the Guttmann Institute.

We previously mentioned that the company Technaid [42] develops lower-limb exoskeletons. In addition, they also develop a wireless system based on magnetic inertial units to capture and analyze human biomechanics.

Werium also develops a similar system called Pro Motion Capture for measuring the joint range of motion based on inertial magnetic units. Werium Assistive Solutions SL. is a company born from the Superior Council of Scientific Research (CSIC) in 2015 [49].

9.3.4   Functional electrical stimulation

In the functional electrical stimulation field, in Spain, there is the company Fesia Technology [50]. Fesia is a technology and scientific-based company engaged in generating advanced functional electrical stimulation (FES) solutions for the rehabilitation of people with diseases or injuries in the nervous system. They have mainly two FES devices, one for hand function rehabilitation called the Fesia Grasp, and the other for the gait rehabilitation called Fesia Walk.

9.3.5   Other

There are other Spanish companies, such as Adamo Robot SL. [51], that use robotic technologies to treat musculoskeletal disorders mimicking the techniques used by physical therapists. The Adamo robot was designed to apply a jet of pressurized air with the correct temperature and without the need for direct contact with the patient. Using the correct air jet flow, the robot imitates manual therapy techniques used by physical therapists, while also increasing the subdermal blood flow and oxygenation of the area.

It is worth highlighting the digital web platform, NeuronUP, widely used by therapists as a tool for cognitive rehabilitation and stimulation processes. It consists of numerous materials and resources for designing treatment sessions in addition to a patient manager for organizing and saving the results of those sessions. NueronUP allows two types of therapeutic interventions: (1) neuropsychological rehabilitation, which is aimed at restoring the degree of the patient's functional condition to the highest possible physical, psychological, and social adaptation level; and (2) cognitive stimulation, which aims at slowing cognitive deterioration and maintaining preserved abilities. It is used for neurodegenerative diseases and normal aging [52].

9.4   Challenges

One of the main challenges of the Spanish health system is the lack of coordination between the 17 regional health systems and the national health system, which increases the differences in health services among the regions. Additionally, most of the regional health systems lack the technologies to deliver intensity rehabilitation therapies due to budget constraints. Therefore the maximization of patient's recovery is compromised by these constraints, and rehabilitation technologies could allow regional health services to increase the duration, intensity, and frequency of rehabilitation therapies and, thus, rehabilitation outcomes.

Another challenge of rehabilitation units will be to adapt or change the work organization and patient management when new technologies are incorporated into these units.

Finally, there is an urgent need to include advanced technologies in the curriculum of the universities educating future professionals. Universities should collaborate with private clinics providing technologies in their treatments to improve therapist's knowledge of how to apply advanced technologies in neurorehabilitation.

9.5 Future opportunities

The scientific literature has shown high evidence of outcomes, efficacy, and effectiveness of rehabilitation technologies, and therefore it is a question of time that regional health systems incorporate different rehabilitation technologies into the clinical practice. Because of the COVID-19 pandemic, the national and regional health system has recognized the need to incorporate novel technologies to deliver therapy in clinical settings and/or in the patient's home. Instead, there will be many opportunities for developing and commercializing novel rehabilitation technologies in Spain.

The opportunity for collaboration between public and private institutions, academic and clinical, in the development of new research involving advanced technologies can be very beneficial for the continuous development of technologies for neurorehabilitation, improving patient's recovery and outcomes.

Acknowledgments

Some of the work in this chapter has been funded by Ministerio de Ciencia e Innovacioón, which is part of Agencia Estatal de Investigacioón (AEI), through the project PID2019−108310RB-I00/AEI/10.13039/501100011033 (SPLASH).

Conflict of interest

Nicolas Garcia-Aracil is the Founder of iDRhA. Jose López Sánchez is Clinical Director at European Neurosciences Center.

References

[1] Instituto Nacional de Estadística. Available from: https://www.ine.es/consul/serie.do?d = true&s = CP335; 2022 (accessed 12.04.22).

[2] Data Commons. Available from: https://datacommons.org/place/country/ESP?hl = es; 2022 (accessed 12.04.22).

[3] Instituto Nacional de Estadística. Population Figures at 1 July 2021, Migration Statistics 2020 (provisional). Press Releases, 2021.

[4] Ministerio de Sanidad (Gobierno de España). Available from: https://www.sanidad.gob.es/estadEstudios/sanidadDatos/home.htm; 2022 (accessed 12.04.22).

[5] Budig K, Harding E. Secondary Prevention of Heart Attack and Stroke: Country Profile for Spain. London: The Health Policy Partnership; 2021.

[6] Díaz-Guzmán J, Egido JA, Gabriel-Sánchez R, et al. Stroke and transient ischemic attack incidence rate in Spain: the IBERICTUS study. Cerebrovasc Dis 2012;34(4):272−81. Available from: https://doi.org/10.1159/000342652.

[7] Ministerio de Sanidad y Consumo. National Health System: Spain. Available from: https://www.sanidad.gob.es/en/organizacion/sns/docs/Spanish_National_Health_System.pdf; 2008 (accessed 12.04.22).

[8] Ministerio de Sanidad, Catálogo Nacional de Hospitales 2021. Available from: https://www.sanidad.gob.es/ciudadanos/prestaciones/centrosServiciosSNS/hospitales/docs/CNH_2021.pdf; 2022 (accessed 12.04.22).

[9] Defensor del Pueblo. Daño cerebral sobrevenido en españa: un acercamiento epidemiológico y sociosanitario. Madrid: Informe del Defensor del Pueblo; 2005.

[10] Aza-Hernández A, Verdugo-Alonso MA. Modelos de atención pública a la población con daño cerebral adquirido en España: un estudio de la situación por comunidades autónomas. Rev Neurol 2022;74:245−57. Available from: https://doi.org/10.33588/rn.7408.2021372.

[11] Alonso de Leciñana M, et al. Características de las unidades de ictus y equipos de ictus en España en el año 2018. Proyecto Pre2Ictus. Neurología. 2020. https://doi.org/10.1016/j.nrl.2020.06.012.

[12] Estrategia en Ictus del Sistema Nacional de Salud, Informe de evaluación y líneas prioritarias de actuación. Informes, Estudios E Investigación 2022 Ministerio de Sanidad. https://www.sanidad.gob.es/bibliotecaPub/repositorio/libros/29231_estrategia_en_ictus_del_Sistema_Nacional_de_Salud_-_informe_de_evaluacion_y_lineas_prioritarias_de_actuacion.pdf (accessed 12.04.22).

[13] Moore JL, Roth EJ, Killian C, Hornby TG. Locomotor training improves daily stepping activity and gait efficiency in individuals poststroke who have reached a "plateau" in recovery. Stroke 2010;41(1):129−35. Available from: https://doi.org/10.1161/STROKEAHA.109.563247.

[14] Kleim JA, Jones TA. Principles of experience-dependent neural plasticity: implications for rehabilitation after brain damage. J Speech Lang Hear Res 2008;51(1):S225−39. Available from: https://doi.org/10.1044/1092-4388(2008/018).

[15] Leech KA, Roemmich RT, Gordon J, Reisman DS, Cherry-Allen KM. Updates in motor learning: implications for physical therapist practice and education. Phys Ther 2022;102(1):pzab250. Available from: https://doi.org/10.1093/ptj/pzab250.

[16] Hart T, Dijkers MP, Whyte J, et al. A theory-driven system for the specification of rehabilitation treatments. Arch Phys Med Rehabil 2019;100(1):172−80. Available from: https://doi.org/10.1016/j.apmr.2018.09.109.

[17] Åstrand A, Saxin C, Sjöholm A, et al. Poststroke physical activity levels no higher in rehabilitation than in the acute hospital. J Stroke Cerebrovasc Dis 2016;25(4):938−45. Available from: https://doi.org/10.1016/j.jstrokecerebrovasdis.2015.12.046.

[18] English C, Healy GN, Coates A, Lewis L, Olds T, Bernhardt J. Sitting and activity time in people with stroke. Phys Ther 2016;96(2):193−201. Available from: https://doi.org/10.2522/ptj.20140522.

[19] Tieges Z, Mead G, Allerhand M, et al. Sedentary behavior in the first year after stroke: a longitudinal cohort study with objective measures. Arch Phys Med Rehabil 2015;96(1):15−23. Available from: https://doi.org/10.1016/j.apmr.2014.08.015.

[20] Kaur G, English C, Hillier S. How physically active are people with stroke in physiotherapy sessions aimed at improving motor function? A systematic review. Stroke Res Treat 2012;2012:820673. Available from: https://doi.org/10.1155/2012/820673.

[21] Plautz EJ, Milliken GW, Nudo RJ. Effects of repetitive motor training on movement representations in adult squirrel monkeys: role of use versus learning. Neurobiol Learn Mem 2000;74(1):27−55. Available from: https://doi.org/10.1006/nlme.1999.3934.

[22] Nudo RJ, Milliken GW. Reorganization of movement representations in primary motor cortex following focal ischemic infarcts in adult squirrel monkeys. J Neurophysiol 1996;75(5):2144−9. Available from: https://doi.org/10.1152/jn.1996.75.5.2144.

[23] Lang CE, Macdonald JR, Reisman DS, et al. Observation of amounts of movement prac-
      tice provided during stroke rehabilitation. Arch Phys Med Rehabil 2009;90(10):1692−8.
      Available from: https://doi.org/10.1016/j.apmr.2009.04.005.

[24] Birkenmeier RL, Prager EM, Lang CE. Translating animal doses of task-specific training
      to people with chronic stroke in 1-hour therapy sessions: a proof-of-concept study.
      Neurorehabil Neural Repair 2010;24(7):620−35. Available from: https://doi.org/10.1177/
      1545968310361957.

[25] Waddell KJ, Birkenmeier RL, Moore JL, Hornby TG, Lang CE. Feasibility of high-
      repetition, task-specific training for individuals with upper-extremity paresis. Am J Occup
      Ther 2014;68(4):444−53. Available from: https://doi.org/10.5014/ajot.2014.011619.

[26] Barrett M, Snow JC, Kirkland MC, et al. Excessive sedentary time during in-patient stroke
      rehabilitation. Top Stroke Rehabil 2018;25(5):366−74. Available from: https://doi.org/
      10.1080/10749357.2018.1458461.

[27] Eng JJ, Tang PF. Gait training strategies to optimize walking ability in people with stroke:
      a synthesis of the evidence. Expert Rev Neurother 2007;7(10):1417−36. Available from:
      https://doi.org/10.1586/14737175.7.10.1417.

[28] Teasell R, Salbach NM, Foley N, et al. Canadian stroke best practice recommendations:
      rehabilitation, recovery, and community participation following stroke. Part One:
      Rehabilitation and Recovery Following Stroke; 6th Edition Update 2019. Int J Stroke
      2020;15(7):763−88. Available from: https://doi.org/10.1177/1747493019897843.

[29] Hornby TG, Reisman DS, Ward IG, et al. Clinical practice guideline to improve locomotor
      function following chronic stroke, incomplete spinal cord injury, and brain injury. J Neurol Phys
      Ther 2020;44(1):49−100. Available from: https://doi.org/10.1097/NPT.0000000000000303.

[30] Krakauer JW, Carmichael ST, Corbett D, Wittenberg GF. Getting neurorehabilitation
      right: what can be learned from animal models? Neurorehabil Neural Repair 2012;26
      (8):923−31. Available from: https://doi.org/10.1177/1545968312440745.

[31] Neurorehabilitation Hospital. Institut Guttmann. Available from: https://www.guttmann.
      com/en; 2022 (accessed 12.04.22).

[32] The European Center of Neurosciences. Available from: https://www.eneurocenter.com/
      en/home/; 2022 (accessed 12.04.22).

[33] Sisters Hospitallers. Available from: https://www.sistershospitallers.org/; 2022 (accessed
      12.04.22).

[34] Neurorehabilitation services for brain injury in vithas Hospitals. Available from: https://
      neurorhb.com/en/; 2022 (accessed 12.04.22).

[35] Neuron. Available from: https://neuronrehab.es/; 2022 (accessed 12.04.22).

[36] TyroMotion. Available from: https://tyromotion.com/en/; 2022 (accessed 12.04.22).

[37] Movex Clinics. Nerorehabilitation robotics. Available from: https://www.movexbilbao.
      com/; 2022 (accessed 12.04.22).

[38] Hospital Los Madroños. Advanced neurorehabilitation unit. Available from: https://hospi-
      tallosmadronos.es/unidades-especializadas/unidad-avanzada-de-neurorrehabilitacion/#;
      2022 (accessed 12.04.22).

[39] Marsi Bionics. Available from: https://www.marsibionics.com/en/mb-active-knee/; 2022
      (accessed 12.04.22).

[40] ABLE Human Motion. Available from: https://www.ablehumanmotion.com/; 2022
      (accessed 12.04.22).

[41] GOGOA Mobility Robots. Available from: https://en.gogoa.eu/; 2022 (accessed 12.04.22).

[42] Technaid. Available from: https://www.technaid.com/; 2022 (accessed 12.04.22).

[43] iDRhA. Innovative Devices for Rehabilitation and Assistance, Available from: https://idr-ha.es/; 2022 (accessed 12.04.22).

[44] RAI. Robotics and Artificial Intelligence, Available from: https://rai.umh.es/; 2022 (accessed 12.04.22).

[45] Eodyne. Available from: https://www.eodyne.com/; 2022 (accessed 12.04.22).

[46] Evolv Rehabilitation TEchnologies. Available from: https://evolvrehab.com/; 2022 (accessed 12.04.22).

[47] DyCare. Clinical assessment by wearable sensors solutions. Available from: https://www.dycare.com/; 2022 (accessed 12.04.22).

[48] B.J. Adaptaciones. Available from: https://bjadaptaciones.com/; 2022 (accessed 12.04.22).

[49] Werium Solutions. Available from: https://www.weriumsolutions.com/; 2022 (accessed 12.04.22).

[50] Fesia Technology. Available from: https://fesiatechnology.com/en/; 2022 (accessed 12.04.22).

[51] Adamo Robot. Collaborative robotics at the service of people. Available from: https://adamorobot.com/; 2022 (accessed 12.04.22).

[52] NeuronUP. Available from: https://www.neuronup.com; 2022 (accessed 12.04.22).

Further reading

OECD/European Observatory on Health Systems and Policies. Spain: country health profile 2021, State of Health in the EU. Paris: OECD Publishing; 2021. Available from: 10.1787/7ed63dd4-en.

Chapter 10

Asia-Pacific region: Japan

Yasuhisa Hirata[1], Tomoyuki Noda[2] and Ichiro Miyai[3]

[1]*School of Engineering, Department of Robotics, Tohoku University, Sendai, Japan,*
[2]*Computational Neuroscience Laboratories (CNS), Advanced Telecommunication Research Institute (ATR), Kyoto, Japan,* [3]*Neurorehabilitation Research Institute, Morinomiya Hospital, Osaka, Japan*

Learning objectives

At the end of this chapter, the reader will be able to:
1. Describe the healthcare structure for stroke rehabilitation in Japan.
2. Describe the guidelines for stroke treatment in Japan and rehabilitation using robotics and mechatronics technologies.
3. Articulate the future trends of medical care and welfare in Japan and the future possibilities of robots for care support and rehabilitation.

10.1 Introduction

The ratio of people aged 65 and over to the total population in Japan is the highest in the world and is estimated to increase in the future. The current population of Japan aged 65 and over is 36.19 million, and the aging rate is 28.8% [1]. The population aged 65−74 is 17.47 million, accounting for 13.9% of the total population, and the population aged 75 and over is 18.72 million, accounting for 14.9% of the total population. This indicates that the population aged 75 and above exceeds the population aged 65−74. Furthermore, by 2065, about 1 in 2.6 people will be 65 years old or older, and about 1 in 3.9 people will be 75 years old or older. This is an urgent issue, especially in the fields of medicine and welfare.

This change in the proportion of the aging population will lead to a rapid increase in the number of people requiring care. In addition, Japan is currently experiencing an increase in the number of nuclear families, and the number of single-person households, married-couple-only households, and households in which both spouses are 65 years old or older. In urban areas, in particular, there is a large proportion of individuals living alone

Rehabilitation Robots for Neurorehabilitation in High-, Low-, and Middle-Income Countries.
DOI: https://doi.org/10.1016/B978-0-323-91931-9.00011-6

either due to being unmarried or divorced. As a result, there are many households with no one to care for them or households where the elderly care for the elderly. As a result, it is becoming increasingly difficult to provide care at home.

In addition, the term "healthy life expectancy" has been attracting attention. The Ministry of Health, Labor and Welfare (MHLW) in Japan defines healthy life expectancy as the period during which a person can live without having their daily life restricted by health problems, while life expectancy is the period until a person dies. This is based on a concept originally proposed by the World Health Organization (WHO) in 2000. According to this concept, even if a person has a chronic disease, as long as he or she is able to live independently without the help of others, he or she is considered healthy. According to a survey by the MHLW, the healthy life expectancy in 2016 was 72.14 years for men and 74.79 years for women in Japan [2]. The difference from the average life expectancy (80.98 years for men and 87.14 years for women) is about 9 years for men and 12 years for women. During this period, the quality of life often declines due to conditions that cause disability, and medical and nursing care costs also increase.

With the advent of this aging society, it is expected that various cutting-edge technologies, including robotics, will be used to solve the problems of elderly care and healthy life expectancy so that people can lead rich and fulfilling lives even if they have disabilities. In Japan, robotics research is flourishing, and various robotic technologies such as life support robots and disaster response robots have been developed, and many of them have been put to practical use. In particular, there are high expectations for the use of robotics technology in the field of nursing care and rehabilitation, for which research and development is underway.

The purpose of this chapter is to introduce the trend of neurorehabilitation in Japan, in particular, the statistics of stroke in Japan and rehabilitation for stroke. In addition, we will introduce the Japanese guidelines for stroke treatment and the treatment using robotics and mechatronics technology. In addition, we will explain how COVID-19 has changed stroke rehabilitation in Japan, and the possibility of telerehabilitation using robotic technology. Finally, we will explain the future trends of medical care and welfare in Japan and the possibilities of assistive and rehabilitation robots.

10.2 Stroke incidence and prevalence in Japan

10.2.1 Incidence of stroke and its social impact in Japan

The number of patients with stroke in Japan is approximately 1.5 million based on data from the MHLW [3] of the Japan government, and the yearly occurrence of stroke is $160 \sim 180$ per 100 thousand population based on annual statistics of patients [4]. According to the data from MHLW, stroke is the third highest cause of death behind cancer and heart disease with approximately 110,000/year dying from stroke (8% of total death). The prevalence of stroke is the fourth highest with 1,170,000 patients currently living with a stroke. In terms of medical costs, stroke is the second highest cause behind cancer accounting for $170 billion per year out of a total medical cost of $4,000 million. Specifically, it is the number one cause of medical costs for the elderly. In Japan, tissue plasminogen activator (tPA) and mechanical thrombectomy as treatments for acute ischemic stroke have been available since 2006 and since the mid-2010s, respectively. Despite such therapeutic developments, stroke is still the number one cause of disability in the elderly (17.2%). In addition to the national health insurance, the national long-term care insurance system was introduced in 2000 in Japan, and since then, stroke and dementia have been the two major conditions covered by insurance with increasing social impact. Thus, along with the progression of aging in Japanese society, stroke management has been regulated by the Cerebrovascular and Cardiovascular Disease Control Act of Japanese national law since 2019 [5].

10.3   Healthcare structure for stroke rehabilitation in Japan

10.3.1   Stroke care unit

Based on the international consensus on the efficacy of stroke treatment by multidisciplinary teams in stroke units [6], the Japanese insurance system introduced stroke care units for acute stroke with the mandatory full-time placement of nurses, physicians, and therapists. This was introduced at the same time as tPA approval in 2006. In response to the Cerebrovascular and Cardiovascular Disease Control Act in 2019, the Japan Stroke Society accredited approximately 1000 primary stroke centers in 2020 that provided 24-hour administration of thrombolytic treatment for acute stroke. Patients with a residual disability even after acute treatment and acute rehabilitation are transferred to the hospitals with Kaifukuki (Convalescent) Rehabilitation Ward (KRW) (Fig. 10.1). Consequently, 40% of acute stroke patients are transferred to KRW. In most regions, the seamless transfer of stroke patients from acute to KRW hospitals is managed by using a local clinical pathway sheet shared and optimized by the hospitals in the region [7].

10.3.2   Kaifukuki (Convalescent) rehabilitation ward

KRW is a unique post-acute rehabilitation system covered by national health insurance [8]. KRW and the national long-term care insurance system were both introduced at the same time in 2000 to prepare for the upcoming "super-aged" society when the proportion of elderly over 65 years would exceed 30% in 2025. For patients with a residual disability after acute

National policies for stroke rehabilitation in Japan

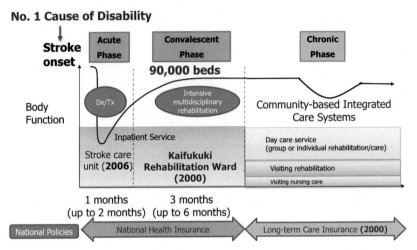

FIGURE 10.1   National policies for stroke rehabilitation in Japan. Stroke rehabilitation services are provided based on national health insurance system and long-term care insurance system in acute and convalescent phases, and in chronic phase, respectively. In acute phase, early rehabilitation is provided in stroke care unit normally within 1–2 days after the onset. Approximately, 40% of acute stroke patients are transferred to Kaifukuki (Convalescent) Rehabilitation Wards if they still have residual disability after 1 month stay in acute hospitals. In Kaifukuki Rehabilitation Wards, approximately 70% of patients go home after 2–3 months of intensive rehabilitation up to 3 hours each day. After discharge home, community-based rehabilitation is provided based on disability assessment according to the protocol determined by long-term care insurance.

treatment of stroke, KRW provides high dose (3 hours per day including weekends) inpatient rehabilitation for up to 150 days. Furthermore, for stroke patients with cognitive disorders, national insurance covers inpatient rehabilitation for up to 180 days. These aim to decrease impairment and disability of post-acute stroke patients and consequently reduce costs of long-term care insurance after discharge to home.

In KRW, multidisciplinary specialists including physicians, nurses, care workers, physical therapists, occupational therapists, speech-language pathologists, registered dietitians, and pharmacists provide intensive rehabilitation and care and disease management. In 2021, the number of KRW beds was approximately 90,000, and nearly half of the beds were allocated to patients with stroke. According to the annual report of the KRW Association [8] in 2020, the mean length of stay in acute hospitals and KRWs of patients with stroke were 36.2 and 83.0 days, respectively. The period of 2–3 months after stroke onset is considered to be the optimal time for intensive task-specific rehabilitation [9].

To be eligible for insurance reimbursement for KRW, there are several required levels of rehabilitation outcomes including motor gains on the Functional Independence Measure (FIM) (FIM motor scores range from 13, totally dependent, to 91, independent), length of hospital stay, and proportion of discharges home. According to KRW Association's annual survey in 2020, mean admission motor FIM, discharge motor FIM, and motor FIM gains of patients with stroke were 39.6, 60.1, and 20.5, respectively, and 72.9 % are discharged home.

10.3.3   Community-based rehabilitation

Long-term care insurance provides helper services, visiting nursing care, and visiting rehabilitation based on the assessed degree of care that needs to lessen the burden of patients and their caregivers. Care need status comprising five stages (1: minimal care needs to 5: maximal care needs) is officially renewed by registered assessors every 6 months for each patient. Individualized care plans are also renewed by registered care managers based on the renewed care need status. For the promotion of early supported discharge [10], insurance started to cover visiting and outpatient rehabilitation up to 12 times per week until 3 months after discharge in 2021. To maintain community life for a longer period after discharge from KRW, each local government is required to build up a community-based integrated care system by 2025, when one-third of the population is expected to be 65 years of age or older.

10.4   Stroke guideline and emerging new treatment

10.4.1   Stroke treatment guideline in Japan

Stroke treatment guidelines written in Japanese were edited by the Japan Stroke Society, were first published in 2004, and then renewed in 2009, 2015, and 2021 [11]. The renewed guidelines reflect recent publications in the stroke field as well as recommendations from other major guidelines such as the World Health Organization and American Heart Association. Consequently, treatment recommendations in the Japanese guidelines are similar to those from other guidelines in the United States and Europe. According to the latest guidelines in 2021, recommendations for the following topics are described in the rehabilitation section. These include recommended outcome measures of stroke rehabilitation, outcome prediction, timing and programs for acute stroke rehabilitation, and post-acute rehabilitation and management for various conditions including motor impairment, disability in activities of daily living (ADLs), gait disorders, paretic upper extremity, spasticity, pain management, dysphagia, malnutrition, bladder incontinence, aphasia, dysarthria, cognitive disorders, poststroke depression, psychiatric symptoms, impaired physical fitness, and seizures.

Specifically, rehabilitation strategies with recommendation A (strong evidence base: two or more high-quality studies) that comprise standard and principal treatments for motor impairment and disability are as follows:

1. Intensive and repetitive walking with either a treadmill (with or without body weight support) or overground walking exercise training combined with conventional rehabilitation for recovery of walking function.
2. Robot-assisted movement training to improve mobility after stroke in combination with conventional therapy.
3. Repetitions of graded and progressive task-specific upper extremity training in accordance with individual capabilities.
4. ADL training tailored to individual needs and eventual discharge setting.

10.4.2 Status of emerging new treatment options

In terms of new treatment options regarding rehabilitation in the latest guidelines, robotic use for mobility training is graded as A as described above. The recommendation level of robot-assisted upper extremity rehabilitation for improving impairment and disability is graded as B (moderate evidence base: at least one high-quality study or multiple moderate-quality studies). To promote technology development in the rehabilitation field, national insurance systems approved the use of assistive robots for upper extremity and gait training as well as functional electrical stimulation for rehabilitation sessions under the supervision of physicians or therapists in 2020. Neuromodulation using noninvasive brain stimulation coupled with rehabilitation is also a treatment option to improve ADLs, and the recommendation level is graded as B. Neurofeedback using functional near-infrared spectroscopy (fNIRS) is also a promising option for neuromodulation. This technology enables brain activity measurement during dynamic movement, and its application to rehabilitation medicine was developed in the early 2000s in Japan [12−14]. Although there are some studies showing the efficacy of fNIRS neurofeedback [15,16], further investigation is necessary for its widespread clinical use. Approval of local IRB is necessary for these treatment options since these are not covered by the insurance.

Several Japanese companies with a background in leading technologies in automotive and industrial robots have released assistive robots to the Japanese market in the last decade. Mechatronics solutions developed in such fields are utilized to manufacture compact robots to operate in the limited space of rehabilitation settings. However, from April 2018, the new Clinical Trials Act (established in April 2017) was adopted, and applicable specific clinical trials, such as those commissioned by a private company, must be registered, creating a new barrier in the development of medical devices.

In gait rehabilitation assistive exoskeleton robots such as Lokomat (Hocoma) [17], the root of the link structure of the robot is connected to a stationary structure with a closed kinematic chain. The gait pattern is constrained by the robot kinematics. In Japan, exoskeletons with open kinematic chains (OKCs) are popular; for example, exoskeleton link structures are not fixed to stationary structures. While OKCs do not constrain the user's body and may contribute to natural gait patterns, the whole exoskeleton structure is attached to the user. The challenge is to ensure the prevention of falls during gait rehabilitation and to reduce the weight of the exoskeleton, especially the weight of the motor. For example, the Walking Assist (Honda), Welwalk (Toyota), and Ankle-Assist Walking Device (Yasukawa) are assistive exoskeletons with OKC for the hip, knee, and ankle, respectively. HAL (Cyberdyne) was originally designed for neurological and muscular intractable diseases, but recently released a knee joint model for patients who have experienced a stroke. As stated above, some of these devices are released as medical devices covered by additional national insurance as well as FES applied to prevent foot drop in patients with abnormal gait.

In upper extremity robots, there are two categories: end-effector and exoskeleton [18], with end-effectors being more popular (refer to Chapter 4 for more details). ReoGo-J (Teijin Pharma), a customized version of ReoGo (Motorika) made for the Japanese context, and CoCoroe AR2 (Yasukawa) are end-effector robots.

In the above-mentioned lower/upper extremity assistive robots, the mechatronics solutions mainly rely on geared motors and are limited by the power-to-weight ratio of the electric motors. Motor systems tend not to be able to backdrive in slow assisted movements or provide enough torque for patients with severe paralysis. It is known that paralysis remains in the distal joints in the lower extremity and the proximal joint in the upper extremity. For the upper extremity, the challenges are shoulder extension associated with upward movements and in the lower extremity, ankle push-off in the stance phase. Shoulder exoskeletons driven with a pneumatic piston [18] can assist severe upper extremity paralysis. As yet another solution, pneumatic actuators, such as pneumatic artificial muscles, may be a promising mechatronics solution both for upper and lower extremity exoskeletons for the shoulder [19] and ankle [20] (Fig. 10.2). Although these exoskeletons are lightweight and inherently compliant, further clinical investigation is necessary. There are also clinical studies of brain−machine interface-based shoulder treatment, assisted movement, and FES associated with motor imagery captured by EEG [21]. There are promising reports on neuromodulation of paralytic lower extremity and gait, such as peripheral nerve electrical stimulation with patterned electrical stimulation [22] and brain stimulation synchronized with gait with transcranial alternating current stimulation [23]. Although these are promising approaches, approval of local IRB is necessary as mentioned above, and nonclinical trials including quality management system in manufacturing and designing test for assured safety of these new technologies will be hurdles for commercialization.

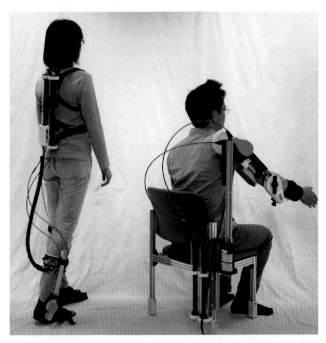

FIGURE 10.2   Lower/upper extremity exoskeletons driven by pneumatic artificial muscles. This figure shows the lightweight, powerful, and compliant exoskeletons for shoulder and ankle rehabilitation. The picture shows two people wearing the ankle exoskeleton robot and the shoulder exoskeleton robot in the picture. On the left of the picture, the ankle exoskeleton robot is ware by a user's right side of the ankle and foot. Force-sensing sensors embedded in the right/left shoes capture the gait pattern to estimate the user's gait cycle. The ankle exoskeleton robot is driven by pneumatic artificial muscles attached to the user's back via Bowden cables. The joint is compliant yet backdrivable to provide transparency of gait assist, and the total weight is lightweight at 2.9 kg. Clinical trials underway to establish medical devices are ongoing, supported by the Japanese AMED project under Grant Number #JP22he2202017. On the right of the picture, the shoulder exoskeleton robot is ware by a user's upper right extremity. The shoulder exoskeleton is lightweight at 5 kg, and its base is fixed on the chair mechanically. In addition to the active joints at the shoulder joints, multiple passive joints and mechanical adjustment mechanisms for height and length are designed to fit a wide range of body sizes easily customized at the clinical scene. Like the ankle exoskeleton, the shoulder joint is driven by pneumatic artificial muscles attached backside of the chair.

10.5   COVID-19 and stroke

10.5.1   Influence of COVID-19 pandemic on stroke rehabilitation service

Starting from the first case in January 2020 to October 2021, Japan intermittently experienced five waves of the COVID-19 pandemic, affecting more than 1.7 million individuals. As a consequence of decreased admissions of acute stroke patients due to the temporary conversion of acute beds to

COVID-19-specified beds, the number of stroke patients admitted to KRW also decreased by approximately 5% in 2020 compared to 2019. Outpatient rehabilitation clinics and visiting rehabilitation services in the community were also restricted in the periods of wave peaks to prevent infection spread. Vaccination started in March 2021, and 75% of the Japanese population received the first vaccine, and 67% received the second vaccination as of October 2021. Consequently, the number of newly infected patients markedly decreased at the end of October 2021, and the inpatient and outpatient service activities gradually rose to previous levels.

To prevent undesirable infections, most practitioners have become aware of the need for online remote rehabilitation services in the future. For example, several trials of remote physical therapy for chronic stroke patients started on a commercial basis before the acquisition of valid data for its efficacy. For cardiac remote rehabilitation, the first clinical trial was supported by the Project for Medical Device Development of the Japan Agency for Medical Research and Development (AMED) [24]. Robot-assisted stroke telerehabilitation might also be suitable for situations that require infection control.

In the training of skilled therapists, face-to-face communication is important to transfer skills. Since it is difficult to transfer skills including tacit knowledge in some areas through videoconferencing, quantitative analysis of rehabilitation, such as assessment of patient's function, becomes more important. In addition, in terms of digitization of therapists' skills, assistive robots do more than just perform physical assistive actions. Assistive robots can be used as a parametrization of the use of assistive devices. If such technical know-how is digitized as big data, AI such as deep learning and machine learning can be applied to find better treatment methods and provide menus for tailor-made rehabilitation.

10.6   Future care using assistive and rehabilitation robots

As urbanization progresses in Japan, changes in the social structure have led to the concentration and dispersion of the population, changes in the population composition, and changes in family structure to nuclear families. In addition, changes in the structure of daily life have led to the dilution of neighborhood relationships, and the simplification and individualization of family relationships. Furthermore, changes in the structure of consciousness have led to a focus on individualism. The changes in these structures have resulted in weakened ties among local residents, and many people have become isolated, leading to the problem of isolated deaths. In order to solve these problems, it is necessary for the entire local community to support the super-aging society, and in Japan, "the community-based integrated care system" that supports the elderly in the community is attracting attention. The community-based integrated care system refers to a system in which housing, medical care, nursing care, prevention, and lifestyle support are provided in an integrated manner so that people can continue to live their own lives in their familiar communities until the end of their lives, even if they require dedicated nursing care. The purpose of this system is to preserve the dignity of the elderly and support independent living.

If telemedicine and telecare, including telerehabilitation, can be provided using robotic technology as described in Section 10.5, it will be possible to provide uniform treatment, rehabilitation, and care anywhere, regardless of the region. In order to achieve this goal, it is necessary to share robots and other systems so that anyone can use them freely. If this can be achieved, cutting-edge technologies such as robots can be used as a social infrastructure at a low cost.

Japan is said to be an advanced country in terms of challenges, with a number of problems at a level that other countries have yet to face, such as a declining birthrate, an aging population, and regional depopulation. Since these issues that Japan faces will eventually affect other countries, there is a high possibility that Japan can provide new solutions. Currently, Japan is promoting a moonshot R&D program in which the government formulates ambitious goals and concepts for social issues and other problems that are difficult to address but are expected to have a large impact if realized and promote challenging R&D while allowing for failure. Under the title of "Adaptable AI-enabled Robots to Create a Vibrant Society" [25], the authors aim to create a vibrant society in which the coexistence and coevolution of a wide variety of robots and people will create a society in which all people can participate by 2050. We are developing a collective of shared AI robots that can be used by anyone, at any time, and that provide appropriate services by changing their shape and functions to suit individual users. In particular, we aim to develop adaptable AI robots that can be used in nursing care and rehabilitation by 2030.

References

[1] Statistical Handbook of Japan. Statistics Bureau, Ministry of Internal Affairs and Communications Japan, 2021. Available from: https://www.stat.go.jp/english/data/handbook/index.html; 2021 (accessed 01.06.22).

[2] Yokoyama T. National health promotion measures in Japan: Health Japan 21(the second term). J Natl Inst Public Health 2020;69(1):14−24.

[3] Patient Survey. Ministry of Health, Labour and Welfare in Japan, Available from: https://www.mhlw.go.jp/toukei/saikin/hw/kanja/17/index.html; 2017 (accessed 01.06.22).

[4] Takashima N, Arima H, Kita Y, Fujii T, Miyamatsu N, Komori M, et al. Incidence, management and short-term outcome of stroke in a general population of 1.4 million japanese-shiga stroke registry. Circ J 2017;81:1636−46.

[5] Kuwabara M, Mori M, Komoto S. Japanese national plan for promotion of measures against cerebrovascular and cardiovascular disease. Circulation. 2021;143:1929−31.

[6] Indredavik B, Bakke F, Solberg R, Rokseth R, Haaheim LL, Holme I. Benefit of a stroke unit: a randomized controlled trial. Stroke. 1991;22:1026−31.

[7]   Fujino Y, Kubo T, Muramatsu K, Murata A, Hayashida K, Tomioka S, et al. Impact of regional clinical pathways on the length of stay in hospital among stroke patients in japan. Med Care 2014;52:634−40.

[8]   Miyai I, Sonoda S, Nagai S, Takayama Y, Inoue Y, Kakehi A, et al. Results of new policies for inpatient rehabilitation coverage in japan. Neurorehabil Neural Repair 2011;25:540−7.

[9]   Dromerick AW, Geed S, Barth J, Brady K, Giannetti ML, Mitchell A, et al. Critical period after stroke study (cpass): a phase II clinical trial testing an optimal time for motor recovery after stroke in humans. Proc Natl Acad Sci U S A 2021;118: e2026676118.

[10] Langhorne P, Taylor G, Murray G, Dennis M, Anderson C, Bautz-Holter E, et al. Early supported discharge services for stroke patients: a meta-analysis of individual patients' data. Lancet. 2005;365:501−6.

[11] Japanese Guideline for the Management of Stroke 2021. Tokyo: Kyowa Kikaku; 2021.

[12] Miyai I, Tanabe HC, Sase I, Eda H, Oda I, Konishi I, et al. Cortical mapping of gait in humans: a near-infrared spectroscopic topography study. NeuroImage. 2001;14:1186−92.

[13] Miyai I, Yagura H, Oda I, Konishi I, Eda H, Suzuki T, et al. Premotor cortex is involved in restoration of gait in stroke. Ann Neurol 2002;52:188−94.

[14] Miyai I, Yagura H, Hatakenaka M, Oda I, Konishi I, Kubota K. Longitudinal optical imaging study for locomotor recovery after stroke. Stroke. 2003;34:2866−70.

[15] Mihara M, Hattori N, Hatakenaka M, Yagura H, Kawano T, Hino T, et al. Near-infrared spectroscopy-mediated neurofeedback enhances efficacy of motor imagery-based training in poststroke victims: a pilot study. Stroke. 2013;44:1091−8.

[16] Mihara M, Fujimoto H, Hattori N, Otomune H, Kajiyama Y, Konaka K, et al. Effect of neurofeedback facilitation on poststroke gait and balance recovery: a randomized controlled trial. Neurology. 2021;96:e2587−98.

[17] Kammen KV, Boonstra AM, van der Woude LHV, Visscher C, Reinders-Messelink HA, den Otter R. Lokomat guided gait in hemiparetic stroke patients: the effects of training parameters on muscle activity and temporal symmetry. Disabil Rehabil 2020;42(21):2977−85. Available from: https://doi.org/10.1080/09638288.2019.1579259 October 8.

[18] Maciejasz P, Eschweiler J, Gerlach-Hahn K, Jansen-Troy A, Leonhardt S. A survey on robotic devices for upper limb rehabilitation. J Neuroeng Rehabilitation 2014;11(1):3. Available from: https://doi.org/10.1186/1743-0003-11-3 December.

[19] Noda T, Morimoto J. Development of upper-extremity exoskeleton driven by pneumatic cylinder toward robotic rehabilitation platform for shoulder elevation. In 2015 IEEE International Conference on Rehabilitation Robotics (ICORR). Singapore, Singapore: IEEE, 2015, pp. 496−501. Available from: https://doi.org/10.1109/ICORR.2015.7281248.

[20] Noda T, Teramae T, Furukawa J-L, Ogura M, Okuyama K, Kawakami M, et al. Development of shoulder exoskeleton toward bmi triggered rehabilitation robot therapy. In 2018 IEEE International Conference on Systems, Man, and Cybernetics (SMC). Miyazaki, Japan: IEEE, 2018, pp. 1105−1109. Available from: https://doi.org/10.1109/ SMC.2018.00195.

[21] Noda T, Takai A, Teramae T, Hirookai E, Hase K, Morimoto J. Robotizing double-bar ankle-foot orthosis. In 2018 IEEE International Conference on Robotics and Automation (ICRA). Brisbane, QLD: IEEE, 2018, pp. 2782−2787. Available from: https://doi.org/ 10.1109/ICRA.2018.8462911.

[22] Takahashi Y, Fujiwara T, Yamaguchi T, Matsunaga H, Kawakami M, Honaga K, et al. Voluntary contraction enhances spinal reciprocal inhibition induced by patterned electrical stimulation in patients with stroke. Restor Neurol Neurosci 2018;36:99−105.

[23] Koganemaru S, Kitatani R, Fukushima-Maeda A, Mikami Y, Okita Y, Matsuhashi M, et al. Gait-synchronized rhythmic brain stimulation improves poststroke gait disturbance: a pilot study. Stroke 2019;50(11):3205−12. Available from: https://doi.org/10.1161/STROKEAHA.119.025354.

[24] First in Japan Clinical Trial for an Online Cardiac Rehabilitation System. Graduate School of Medicine, Osaka University. Available from: https://www.med.osaka-u.ac.jp/eng/archives/5996; 2020 (accessed 01.06.22).

[25] Adaptable AI-enabled Robots to Create a Vibrant Society in Moonshot R&D. Available from: https://srd.mech.tohoku.ac.jp/moonshot/en/; (accessed 22.01.15).

Chapter 11

Asia-Pacific region: Australia

Marlena Klaic[1] and Vincent Crocher[2]
[1]Melbourne School of Health Sciences, University of Melbourne, Parkville, VIC, Australia,
[2]Department of Mechanical Engineering, University of Melbourne, Parkville, VIC, Australia

Learning objectives

At the end of this chapter, the reader will be able to:
1. Understand the implications of Australia's geography on stroke outcomes.
2. Appreciate demographic and regional trends in stroke secondary to changing lifestyle patterns.
3. Consider the potential of technology to address current barriers to accessing stroke rehabilitation services across Australia.

11.1 Australia geography, demography, and stroke incidence

11.1.1 Geography and general demography

Australia has one of the lowest population densities in the world with more than 25 million inhabitants across 7.6 million square kilometers (3 inhabitants/km^2). Most of the population (68%) reside in the 8 capital cities, which are distributed in coastal areas. Melbourne and Sydney account for 10 million inhabitants with the remaining population distributed across the rest of the country, including large central areas with densities as low as 0.1 inhabitants/km^2 (Fig. 11.1). This distribution remains relatively constant with similar growth rates (around 1%) reported by the Australian Bureau of Statistics for both regional areas and major cities [1].

Distribution patterns of Aboriginal and Torres Strait Islander peoples differ significantly with two-thirds residing in rural areas. They also constitute the main part of the inhabitants in remote and very remote Australia, located mostly in New South Wales, Queensland, and Western Australia [2].

Due to an increase in life expectancy and a reduction in the fertility rate over the last decades, the Australian population is aging, with an increasing proportion of people aged 65 and older [3]. Older Australians

Rehabilitation Robots for Neurorehabilitation in High-, Low-, and Middle-Income Countries.
DOI: https://doi.org/10.1016/B978-0-323-91931-9.00019-0

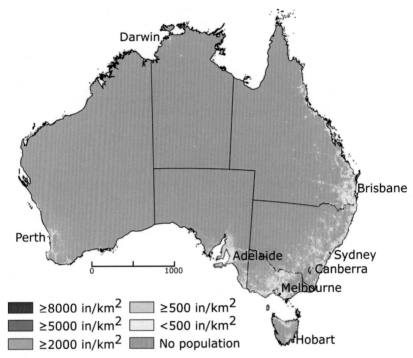

FIGURE 11.1 Australian population distribution in capital cities in 2020. Short text description: This illustration shows a map of Australia with population density highlighted for the capital cities. There are higher rates of people living along the eastern coast of Australia with the highest numbers located in Melbourne and Sydney.

are overrepresented in regional locations (Fig. 11.2) with 39% being over 50 years old compared to 32% in capital cities.

More recently, the COVID-19 pandemic has had a significant impact on Australia's population growth. Restrictions in international travel and other pandemic effects have contributed to the decline in population growth from 1.3% to 0.1%, the lowest since World War I [3]. This has further exacerbated the current challenges Australia faces with an aging population.

11.1.2 Stroke incidence, prevalence, and socioeconomic impact

The population distribution across Australia is meaningful as the incidence of stroke in regional locations is 17% higher than in metropolitan areas (Fig. 11.3), primarily due to the older age demographic, overall poorer health status, and limited access to specialist stroke care [4]. Stroke prevalence in Australia is 1.3%, and incidence was approximately 28,000 in 2020 with incidence predicted to increase to more than 50,000 per annum by 2050 due

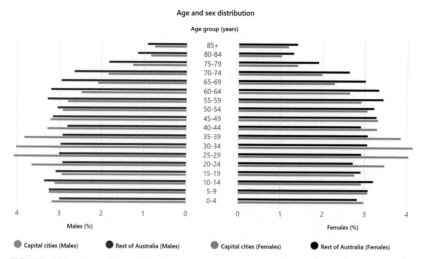

Age and sex distribution

Age group (years)

FIGURE 11.2   Australian population age and sex distribution. Short text description: This figure is an illustration of where Australians live according to their age groups and sex. For example, there is a higher rate of males aged between 0 and 4 years old living in capital cities compared to the rest of Australia.

to the aging population and increased prevalence of chronic diseases [5]. Furthermore, there is an increasing incidence among younger Australians with 24% of stroke survivors being under the age of 54 in 2020 in comparison to 14% in 2012. This concerning evolution seems to be due to higher risk factors (tobacco consumption, overweight and obesity rates, and physical inactivity) among the younger population.

Aboriginal and Torres Strait Islander peoples are also significantly more affected by stroke with an age-standardized incidence up to three times higher than non-Indigenous Australians [6,7]. The factors contributing to this include higher rates of risk factors for stroke, such as smoking and physical inactivity, and poorer access to specialist stroke care and secondary prevention, particularly for those who reside in regional or remote Australia. The higher incidence among this group, coupled with an average older population in rural areas, contributes to the overrepresentation of stroke survivors in regional Australia which accounts for 43% of the country [4].

The mortality rate for stroke in 2020 was 32%, and the number of Australian stroke survivors reached 445,087, corresponding to 1.7% of the population [5]. One-third of stroke survivors have a disability which impacts their ability to independently complete a range of activities of daily living [8]. Many stroke survivors report an overall reduced quality of life with rates of depression and anxiety estimated at 25%−79% [9]. Aboriginal and Torres Strait Islander peoples have a burden of disease that is 2.3 times higher than the rate for non-Indigenous Australians [10].

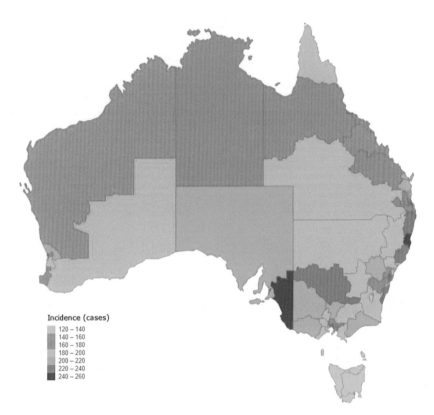

FIGURE 11.3   Australian stroke incidence in 2020. Short text description: This is a map of Australia with different areas colored according to how many cases of stroke occurred. For example, the most southeastern area of South Australia is shaded in dark blue, indicating an incidence of 240–260 strokes, one of the highest rates in all of Australia.

The direct healthcare cost incurred by stroke is estimated to be 1.3 billion AUD a year (0.95 billion USD), of which 46% is funded by the federal government, 28% by the state governments, and the rest by individuals and their families [5]. The overall financial burden, including loss of productivity (due to early death or disability), carer costs and deadweight cost, is estimated to be $6.2 billion AUD a year (4.5 billion USD), equivalent to 0.3% of Australia's GDP in 2020.

11.2   Australian healthcare system

The Australian healthcare system has been ranked as the third best performing out of 11 developed nations and ranked highest for equity [11]. This is largely a reflection of Australia's universal healthcare scheme, Medicare, which ensures that eligible individuals can afford healthcare, irrespective of their income. Australia spent approximately \$185 billion AUD on health in 2017−2018, which is 10% of the GDP and included acute and rehabilitation admissions, primary healthcare, aids and equipment, and capital expenditure. The Australian health system includes:

- Primary healthcare delivered in several settings including medical and allied health practices and community health centers. Primary health is typically an individual's first contact point with the health system and includes care that is not delivered in a hospital setting. Primary health accounts for approximately 38% of Australia's total health expenditure.
- Secondary health services which includes care provided by a specialist or service or facility often in a hospital setting (public and private). Secondary health accounts for approximately 31% of Australia's total health expenditure.

11.2.1   Medicare

Introduced in 1984, Medicare is primarily funded through the taxation system which includes a levy on taxable income and a surcharge payable by high-income earners who do not have private health insurance. Medicare subsidizes or fully funds the following aspects of healthcare in Australia:

- Medical services: Costs associated with accessing primary health, secondary health, specialists, allied health professionals, and some dental services are either fully funded by Medicare through a bulk billing system or subsidized according to a schedule of fees for the relevant service.
- Hospitals: Medicare funds the cost of admission to a public hospital including, but not limited to surgery, acute stay, and rehabilitation. A portion of the costs associated with admission to a private hospital is also covered by Medicare.
- Medicines: Prescription drugs listed on the Pharmaceutical Benefits Scheme (PBS) are greatly subsidized through Medicare with health consumers paying a small portion of the cost, or none at all, if they are deemed to have a low income. The PBS also has a safety net which lowers the cost of prescription medicines once an individual spends above a threshold amount.

11.2.2   Private health insurance

More than 50% of Australians have private health insurance, the majority of who live in a major city [12]. There are 38 private health providers in Australia, and the industry spent an estimated \$8.6 billion AUD on private hospital costs in 2016−2017 [13]. Although private health insurance is voluntary in Australia, individuals above an income threshold will incur a Medicare levy surcharge if they do not purchase private health insurance.

Australians are incentivized to acquire private health insurance before the age of 31 with a "lifetime health cover" initiative that avoids payment of additional loading.

11.2.3 Acute hospital stroke care

In 2020, almost 70,000 hospital admissions were attributable to a stroke diagnosis with an average length of stay in the acute setting of 7 days [14]. The costs to health services for these admissions were over $700 million AUD. Stroke care units are described as an organized model of care within a dedicated ward of the hospital where care is provided by a multidisciplinary specialist team [15]. Stroke care units and time-critical stroke treatments, such as thrombolysis and endovascular clot retrieval, have been shown to reduce mortality and morbidity and increase the likelihood of the stroke survivor returning home, independent of stroke type or severity, patient sex, or age [15]. However, access to time-critical treatments and stroke unit care is limited in regional and remote Australia [16]. This has significant implications for Aboriginal and Torres Strait Islander peoples as they are more likely to be residing in regional and remote Australian locations than non-Indigenous Australians. Inequitable access to the best evidence of acute stroke care contributes to the overrepresentation of stroke and stroke-related disability in Aboriginal and Torres Strait Islander peoples and those who live in regional or remote Australia (Table 11.1).

This difference in acute stroke care between urban and remote Australia is well illustrated by the existence of the Melbourne Mobile Stroke Unit (MSU) in metropolitan areas and the reliance on the Royal Flying Doctor Service (RFSD) in regional and/or rural locations. While the first MSU, initiated in Melbourne in 2017, reduced the "door-to-needle" time to a median of 90 minutes [17], the median time for aeromedical retrieval alone for stroke patients relying on the RFSD was 238 minutes.

TABLE 11.1 Australian stroke treatment facilities by location from [5].

	Major cities	Inner regional	Outer regional
Stroke unit	61	25	5
Clot dissolving treatment	54 (49 accessible 24/7)	33 (31 accessible 24/7)	11 (all accessible 24/7)
Clot retrieval treatment	19 (13 accessible 24/7)	0	0
Education on secondary prevention	75%	69%	66%

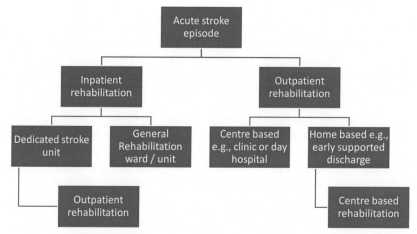

FIGURE 11.4   Potential discharge pathways following a stroke. Short text description: This is a diagram illustrating the different pathways for stroke survivors following their hospital admission. On the left is a box with inpatient rehabilitation followed by two boxes indicating this can happen in a dedicated stroke unit or a general rehabilitation ward or unit. On the right is a box with outpatient rehabilitation followed by two boxes indicating this can happen in a center or the home.

11.3   Structure of rehabilitation in Australia

Rehabilitation after a stroke should commence as soon as a patient is medically stable, ideally in the acute setting. Individuals who have more significant impairments or ongoing goals may require rehabilitation beyond the acute hospital episode, which can be delivered in one or more modes including inpatient, community-based (center or home), and/or early supported discharge such as "rehabilitation in the home" programs (Fig. 11.4) [18]. A recent report exploring access to rehabilitation for stroke survivors in Australia found that although 81.3% were identified as having rehabilitation needs, only 52.2% were referred for further rehabilitation beyond the acute episode [19]. Rehabilitation can be delivered in either the private or publicly funded sector, depending on whether the individual has purchased private health insurance and/or has access to the National Disability Insurance Scheme (NDIS).

11.3.1   Inpatient stroke rehabilitation

Stroke is one of the top three primary impairment types reported for Australian rehabilitation care episodes with 13,364 patients discharged from a public health rehabilitation service in 2017−2018 [14]. The median length of stay for an admitted inpatient rehabilitation episode following a stroke was 22 days, and approximately two-thirds of patients receive rehabilitation in a public hospital [20].

A recent audit of 111 stroke rehabilitation services (15 private and 96 public) throughout Australia, including metropolitan, regional, and remote locations, found that 21% of patients admitted for inpatient rehabilitation after stroke were treated in a dedicated stroke care unit [19]. Most services (73%) included in the audit provided an estimated two or more hours of active therapy to stroke survivors, at least 5 days per week. This was delivered in multiple modes including one-on-one treatments and circuit classes in a group setting. The most frequently documented impairment was difficulties with activities of daily living (87%) followed by mobility deficits (76%) and upper-limb impairment (72%). Therapy delivered to address these impairments included task-repetitive practice and "other" therapy. Most (65%) patients were discharged to the usual residence following their inpatient rehabilitation episode, and 64% of patients were referred for further rehabilitation in the community.

11.3.2   Outpatient stroke rehabilitation

Outpatient stroke rehabilitation may be delivered in a center or day hospital or within the stroke survivor's home, either as an early supported discharge model or home-based rehabilitation service. The annual costs associated with outpatient stroke care, including but not limited to allied health services, specialist care, and diagnostic imaging, were estimated at $47.8 million AUD [5].

11.3.2.1   Center-based

Ongoing rehabilitation following a stroke may be delivered in a community center or day hospital, often affiliated with a health service. Termed ambulatory rehabilitation, these multidisciplinary programs aim to improve or maintain function, are goal-directed, and time-limited (6−8 weeks). Survivors of stroke may be offered annual bursts of therapy via a center-based program and can be referred to these programs directly from the hospital or the community.

More recently, dedicated stroke support centers have been established by stroke-specific organizations, such as the Stroke Association of Victoria [21]. With a focus on inclusion, these support centers offer access to a variety of programs including vocation support, skill development (including upper-limb recovery), and social support.

11.3.2.2   Home-based

Early supported discharge (ESD) services include home-based rehabilitation provided by a multidisciplinary team of clinicians with expertise in stroke rehabilitation, though there is considerable variability in how this model of care is delivered. The Australian clinical guidelines for stroke management recommend that patients with mild to moderate impairment following a stroke should be offered ESD [22]. There has been a marked increase in health services offering ESD services from 16% (2016) to 42% (2020) which is likely due to growing evidence of its effectiveness [15].

In addition to publicly funded models, allied health professionals delivered an estimated 850,000 hours of private home-based rehabilitation in 2020 to survivors of stroke, at a cost of $185.7 million AUD. This was in addition to any rehabilitation offered as part of a community rehabilitation program.

11.3.3   The National Disability Insurance Scheme

Stroke has the highest associated rate of disability in comparison to all other chronic conditions with 42% of Australian survivors reporting a profound limitation in activities of daily living [23]. Commonly reported impairments following a stroke are limitations in physical activities, incomplete use of the upper and lower limbs, fine motor impairment and difficulties with speech [24]. The National Disability Insurance Scheme (NDIS) was established to support individuals with a permanent and significant disability (under 65 years old) to achieve the goals and objectives they set as part of their care plan. Funding can be provided within three types of support budgets including core supports (e.g., assistance with daily living), capability-building supports (e.g., increased social and community participation) and capital supports (e.g., assistive technology). Approximately 5161 individuals with a primary disability secondary to stroke were supported by NDIS in 2020, at an annual cost of $388.2 million AUD [5].

11.4   Robotic-assisted stroke rehabilitation

11.4.1   Guidelines and current use

Emerging technologies such as virtual reality, interactive gaming, and robotics have been shown to improve patient outcomes following a stroke. In reflection of this growing body of evidence, the Australian and New Zealand Clinical Guidelines for Stroke Management have recently included a weak recommendation in favor of using emerging technologies to improve standing, walking, arm activity, and activities of daily living [22].

A dedicated report for the Australian Government concluded there was evidence that robotic-assisted therapy had superior outcomes for UL motor function and speed of movement compared to usual care for chronic stroke survivors [25]. However, they recommended further cost analysis comparing manual intensive therapy with robotic-assistive therapy. Additionally, the use of robotics for both LL and UL training has been included in the Stroke Foundation guidelines since 2010 and was then considered as one of the research priorities to increase UL activity after stroke.

Beyond these guidelines and recommendations, there is little published evidence regarding the actual penetration and use of emerging technologies for stroke rehabilitation in Australia. A national audit completed in 2012 with 111 stroke hospital rehabilitation services (both public and private hospitals) found that 9% had access to robotic equipment and 76% had access to commercially available gaming technology [26]. Anecdotally, there does appear to be growing interest in using emerging technologies with social media reports describing launches of robots, stroke-specific games, and virtual reality applications in both private rehabilitation settings and public hospitals [27−30].

An online survey conducted by the authors in 2021 with OTs and PTs in Australia (n = 42 respondents, refer to Appendix A) seems to indicate a growing penetration with 23% of respondents using, or having used, robotic devices for neurorehabilitation. Although the small number of respondents does not allow for definitive conclusions, it is interesting to note a predominance of devices for UL training (N = 9) compared to ones for LL (N = 3). Not surprisingly, the robots used correspond primarily to the commercially available devices from the main players in the field (refer to Chapter III) with the Tyromotion devices (N = 6), Hocoma (N = 1), InMotion (N = 1), H-MAN (N = 1), and Ekso Bionics exoskeleton (N = 1).

11.4.2 Limits and challenges to robotic adoption

Despite the potential of emerging technologies to improve patient motor outcomes and anecdotal evidence suggesting growing interest in this form of therapy, adoption remains limited. One of the most significant barriers to adoption in Australia is the high cost of the devices, and the lack of dedicated funding, particularly in publicly funded services, to purchase such equipment. Media reports suggest that funding relies on grants, often with a research component, or community fundraising. It is thus not surprising to see growing adoption by privately funded health organizations, which may perceive emerging technologies as a possible differentiator, helping them to provide added value to their customers.

The adoption might also be limited by the availability of robotic systems in Australia. Indeed, Australia represents a relatively small market but requires a dedicated approval process for these devices by the Therapeutic Goods Administration (TGA). If this regulatory approval benefits from "bridges" with its European counterpart, it may be an additional constraint to some manufacturers to provide their devices in Australia.

Additionally, and not specific to Australia, a recent systematic review found that end-user's decision to use emerging technologies was influenced by multiple factors including attitude toward technology, funding to purchase the device, usability, and applicability in practice [31]. Broader research on the implementation of technology often uses the technology acceptance model (TAM) to explore barriers and facilitators to implementation. The TAM suggests that an individual's intention to use technology, such as robotic devices for rehabilitation, can largely be predicted by perceived usefulness and perceived ease of use [32]. In turn, these factors can be mediated through increased practice using the technology and sufficient resources to support implementation, including adequate dedicated training. However, the high turnover of clinical staff across the health sector suggests that retaining knowledge and skills in using these often-complex devices may remain a significant barrier.

11.4.3   Research and future opportunities

The ANZCTR reports only 8 trials involving the use of robotics for stroke rehabilitation since 2008 in Australia and New Zealand, and overall, 15 trials using robotics for different neurological pathologies. By comparison, there were 236 trials registered for stroke rehabilitation over the same period. Dedicated research projects may not only contribute to the development of evidence and new devices but also help the dissemination of practices and better integration of robotic devices in existing practices. This typically should involve codesign processes between the end users and those developing such devices to ensure they adequately address both therapists and patients' needs.

In terms of adoption, the NDIS rollout may act as a game changer in the use of these systems, both through an additional opportunity given to patients to select robotic-assisted therapy over traditional, and through a potential uptake of devices for individual home-based practice or as assistive devices which can be funded by the NDIS.

11.5   Conclusion

Australia's greatest health system challenge is the aging population and increasing rates of noncommunicable diseases such as stroke, which is driving the need to implement emerging technologies [32]. Access to rehabilitation following a stroke is affected by multiple factors in Australia. Research has found that less than 40% of patients are assessed for rehabilitation, despite 75% having ongoing goals best achieved through rehabilitation [33]. This inequitable access is a result of both individual and organizational factors, including limited available rehabilitation beds requiring a prioritization approach. Australia also has geographical limitations which influence access to evidence-based stroke care, including rehabilitation. Furthermore, the recent COVID-19 pandemic has required health services to consider alternative models of healthcare delivery, including telehealth and other home-based technologies. Given the limited number of rehabilitation beds available, growing numbers of stroke survivors, and geographical barriers, it is imperative that health services consider how emerging technologies can best be used to support rehabilitation efforts, particularly through outpatient services and private practices.

Although recent audit results suggest that most Australian rehabilitation services are providing 2 hours per day of scheduled therapy, published studies have found that patients spend as little as 4—11 minutes engaged in practice, particularly in relation to rehabilitation of the upper limb [34—36]. Yet, research suggests that the dosage of therapy is important, and this includes frequency of attendance, length of session, and number of repetitions [37,38]. Emerging technologies, particularly robotics and serious gaming, have the potential to assist patients in achieving recommended amounts of practice and can be used in outpatient settings, including the patient's home, but this technology has not reached its full potential in Australia.

Conflict of interest

VC is a coinventor of the EMU rehabilitation device and has financial interests in the commercialized version of the device.

As part of his research activities, VC is partially funded by industry grants (ARC-LP scheme) co-awarded to the University of Melbourne and Fourier Intelligence (Shanghai, China).

Appendix A

Online survey questions

Exploring the use of robotic devices in Australian neurorehabilitation settings

Q1 What is your profession?

- Occupational therapist
- Physiotherapist
- Other _____

Q2 Where is your practice located where you primarily work?

- Metropolitan
- Regional
- Rural
- Other _____

Q3 What is the nature of your practice?

- Inpatient rehabilitation, for example, rehabilitation ward/hospital
- Outpatient rehabilitation, for example, day therapy, rehabilitation clinics
- Home-based therapy, for example, rehabilitation in the home
- Other _____

Q4 How many years have you practiced clinically?

- < 1 year
- 2 years
- 3 years
- 4 years
- 5 years
- 6−10 years
- 11−15 years
- 16−20 years
- > 20 years

Q15 In this survey, we consider robotic devices as devices which provide physical assistance (or resistance) to the patient. They are often, but not necessarily, coupled to a gaming system. Passive devices and/or gaming systems alone (e.g., Wii, Wii board, AbleX, Rejoyce, ArmeoSpring, SaeboMAS, etc.) are here NOT considered robotic devices. If you are unsure if a device falls within this definition of a robotic device, please consider it and provide the name or details about the device if possible.

Q7 Do you have any robotic devices in your current setting?

- No
- Yes

Q14 What is the name of the organization where you mainly practice?
Q18 What is the name of the organization where you used these devices?
Q10 How many robotic devices do you have in your setting?
Q9 Which devices do you have in your setting (name or description)?
Q11 How often do you use robotic devices when delivering neurorehabilitation?

- Daily
- 2−4 times a week
- Twice a month
- Once a month
- Never

Q12 Have you ever used a robotic device in your clinical practice, for example, at another workplace in Australia?

- Yes
- No

Q13 Which device(s) did you use (name or description)?
Q15 Did you use the device(s) for neurorehabilitation?

- No
- Yes

Q14 What was the name of the practice where you used the device(s)?

References

[1] Australian Government. *Our country*. About Australia. Available from: https://info.australia.gov.au/about-australia/our-country. n.d. (cited 9 December 2021).

[2] Australian Bureau of Statistics. *Estimates of Aboriginal and Torres Strait Islander Australians*. Aboriginal and Torres Strait Islander Peoples. Available from: https://www.abs.gov.au/statistics/people/aboriginal-and-torres-strait-islander-peoples/estimates-aboriginal-and-torres-strait-islander-australians/latest-release. 2018 (cited 14 December 2021).

[3] Australian Institute of Health and Welfare. Deaths in Australia. Life expectancy & death. Available from: https://www.aihw.gov.au/reports/life-expectancy-death/deaths-in-australia/contents/life-expectancy. 2021 (cited 29 November 2021).

[4] Deloitte. *No postcode untouched. Stroke in Australia 2020*. Available from: https://strokefoundation.org.au/What-we-do/Research/No-Postcode-Untouched. 2020 (cited 9 December 2021).

[5] Deloitte. *The economic impact of stroke in Australia 2020*. Available from: https://strokefoundation.org.au/What-we-do/Research/Economic-impact-of-stroke-in-Australia. 2020 (cited 9 December 2021).

[6] Balabanski AH, et al. Stroke incidence and subtypes in Aboriginal people in remote Australia: a healthcare network population-based study. BMJ Open 2020;10(10):e039533.

[7] Australian Institute of Health and Welfare. *Heart, stroke and vascular disease - Australian facts*. Heart, stroke & vascular disease. Available from: https://www.aihw.gov.au/reports/cvd/092/hsvd-facts/contents/disease-types/stroke. 2021 (cited 9 December 2021).

[8] Better Health Channel. Effects of stroke. *Stroke*. Available from: https://www.betterhealth.vic.gov.au/health/conditionsandtreatments/effects-of-stroke. n.d. (cited 9 December 2021).

[9] Khan F. Poststroke depression. Aust Fam Physician 2004;33(10):831−4.

[10] Australian Institute of Health and Welfare. *Indigenous health and wellbeing*. Australia's health 2020. Available from: https://www.aihw.gov.au/reports/australias-health/indigenous-health-and-wellbeing. 2020 (cited 29 November 2021).

[11] Australian Institute of Health and Welfare. *International comparisons of health data*. Australia's health 2020. Available from: https://www.aihw.gov.au/reports/australias-health/international-comparisons-of-health-data. 2020 (cited 9 December 2021).

[12] Australian Institute of Health and Welfare. *Private health insurance*. Australia's health 2020. Available from: https://www.aihw.gov.au/reports/australias-health/private-health-insurance. 2020 (cited 9 December 2021).

[13] Australian Competition & Consumer Commission. *Report to the Australian Senate on anti-competitive and other practices by health insurers and providers in relation to private health insurance*. Available from: https://www.accc.gov.au/system/files/Private%20Health%20Insurance%20Report%20-%202019-2020.pdf. 2020 (cited 9th December 2021).

[14] Australian Institute of Health and Welfare, Admitted patient care 2017−18: Australian hospital statistics, in Health services series no. 90. 2019, AIHW: Canberra.

[15] Langhorne P, Ramachandra S. Organised inpatient (stroke unit) care for stroke: network meta-analysis. Cochrane Database Syst Rev 2020;4(4):Cd000197.

[16] Walter S, et al. Stroke care equity in rural and remote areas - novel strategies. Vessel Plus 2021;5(27).

[17] University of Melbourne. *Reaching stroke patients in time to give life-saving treatment*. Research at Melbourne. Available from: https://research.unimelb.edu.au/research-at-melbourne/impact/reaching-stroke-patients-in-time-to-give-life-saving-treatment. n.d. (cited 14 December 2021).

[18] Walters R, et al. Exploring post acute rehabilitation service use and outcomes for working age stroke survivors (≤ 65 years) in Australia, UK and South East Asia: data from the international AVERT trial. BMJ Open 2020;10(6):e035850.

[19] Lynch EA, et al. Access to rehabilitation for patients with stroke in Australia. Med J Aust 2019;210(1):21−6.

[20] Stroke Foundation. *National stroke audit - Rehabilitation services report 2020*. Available from: https://informme.org.au/stroke-data/Rehabilitation-audits. 2020 (cited 9 December 2021).

[21] Stroke Association of Victoria. *Stroke Support*. Available from: https://www.strokeassociation.com.au/. n.d. (cited 9 December 2021).

[22] Stroke Foundation. *The Australian and New Zealand Clinical Guidelines for Stroke Management*. Available from: https://informme.org.au/en/Guidelines/Clinical-Guidelines-for-Stroke-Management. n.d. (cited 9 December 2021).

[23] Australian Institute of Health and Welfare. *Australian burden of disease study: Impact and causes of illness and death in Australia 2018*. Burden of disease. Available from: https://www.aihw.gov.au/reports/burden-of-disease/abds-impact-and-causes-of-illness-and-death-in-aus/summary. 2021 (cited 9 December 2021).

[24] Australian Institute of Health and Welfare. How we manage stroke in Australia. Heart, stroke & vascular disease. Available from: https://www.aihw.gov.au/reports/heart-stroke-vascular-disease/how-we-manage-stroke-in-australia/contents/table-of-contents. 2006 (cited 7 December 2021).

[25] Australia and New Zealand Horizon Scanning Network. *Robot-assisted therapy for long-term upper limb impairment after stroke*. Available from: http://www.horizonscanning.gov.au/internet/horizon/publishing.nsf/Content/prioritising-summaries-2010. 2010 (cited 14 December 2021).

[26] Stroke Foundation. *National Stroke Audit - Rehabilitation Services Report 2012*. Available from: https://informme.org.au/stroke-data/Rehabilitation-audits. 2012 (cited 9 December 2021).

[27] Wellings H. *The new touch-screen game therapy helping patients recover from stroke*. Available from: https://7news.com.au/lifestyle/health-wellbeing/the-new-touch-screen-game-therapy-helping-patients-recover-from-stroke-c-600802. 2019 (cited 14 December 2021).

[28] RMIT. *Game changer: New software for stroke rehab*. Available from: https://www.rmit.edu.au/news/all-news/2019/dec/new-software-for-stroke-rehab. 2019 (cited 9 December 2021).

[29] Sheehan A. *Cooroy hospital robotic rehab program uses virtual-reality technology to help stroke patients*. Available from: https://www.abc.net.au/news/2020-12-26/robotics-rehab-program-at-cooroy-hospital-helps-stroke-victims/13006646. 2020 (cited 9 December 2021).

[30] HCF. *Robots Helping Patients Recover From A Stroke*. Health Agenda Research and Insights. Available from: https://www.hcf.com.au/health-agenda/health-care/research-and-insights/robots-helping-recover-from-stroke. 2019 (cited 9 December 2021).

[31] van Ommeren AL, et al. Assistive technology for the upper extremities after stroke: systematic review of users' needs. JMIR Rehabil Assist Technol 2018;5(2):e10510.

[32] Venkatesh V, Bala H. Technology acceptance model 3 and a research agenda on interventions. Decis Sci. 2008;39(2):273−315.

[33] National Stroke Audit − Acute Services Report 2019. Stroke Foundation, Editor. 2019: Melbourne, Australia.

[34] Lang CE, et al. Observation of amounts of movement practice provided during stroke rehabilitation. Arch Phys Med Rehabil 2009;90(10):1692−8.

[35] Lang CE, MacDonald JR, Gnip C. Counting repetitions: an observational study of outpatient therapy for people with hemiparesis post-stroke. J Neurol Phys Ther 2007;31 (1):3−10.

[36] Hayward KS, Brauer SG. Dose of arm activity training during acute and subacute rehabilitation post stroke: a systematic review of the literature. Clin Rehabil 2015;29 (12):1234−43.

[37] Michaelsen SM, Dannenbaum R, Levin MFJS. Task-specific training with trunk restraint on arm recovery in stroke: randomized control trial. Stroke 2006;37(1):186−92.

[38] Birkenmeier RL, et al. Translating animal doses of task-specific training to people with chronic stroke in 1-hour therapy sessions: a proof-of-concept study. Neurorehabil Neural Repair 2010;24(7):620−35.

Chapter 12

Asia-Pacific region: Republic of Korea

Bong-Keun Jung[1], Inhyuk Moon[2], JiHyun Kim[3], Jin-Hyuck Park[4] and Won-Kyung Song[5]

[1]*Department of Mechanical Engineering, Seoul National University, Kwanak-gu, Seoul, Republic of Korea,* [2]*Department of Robots and Automation Engineering, Dong-Eui University, Busanjin-gu, Busan, Republic of Korea,* [3]*Department of Occupational Therapy, Far East University, Eumseung-gun, Chungcheongbuk-do, Republic of Korea,* [4]*Department of Occupational Therapy, Soonchunhyang University, Asan, Chungcheongnam-do, Republic of Korea,* [5]*Department of Rehabilitation and Assistive Technology, National Rehabilitation Center, Gangbuk-gu, Seoul, Republic of Korea*

Learning objectives

By the end of this chapter, readers will be able to:

1. Understand the typical neurorehabilitation strategy for the South Korean stroke population and how rehabilitation robots are treated as an essential mode for the continuum of rehabilitation.
2. Implement government-led rehabilitation robotics R&D to improve patient quality of care in rehabilitation settings.

12.1 Introduction

12.1.1 Overview of country

The Republic of Korea (ROK) is a country in East Asia. The Korean population is highly homogeneous. Almost the entire population is ethnically Korean, and there is a small minority of ethnic Chinese permanent residents. All Koreans speak the Korean language and use an indigenous writing system with the Korean script alphabet known as Hangul. ROK occupies the southern part of the Korean peninsula and borders the Democratic People's Republic of Korea (North Korea) to the north. The capital of ROK is Seoul, the fifth most significant metropolis in the world.

Rehabilitation Robots for Neurorehabilitation in High-, Low-, and Middle-Income Countries.
DOI: https://doi.org/10.1016/B978-0-323-91931-9.00021-9

The Korean economy has grown remarkably in the past few decades. It is a developed country with a high-income economy. ROK joined the Organization for Economic Co-operation and Development (OECD) in 1996 and is now the most industrialized member among the OECD countries. The current population of ROK is over 51 million. The population growth rate has decreased recently, but the population density is high and concentrated because most citizens live in urban areas, with roughly half the total population livings in the Seoul capital area. The number of foreigners is also growing, especially in the major metropolitan areas. ROK is a fast-aging country with a low fertility rate and has one of the highest life expectancies (83.5 years) in the world [1].

12.2 Neurorehabilitation for stroke in Republic of Korea

According to the Epidemiologic Research Council of the Korean Stroke Society (2018), currently, one in 40 adults have stroke, 232 out of every 100,000 people will have a stroke each year, and the proportion of cerebral infarction cases is higher than cerebral hemorrhage-type stroke [2]. Yet, stroke mortality is gradually declining, increasing the need for neurorehabilitation for stroke survivors [2].

12.2.1 Continuum of care and service delivery

A multidisciplinary team of experts (e.g., physicians, rehabilitative nurses, OTs, PTs, and STs) perform stroke neurorehabilitation at comprehensive stroke rehabilitation units. Stroke rehabilitation focuses on mat mobility exercises and activities of daily living (ADLs) after the acute stage of stroke and as soon as the patient is medically stable (approx. 72 hours). Rehabilitation is usually performed for 1−2 months at a tertiary/general hospital. After acute stroke rehabilitation, patients with stroke are generally transferred to a subacute rehabilitation hospital, where intensive rehabilitation treatment is conducted for 6 months. Finally, patients are transferred to a nursing hospital where conservative rehabilitation focuses on gait training. Patients with stroke who have been discharged continue to receive rehabilitation through outpatient clinics or community-based rehabilitation service centers. However, the post-discharge treatment heavily depends on the individual hospital protocol [3,4].

In ROK, the criteria for stroke rehabilitation accreditation follow the standards outlined by the National Health Insurance Service (NHIS) and the National Health Insurance Act and consists of 48 rehabilitation treatment items. Rehabilitation costs are divided into fees paid by the NHIS (via reimbursements) and paid by patients (nonreimbursable). Rehabilitation costs covered by the NHIS are approved for the first 2 years of care after the onset of a stroke. Still, they may be continuously approved depending on the patient's functional recovery [3,4]. The cost of rehabilitation services provided by medical institutions is based on a fee-for-service system that follows a relative value point system [5]. Yet, there is no fee system for community-based rehabilitation via home treatment [6].

12.2.2   Stroke assessment and treatment

In the early stages of a stroke, physical and cognitive functions are heavily monitored, patient recovery and treatment responses are predicted, and the basis of a post-discharge recovery plan is provided. The Korean version of the Modified Barthel Index (K-MBI) and functional independence measure (FIM) are the most commonly used tools for ADL assessment for acute and subacute stroke stages, followed by the Korean version of the Mini-Mental Status Examination to evaluate the severity of cognitive decline. During the chronic stroke stage, ADL and instrumental ADL (IADL) assessments are also used [7].

Rehabilitation developmental therapies for disorders of the central nervous system, such as proprioceptive neuromuscular facilitation (PNF) and neurodevelopmental treatment (NDT), are the most frequently performed treatments in medical institutions, followed by functional electrical stimulation (FES) treatment [8]. Mat exercises, mobility treatments, and ADL training are frequently implemented in advanced/general hospitals, and occupational and gait therapy is commonly performed in rehabilitative and nursing hospitals [8]. Within communities, rehabilitation is mainly implemented to alleviate complications or pain caused by stroke. Specifically, community-based care may take the form of range of motion (ROM) exercises to prevent pressure sores or contractures, home modifications to ensure ADL independence, and the provision of assistive devices [8].

12.2.3   Outcome and limitations

A variety of rehabilitation treatments are provided for dysfunctions caused by stroke. Rehabilitation for motor function recovery is primarily divided into upper-extremity (UE) and lower-extremity (LE) functions. In addition, treatments for sensory impairments, swallowing, and cognitive disorders are mainly conducted in medical settings, and their clinical efficacy has been proven. However, the recommendation for treatments depends on their effectiveness [7]. After discharge, no suitable program for continuous rehabilitation exists, so discharge is avoided by increasing the length of a patient's hospital stay. Therefore a home care or community-based service that maintains the continuity of rehabilitation treatment needs to be established. As a result of these efforts, community-based rehabilitation has been implemented in 254 public centers across ROK since 2017, and the demand for home-based rehabilitation services is increasing, but still with limited accessibility [9].

12.2.4 Technology's place in the healthcare process

The degree of neuroplasticity is a fundamental mechanism underlying functional movement to improve independence in a patient with a stroke. Therefore the effective use of neuroplasticity is targeted by stroke rehabilitation. A variety of rehabilitation treatments for patients with stroke have focused on high-intensity and repetitive practices sufficient to induce neuroplasticity [10]. Recently, these treatments have been successfully supplemented by emerging robotic technologies [11]. With the recent advancements in robotic technologies, the implementation of these devices in stroke rehabilitation in ROK has only accelerated, following similar trends observed in other developed countries [11]. For example, AMADEO (Tyromotion GmbH, Graz, Austria), designed explicitly for hand rehabilitation, was used for patients with chronic stroke to improve hand function in ROK [12]. According to the most recently released clinical practice guideline for stroke rehabilitation in ROK, other types of technology, such as repetitive transcranial magnetic stimulation (rTMS), virtual reality (VR), and telerehabilitation (TR) are also being implemented into stroke rehabilitation protocols, and their clinical efficacy has been proven [12]. These technologies are advantageous since they can be used in medical institutions, homes, or community care centers. These technologies will be a critical part of the continued rehabilitation after discharge, minimizing unnecessary hospitalization and dependence on medical institutions and reducing the financial and psychosocial burden on patients with stroke and their families. A plan to reform ROK's insurance system to provide coverage for community-based rehabilitation services is currently being extensively debated [13].

12.3 Robotics rehabilitation trends in Republic of Korea

ROK has invested funds into rehabilitation-related robotics research and development projects since 1999 that have expanded to an annual budget of 10 million dollars (USD), including service-oriented robots. As of 2022, over 15 domestic rehabilitation robotics companies focus on physical and cognitive rehabilitation. The market for rehabilitation robots in ROK is expected to grow [14,15].

With the constantly increasing demand for technology-driven medical procedures and evidence-based practices among elderly individuals in ROK, rehabilitative robots have become a critical part of the path to recovery for stroke survivors. This has inspired technology companies and universities to enter the field to address the demand and develop more advanced rehabilitation robots.

More broadly, rehabilitation robots can transform the patient experience in the healthcare system from hospital-based intensive rehabilitation to home-based telerehabilitation. With these innovations, the future trajectory of rehabilitation treatments appears to be increasingly digital and individualized. Robotic rehabilitation technologies represent a new way to serve patients quickly, comprehensively, and more efficiently, so hospitals can cope with the ever-changing demands.

12.3.1   Issues in rehabilitation robotics

For the past two decades, ROK has experienced rapid and vast development in rehabilitation robotics. Robot-assisted therapies have been applied in inpatient and outpatient rehabilitation clinics. Many neuroscience-related studies have shown that repeated robot-assisted therapy (RAT) can partially restore impaired brain function. Despite the advances in ROK's rehabilitation robotics field, several issues remain: (1) rehabilitation robots should be treated as an extra element of treatment rather than as a competitive element; the perception of rehabilitation robots should shift to one where they are considered as a technology that subsidizes services rather than a competitor meant to replace services currently provided by clinicians. (2) Robots should not only support basic repetitive motions but also provide stimulation and feedback to patients. (3) The distribution and implementation of rehabilitation robots should be expanded to various clinical environments, including hospitals, community-based rehabilitation facilities, and patients' homes. (4) More published testing and evaluation of these devices are necessary for translational research. (5) Affordable access to care is critical for the widespread implementation of RAT; the current medical insurance infrastructure does not sufficiently support this form of care.

12.3.2   Republic of Korea's recent funding endeavors in rehabilitation robotics

The Korean government has recognized the importance of robotics as a critical axis of the Fourth Industrial Revolution. Therefore numerous government ministries (e.g., Ministry of Trade, Industry, and Energy, Ministry of Science and ICT, MOH and Welfare, and Ministry of SMEs and Startups) have developed their own funded projects targeting rehabilitation robotics, including cross-ministry projects and other collaborative endeavors.

The Ministry of Health and Welfare (MHW) is one of the biggest proponents pushing forward the implementation of rehabilitation robotics nationwide. The MHW has secured billions of won in funding to support the development of rehabilitation and care robots for the elderly and disabled individuals, with neurorehabilitation robots as one primary area of research and development.

The National Rehabilitation Center (NRC), one of ROK's leading rehabilitation institutes, has established two primary multiphase research and development projects for rehabilitation robots since 2013; one is the

Translational Research Program for Rehabilitation Robots (TRPRR), and the other is the Pilot Supply Program for Rehabilitation Robots (PSPRR).

NRC's TRPRR consists of four stages: (1) technology enhancement, (2) testing, (3) clearance/approval, and (4) clinical trials. This all-in-one process was aimed to reflect the perspectives of clinical experts during early-stage development and to produce commercially and clinically applicable products for the marketplace. TRPRR has linked various robot technologies to clinical research over the last decade (Fig. 12.1). Most of the supported technologies targeted by the TRPRR were shown to be clinically feasible for rehabilitation through prior R&D programs. Still, they had previously needed help to enter clinical trial stages [16].

The PSPRR supports rehabilitation robotics companies entering the market with successful sales records supported by clinical evidence. Hospitals in Korea have a high standard of medical care, and barriers to adopting newly developed medical devices without reference records of success are high. PSPRR supports hospitals in purchasing newly developed rehabilitation robots and providing published clinical evaluations of their effectiveness,

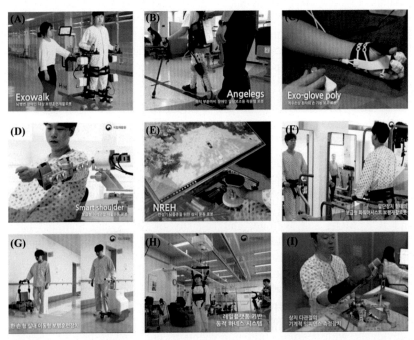

FIGURE 12.1 (A) Exowalk: powered gait training robot; (B) Angelegs: exoskeleton gait-assist robot; (C) Exo-glove Poly: soft wearable robot for hand; (D) Smart Shoulder: powered-assist shoulder rehabilitation robot; (E) NREH: interactive upper-extremity tiltable planar exercise robot; (F) End-effector-based power-assist gait robot; (G) unimanual gait-assist robot; (H) rail-type mobility-assist robot; (I) upper-extremity impedance measuring robot.

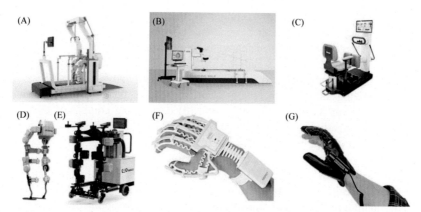

FIGURE 12.2 (A) Walkbot: total body weight support treadmill; (B) Morning Walk: end-effector-type gait training robot platform; (C) 3DBT-33: trunk stability training robot; (D) Angelegs: walk-assist robot; (E) Exowalk: gait training robot; (F) Rapael Smart Glove: soft wearable glove type for neurorehabilitation; (G) Neomano: soft glove for functional hand training.

benefiting both industry and hospitals. Fig. 12.2 highlights some of the rehabilitation robots supported by PSPRR. Through PSPRR, the industry has collected valuable clinician feedback for developing future iterations with improved device function.

The MHW and the Ministry of Trade, Industry, and Energy (MTIE) have collaborated to support the PSPRR; the MTIE expanded its support to more local businesses for additional facilities.

12.3.3   Rehabilitation robot-related policies in Republic of Korea

Rehabilitation robots can be divided into two categories: (1) therapeutic rehabilitation robots and (2) assistive robots. Therapeutic rehabilitation robots, under the jurisdiction of the Medical Device Act (MDA), are classified as *medical devices* and need to receive a medical device certification from the Ministry of Food and Drug Safety (MFDS) (e.g., robotic guidance rehabilitation system: A67000 is a classification for a medical device for orthopedics and restoration). In partnership with global regulations, Korean medical device regulatory systems have given the MFDS oversight of medical devices to improve efficiency [17]. Most neurorehabilitation robots are classified as Class II (A.67020.02, exerciser, orthopedic, electrically powered) or Class III (A.67080.01, robotic guidance rehabilitation system) medical devices (e.g., a medical device for orthopedics and restoration), requiring more comprehensive testing and clinical trials. Industries can apply for a New Medical Technology Evaluation (NMTE) or an Innovative Medical Technology Evaluation (IMTE) for their newly developed products through the NECA (Center for New Technology Assessment), which results in the establishment of updated medical insurance coverage guidelines. New legislation has emerged in support of innovative medical devices and the promotion of the rehabilitation robotics industry: (1) designating effective devices with cutting-edge technologies and safety standards as innovative medical devices, (2) prioritizing the overall review of such regular devices, and frequent reviews of each stage of innovation, and (3) designating companies that are proactively investing in R&D as innovative device companies, and offering them the first choice in bidding for government-funded projects [18].

Contrary to therapeutic rehabilitation robots, assistive robots are not treated as medical devices and are exempted from medical device certifications, resulting in easy access to the market without clinical trials. Assistive products, including prosthetics and orthoses, are listed as nonmedical devices, instead, as welfare technology for individuals with disabilities and fall under the purview of the disability and welfare law. In some instances, rehabilitation robot companies use this classification to commercialize their products rapidly.

The NHI's selective coverage system for gait rehabilitation robots for patients with stroke has been in effect since February 1, 2022. Coverage systems can be applied in the following cases: (1) when economic feasibility or treatment effectiveness is uncertain and additional evidence is needed for verification, (2) even if economic feasibility is low, there is a potential benefit to the health recovery of the insured and their dependents, and (3) when there is a social demand for medical care benefits, or it is recognized as particularly necessary for strengthening public health. The functional ambulation category at the start of rehabilitation treatment should be less than 2 (based on levels 0−5), and the training period is up to 6 months after the onset. Applied equipment is limited to robotic guidance rehabilitation systems (Class III, Item No. A67080.01). Robotic guidance rehabilitation system refers to a robot automation system mechanism used for muscle reconstruction and joint motion recovery [19].

12.4 Korean rehabilitation robotics research highlights from universities, research institutes, and industry

Initially, rehabilitation robotics in ROK emerged as prosthetics and orthotics (P&O) in the 1940s [20]. By the 1990s, mechanical, electrical, and robotics engineers began to expand their development of robotic technologies to include products for individuals with disabilities (IWDs); however, their focus was on human performance assistance rather than clinical rehabilitation [21]. Presently, most original research is performed at research-intensive universities and research institutes. With the help of government funding, the number of university−industry collaborative research projects has increased, including hospital-based research initiatives. This boom in university-based robotics research has drawn the attention of major companies like Samsung and Hyundai, who are now looking into developing their own commercial rehabilitation robots.

There are two big themes in rehabilitation robotics research in ROK: advancing the development of rehabilitation robotics for neurorehabilitation and advancing wearable robots to assist ADLs (aka care robot). The former theme requires comprehensive research collaborations between clinicians and engineers. For example, sensory and motor function recovery after CNS damage depends on the patient's degree of neuroplasticity, requiring physiological limb muscle activation and appropriate sensory receptors. Researchers previously focused on developing robot technologies that mimic human movement without the support of comprehensive clinical knowledge, resulting in their inability to be commercialized. Consequently, developing neurorehabilitation robots requires specialized engineering and expert clinical knowledge. This demand for interdisciplinary research facilitated the establishment of Korean Society of Rehabilitation Robots (KORERO) in 2019.

12.4.1   University-based rehabilitation robotics development

Rehabilitation robotics developed by research-intensive universities began in the early 2000s with experimental exoskeleton robots that assist patients with amputations or patients in need of musculoskeletal assistance. In ROK, this research initially targeted LE gait training and the promotion of optimal robot−patient kinematic synchronization. Walking Assist for Hemiplegia (WA-H) is an example of one of the first exoskeleton gait training solutions [22,23]. Researchers are attempting to improve the robot−patient interfaces, reduce device weight and bulk, minimize the number of actuators, and provide the power necessary for full gait assist. Research improvements thus far have produced advancements in robot control algorithms and the wearability of exoskeleton robots, like the Sogang University Biomedical Assistive Robot (SUBAR), implementing flexible power transmission systems via robust control methods. SUBAR now includes an intelligent control system and has been rebranded to Sliding Mechanism AI-Robotic Therapist (SMART)—now available in Korean markets as a Class II certified medical device [24,25]. At the Korean Advanced Institute of Science and Technology (KAIST), alongside clinicians at Yonsei University Medical Center (YUMC), a series of exoskeleton rehabilitation robots focused on motor-learning-based gait training have been developed by researchers (Angel Legs for adults in 2016 and Angel Suit for children 2018) [26].

Korean university researchers have developed several rehabilitation robots for upper extremities: DULEX-II (2008), a linear-actuated wearable hand robot for stroke survivor hand function assistance [27]; Smart Shoulder

(2017), a shoulder rehabilitation exoskeleton robot that employs a passive shoulder joint tracking mechanism [28]; Easy-Flex (2013), a lightweight, low-cost hand robot that uses a spring tension tool for repetitive hand exercises and fingertip force measurements [29]. Exo-Glove (2015) is a soft tendon-driven wearable glove using a soft tendon routing system [30]. Exo-Glove Poly (EGP), a soft wearable polymer glove that assists patients with hand impairments, was commercialized in 2010 under the brand name NEOMANO [31].

Recently, the applications within the neurological field have been increasing with the advancements in robotics. Especially soft robotics has been a fast-growing field in rehabilitation robotics research in ROK. This soft robotics technology has enhanced rehabilitation robots' user adaptability and wearability and is expected to bring a new era of future neurorehabilitation. With the emergence of the COVID-19 pandemic, the demand for telerehabilitation has increased, resulting in new government—university partnerships to develop teleoperating rehabilitation robot systems.

12.4.2 Research institute-based rehabilitation robotics development

Research institutes in ROK are government-funded centers that conduct many collaborative research ventures and longitudinal studies related to rehabilitation robotics. The NRC is a representative example of a research institute developing these robots. NRC's Robotic Exoskeleton (NREX) is an upper-limb exoskeleton robot designed to assist the arm movements of patients with weak muscular strength and motor control [32]. Cowalk is a gait rehabilitation exoskeleton developed by the Korea Institute of Science and Technology (KIST) that can provide simultaneous pelvis movements during gait training [33]. Beyond the development of exoskeleton robots, kinetic and kinematic data collected by wearable sensors embedded into rehabilitation robots have been highly informative. KIST has undertaken efforts to develop Human-Robot Interface (HRI) technologies, and Cowalk's human—robot interface detects a user's brain signals and movement intentions via EEG and EMG. These signals are applied to exoskeletons' synchronized and interactive movements [34]. The Korea Institute of Machinery and Materials (KIMM) is another center of significant robotics development in rehabilitation robotics; they have focused on both upper- and lower-limb rehabilitation therapy robots for stroke patients.

Along with investments in a research capacity, the Korean government has expanded its scope to policies and R&D expenditures necessary for building an ecosystem that supports technology transfer and commercialization. Many financial support programs have already begun backing the transfer of research institute technologies to the rehabilitation industry via commercialization.

12.4.3   Industry-based rehabilitation robotics developments

Most robotics companies developing and manufacturing rehabilitation robotics solutions in ROK are small- to medium-sized. Walkbot, a robot-assisted gait training system designed for the rehabilitation of patients with neurological disorders, is the first commercialized rehabilitation robot in ROK by P&S Mechanics (2006). Walkbot™ is on sale in European countries, North America, and East Asia. Morning Walk, developed by Curexo, is the first end-effector rehabilitation robot for the treatment of lower-limb conditions (foot and ankle) via walking on a machine; this device is connected to a robotic arm that supplies the force to the user while sensors measure the patients' performance. Exowalk, an overground-type gait rehabilitation robot, was developed by HMH Inc. in 2017 and approved as a medical device.

In addition to small-sized robot companies' efforts, investment in developing rehabilitation robots by large corporations has attracted attention recently. Gait Enhancing and Motivating System-Hip (GEMS-H), developed by Samsung Electronics, assists the movement of the hip joint for patients with stroke and the elderly and has been studied in the United States (Shirley Ryan Ability Lab) and ROK (Samsung Medical Center). In 2020, an ISO 13482 certification was obtained, and GEMS is expected to be on sale in 2022. Hyundai Motor Company has developed several robotic solutions for patients with neurological conditions. H-MEX is a wearable walking assistance robot, and H-LEX is a more versatile device for daily walking. RearMEX, a seven DoF upper-extremity exoskeleton, was also developed to provide individualized robot-assisted therapy (RAT) for stroke survivors [35].

12.5   Post R&D studies

After the R&D process, clinical trials are a critical part of the approval process to classify rehabilitation robots as medical devices. Many rehabilitation hospitals in ROK perform clinical studies to assess the effectiveness of RAT. However, the effects of RAT for stroke survivors compared to conventional treatment have not been fully supported by the literature [33,36,37]. Instead, several studies have reported higher efficacy with end-effector RAT than with exoskeleton-type robot-assisted treatment [38].

The NECA has been performing national-level clinical studies since 2020. They are collecting information on standardized lower-extremity rehabilitation RAT protocols and measuring outcomes from eleven hospitals for the past 3 years. The results of this longitudinal study will establish the clinical effectiveness and value of RAT and hopefully generate sufficient clinical evidence of their use, ultimately supporting the push for insurance-covered compensation for robot-assisted treatment via reimbursement. Nevertheless, the TRPRR from the NRC has played a pivotal role in the clinical studies conducted for rehabilitation robotics commercialization. Some companies have performed their clinical studies in collaboration with international research institutes (e.g., Rehabilitation Institute of Chicago, Stanford University) to expand their market share abroad.

12.6 Conclusion

This chapter provides an overview of the emerging field of rehabilitation robotics and neurorehabilitation in ROK, including current achievements and the future direction of neurorehabilitation. Robotics has the potential to transform current rehabilitation practices into technology-driven evidence-based practices. However, for these treatments to be widely accessible, accessibility to robotic solutions in less-supervised environments outside the hospital (e.g., community-based rehabilitation centers) will be essential. The Korean government has recognized the importance of the continuum of care, particularly for stroke rehabilitation. It has begun to discuss healthcare system reforms, including remote home healthcare and home-visit rehabilitation treatment via national health insurance coverage. Rehabilitation robots can power home-based rehabilitation services to assist with repetitive training regimes and patient performance monitoring.

Interest in rehabilitation robotics R&D is much higher in ROK than in any other country. However, several challenges, including cost, efficacy, and reimbursement issues, must be addressed to make RAT more accessible and marketable. Significant progress was initially possible due to appropriate government support and innovative technology policies; however, the collaborations between engineers and clinicians have had the most beneficial effect on translational research and market-ready products. The following steps for ROK will be to continue developing and fostering an ecosystem where clinicians, engineers, and users can communicate with each other to innovate rehabilitation robotics.

Acknowledgments

This work was supported by the Technology Innovation Program (A lightweight wearable upper limb rehabilitation robot system and untact self-training and assessment platform customizable for individual patients — No. 1415178513) funded by the Ministry of Trade, Industry and Energy, Korea, and the Research Program of National Rehabilitation Center, Korea [NRCTR-IN22004].

References

[1] Steinberg DI. The Republic of Korea: Economic Transformation and Social Change. Routledge; 2019.

[2] Kim JY, Kang K, Kang J, Koo J, Kim DH, Kim BJ, et al. Executive summary of stroke statistics in Korea 2018: a report from the epidemiology research council of the Korean Stroke Society. J Stroke 2019;21:42−59.

[3]   Kim WS, Jung YS, Paik NJ, Shon MK, Jee SJ, Shin YI, et al. Multidimensional approach for rehabilitation status and social adaptation in stroke patients after discharge. Public Health Wkly Rep 2020;13(42):3009−26.

[4]   Lee HS, Jung MY. A study on the use of rehabilitation treatment for stroke patients. Korean J Occup Ther 2021;29(2):1−14.

[5]   National Rehabilitation Center. Research on rehabilitation medical delivery system and policy improvement plan, 2018, 17-S-02.

[6]   Song YJ, Cha YJ. Occupational therapy medical insurance review of issues and improvement of the system in Korea. Korean J Occup Ther 2015;23(1):123−35.

[7]   Kim DY, Kim YH, Lee J, Chang WH, Kim MW, Pyun SB, et al. Clinical practice guideline for stroke rehabilitation in Korea 2016. Brain Neurorehabil 2017;10(suppl 1):e11.

[8]   Song WJ. A study on medical utilization in acute first-ever stroke patient. Doctorate dissertation. Pochun, South Korea: Cha University; 2017.

[9]   Moon KT, Park HY, Kim J-B. The effects of occupation-based community rehabilitation for improving occupational performance skills and activity daily living of stroke home disabled people: a single subject design. Ther Sci Rehabil 2020;9(2):99−117.

[10] Pekna M, Pekny M, Nilsson M. Modulation of neural plasticity as a basis for stroke rehabilitation. Stroke 2012;43:2819−28.

[11] Chang WH, Kim Y-H. Robot-assisted therapy in stroke rehabilitation. J Stroke 2015;15 (3):174−81.

[12] Park JH. Effect of robot-assisted hand rehabilitation on hand function in chronic stroke patients. J Korean Robot Soc 2013;8(4):273−82.

[13] Lee Y, Kang E, Kim S, Byun J. Suggestion of long-term care system reform in view of aging in place. Korean Institute for Health and Social Affair; 2017.

[14] Kim M. New Legislation and the Reform of the Rules on Robots in Korea. Robotics, Autonomics, and the Law. Nomos eLibrar, 2017.

[15] Moon S, Park KH, Lee SG, Cho YJ, Ryu HY. Standardization activities for service robots in Korea. In: ICCAS-SICE 2009 - ICROS-SICE International Joint Conference 2009, Proceedings; 2009.

[16] Song WK. Trends in rehabilitation robots and their translational research in National Rehabilitation Center, Korea. Biomed Eng Lett 2016;6(1):1−9.

[17] Moon IH. Development of a Guideline for Evaluating and Testing Safety and Performance of Robot-assisted Rehabilitation System: the Ministry of Food and Drug Safety; 2015.

[18] Lee M, Ahn J. The current status and future direction of Korean health technology assessment system. J Korean Med Assoc 2014;57(11):906−11.

[19] Yon JH. Controversy related to the preliminary coverage system of health insurance. J Korean Med Assoc 2018;61(6):332−5.

[20] Kim MH. A literature review on the industrial history of assistive technology (rehabilitation engineering) and the developmental plan of industry. Physical Therapy Korea 1997;4 (3):84−96.

[21] Chang PH, Park HS. Development of a robotic arm for handicapped people: a task-oriented design approach. Auton Robot 2003;15(1):81−92.

[22] Rosen J, Ferguson PW. Wearable robotics: systems and applications. 1st ed. Elsevier; 2020, 302.

[23] Sung J, Choi S, Kim H, Lee G, Han C, Ji Y, et al. Feasibility of rehabilitation training with a newly developed, portable, gait assistive robot for balance function in hemiplegic patients. Ann Rehabil Med 2017;41(2):178−87.

[24] Kong KC, Jeon DY. Design and control of an exoskeleton for the elderly and patients. IEEE/ASME Trans Mechatron 2006;11(4):428−32.

[25] Kong K, Moon H, Hwang D, Jeon D, Tomizuka M. Impedance compensation of SUBAR for Back-drivable force-mode actuation. IEEE Trans Robot 2009;25(3):512−21.

[26] Choi H, Na B, Kim S, Lee J, Kim H, Kim D, et al. Angel-suit: a modularized lower-limb wearable robot for assistance of people with partially impaired walking ability. In 2019 Wearable Robotics Association Conference (WearRAcon) 2019;51−6.

[27] Kim YM, Park CY, Jung SY, Moon I. Design of a Rehabilitation Exercise Robot for Upper-extremity Paralysis in Stroke Patients. In Proceedings of the Korean Society of Precision Engineering Conference. 2008;(3):181−2.

[28] Lee KS, Park JH, Park HS. Compact design of a robotic device for shoulder rehabilitation, In 14th International Conference on Ubiquitous Robots and Ambient Intelligence (URAI), 2017;679−682.

[29] Jung BK, Gu WH, Cha YR, Kim JY. The upper extremity rehabilitation exercise device "easy-flex" for people with stroke: a usability test and its results. J Integr Des Res 2017;16(3):71−80.

[30] In H, Kang BB, Sin M, Cho K. Exo-glove: a wearable robot for the hand with a soft tendon routing system. IEEE Robot Autom Mag 2015;22(1):97−105.

[31] Kang BB, Lee H, In H, Jeong U, Chung J, Cho KJ. Development of a polymer-based tendon driven wearable robotic hand. In Robotics and Automation (ICRA), IEEE International Conference on 2016:3750−3755.

[32] Song JY, Lee SH, Song WK. Improved wearability of the upper limb rehabilitation robot NREX with respect to shoulder motion. J Korea Robot Soc 2019;14(4):318−25.

[33] Lee SH, Park G, Cho DY, Kim HY, Lee JY, Kim S, et al. Comparisons between end-effector and exoskeleton rehabilitation robots regarding upper extremity function among chronic stroke patients with moderate-to-severe upper limb impairment. Sci Rep 2020;10(1):1806.

[34] Choi J, Kim KT, Jeong JH, Kim L, Lee SJ, Kim H. Developing a motor imagery-based real-time asynchronous hybrid BCI controller for a lower-limb exoskeleton. Sensors 2020;20:7309. Available from: https://doi.org/10.3390/s20247309.

[35] Kim B, Ahn K, Nam S, Hyun D. Upper extremity exoskeleton system to generate customized therapy motions for stroke survivors. Robot Autonomous Syst 2022;154:104128.

[36] Kang CJ, Chun MH, Lee J, Lee JY. Effects of robot (SUBAR)-assisted gait training in patients with chronic stroke Randomized controlled trial. Medicine (Baltimore) 2021;100 (48):e27974.

[37] Lee HY, Park JH, Kim TW. Comparisons between Locomat and Walkbot robotic gait training regarding balance and lower extremity function among non-ambulatory chronic acquired brain injury survivors. Medicine (Baltimore) 2021;100(18):e25125.

[38] Kim HY, You JSH. A review of robot-assisted gait training in stroke patients. Brain Neurorehabil 2017;10(2):e9.

Chapter 13

Middle East region: Israel

Ronit Feingold-Polak[1], Patrice L. Weiss[2,3] and Shelly Levy-Tzedek[1]

[1]Department of Physical Therapy, Ben-Gurion University, Be'er Sheva, Israel, [2]Pediatric and Adolescent Rehabilitation Center, ALYN Hospital, Jerusalem, Israel, [3]Department of Occupational Therapy, University of Haifa, Haifa, Israel

Learning objectives

At the end of this chapter, the reader will:

1. Be aware of the main barriers and facilitators to the incorporation of technology into patient care.
2. Have an operational set of guidelines for the "dos and don'ts" for the successful incorporation of technology into patient care.

13.1 Introduction

13.1.1 Country history and demographics

Israel is a small, young country (established in 1948) located at the juncture of Africa, Asia, and Europe [1]. Its population density is among the highest in the Western world, with 9.4 million people living in it in 2021. The largest population groups are Jews (74%) and Arabs (21%) [2]. Among the Organization for Economic Co-operation and Development (OECD) countries, Israel has the highest proportion of residents who are immigrants (60% vs the OECD average of 18%) [3]. It has a modern market-based economy with a substantial high-technology sector. The health status in Israel is similar to that of other developed countries, even though the share of gross domestic product (GDP) spent on health is relatively low (in 2020, 7.6% of the GDP were spent on health). Life expectancy in Israel is slightly above the average for the EU member states (EU-15) for both men (80.8 years, compared with 79.1 for the EU-15) and women (84.4 years, compared with 84.2 for the EU-15); life expectancy for Israeli men is among the highest for OECD countries [1]. Acute cerebrovascular event is the third leading cause of death in Israel, accounting for approximately 5.5%

Rehabilitation Robots for Neurorehabilitation in High-, Low-, and Middle-Income Countries.
DOI: https://doi.org/10.1016/B978-0-323-91931-9.00033-5
209

of all deaths, and is the primary neurological cause of disability in adults [4]. Age-adjusted mortality from cerebrovascular disease is highest in the southern region [3].

13.1.2 Healthcare structure

Israel has an advanced health system [3] and a National Health Insurance (NHI) system that provides universal coverage [1]. Every Israeli citizen or permanent resident may choose from among four competing, nonprofit health funds, called health plans or health maintenance organizations (HMOs). These HMOs must provide their members with access to a statutory benefits package [1]. To address inequalities in availability and access to healthcare, universal healthcare insurance for all Israeli citizens was introduced in 1995. This law provides a broad basket of high-quality preventive, curative, and rehabilitation healthcare services [3]. The Ministry of Health (MOH) owns and operates about half of the nation's acute care hospital beds, although they function with increasing autonomy. The largest HMO operates another third of the beds, and the remainder of the beds are operated through a mix of nonprofit and for-profit organizations. Within HMOs, patients have a great deal of freedom in choosing their community-based physicians—both primary and specialist—from among the physicians affiliated with the HMO. All HMOs and hospitals have sophisticated information systems that include electronic medical records and data on activity levels, services provided, and quality of care; there are also several systems for aggregating data across providers, including national registries for conditions such as stroke and reporting of cases of infectious disease. Health outcomes in Israel are comparable with the best health outcomes in the world, as reflected, for example, in the high life expectancy [3].

13.1.3 The national stroke registry

In 2014, an Israeli National Stroke Registry (INSR) was established in the Israel Center for Disease Control (ICDC). The registry enables the identification of needs in the treatment and prevention of acute cerebrovascular events, the monitoring of changes in incidence rates and treatment quality, as well as the planning of interventions and the assessment of their efficacy [4]. In recent years, the Medical Directorate in the Israeli MOH launched a National Plan for the Treatment and Prevention of Stroke Damage [5]. The components of the national plan include raising public awareness of acute cerebrovascular events, training specialized medical personnel, and establishing stroke units, and introducing quality measures for the treatment of acute cerebrovascular events in emergency departments and hospitalizing wards [4].

13.1.4   Stroke incidence and prevalence

According to the ICDC, in 2019, 19,244 new stroke cases were reported from 26 hospitals in Israel, of whom 65.5% were ischemic strokes, 26.4% were transient ischemic attacks, and 8.2% were hemorrhagic strokes. The age-adjusted incidence was 3.2 cases for 1000 people. The incidence rates were 56% higher among men than women and 69% higher among Israeli Arabs than Israeli Jews. The average stroke age was 71.8 years. About 19% of the cases were under 60 years old, and 31% of the cases were above 80 years old; there is only a small gender-based asymmetry with men accounting for 55.6% of cases. The average age among men was younger (69.7 ± 13.1 years) than women (74 ± 13.7 years) and among Arabic-speaking people (65.0 ± 13.5 years) than Hebrew-speaking people in Israel (73.2 ± 13.2 years); the former group has a higher prevalence of underlying risk factors [6]. The most common underlying risk factors were high blood pressure (61.8%), hyperlipidemia (47.3%), diabetes (37.4%), and ischemic heart disease (22.2%). Among those aged 80 years and older, atrial fibrillation was highly prevalent (27.9%). The survival rate in the first-month poststroke was 91% postischemic stroke and 73% posthemorrhagic stroke [6].

13.1.5   Stroke rehabilitation standard of care

According to the universal healthcare law, which was established in 1995, every Israeli citizen who has had a stroke is entitled to rehabilitation, following the recommendation of either a doctor specialized in physical medicine and rehabilitation or a geriatric doctor [7]. Post-acute rehabilitation treatment can be provided in either an inpatient setting or in the community [8]. Israel has a limited number of rehabilitation day centers that offer multidisciplinary treatment in an outpatient setting. Therefore post-acute care usually starts in a specialized rehabilitation inpatient facility [8] or at the patient's home. The at-home rehabilitation services have greatly developed in recent years. In 2014, the MOH established the rehabilitation department, whose aims include: to delineate a national rehabilitation policy; to lead the development of rehabilitation processes at hospitals, in the community, and at patients' homes; to encourage the establishment of rehabilitation units in the periphery of the country; and to regulate and supervise the rehabilitation services in these different settings.

The availability of rehabilitation services differs according to geographic location. The availability of rehabilitation services is greater in the center of the country and more limited in the periphery of the country, especially in the northern district [8]. Rehabilitation services are composed of multidisciplinary services which include doctors, nurses, physiotherapists, occupational therapists, speech therapists, social workers, and psychological treatment. According to the ICDC, in 2019, 12.9% of the stroke cases received rehabilitation only in an inpatient setting, 21.8% of the cases received rehabilitation only in the community, 16.4% received rehabilitation in both settings, and 48.9% did not receive rehabilitation in either setting. While the reasons for not receiving treatment are not detailed in the ICDC report, they include minor events that do not require treatment (transient ischemic attacks—TIAs), treatment that was not reported by the HMO, and lack of available services in underserved areas.

13.1.6 Rehabilitation robotics and assistive technology

To the best of our knowledge, there is no aggregated knowledge base as to the extent of the use of robotics and assistive technology in rehabilitation centers in Israel. Based on anecdotal information gathered in the current study, as well as in conversations with other clinicians in Israel, it appears that the main type of robotic devices used in Israeli rehabilitation centers is the physically assistive kind, such as the Lokomat for gait training, or Gloreha for the upper-limb training, in addition to other assistive technologies, such as the Myro, the Pablo, and other virtual reality technologies. While academic research in Israel does show a benefit for patients using a socially assistive robot [9], this is a rather recent development, which is not yet been implemented in the clinic. In another line of research in Israel, a recently developed assistive robot is currently being tested in the clinic [10,11]. The development of a robot with neuromorphic nonlinear adaptive control was inspired by the brain's spiking neural network to control object manipulation during real-time machine learning-driven control with simpler (and cheaper) hardware. Neuromorphic control, therefore, holds the key to providing robotic assistive technologies at an affordable price.

13.1.7 SWOT experiment

As there are no official data on the prevalence of the use of assistive technology for neurorehabilitation in Israel, or on the factors that could account for the successful or failed integration of such technology as a clinical tool, we conducted focus groups with stakeholders to identify facilitators and barriers to such integration.

13.2 Methods

13.2.1 Focus group

A focus group is a group interview involving a small number of demographically similar people or participants who have other common traits/experiences [12]. Their reactions to specific questions are studied. Focus groups are used to better understand people's reactions to products or services or participants' perceptions of shared experiences. The discussions can be guided or open.

13.2.2   Protocol and participants

We performed two 90-minute focus groups to gather insights related to factors that could account for the successful integration of technology as a clinical tool and to identify barriers to this integration. One focus group included six participants, and the other included seven from across the country representing clinical expertise (physiatry [one], occupational therapy [four], physical therapy [five], technology developer [one], researchers [three], distributors of rehabilitation technology [one], and hospital or governmental administrators [two]); the numbers add up to more than 13 since several participants hold more than one role. There were no differences in the makeup of the two groups, rather participants were assigned to achieve equivalence in terms of professional background, technology exposure, and seniority.

The three authors of this chapter, each of whom has considerable experience in clinical and research applications of rehabilitation technologies, served as moderators for both groups, with author RF acting as the principal moderator. They prepared guiding questions, shown in Appendix 1, which were used to facilitate the discussion and ensure that the major areas of interest were addressed.

Due to the COVID-19-related restrictions as well as logistical convenience, the focus groups were held via Zoom. The sessions were video-recorded and transcribed. The study protocol was approved by the Ethics Committee of Ben-Gurion University, and all participants gave their informed consent to participate.

13.2.3   Data analysis

The three chapter authors independently appraised the transcripts and conducted a thematic analysis in order to analyze the qualitative data by systematically reviewing the participant's experiences, opinions, and views [13−16]. The themes were compiled, and any redundancies were removed via consensus of the three authors.

The themes were then categorized via a SWOT analysis to identify the strengths, weaknesses, opportunities, and threats, a method designed to support the strategic planning of projects [17]. It is typically initiated by specifying the objective of a particular project and then identifying the internal and external factors that support or detract from achieving it. This has been used to great effect in understanding, for example, the state of the art in applications of virtual reality to rehabilitation at the early-to-mid stages of technology adoption [18]. More recently, Nwosu et al. [19] used a SWOT analysis to examine the future impact of medical robotics on palliative, supportive care, and end-of-life care. In this chapter, a SWOT analysis is used to map the strengths, weaknesses, opportunities, and threats related to current clinical applications of rehabilitation technology.

13.3 Results

Table 13.1 lists the themes that were identified for the SWOT analysis's strengths and weaknesses for clinical applications of rehabilitation technology.

TABLE 13.1 Themes identified for the SWOT analysis's strengths and weaknesses for clinical applications of rehabilitation technology.

Strengths	Weaknesses
• Prestige value in clinic being associated with advanced technologies	• Current technologies are not robust or user-friendly enough
• Literature supports the use of technology	• Robotic technologies represent a greater technology challenge
• Provides objective documentation of performance and results	• Difficulty in matching technology to patient's abilities; often personalization is not available
• Increase treatment time	• Some technologies require time and skills at a higher level
• Technology can be an additional therapeutic tool	• Turnover of trained clinicians
• When used together with gamification, it can increase motivation and participation	• Inconvenient physical location of technology
• Treatment can be much more varied	• Personal characteristics and attitudes of clinicians
• Client can be more independent in therapeutic process	• Fear of abandonment of the patient by the clinician
• Provides options that are not available during conventional treatment	• Patients need a certain cognitive level to be able to use technology
• Potential to achieve more effective/ efficient therapy	• Need for extra time for technology setup
• Technology can be used as priming to be followed up by conventional tools	• Technology should not be used just as a game without sufficiently taking advantage of its full rehabilitation potential (e.g., provision of meaningful feedback)
• Possibility of using more than one technology at a time can be powerful	
• Quantitative outcome measures	
• More positive image of disability	

13.3.1   Strengths

The following key strengths were raised by participants: the prestige value in clinics that are associated with advanced technologies which they suggested would lead to administrative support to acquire and use them. Many technologies are supported by studies in the literature that show their effectiveness. Technology can provide objective documentation of the performance and results of therapy for both therapists and patients, and they can facilitate an increase in therapy time especially if one-to-one treatment is followed by practice time. Therapists view technology as an additional, powerful therapeutic tool depending on a therapist's knowledge of its capabilities and a patient's willingness to use it. Technology, in particular when gamification is applied, increases motivation and participation by the patient even though they may be initially apprehensive about using the devices. Treatment can be much more varied, especially for those in long-term treatment. Patients can be more independent in their own therapeutic process, for example, if the use of technology at home is supported via remote therapy.

In summary, technologies provide options that are not available during conventional treatment, e.g., gait training using antigravity (weight support) devices. There is potential to achieve more effective/efficient therapy due, for example, to increasing the intensity of therapy, documentation of outcomes, and more time for handling by the therapist, all while presenting patients with more challenging activities. Feedback on results and performance can be accomplished, in part, without the therapist, in either clinical or home settings. Therapists with expertise in specific technologies learn to manipulate difficulty levels and stimuli to achieve specific therapeutic objectives. Technology can be used as a priming tool to work on abilities that are then followed up using conventional tools. Outcome measures focus on objective, quantitative data providing documentation of treatment progress. In general, outcomes can be used to greatly improve treatment options and support clinical decision-making. Finally, many patients feel that the use of technology broadcasts a message of ability, rather than focusing on their disability.

13.3.2   Weaknesses

With regard to weaknesses, the key themes included a view that many current technologies are not robust or user-friendly enough and therefore need technical support in order to provide consistency during treatment. Robotics technologies represent a greater challenge than other devices due, in part, to their moving parts and greater complexity. Indeed, several therapists reported not using the Lokomat (Hocoma DIH brand) due to its setup time; they use it just for weight support rather than as a robot. Difficulty in matching technology affordances to patient's abilities was a common theme, especially with some of the cognitive aspects. In some cases, the provision of feedback is not finely tuned to a patient's performance which leads to frustration. In general, care must be taken to provide adequate supervision of patients in order to prevent compensatory movements.

Focus group participants reported a tendency to rely on "lighter" technologies such as the Wii and low-immersion virtual reality games since they present fewer technical challenges. They remarked on a relatively high level of turnover of trained clinicians resulting in a difficulty in maintaining a roster of specific therapists who are responsible for the widespread use of technology within the clinic. Difficulties tended to arise when the technologies are not located sufficiently close to treatment rooms; many remarked that devices will be used far less frequently whey they are too far away, if it is too complex to set them up, or if training time is too long or complex. Personal characteristics and attitudes of therapists were viewed as potential weaknesses, e.g., in the case of people with technophobia. In some cases, this was attributed to the age of the therapist; however, overall, age was not a major criterion for not using technology. Some patients fear they may be "abandoned" by the therapist (left on their own to use the technology); they want and need the therapist's attention and handling and fear that the use of technology will take away from one-to-one time with the therapist. Patients need a certain cognitive level to be able to use technology.

Table 13.2 lists the themes that were identified for the SWOT analysis's opportunities and threats for clinical applications of rehabilitation technology.

13.3.3 Opportunities

With regard to opportunities, recruitment or training of additional therapists with technology experience can serve to support the rest of the staff; the participants agreed that there should be two or more therapists who are given time to support technology identification and usage. Recognition of such roles was deemed to be a primary responsibility of the clinical administration. It is important not to be too "democratic" in letting therapists decide whether to use a given technology since, overall, response to technology is positive as long as management perceives that the use of devices is not designed to replace conventional therapy. Indeed, technology broadens our understanding of what clinicians can do, especially their ability to think of and create alternate therapeutic possibilities. It is particularly important to promote interaction with technology developers to enhance the clinician's role as a design partner to ensure that technologies are more usable, accessible, and functional (i.e., similar to the demands of daily life activities). Greater technology support leads to more willingness to use technology. Nevertheless, it is equally important to recognize that not all technologies are suitable for use by all therapists for all patients; it is not a universal

TABLE 13.2 Themes that were identified for the SWOT analysis's opportunities and threats for clinical applications of rehabilitation technology.

Opportunities	Threats
• Select and train a few therapists to support the rest of the team	• Therapist "burnout" attitudes: "Been there, done that"
• Define and promote the role of administration in supporting technology usage	• Trade-off between the time to reach the location where technology is located and/or set it up and the actual time left for therapy
• Work with technology developers as design team partners	• Lack of clarity of what device contributes to therapy
• Greater technology support at all levels leads to more willingness to use technology: 1. incentives, 2. training, 3. encourage demand by patients	• Timing and circumstances of training in device usage are essential
• Recognize that not all technology is suitable for use by all therapists for all patients	• Risk to patients if technologies are used unsafely
• Recognize other specializations that can perhaps support OTs and PTs when using technology	• Physical location of technology may not sufficiently support device usage
• Technology should not usurp therapy	• Technology adoption is a process that needs time to accommodate
• Optimal timing of training to clinician on device usage (should be close to when used clinically)	• Cost of technology is still very high
• Promote more focus on technology in academic courses	
• Encourage developers to make applications as functional as possible	

mandate for all. One participant who had extensive experience in several European countries recommended the use of new professional specializations such as robotics specialists who combine clinical and technical skills.

The greatest opportunities occur when ways to promote greater use of technology are considered, such as various incentives to encourage therapists to use technology, giving therapists time to train in the use of technologies, and encouraging patients to demand to use the technology (when clinically appropriate). The promotion of technology usage, for example, can be achieved by seeing other therapists successfully use the technology, by the availability of usage tips (online or via workshops), and by greater focus on technology during academic training and continuing education.

13.3.4 Threats

With regard to threats, overall, technology devices are developed and marketed on an ongoing basis. In addition to the cost of acquiring and maintaining technology, rehabilitation professionals, to a large extent, have not yet formulated "best practices" for how to best integrate it into therapy. Negative or pessimistic attitudes of therapists (e.g., "Been there, done that") can lead to "burnout" for those who have been disappointed with technologies and hesitate to give them another chance. Moreover, if other therapists in the department see that it is not used, then they will also not use it. There is a need to consider the trade-off between the time to get to where a given technology is located (when it is not near conventional treatment sites) and the time to set it up (if it cannot remain "active" between uses) and the actual time left for therapy. If either travel or setup time is too much, then therapists will not consider using the technology. It is important to consider the overall layout of rehabilitation departments to both accommodate and facilitate technology usage. If it is not near or close to where therapy normally takes place, then it will be less used; in addition, some technologies require a quiet setting that is free from distraction.

It was clear from the discussion that there is not always a consensus on technology-related terminology (e.g., virtual gaming versus immersive virtual reality), and this may lead to misunderstandings about which devices are suitable for a range of treatment protocols. Although all participants agreed that training in correct technology usage is very important, the timing and circumstances of the training are equally important. In general, training is less helpful when it is given too early (at a time when the therapist cannot yet use the device, e.g., if it has not yet been purchased for the clinic) or when it is given to too large a group at once so that there is not enough time for hands-on practice.

13.4 Emergent guidelines

Based on the discussions we held with the stakeholders, we complied the following guidelines for the successful incorporation of technology for neurorehabilitation, including assistive robotics; we detail those at four different levels: the technology development level, the institutional level, the clinical team level, and the clinician–patient level.

13.4.1 Technology development level

The clinicians were adamant about two main factors, which were the make-or-break for using technology in the clinic: (1) the user-friendliness of the interface, it should be very simple to operate and adjust the settings per patient; and (2) the setup time per patient should be 5 minutes or less. In addition, they indicated (3) the importance of using participatory design when developing assistive technology; they said that devices that were developed without clinicians' input are often quickly abandoned in the clinic.

13.4.2   Institutional level

Participants made it clear that the medical institution should facilitate the incorporation of technology in the following ways: (1) dedicating physical space for the technologies, which is located in very close proximity, or integrated with, the space used for one-on-one physical and occupational therapy; this location needs to have a reliable internet connection, for devices that rely on online accessibility; (2) dedicating information technology (IT) personnel for timely technical support of the devices.

13.4.3   Clinical team level

(1) Appointing a team member to be the go-to person per technological device; they will be in charge of the day-to-day operation of the device; it is advisable that they will not be technology-averse; (2) instituting a rewarding structure for the use of technology (if only by noting it as part of the expectations, and following up on it by quantifying the actual extent of use); (3) clearing out time in the schedule of therapists to work with the technology, and get acquainted with it, with no patients scheduled for this time.

13.4.4   Clinician−patient level

(1) Identifying the patients who will benefit most from working with the technology is key; for the first instances of the therapist using a new device, it is recommended to also choose patients that will be excited about using technology; a positive experience for both patient and therapist will likely lead to further use, and adoption of this approach by nearby clinicians and patients. (2) Managing expectations from the technological device is important to avoid disappointment and resentment on the side of the patient. (3) Framing independent usage of technology: having the patient work independently with the device can significantly increase therapy time; this must not come at the expense of time spent with a therapist. Many participants were adamant about technology being a tool in the hands of the therapist and not a replacement of the therapist. Independent use of technologies before, after, or in between therapy sessions can be beneficial.

13.5 Conclusion

The implementation of many of the guidelines listed above will require dedicated funds for this cause. It is also dependent on a shift in the mindset of several actors: from technology-oriented curricula at the university training programs, through clinical institutions dedicating specific resources (space, personnel, and working hours) to promote the use of technology, to companies incorporating clinical advisors on their team already at early stages of product development.

While these insights were brought up in focus group discussions that were held to capture the state of affairs in Israel, we anticipate that these guidelines will be of universal use.

Acknowledgments

We appreciate the contributions of the participants in the focus groups, which included Ilanit Baum-Cohen, Ahmi Ben Yehuda, Sigal Berman, Atara Blaunstein, Hila Dahan, Nitzan Hartal, Yifat Koren-Levin, Micha Kutner, Richard Levi, Justine Raz, Alexandra Saad-Daniel, Liraz Sammet, and Iuly Treger. Financial support was provided by the Rosetrees Trust, by grant no. 3000017258, from the Chief Scientist Office of the Israeli Ministry of Health, and by the National Insurance Institute of Israel.

Conflict of interest statements

The authors declare no conflict of interest.

Appendix 1

The guiding questions used during the focus group discussions are as follows.

Question 1: Do you think the use of rehabilitation technology can promote the rehabilitation of your patients? In what aspects and how? What do you think are the advantages and disadvantages of using technology in rehabilitation, specifically robotics? Do you see a difference between robotic devices and other technologies—for you and the patient?

Question 2: What enables you to use technology (and specifically—robotics) in everyday rehabilitative care? What makes it difficult for you to use technology on a daily basis? What do you think would have allowed you to use more technology in rehabilitative care?

Question 3: When you want to use technology in the workplace, what are the difficulties and barriers you face? What helps and promotes the use of rehabilitation technology in your workplace?

Question 4: Is there a type of technology that you prefer to use, or actually use more frequently? If so, what is it? Why do you think you use it more than other types?

Question 5: Is there technology in your workplace that you do not use? What is it, and why?

Question 6: Do you think the world of rehabilitation has changed as a result of the COVID-19 pandemic? In what ways? Has there been a change in the use of technologies as a result of the limitations that the pandemic placed on us?

Question 7: With regard to technology and specifically robotics, where do you think the rehabilitative world will be in 5 years?

Question 8: Regarding technology and specifically robotics, where would you like to see the world of rehabilitation in Israel in 5 years?

References

[1] Rosen B, Waitzberg R, Merkur S. Israel: health system review. Health Syst Transit 2015;17(6):1–212.

[2] CBS. Israeli Central Bureau of Statistics, 2021.

[3] Muhsen K, Green MS, Soskolne V, Neumark Y. Inequalities in non-communicable diseases between the major population groups in Israel: achievements and challenges. Lancet 2017;389(10088):2531–41. Available from: https://doi.org/10.1016/S0140-6736(17)30574-3.

[4] Ben Shoham A, Liberant-Taub S, Sharon M, Zucker I. The number of acute cerebrovascular events in Israel: a forecast until 2040. Isr J Health Policy Res 2019;8(1):67. Available from: https://doi.org/10.1186/s13584-019-0337-1.

[5] Ministry of Health. I.M.o.H. Stroke (Hebrew). Available from: https://www.health.gov.il/Subjects/disease/Pages/stroke.aspx [Internet] cited 2021 (accessed on 08.02.2021).

[6] Zucker I, Libroder C, Ram A, Hershkovitz Y, Tanne D. Israeli National Stroke Registry 2019 Report. Israel Center of Disease Control (ICDC). 2021 Apr:38.

[7] Clarfield AM, Manor O, Nun GB, Shvarts S, Azzam ZS, Afek A, et al. Health and health care in Israel: an introduction. Lancet. 2017;389(10088):2503–13. Available from: https://doi.org/10.1016/S0140-6736(17)30636-0.

[8] Zucker I, Laxer I, Rasooli I, Han S, Cohen A, Shohat T. Regional gaps in the provision of inpatient rehabilitation services for the elderly in Israel: results of a national survey. Isr J Health Policy Res 2013;2:27. Available from: https://doi.org/10.1186/2045-4015-2-27.

[9] Feingold-Polak R, Barzel O, Levy-Tzedek S. A robot goes to rehab: a novel gamified system for long-term stroke rehabilitation using a socially assistive robot-methodology and usability testing. J Neuroeng Rehabil 2021;18(1):122. Available from: https://doi.org/10.1186/s12984-021-00915-2.

[10] Tsur EE. Neuromorphic engineering: the scientist's, algorithm designer's, and computer architect's perspectives on brain-inspired computing. CRC Press; 2021.

[11] Ehrlich M, Zaidel Y, Weiss PL, Melamed Yekel A, Gefen N, Supic L, et al. Motion guidance and adaptive control of a wheelchair mounted robotic arm with neuromorphically integrated velocity readings and online-learning. Unpublished data.

[12] Restall G, Diaz F, Wittmeier K. Why do clinical practice guidelines get stuck during implementation and what can be done: a case study in pediatric rehabilitation. Phys Occup Ther Pediatr 2020;40(2):217–30. Available from: https://doi.org/10.1080/01942638.2019.1660447.

[13] Braun V, Clarke V. Using thematic analysis in psychology. Qual Res Psychol 2006;3 (2):77−101.

[14] Terry G, Hayfield N, Clarke V, Braun V. Thematic analysis. The sage handbook of qualitative research in psychology,. SAGE; 2017. p. 17−37.

[15] Chapman AL, Hadfield M, Chapman CJ. Qualitative research in healthcare: an introduction to grounded theory using thematic analysis. J R Coll Physicians Edinb 2015;45 (3):201−5. Available from: https://doi.org/10.4997/JRCPE.2015.305.

[16] Clarke V, Braun V. Thematic analysis. In: Teo T, editor. Encyclopedia of critical psychology. New York, NY: Springer; 2014. Available from: https://doi.org/10.1007/978-1-4614-5583-7_311.

[17] Gürel E, Merba T. SWOT analysis, a theoretical review. J Int Soc Res 2017;10(51).

[18] Rizzo A, Kim GJ. A SWOT analysis of the field of virtual reality rehabilitation and therapy. Presence 2005;14(2):119−46.

[19] Nwosu AC, Sturgeon B, McGlinchey T, Goodwin CD, Behera A, Mason S, et al. Robotic technology for palliative and supportive care: strengths, weaknesses, opportunities and threats. Palliat Med 2019;33(8):1106−13. Available from: https://doi.org/10.1177/0269216319857628.

Part III

Background on healthcare systems, rehabilitation, stroke rehabilitation, and rehabilitation robotics in selected LMICs

Chapter 14

North America and Caribbean region: México

Daniel Comadurán Márquez[1,2], Pavel Loeza Magaña[3] and
Karla D. Bustamante Valles[1,4]

[1]*Research Department, Centro de Investigación en Bioingeniería A.C., Chihuahua, Mexico,*
[2]*Cumming School of Medicine, University of Calgary, Calgary, AB, Canada,* [3]*Rehabilitation
Medicine, Sports Science, Universidad Nacional Autónoma de México, Ciudad de México,
Mexico,* [4]*Orthopaedic & Rehabilitation Engineering Center, Marquette University, Milwaukee,
WI, United States*

Learning objectives

At the end of this chapter, the reader will be able to:

- Understand the organization and challenges of the rehabilitation healthcare system in Mexico.
- Recognize the standard of rehabilitation care, and the limitations and challenges faced by healthcare personnel in Mexico.
- Understand the current and future mechatronic and robotic rehabilitation practices in Mexico.

14.1 Current healthcare situation

14.1.1 Geographical and economic situation

Mexico, part of North America, is often mislabeled as a Central or South American country as it is a Spanish-speaking and developing country. According to the World Bank Organization, Mexico has a population of about 130 million people, and it is the second-largest economy in Latin America, now considered an upper-middle income economy [1,2]. However, the country's economic growth has been only about 2% over the last three decades; 48.8% of the total population live under the poverty line, and 45% of the population with a disability live in poverty [3]. Mexico has the lowest shared prosperity indicator in Latin America with great inequalities in income, access to education, and access to health services [1]. Disability is similarly distributed in

Rehabilitation Robots for Neurorehabilitation in High-, Low-, and Middle-Income Countries.
DOI: https://doi.org/10.1016/B978-0-323-91931-9.00034-7

rural and urban areas (27.4% vs 21.6%, respectively); the Instituto Nacional de Estadística y Geografía (INEGI) estimates that about 7.5% of the population are disabled (i.e., 9.17 million) [4]. From the different disability categories, mobility disability accounts for 45.3% of disability in Mexico [3]. However, according to the United Nations Developed Program Proportions, disability in Mexico should be at least 9.9% [5]. This discrepancy may be due to the lack of, or inefficient, data collection, which has been reported in many developing countries [6]. INEGI reports the prevalence of disability only according to seven categories related to activity limitation, such as walking, listening, or learning, and does not provide information on the incidence or prevalence of diseases causing disability [4]. INEGI also reports the inequalities people with disabilities face in access to education, jobs, and healthcare [7].

14.1.2 Healthcare system overview

The Mexican public healthcare system is inadequate with severe limitations in its human and physical resources due to a highly fragmented structure, widespread poverty, and scarce financial resources. About 13% of the population do not have access to healthcare despite the efforts to increase service over the past several years [8]. Quality of healthcare directly correlates with income levels. For example, infant mortality rates are 20 times higher in poorer and disadvantaged zones. This fragmentation also reflects the discrepancy in data collection. The lack of proper statistics circumvents the implementation of programs to solve health problems [8].

The Mexican healthcare system is divided into private and public sectors (Fig. 14.1). The public system is further divided into two main categories. As described by Gomez et al. [8], the first category is the Social Security Institutes which are the public healthcare systems providing services to persons working in a formal economy with job benefits. Their main income comes from employers' and employees' fees. The Social Security Institutes are Instituto Mexicano del Seguro Social (IMSS), Instituto de Seguridad y Servicios Sociales de los Trabajadores del Estado (ISSSTE), Petróleos Mexicanos (PEMEX), Secretaría de la Defensa Nacional (SEDENA), and Secretaría de Marina (SEMAR). Each institute provides services to a segment of the population. For example, SEMAR provides services exclusively to marine employees, ISSSTE to teachers, etc. These institutes are separate entities that work independently. The second category under public services is the healthcare systems that deliver services to people working in a nonformal economy, without job, legal benefits. These institutions provide services with resources from the federal government and, in a small percentage, from state authorities and co-payments that patients must pay when they undergo treatment. Under the federal government category are the Secretaría de Salud (SSA), Servicios Estatales de Salud (SESA), and Programa IMSS-Oportunidades (IMSS-O). As with the Institutes of Social Security, these

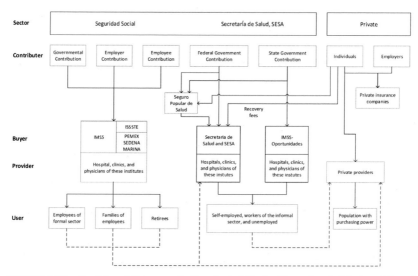

FIGURE 14.1 Short description: organization chart of the health services in Mexico. *Adapted from O.G. Dantés, S. Sesma, V.M. Becerril, F.M. Knaul, H. Arreola, J. Frenk, Sistema de salud de México. Salud Publica de México 53 (2011) s220–s232. Long description: Organization chart of the health services in Mexico. The chart is divided into sector, contributor, buyer, provider, and user categories. The sector category has two main divisions: private and public, with the public division containing Seguridad Social (social security), and Secretaría de la Salud (bursary of health). Contributors include governmental (state and federal) bodies, employers, and individuals. Buyers include different governmental institutes and private entities. Lastly, users include employees, their families, or self-employed individuals.*

entities work independently from each other. Private healthcare systems are divided into private healthcare insurance and out-of-pocket payments [9]. The healthcare system in Mexico is a complicated system with multiple sectors, contributors, buyers, and providers (Fig. 14.1).

According to SSA, there are about 164,000 physicians at an approximate rate of 1.25 physicians per 1000 people [10]. This is a low rate when compared to countries like Canada and the United States with rates of 2.31 and 2.58, respectively [11]. These numbers are for general and specialized physicians. The Physical Medicine and Rehabilitation Mexican Board reports a total of 2033 physicians [12]. The lack of medical personnel leaves an overwhelmed healthcare system that struggles to meet demand. The 2020 national census estimates that there are over 6 million people with disabilities in Mexico; thus there is approximately 1 rehabilitation physician per 2721 people with disabilities [13], and the number of persons with disability is believed to be higher due to a lack of proper data collection systems. Additionally, Mexico with one of the highest incidences of obesity and diabetes, compounded by the lack of preventive measures, as well as a high-risk population, has made stroke a major health problem [14].

It is estimated that stroke was the fifth leading cause of death in Mexico in 2018 [10]. As previously mentioned, due to the fragmentation of the health system there is no reliable data on stroke incidence or prevalence. The information available comes from surveys or studies published, and estimations are made based on those numbers. The Brain Attack Surveillance study made in the community of Durango estimates there are 8 stroke survivors per 1000 people [15].

The National Mexican Registry of Cerebral Vascular Disease (RENAMEVASC) was created with the purpose of estimating the incidence by age and gender of stroke as well as the risk factors. This study found that hemorrhagic stroke is the most frequent, and it is due mainly to uncontrolled high blood pressure [16]. This correlates with the prevalence of high blood pressure in Mexico, a major healthcare problem with a prevalence of 42.3% for adults >20%, 60% for adults between 50% and 60%, and 70% for adults over 60 years of age. Additionally, 38% of the strokes were classified as unknown due to limited diagnostic equipment. For patients with ischemic stroke, data show that 20% presented serious disability with a modified Rankin scale of 4−5. This percentage is 21% for intracerebral hemorrhagic stroke.

Studies highlight that, while the mortality rate for stroke patients has diminished in developed countries, it is increasing in Mexico [17]. Data report that the mortality rate rose from 25.1% in the year 2000 to 28.3% in 2008, emphasizing the sub-registry that may occur due to different healthcare systems or other issues (e.g., there is not a universal word for stroke in Spanish, so patients may be diagnosed differently by region). Regarding the sequela of stroke, the records show that patients with moderate to severe functional impairments range from 24% to 59% [18].

14.2   Standard of rehabilitation care

As mentioned before, Mexico's healthcare system is divided into public and private sectors. Quality of care varies significantly among these, even though many physicians practice in both sectors. Furthermore, the standard of care for stroke patients or other persons with disabilities may vary because each major healthcare provider operates independently.

The private system offers inclusive and comprehensive care to the population with economic solvency, while the public system provides care to a larger sector of the population. Thus the public systems' reach is impacted by saturation and budget constraints. The federal government's public healthcare system intends to protect the poorest sector with a restricted package of services, with limited benefits for temporary or permanent disability [19].

Efforts have been made to provide rehabilitation services all over the country. However, the largest number of healthcare providers remains in urban areas. Rehabilitation in Mexico is mainly based on public healthcare institutions with the IMSS being the largest healthcare provider [20]. The Mexican social security law requires that every person affiliated with IMSS has the right to medical, rehabilitation, and all necessary services to get integral care. IMSS has four tertiary health service units, three in Mexico City and one in Monterrey. Additionally, it has 14 high-specialty rehabilitation units and 101 regional rehabilitation services in hospital areas. Most of these units are composed of one physiatrist, four physiotherapists, a social worker, and a nurse. The tertiary level of healthcare can be found in both the public and private sectors with institutions offering high-intensity services in rehabilitation, medical education, and medical research. These units provide the main source of research currently available in Mexico, but mostly in Mexico City or a large metropolis (Fig. 14.2). IMSS provides 164 services to more than 50 million inhabitants [20]. Despite its efforts to provide quality services, IMSS is known for its insufficiency to meet demand in overall care.

The Sistema Nacional para el Desarrollo Integral de la Familia (DIF) is a public organization that provides care to people with disabilities. DIF has 128 centers around the country. At the same time, the influence of the private (nongovernmental) sector in rehabilitation has become more influential

Chihuahua*
Nuevo León
Jalisco
Ciudad de México

The image shows a map of Mexico where the states with the larger metropolis are highlighted

FIGURE 14.2   Short description: Map of Mexico with state divisions. The map highlights our home state, Chihuahua, as well as Nuevo Leon, Jalisco, and Ciudad de Mexico; the states with the major metropolis in the country: Monterrey, Guadalajara, and CDMX, respectively. Long description: Map of Mexico with state divisions. The map highlights our home state, Chihuahua, as well as Nuevo Leon, Jalisco, and Ciudad de Mexico; the states with the major metropolis in the country: Monterrey, Guadalajara, and CDMX, respectively.

in recent years in Mexico. Teleton is a private organization in Mexico that has 20 centers throughout the country with a specialized focus on pediatric rehabilitation [21].

14.3 Guidelines for stroke care

The main institutions developing clinical guidelines of care are IMSS and SSA through the Centro Nacional de Excelencia Tecnológica en Salud (CENETEC). CENETEC assists Mexican healthcare providers in the decision-making process through scientific evidence-based guidelines [22]. The objective of these guidelines is to establish the standard of care for stroke survivors to align with the National Health Program (Programa Nacional de Salud). IMSS clinical practice emphasizes prevention as a key factor to reduce the incidence of stroke. Diabetes, hypertension, alcohol, and tobacco consumption are highlighted as the highest risk factors for strokes [23]. Most of the treatments mentioned in the IMSS clinical guidelines focus on antithrombotic and antiplatelet therapy to reduce the risk of a recurring stroke.

CENETEC guidelines refer to the diagnostic and early treatment of ischemic stroke on the second and third levels of attention [22]. The guide provides information about the intensity and time of rehabilitation; it cites the prognosis will be better with an early rehabilitation treatment endorsing mobilizations and cognitive evaluation. Other recommendations include the use of the Barthel scale, the use of Botox for spasticity, and speech therapy. These guidelines were updated in 2015 adding that the rehabilitation programs should include the needs of the patient and their environment and also adding the use of aerobic and resistance exercises considering safety due to balance issues and supervision by physical therapy or cardiac rehabilitation, and the evaluation of cardiac function. The guides now recommended an intensity of 2−3 sessions per week of at least 40 minutes per session. Other suggestions include a low-fat diet, physical agent modalities such as heat or cold to manage shoulder pain, and psychological assessment. The guidelines come from a systematic review of the literature and not from studies in the Mexican population.

There are many efforts being made for the appropriate care of stroke patients, but the improvement of functional prognosis has not been impacted [18]. A study on the burden of stroke in Latin America with limited available information found the rates of severe disability in rural and urban Mexico were between 29.8% and 18.5%, respectively [24].

Mexican legislature has promoted recent laws (general law for the incorporation of the people with disability, June 2005) aimed to protect the rights of people with disabilities and to ensure equal access to social services, and human rights. However, the real situation of people with disabilities in Mexico is not well known or documented [20].

14.3.1   Current standard of care survey

Because of the limited and inaccurate information, we implemented an online survey to determine the knowledge of physicians with regard to the standard of care for stroke, specifically regarding robotic rehabilitation. The survey was distributed via email and social media. In total, 47 responses were recorded from different types of physicians (Fig. 14.3). Most answers were form Mexico City (n = 15) and Chihuahua (n = 9). The years of practice of the responders were 13.7 ± 9.6 (Mean ± SD). Physicians reported to practice in public (n = 21), private (n = 13), or both sectors (n = 13).

Almost half of the physicians surveyed were unaware of the number of specialized rehabilitation hospitals in Mexico (n = 20), and the other half mentioned between 1 and 5 rehabilitation hospitals (n = 22). Most physicians (n = 43) mentioned that a multidisciplinary team could improve the treatment of the patients. However, many of them mentioned that this is not possible to have multidisciplinary teams due to a lack of proper communication channels and a collaborative culture. Additionally, the physicians reported that, since stroke can cause temporal or permanent disabilities, the patient's needs might change over time and different strategies must be used to improve the patient's quality of life.

Most physicians reported that the clinics to which they refer their stroke patients do not use high-technology approaches for stroke rehabilitation (n = 31). Those that reported the use of technology, reported equipment such as functional electric stimulators, MotoMed, virtual reality, and mobilization robots (e.g., Lokomat, and Armeo-Spring). The physicians also reported that the biggest barrier to adapt these technologies was economic; the human resource factor might also play an important role.

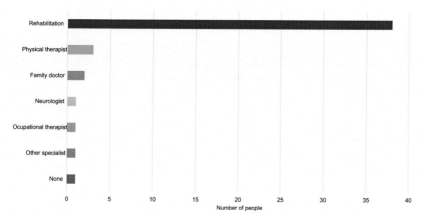

FIGURE 14.3 Short description: Medical specialty of health professionals surveyed. Long description: Horizontal bar graph showing the medical specialty of health professionals surveyed. Rehabilitation (n = 38), physical therapist (n = 3), family doctor (n = 2), neurologist (n = 1), occupational therapist, other specialist, and none (n = 1, each).

TABLE 14.1 Limitations and challenges found in the public and private healthcare sectors according to the implemented survey.

Public sector	Private sector
• Difficulty for patient mobilization (i.e., attend physician and rehabilitation appointments)	• Lack of continuity of treatment (i.e., short rehabilitation sessions, high session costs)
• Lack of integral rehabilitation services	• Lack of availability from family members and adherence to treatment
• Limited resources (e.g., treatment time, equipment, etc.)	• Lack of treatment for acute cases

The physicians surveyed were aware of the limitations of the knowledge of the incidence and prevalence of stroke in Mexico. There is a need for rehabilitation and the development of human resources specialized to treat neurological patients. Additionally, the lack of adequate technology for treatment and rehabilitation has a direct impact on the quality of life of patients. Limitations in the public and private sectors are summarized in Table 14.1.

14.4   Current robotic and mechatronic solutions

As mentioned before, Mexico's health system is insufficient to provide quality care to its growing population [25]. In addition to the low medical personnel-to-population ratio, the cost of medical equipment can be higher in Mexico than in developed countries considering the individuals' purchasing power, importation cost, and taxes [25]. These conditions give rise to the need for affordable technologies for poststroke rehabilitation. Previous studies have looked at the implementation of technology-assisted gymnasiums for stroke rehabilitation [26]. The concept of technology-assisted gym is a boot camp-style gym that allows one therapist to provide therapy to up to six patients simultaneously [26]. The study (n = 10) showed that technology-assisted therapy was equally or more effective than conventional therapy schemes. Additionally, technology-assisted therapy is cost- and labor-effective, which makes it a feasible option to be implemented in developing countries, such as Mexico [26].

The concept and results for the technology-assisted gym were translated into clinical practice by adopting a social business model; the detailed operations of the organization, CIBAC in Chihuahua Mexico, are described in [27]. In summary, patients pay according to their capabilities through a socioeconomic assessment. The therapy costs are kept low by using the technology-assisted gym where one therapist can oversight the treatment of several patients assisted by the technology, with the help of national and international donors, as well as funding campaigns. Many patients can only afford to pay $2 US dollars per session or less. The daily minimum wage set by the Mexican government for 2022 is $172.87 Mexican pesos (MXN), around $8.64 USD. The therapy sessions are about 2-hours long, three to five times per week. Patients receive a 2-hour long session, which is not comparable in efficacy to inpatient hospital therapy, but inpatient hospital therapy is usually only available in Mexico in larger cities. The 2-hour sessions are more than double the time that patients usually receive at public institutions being 1-hour session with a frequency of three times a week [27]. Compared to traditional therapy ($250 MXN, $19.21 USD per session in 2012 [26]), our implementation costs are lower (i.e., $6.99 USD per session [26]) in the short (<2 years) and (i.e., $4.29 USD per year) in the long term (>2 years) [28]. There are still many factors to be considered to make the business model sustainable (e.g., treating non-neurological disorders, agreements with private insurance companies, etc.). However, this implementation is, to the best of our knowledge, a feasible approach for neurological rehabilitation services to provide an integral service. It is important to note that the exchange rates MXN−USD vary over time, e.g., 13 MXN per one USD in 2012 and 20 MXN per one USD in 2022.

There are other healthcare centers implementing robotics solutions in Mexico. Hocoma, a Swiss company that develops robotics for movement and neurological rehabilitation, reports having neurorehabilitation devices in 30 rehabilitation facilities in Mexico [29]. All these rehabilitation facilities are in urban areas and seem to be equally distributed among the public (e.g., IMSS, ISSSTE, SSA) and not-for-profit organization sectors. The Mexico City metropolitan area has the most Hocoma systems, with 10 rehabilitation centers, followed by Guadalajara (3), Puebla and Aguascalientes (2 each) [29]. Another important initiative in robotics is the Cerebro clinic in Mexico City, using a for-profit business model and providing neurorehabilitation services with robotic equipment from Hocoma, Tyromotion, and TheraTraining. The Cerebro clinic has published an extensive review in which they established several criteria to select candidates for robotic rehabilitation [30].

Stroke rehabilitation in Mexico is hampered by a lack of clinical resources and funding. Robotics rehabilitation has shown to be as effective as traditional care while presenting a cost- and labor-efficient option for developing countries if used with the methodology explained where one therapist could provide therapy to up to six patients [26].

14.5   Barriers to implement robotic and mechatronic solutions

There are three key qualitative factors identified to ensure that a robotic therapy solution can be implemented and adopted in Mexico for rehabilitation: technological, cultural, and economical [31]. It is suggested that the three factors are at the same hierarchical level and that every proposed solution should try to fulfill all requirements. The factors are:

1. **Technological:** defined as the qualitative factors directly linked to the therapy requirements (i.e., the use of the technology addressing the problem in an efficient manner).
2. **Cultural:** defined as the barrier to the adoption of technology from the end user or the healthcare provider (i.e., physician or therapist).
3. **Economic:** this is the qualitative factor that restricts the design of the technology according to the assigned budget.

As with most healthcare technologies, robotic solutions must start in a research environment to ensure their efficacy and safety. Further, there must be a translation stage to go from the laboratory to the clinical setting [30]. In our survey, physicians and therapists reported two barriers to adopt new technologies: for the patients, the elevated costs of therapy in private institutes; and for the healthcare personnel, the cultural factor and the lack of awareness of such therapies. To address the former, the Cerebro clinic has reached an agreement with private insurance companies to offer their services to a broader audience. While IMSS and ISSSTE hospitals have access to robotic rehabilitation, most of these facilities are in large urban areas or regional hospitals, thus inaccessible to a large sector of the population. For example, in the state of Chihuahua, with a population of 3.7 million, there are no major hospitals with such services [27].

The economic and cultural factors seem to be the most challenging when translating research into clinical practice [28]. Regarding the economic factor, the chronic nature of motor disability caused by stroke can make treatment unaffordable for many patients in Mexico. Additionally, insurance companies usually cover a limited number of therapies [28]. In the cultural factor, neurologists in Mexico often do not prescribe intensive physical rehabilitation. This could be because of a lack of facilities that provide this service, lack of awareness of the benefits of this treatment, or the ability of the patient to pay for such services [28]. To ensure a successful translation from the laboratory into clinical practice, technological, cultural, and economic factors should be taken into consideration. The technological portion of this adequate translation scenario is not met by the basic equipment recommended by the National Health Council (Consejo de Salubridad Nacional) which does not include the equipment needed for robotic stroke rehabilitation to be in every major public hospital [32]. One option could be a universal private rehabilitation provider that could treat patients from both, the public and private sectors of the system. This approach could bring the same quality of care to both sectors while maintaining low costs, if supported by private, federal, and state funding. However, to the best of our knowledge, such implementation is yet to be seen.

Another barrier is that robotic equipment might not be used to its fullest capacity. For example, most robotic rehabilitation equipment offers quantitative measurements that allow the clinician to evaluate patient progress. We found that many clinicians do not use the quantitative features of the devices, likely due to lack of training. In Mexico, clinical personnel often has limited research training in their education, especially in cities outside the main urban areas. The lack of training could be related to costs as an on-site biomedical engineer or external training courses are needed to provide training.

14.6   Stroke care during the COVID-19 pandemic

Because of the strict lockdowns between March and July 2020, most clinics closed or operated at a reduced capacity. The only services that continued were those deemed essential (e.g., oncology, emergencies). The lockdowns were implemented to slow the spread of COVID-19 to protect the vulnerable sectors of the population. Stroke survivors are among the most vulnerable individuals (i.e., elderly individuals with chronic preexisting conditions) [33]. Physicians in Mexico have reported that, even after the lockdowns are lifted, the patients are not returning to the clinics as expected. This could be because the patients might not feel safe to return to therapy. Additionally, the lockdowns had an impact on the economy; thus patients may not be able to pay for the rehabilitation services anymore.

Clinics in Mexico are currently changing their operations to be able to maintain the recommended guidelines of social distancing, the number of concurrent patients, as well as improving the sanitization protocols. Some clinics started opting for telerehabilitation programs, as well as reducing the time of therapy. However, long-term measures are yet to be defined to ensure that patients will be able to return in a safe manner. Our survey indicated that physicians had to adapt their practice to safely accommodate patients. Strict cleaning protocols and reduced operation have affected most institutions. Additionally, the physicians report that the patients have been affected economically, thus affecting their ability to pay for treatment and mobility to health institutions. The patients' caretakers (mostly family members) are also hesitant to take the patients to therapy since most of them are part of the vulnerable population.

About half of the physicians reported to have tried telemedicine approaches to continue their practice (Fig. 14.4). In the CIBAC clinic, we found many patients hesitant to use telemedicine. Many patients reported that they think the quality of care is inferior with telemedicine compared to on-clinic care. Patients are interested in using the devices found in the clinic; however, these devices are expensive and difficult to take to the patients' home for individualized care. Despite the hesitation, and because on-site therapy was not an option, few of our patients' tried telemedicine for a few months with good clinical outcomes. Some family members reported

There are three photographs of three patients with their corresponding therapist during telerehabilitation sessions. The photographs show screenshots on the mobile phones used for session where you can see the face of the therapist in a small square and the patient on the main area of the photographs. On the first photographs the patient is trying to lift his leg to pass an obstacle. Second photograph shows the therapist giving indication to the patient to give few steps without the use of her walker. Third photograph a patient can be observe trying to stand up.

FIGURE 14.4 Short description: Screenshots of the telerehabilitation process. Long description: Three screenshots of video calls between patients and their therapists to illustrate the telerehabilitation process. First screenshot shows a video call between a male patient with hemiparesis and his therapist. The patient is lifting his left foot to step on a jug of water, and his therapist appears in the top right corner supervising the exercise. Second screenshot shows a video call between a female patient with cerebral palsy and her therapist. The patient is outside her house doing an unassisted balance test. Her therapist appears on the top left corner supervising the exercise. Third screenshot shows a subject performing an unassisted squat in their bedroom. Their therapist appears in the bottom right corner supervising the exercise.

improved personal hygiene and good mood on the day the patient had a telemedicine session. The COVID-19 pandemic has been an opportunity for all healthcare providers to consider and adopt new treatment options. We believe new generations might be more receptive to the use of telemedicine.

14.6.1 The future of rehabilitation in Mexico

Developing countries seem to be the most impacted by COVID-19, both, in economic and health perspectives [34]. Among the guidelines to slow the spread of the virus, there are vaccination wherever available, avoiding crowds in closed spaces, the use of facemasks or protective equipment when indoors, and constant disinfection of hands and highly touched surfaces [35].

Clinics and hospitals must adapt to these guidelines to ensure that rehabilitation services are not impacted by COVID-19. Telerehabilitation hybrid models (i.e., telerehabilitation with some on-site services) and robotics rehabilitation allow for a service with less physical interaction between therapists and patients. Studies have shown that telerehabilitation can be successfully implemented even when broadband internet is not available [36].

A systematic review analyzing the cost of robotic rehabilitation for stroke patients concluded that robotic therapy seems to offer a better cost−benefit relationship for patients with severe impairments with a JBI Grade B with a recommendation to consider for practice with the number of patients that could be treated and the time required of a therapist at a robotic session [37]. Additionally, a systematic review on robotic-assisted gait rehabilitation for stroke patients suggests that the future of robotic rehabilitation includes the use of robots as part of the standard rehabilitation treatment, but clinical guidelines still need to be improved as recommendations for actual practice are still a grade B or C [38]. A systematic review recommended robotic rehabilitation for the upper-limb robotic, but the clinical guidelines about the specificity of the treatment such as the length of the treatment and patient characteristics need to be improved [39].

There are many factors affecting the future of robotic care for stroke patients. The rehabilitation process for a stroke patient needs to be multidisciplinary, engaging, intensive, and effective. The time required for a person to recover independence and reach their maximum ability is usually long. The economic implications of these types of treatments are high; robotic care has the potential to reduce the cost of treatment, improve patient engagement and intensity of treatment, and provide quantitative metrics of patient improvement for better monitoring. To aid technology adoption, research into the development of clinical guidelines is essential [39]. We must find ways to develop and maintain equipment with locally available resources that could provide lower implementation costs. Additionally, we need to update the standard of care to adapt to emergent situations, such as the COVID-19 pandemic (Fig. 14.5).

Two photographs appear in figure 5 where you can observe the patient and the therapist with mask on other protective gear due to the COVID-19 pandemic.

FIGURE 14.5 Short description: The "new normal" of physical rehabilitation. (A) The therapist has a face shield and face mask while providing service to the patient (notice the sign in the background that states the mandatory use of hand sanitizer). (B) Therapists and patient with face masks. Long description: Two photos of in-person therapies to illustrate the new standards of care for physical rehabilitation. First photo shows a female patient lying face down, in a sphynx pose. Her therapist is assisting the exercise by holding the patient from the elbows, where the elbows contact the support structure. The therapist is using a face shield as well as a cloth face mask. There is a sign on the background that states the mandatory use of hand sanitizer. The second photo shows two therapists and one patient sitting in a Bobath therapy table, and all three individuals have cloth facemasks.

References

[1] Mexico Overview. Available from: https://www.worldbank.org/en/country/mexico/overview (accessed on 12 September 2020).

[2] World Bank Country and Lending Groups. World Bank Data Help Desk [Internet]. Available from: https://datahelpdesk.worldbank.org/knowledgebase/articles/906519-world-bank-country-and-lending-groups (accessed on 31 October 2021).

[3] Disability in Mexico. Global Disability Rights Now! [Internet]. Available from: https://www.globaldisabilityrightsnow.org/infographics/disability-mexico (accessed on 12 September 2020).

[4] Geografía IN de E y. Estadísticas a propósito del día internacional de las personas con discapacidad. INEGI México; 2012.

[5] Metts R. Disability and development. In: Background Paper Prepared for the Disability and Development Meeting, World Bank; 2004.

[6] World Health Organization. World report on disability 2011. World Health Organization; 2011.

[7] "Estadísticas a Propósito del día internacional de las personas con discapacidad (3 de diciembre)" Datos Nacionales.

[8] Systematic Country Diagnostic Mexico [Internet]. Available from: www.worldbank.org; 2018 (accessed on 12 September 2020).

[9]   Dantés OG, Sesma S, Becerril VM, Knaul FM, Arreola H, Frenk J. Sistema de salud de México. Salud Publica de México 2011;53:s220−32.
[10] Sistema de Información de la Secretaría de Salud [Internet]. Available from: http://sinais-cap.salud.gob.mx:8080/DGIS/ (accessed on 12 September 2020).
[11] Physicians (per 1,000 people) - United States | Data [Internet]. Available from: https://data.worldbank.org/indicator/SH.MED.PHYS.ZS?
end = 2016&locations = US&start = 2016&
view = bar (accessed on 12 September 2020).
[12] Consejo Mexicano de Medicina de Rehabilitación A.C. [Internet]. Available from: https://www.consejorehabilitacion.org.mx/. (accessed on 31 October 2021).
[13] Población. Discapacidad [Internet]. Available from: http://cuentame.inegi.org.mx/poblacion/discapacidad.aspx. (accessed on 31 October 2021).
[14] Incidence of type 2 diabetes in Mexico. Results of The Mexico City Diabetes Study after 18 years of follow-up. Salud Pública de México 56(1), 11−17.
[15] Cantu-Brito C, Majersik JJ, Sánchez BN, Ruano A, Quiñones G, Arzola J, et al. Hospitalized stroke surveillance in the community of Durango, Mexico: the brain attack surveillance in Durango study. Stroke. 2010;41(5):878−84.
[16] Carlos C-B, José R-SL, Erwin C, Antonio A, Carolina L-J, Luis M-BM, et al. Factores de riesgo, causas y pronóstico de los tipos de enferme-dad vascular cerebral en México: Estudio Renamevasc. Rev Mexicana de Neurocienc 2011;;20(24):224−34.
[17] Erwin C, José Luis R-S, Luis Manuel M-B, Antonio A, Jorge V-C, Fernando B, et al. Mortalidad por enfermedad vascular cerebral en México, 2000−2008: una exhortación a la acción. Rev Mexicana de Neurocienc 2011;12(5):235−41.
[18] Góngora Rivera F. Perspective on stroke in Mexico. Medicina Universitaria 2015;17 (68):184−7.
[19] Sandoval H, Pérez-Neri I, Martínez-Flores F, Del Valle-Cabrera MG, Pineda C. Disability in Mexico: a comparative analysis between descriptive models and historical periods using a timeline. Salud Publica de México.
[20] Guzman JM, Salazar EG. Disability and rehabilitation in Mexico. Am J Phys Med & Rehabilitation 2014;93:S36−8.
[21] Directorio de CREE y CRI a nivel nacional 2016 [Internet]. Available from: https://www.gob.mx/cms/uploads/attachment/file/153868/Directorio_de_CREE-CRI_Nacional_2016_final.pdf; 2016. (accessed on 18 August 2021).
[22] Guía De Práctica Clínica. GPC Diagnóstico y Tratamiento Temprano de la [Internet]. Available from: http://www.cenetec.salud.gob.mx/contenidos/gpc/catalogoMaestroGPC.html. 2017 (accessed on 18 August 2021).
[23] Guía de Práctica Clínica. Vigilancia y prevención secundaria de la enfermedad vascular cerebral [Internet]. Available from: http://imss.gob.mx/profesionales-salud/gpc. (accessed on 26 September 2020).
[24] Ferri CP, Schoenborn C, Kalra L, Acosta D, Guerra M, Huang Y, et al. Prevalence of stroke and related burden among older people living in Latin America, India and China. J Neurol, Neurosurg Psychiatry 2011;82(10):1074−82.
[25] Schwartz MS. Biofeedback: A Practitioner's Guide (2nd ed.). 1995. Available from: http://ovidsp.ovid.com/ovidweb.cgi?
T = JS&PAGE = reference&D = psyc3&NEWS = N&AN =
1995-98538-000.
[26] Bustamante Valles K, Montes S, Madrigal MDJ, Burciaga A, Martínez ME, Johnson MJ, et al. Technology-assisted stroke rehabilitation in Mexico: a pilot randomized trial

comparing traditional therapy to circuit training in a Robot/technology-assisted therapy gym. J Neuroeng Rehabil 2016;13(1):83. Available from: https://www.ncbi.nlm.nih.gov/pmc/articles/PMC5025604/.

[27] Instituto Nacional de Estadística G e I. Population in Chihuahua - 2020 census [Internet]. Instituto Nacional de Estadística y Geografía. INEGI. Available from: http://en.www.inegi.org.mx/app/areasgeograficas/?ag = 08; 2021 (accessed on 31 October 2021).

[28] Valles KDB, Marquez DC, Johnson MJ, Bustamante Valles KD, Comaduran Marquez D, Johnson MJ. Robotic rehabilitation therapy in chihuahua mexico, challenges from translating a clinical research protocol to clinical practice. In: 2020 IEEE 11th Latin American Symposium on Circuits & Systems (LASCAS) [Internet]. IEEE; 2020 [cited 2020 Apr 23]. p. 1−4. Available from: https://ieeexplore.ieee.org/document/9068966/.

[29] Latin America - Hocoma [Internet]. Available from: https://www.hocoma.com/partners/references/latin-america/. (accessed on 12 September 2020).

[30] Loeza P. Introducción a la rehabilitación robótica para el tratamiento de la enfermedad vascular cerebral: revisión. Rev Mex Med Fis y Rehab 2015;27(2):44−8. Available from: www.medigraphic.org.mxhttp://www.medigraphic.com/medicinafisica.

[31] Ponce P, Lopez EO, Molina A. Implementing robotic platforms for therapies using qualitative factors in Mexico. International Conference on Smart Multimedia. Springer; 2019. p. 123−31.

[32] Consejo de Salubridad General. CUADRO BÁSICO Y CATÁLOGO DE INSTRUMENTAL Y EQUIPO MÉDICO [Internet]. Mexico City; Available from: http://www.csg.gob.mx/descargas/pdf/priorizacion/cuadro-basico/iyem/catalogo/2017/EDICION_2017_TOMO_II_EQUIPO_MEDICO_-_link.pdf. (accessed on 31 October 2021).

[33] WHO, Mackay J, Mensah GA. The atlas of heart disease and stroke. J Hum Hypertension 2004;19:112. Available from: http://www.who.int/cardiovascular_diseases/resources/atlas/en/.

[34] The World Bank. COVID-19 Research [Internet]. Available from: https://www.worldbank.org/en/research/brief/COVID19-research. (accessed on 31 October 2021).

[35] World Health Organization. WHO: Coronavirus [Internet]. Available from: https://www.who.int/health-topics/coronavirus#tab = tab_1. (accessed on 31 October 2021).

[36] Russell TG, Buttrum P, Wootton R, Jull GA. Rehabilitation after total knee replacement via low-bandwidth telemedicine: the patient and therapist experience. J Telemed Telecare 2004;10:85−7. Available from: http://journals.sagepub.com/doi/10.1258/1357633042614384.

[37] Lo K, Stephenson M, Lockwood C. The economic cost of robotic rehabilitation for adult stroke patients: a systematic review. JBI Database Syst Rev Implement Rep 2019;17(4):520−47. Available from: https://doi.org/10.11124/JBISRIR-2017-003896.

[38] Calabro RS, Sorrentino G, Cassio A, Mazzoli D, Andrenelli E, Bizzarini E, et al. Robotic-assisted gait rehabilitation following stroke: a systematic review of current guidelines and practical clinical recommendations. Eur J Phys Rehabil Med 2021;57(3):460−71. Available from: https://doi.org/10.23736/S1973-9087.21.06887-8.

[39] Morone G, Palomba A, Cinnera AM, Agostini M, Aprile I, Arienti C, et al. Systematic review of guidelines to identify recommendations for upper limb robotic rehabilitation after stroke. Eur J Phys Rehabil Med 2021;57(2):238−45. Available from: https://doi.org/10.23736/S1973-9087.21.06625-9.

Chapter 15

North America and Caribbean region: Costa Rica

Beatriz Coto-Solano[1] and Arys Carrasquilla Batista[2]

[1]*Rehabilitation Service, Hospital R. A. Calderón Guardia, Caja Costarricense de Seguro Social, San José, Costa Rica,* [2]*Department of Mechatronics Engineering, Instituto Tecnológico de Costa Rica, Cartago, Costa Rica*

Learning objectives

At the end of this chapter, the reader will be able to:

1. Understand Costa Rica's demographic and epidemiological profile, health situation, and health system structure.
2. Understand stroke prevalence in Costa Rica and the challenges related to neurorehabilitation, including the standards of care for stroke rehabilitation.
3. Recognize attitudes about and/or barriers to effective rehabilitation and the use of technological solutions in stroke therapy.

15.1 Introduction

Costa Rica is uniquely characterized by its extended public health system and its complex epidemiological profile. Despite exporting instruments and devices for medical use [1], the country is struggling to incorporate technology, particularly robotic systems into rehabilitation.

This chapter aims to present Costa Rica's context and provide a better understanding of the stroke rehabilitation management currently provided, the efforts undertaken to incorporate technology as a tool for the improvement of medical care, the challenges faced in this process, and views on the future of rehabilitation technology in the country.

15.1.1 Country history and demographics

Costa Rica is a country in Central America with a population of 5,163,038 inhabitants in an area of 51,100 km^2. Characterized by political stability and

Rehabilitation Robots for Neurorehabilitation in High-, Low-, and Middle-Income Countries.
DOI: https://doi.org/10.1016/B978-0-323-91931-9.00009-8

a large investment in social development during the past century, its success is reflected in strong human development indicators: it has one of the oldest democracies in Latin America, relatively low rates of poverty, low infant mortality, a life expectancy of 78.1 years for men and 83.2 for women and an unemployment rate of 15.3% [1−3].

Social determinants such as equality, education, gender, ethnic group, and migratory situation accompanied by cultural, economic, and environmental circumstances account for the development of noncommunicable diseases (NCDs) [4]. The strong social orientation of Costa Rican governments in past decades is reflected in its health outcomes. However, "good health outcomes should not be confused with good medical care" [5] and the lack of alignment in measures such as healthy lifestyles, environments, and recreation aiming to prevent disease has taken its toll on the health system.

Additionally, as has been the dynamic worldwide, the demographic pyramid has inverted, with a larger group of adults and elders on top of the pyramid which presents a greater risk of NCDs [1].

15.1.2 Communicable and noncommunicable disease prevalence

Costa Rica belongs to upper-middle-income countries, with an epidemiological profile which corresponds to a high-income country in that it has a predominance of morbimortality related to NCDs such as cardiovascular diseases and cancer.

Infectious diseases have given way to noncommunicable chronic diseases as the main cause of morbidity and mortality in Costa Rica. Since 1970, circulatory system diseases represent the first cause of death among the Costa Rican population. Noncommunicable chronic diseases have common risk factors such as adopting a high-fat diet, sedentarism, smoking, alcohol abuse, obesity, hypertension, diabetes mellitus, and dyslipidemia [6].

The Survey of Cardiovascular Risk Factors conducted in 2018 by the Caja Costarricense del Seguro Social (Costa Rican Bursary for Social Insurance, henceforth CCSS) showed concerning trends [7]. It documented a diagnosed diabetes prevalence of 10.9% and non-diagnosed of 3.9%, as well as a diagnosed hypertension prevalence of 32.4% and non-diagnosed of 5%. In the case of dyslipidemia, 27.1% of the survey respondents presented the condition, with an overweight prevalence of 39.5% and 31.2% of obesity. In addition to this, 36.1% of the population do little to no physical activity and 11.1% of those surveyed were active smokers [7].

In 2019, 80.7% of all deaths were caused by NTDs, 31.7% of which were caused by cardiovascular diseases [4]. The National Strategy and Operative Plan for a comprehensive approach to chronic non communicable diseases and obesity (2022−2030) states as its first goal a reduction of 17% in premature mortality caused by NTD, and for the first time introduced the promotion of rehabilitation strategies, social reintegration and palliative care of people with NTDs as a strategic line for the comprehensive approach to NTDs [4].

15.1.3 Stroke incidence and prevalence

The only epidemiological description of the stroke population hospitalized in the National Rehabilitation Center studied inpatients from 2008 to 2012. The average patients were 65 years old or older, and their most frequently associated risk factors included hypertension, diabetes, cardiopathy, and alcohol and tobacco intake. Clinically, they presented ischemic strokes of anterior circulation with hemiparesis, severe functional dependence, and cognitive deficits according to the Functional Independence Measure (FIM), with favorable evolution documented as mild dependence through FIM [8].

The incidence of stroke in 2015 was 6.5 per 100,000 people and the premature mortality rate from cerebrovascular disease in 2020 was 14.59 pero 100,000 people [4,9]. As for circulatory system diseases, strokes accounted for 27.2% of total deaths, exceeded only by coronary ischemic cardiopathy. Strokes have a mortality rate of 6.25 per 100,000 among patients 30 and 69 years old [10] which is relevant as it compromises the economically active population.

15.2 Healthcare structure and resources

15.2.1 Rehabilitation structure

Costa Rica has a healthcare system that covers more than 95% of the country's population. The system is strongly universalized and has achieved higher-than-average health indicators when compared to other Organization for Economic Cooperation and Development (OECD) countries. The entity in charge of healthcare is the national socialized provider CCSS. Traditionally, the institution has provided high-end technology for surgical and medical equipment through the acquisition of already-made commercial devices. It has also used technology to improve its management of healthcare information, but despite its achievements, numerous limitations still hinder technological translation and implementation into the healthcare system.

The CCSS runs an ascending complexity healthcare network, ranging from primary care centers to national specialized centers. The CCSS is made up of a set of health centers organized by regions and levels of care, with different degrees of complexity and solving capacity, and articulated vertically and horizontally. Geographically, it is distributed in regions, which are organized into three major networks: south, east, and northwest [4]. Each region has three levels of care. The first level comprises health areas made up of basic healthcare teams and peripheral clinics, who provide comprehensive care programs. The second level of care provides diagnosis support, specialized outpatient consultation, and basic surgical procedures in central clinics and 13 peripheral hospitals (class C) and six regional hospitals (class B). The third level provides high technology and highly specialized services, and it is made up of four national hospitals (class A), six specialized centers and units, and 6 specialized hospitals, among them the National Rehabilitation Center [11].

15.2.1.1 Rehabilitation access

Costa Rica faces complex geographical challenges. The country is relatively small, but it has abundant geographical barriers which restrict the reach of healthcare. There are many hard-to-reach population groups (e.g., Indigenous communities, rural communities in rainforests, and communities on small islands). Therefore complex rehabilitation care is concentrated in the region with the larger cities, and even though rehabilitation is provided in some smaller urban areas, the service there tends to only include physical therapy. Nationwide, 22 centers hold outpatient rehabilitation consultations. Class A facilities and specialized centers are clustered in the metropolitan area, creating a gap in the access of care between the metropolitan areas and the peripheral regions.

Patient transfers, even within the peripheral regions, demand long and expensive trips for many users, making access prohibitive, particularly in stroke and neurological patients where frequent visits are necessary for appropriate treatment. Some rural regions also have limitations in basic services such as electricity and mobile network coverage, which might become a barrier when implementing community-based rehabilitation strategies.

15.2.1.2 Rehabilitation capacity

The number of physical medicine and rehabilitation physicians (henceforth PM&R) varies by hospital; in class A hospitals, it ranges from 2 to 6, depending on hospital size. However, these specialists are not assigned exclusively to neurological patients, but rather to the whole hospital.

The National Rehabilitation Center provides inpatient care through a stroke unit with 18 beds, while also providing outpatient care. The National Geriatric and Gerontological Hospital also has an inpatient 30-bed functional rehabilitation unit, though it is not specific to strokes, and provides outpatient service through a day hospital. Both specialized facilities employ physical, occupational, and speech therapists. Class A and B hospitals provide PM&R services to hospitalized patients through interconsultation. This interconsultation starts very early in the hospitalization process, sometimes even in the emergency room as the patient is admitted, thanks to the fact that doctors in ERs have become familiar with rehabilitation care. General hospitals offer mostly physical therapy services in inpatient and outpatient care, while a few centers have occupational therapy. Patients can be sent to a more complex rehabilitation service if needed.

The waiting time for assessing and receiving rehabilitation therapy can be more than a year, and the intensity of therapy can be less than five sessions for some patients. The number of patients significantly exceeds the public health system's capacity in all its medical facilities, and insufficient infrastructure and a technological gap add to the problem.

In order to comply with the WHO Global Action Plan for the Prevention and Control of Non-communicable Diseases, a Latin American Ministerial Meeting was held in 2018 in Gramado, Brazil. It brought together ministers of health and stroke experts from 13 Latin American countries to discuss the growing burden of strokes in the region and to identify ways of cooperating to reduce the prevalence of strokes. Some agreed-upon actions include increasing the workforce and improving access to adequate stroke care and rehabilitation, both in hospitals and in the community [12].

Costa Rica is also committed to ensure the protection and full and equal enjoyment of all human rights included in the Convention on the Rights of Persons with Disabilities, which includes access to assistive technology as a human right and has called for international cooperation to improve access to it. Costa Rica has also committed to the WHO's six global leadership priorities, including an "increase in the access to high quality, effective and affordable medical products, including support for innovation in affordable health technology, local production, and national regulatory authorities."

15.2.1.3   Stroke rehabilitation standard of care

The National Guide for Stroke Management and Creation of Unified Stroke Units was published in 2010 [13]. Nevertheless, early rehabilitation was not considered part of the guide. Currently, there is no national guide or protocol for the rehabilitation treatment of stroke patients, and there is limited access to quick rehabilitation after discharge [12]. Each hospital defines its method for working with patients. There is no unified timeframe for starting rehabilitation or a defined number of minimum sessions. Access to physical, occupational, and speech therapy is unequal among hospitals. Patients can be referred to a more complex facility with these services, but equipment differs across centers and follow-up can change as well. Finally, regarding scales and outcome measures, the most used are FIM, Barthel Index, and Montreal Cognitive Assessment, but there is no standardized application of these measures among hospitals which makes it difficult to compare treatment outcomes.

The standard care is usually physical therapy delivered through traditional exercise techniques. Usually, 3–4 patients receive therapy at the same time by the same therapist. Depending on the hospital, the number of sessions might range from 6 to 20. The equipment usually involves therapy balls and resistance bands, weights, and work within the parallel bars.

Cognitive rehabilitation is available at the National Rehabilitation Center and the National Geriatric and Gerontology Hospital. Most of it is carried out through specific software such as Gradior [14], a platform developed in Spain, which allows for individualization of treatment goals. Patients might be assigned homework printed on paper sheets, but currently there are no technological solutions being used for work at home.

The program "Opening Roads" in the city of Liberia brought together the Ministry of Science, Technology and Telecommunications, the National Learning Institute, the National Technical University, and CCSS. It offers a computer laboratory for professional instruction in computer use and English to patients with disabilities, aiming to improve employment opportunities.

15.3   Technology-assisted rehabilitation

In Costa Rica, specifically at the Instituto Tecnológico de Costa Rica (also known as TEC or ITCR), there are several engineering programs that have been working in different areas related to rehabilitation devices. There are projects in which multiple departments have collaborated to solve a specific problem, share research assistants, or develop joint graduation projects to solve their needs. The mechatronics, industrial product, and industrial design programs have been involved in the manufacture of rehabilitation devices; each department has its own research laboratory and research assistants.

ITCR and the University of Pennsylvania (UPenn) signed an MoU in 2018 [15] to adapt and deploy rehabilitation devices and other similar robotic rehab devices already developed at UPenn research labs.

15.3.1   Assistive technology

Assistive technology (AT) in Costa Rica is provided through private payment, or through a prescription given by CCSS or the Ministry of Education if the user is enrolled in the education system. Most stroke patients are out of the education system, so their only options are to have their treatments paid for by CCSS or to pay it themselves.

CCSS holds an AT list of authorized subsided products, focusing mainly on mobility which has not been updated in the past 10 years. There is a proposal for an update waiting to be approved which will help broadening the range of AT that can be prescribed by the public system. However, even after the update, the list will not meet the Minimum AT Product List developed by WHO, and it will be particularly lacking in products for cognition and activities of daily living. CCSS pays different suppliers for AT products, and it owns the prosthesis and orthosis laboratory which produces lower-limb prosthesis and has worked to professionalize its technicians.

The PM&R National Technical Committee and WHO are working together through the initiative Global Cooperation on Assistive Technology (GATE). The committee applied the rapid Assistive Technology Assessment survey (rATA) to measure the access, needs, and unmet needs of AT users in Costa Rica. Results are part of the first WHO-UNICEF Global Report on Assistive Technology [16] and included populations from all demographic and geographic segments. Despite the results that show high usage of ATs (68%), there is also a high percentage of users with unmet needs (47%), where an unmet need is understood as the need for a replacement or the need for a different AT. The major barrier to access was the lack of affordability of the AT. The most used ATs are visual aids, followed by mobility aids, and these are predominantly funded through public/government funding. Access to the actual devices can be a challenge: almost 14% of users must travel more than 100 km to obtain their AT [17].

15.3.2   Robots

Currently, the only robotic device available in CCSS is the MOTOMED 2. It has been used by medical and surgical ICU patients, but there are plans for including it in the neuroICU where stroke patients might benefit from its use. The National Insurance Institute, a public healthcare provider in charge of traffic- and work-related accidents, uses *Diego*® and *Pablo*®, upper-limb solutions from Tyromotion Company.

The ITCR laboratories have robotic/mechatronic solutions under development. The first such project is titled "Wireless control system for a robotic vehicle to verify the compliance of Law 7600"; it aims to perform various measurements, automatically, to verify compliance with the 7600 Law, which describes the legal framework for the rights of persons with disabilities in Costa Rica, with an emphasis on accessibility. The system is controlled using a Wii remote controller. It sends signals to the robotic vehicle through an application, where they are interpreted and translated into instructions that allow the movement of the robot, as well as control and data acquisition of various functions. Upon completion of the measurements, the information gathered by the robot is shown as a map which identifies the places that are accessible (i.e., 7600-compliant) [18].

The second robotic development is aimed at providing autonomy of movement for people in need a wheelchair. The project proposes an affordable and simple way to control a wheelchair through electroencephalogram (EEG) signal processing. Preliminary results have shown differences between involuntary and slow voluntary blink signals in the EEG, which drive specific kinds of wheelchair movement [19]. Fig. 15.1 shows the graphic representation for a proof of concept to control a wheelchair.

The third project for robotic rehabilitation entails the development of a haptic interface to measure hand strength and perform rehabilitation exercises for the upper limbs. A functional prototype was built using 3D printing, and it is used along with an existing haptic controller to collect data about performed rehabilitation exercises. The device is equipped with internet access, so it sends its data to an Internet of Things (IoT) platform. The code was developed in Python, and the IoT platform was used for visualization and control, both local and remote, by a rehabilitation professional. Tests

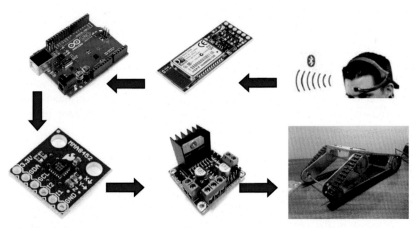

FIGURE 15.1   Graphic representation for a proof of concept to control a wheelchair. *Fig. 15.1 illustrates all required electronic and electrical mechanical devices for a proof of concept to control a wheelchair. First, there is a person using a BrainWave Headset that reads and process electroencephalogram (EEG) signals and sends them through Bluetooth to an electronic printed circuit board (PCB) specially designed to receive Bluetooth signals. The third element is an Arduino board where some filters and algorithms are developed to recognize a slow voluntary blink that sends control signals to a MMA8452 electronic device which is a smart low-power, three-axis, capacitive micromachined accelerometer with 12 bits of resolution. This accelerometer is packed with embedded functions with flexible user programmable options, configurable to two interrupt pins. Embedded interrupt functions allow for overall power savings relieving the host processor from continuously polling data. Finally, a power electronic PCB is responsible for sending power signals to each motor that moves a prototype that represents a possible wheelchair in this proof of concept.*

were carried out to determine the error in the measurements and the data communication between each device. The final device fulfilled all given tasks and is fully functional [20]. Fig. 15.2 shows the hand grip measuring device, whereas Fig. 15.3 shows the graphic interactive interface available to the health professional; the hand grip rehab device has IoT capabilities and is connected to a haptic interface [20].

One unfortunate aspect of these academic projects is that they have yet to be transferred from the laboratories to any clinical sites or communities. There is ongoing work to implement a robotic gym for stroke treatment in one of the national hospitals, but despite being in development for 3 years, the project is still in the early stages, facing numerous technical and bureaucratic barriers. It is being built in an alliance between ITCR and UPenn's GRASP Rehabilitation Robotics Lab. When the project comes to fruition, it will be after the first alliance for technological translation between the CCSS, a public, and a foreign university. As part of the MoU signed between ITCR and UPenn, UPenn's Rehab Robotics Lab has provided the ITCR mechatronics lab with circuit off-the-shelf (COTS) components, motor controllers, microprocessors, servomotors, and other electronic circuits, as well

FIGURE 15.2 First prototype of the hand grip measuring device with Internet of Things capabilities [20]. *Fig. 15.2 shows the first prototype of the hand grip measuring device with IoT capabilities. This device is plastic and has gray color; the front part has a rounded shape that connects to a haptic interface to enable different forces and movements left and right, up and down. In the center, there is an interchangeable spring that allows different forces; finally, there are two plastic bars; and the one on the left has all fingers shapes in order to place a hand around them and start a rehab session.*

as circuit diagrams to replicate robotic devices similar to the ones developed at UPenn's research labs.

A new research proposal started in July 2021 between ITCR, UPenn, and CCSS: the development of a Gym of Mechatronic Devices for Rehabilitation (REHAB-GYM), where multidisciplinary research and development will provide devices for data-driven rehabilitation. This will provide a technology that decreases cost and increases the number of patients that can be treated at the same time by the same professional. It will also contribute to the improvement of the quality of care for people with cerebrovascular events or long hospital stays.

The REHAB-GYM includes several devices working at the same time. The design of mechatronic devices entails two phases: first, the design and

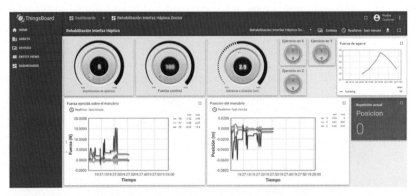

FIGURE 15.3 Graphic interface of hand grip measuring device on the Internet of Things platform [20]. *Fig. 15.3 illustrates a graphic interface of the IoT platform implemented to measure force, quantity of repetitions, distance of movements in X, Y, and Z directions. There are three knobs to set repetitions, force control and distance. In the lower left hand side of the interface, there are two graphics: the first one measures force (N—Newton) vs. time and the next one position (m—meter) versus time. In the upper right corner, there is a graph to show hand grip strength values.*

production of therapeutic devices for the upper limbs. In later stages, it will also include the integration of rehabilitation systems for lower limbs and balance, as well as devices for cognitive rehabilitation. This approach seeks to create an entertaining therapeutic environment that includes social therapy, where patients interact with each other, thereby increasing motivation to complete the given task. Patients will be able to have a more intensive rehabilitation process and enhance their recovery and rehabilitation, thus reducing the severity of their sequels after a cerebrovascular event. The comprehensive equipment responds to multiple requirements, including the track of individualized vital signs, exercise strength, and adjustable forces while using washable materials that are safe for patients and clinical staff.

New prototypes for rehab of upper limbs, specifically for hand strength and grip functionality, have been developed. The first one is a device that measures the angular position of the wrist in the four main directions of movement of this joint. This mechatronic design incorporates the use of a cloth glove, adjustable to various hand sizes, in which flexion sensors are placed, and the reading of these sensors is interpreted through a microcontroller-based control system using a computer, which generates a report of the data. This system has an automated solution for the measurement process and is capable of autonomously measuring the angle at which the wrist is positioned while also saving this data for further analysis. The aim of this device is to reduce the time spent in therapy sessions, and measure and report the patient's progress.

As part of REHAB-GYM research project, a second prototype is under development. This is a low-cost device specifically developed for the rehabilitation of the hand grip function. Five different elastic exercise bands are used to enable levels of resistance for hand grip function as the patient closes and opens their hand. The device uses easily accessible items such as materials available at hardware stores. In addition, it incorporates a micro-controller that allows recording the data of the movement and force made by the patient, for later analysis by a health professional. A user interface for exercise monitoring and a functional prototype were developed and are being upgraded for better performance.

15.4   Challenges

One of the main barriers to technological implementation in Costa Rica is transcending the research level and reaching actual implementation. There are many projects that are never put into service because of weak integration between the clinical and technological sites. The main role of CCSS is as a care provider, and despite aiming for an institutional strategic plan that prior-itizes research, there is a lack of assignment of the proper time for research, resources, incentives, and support. Research on devices and technological devices is even more limited, as there is little relevant experience in bioeth-ics committees. A cultural change is needed to reconceptualize research as a complement to clinical care, necessary for quality improvement.

There is an urgency in increasing clinical staff and engineers' skills to promote a multidisciplinary approach for adequate decision-making in terms of rehabilitation technology selection, adaptation, implementation, and con-tinuous evaluation. The opportunities for collaboration could allow for more local product development, reducing costs, allowing for better coupling to the Costa Rican context, and modifying the trend of sole acquisition of com-mercial devices.

The COVID-19 pandemic has emphasized the need for removing bureau-cratic barriers to biomedical investigation, including shortening the authori-zation processes, encouraging public—private collaborations, and strengthening the legal framework in areas such as intellectual property. It also requires controls on the judicialization of health: In the current system, patients can sue the CCSS so that it is forced to provide them with treat-ments, interventions, or procedures which are not always supported by tech-nical or medical criteria [21].

15.5   Future opportunities

The COVID-19 pandemic has generated growing interest in rehabilitation because of the need for ICU and post-COVID rehabilitation. The implemen-tation of telerehabilitation has also become urgent; this opportunity is further enhanced by the widespread access to smartphones. Nevertheless, the CCSS does not have an institutional platform for telerehabilitation, and it is therefore dependent on clinicians' usage of personal resources to perform rehabilitation tasks.

New barriers have emerged, including economic limitations and growing inequalities in educational access among users, which might complicate appointments and the quality of the care provided. These barriers could be counteracted with new digital solutions and mobile applications designed to simplify tasks for people suffering from cognitive disorders and to improve adherence to cardiac rehabilitation programs, among others. These would reduce the time and resources invested in getting patients to the medical center and would provide real-time information on the fulfillment of rehabilitation objectives. Additionally, health education for family members of rehabilitation patients can be delivered through these devices.

The supply chain crisis of late 2021 has also become an opportunity to produce solutions locally (e.g., the manufacture of locally developed artificial breathing machines). The Mechatronics Engineering Department at ITCR is planning to offer a new academic masters program in bioengineering, focusing on trends such as IoT, eye tracking, and voice-activated devices. There are also opportunities to improve upon older rehab devices, making low-cost upgrades to give them connectivity to the internet or voice activation capabilities.

In 2018, the National Survey on Disability was published by the National Institute of Statistics and Census, providing valuable epidemiological data. This data, together with the rATA survey, will allow for a better understanding of the needs regarding AT, better resource allocation, and research development.

In the context of social security, the needs involve compact, affordable solutions that are capable of performing in urban and rural sites. The goal of Costa Rica's CCSS is to provide equal access to healthcare throughout the territory, and rehabilitation robotics could be a strategy to fill the existing gaps. Its large coverage and data acquisition capabilities provide an ideal head start for research.

Acknowledgments

We would like to thank Dr. Rolando Coto-Solano and Dr. Samantha Wray for their contribution to previous versions of this work.

Conflict of interest statements

None of the authors presents a conflict of interest.

References

[1]   INEC. Costa Rica en cifras 2021. San José, Costa Rica; 2021.

[2] The World Bank. The World Bank in Costa Rica [Internet]. Available from: https://www.worldbank.org/en/country/costarica/overview#1; 2021.

[3] OECD. OECD Economic Surveys: Costa Rica 2020 [Internet]. Paris: OECD. Available from: https://www.oecd-ilibrary.org/economics/oecd-economic-surveys-costa-rica-2020_2e0fea6c-en; 2020. 1−140 p.

[4] Salas Peraza D, Garita Castro A, Méndez Briceño R, Claramunt Garro M, Fernández Esquivel M, Jiménez AA, et al. Estrategia Nacional de Abordaje Integral de las Enfermedades No Transmisibles y Obesidad 2022-2030. 2022;1−156.

[5] Ávila Agüero ML. La problemática de la Caja Costarricense del Seguro Social desde la óptica de los determinantes de la salud. Acta Med Costarric 2020;55(3):139−42.

[6] OPS/CCSS/INCIENSA/Ministerio de Salud Costa Rica. Encuesta de Diabetes Hipertensión y Enfermedades Crónicas en Costa Rica. San José, Costa Rica; 2004.

[7] Caja Costarricense del Seguro Social. Vigilancia de los Factores de Riesgo Cardiovascular, Tercera encuesta 2018. CCSS 2021:2020−p.

[8] Vargas Quesada C. Caracterización epidemiológica de las personas con secuelas de evento cerebrovascular, hospitalizadas en Centro Nacional de Rehabilitación, de enero 2008 a diciembre 2012. [Internet]. Universidad de Costa Rica. Available from: http://repositorio.sibdi.ucr.ac.cr:8080/xmlui/handle/123456789/4216; 2013.

[9] Ministerio de Salud Costa Rica. Boletín Estadístico de Enfermedades de Declaración Obligatoria en Costa Rica del año 2015 [Internet]. San José, Costa Rica. Available from: https://www.ministeriodesalud.go.cr/index.php/vigilancia-de-la-salud/estadisticas-y-bases-de-datos/notificacion-individual/3167-boletin-de-morbilidad-enfermedades-de-declaracion-obligatoria-2015-2/file; 2015.

[10] Evans R, Pérez J, Bonilla R. Análisis de la mortalidad por enfermedades cerebrovasculares en Costa Rica entre los años 1920−2009. Arch Cardiol México 2016;86(4):358−66.

[11] Santacruz J, García R, López M, Picado K, Ramírez A. Perfil del sistema de servicios de salud de Costa Rica. Organ Panam la Salud, Minist Salud [Internet]. Available from: http://www.cor.ops-oms.org; 2004;1−21.

[12] Ouriques Martins SC, Sacks C, Hacke W, Brainin M, de Assis Figueiredo F, Marques Pontes-Neto O, et al. Priorities to reduce the burden of stroke in Latin American countries. Lancet Neurol 2019;18(7):674−83.

[13] Fernández H, Carazo K, Henríquez F, Montero M, Valverde A. Guía Nacional de Manejo del Evento Cerebrovascular y Creación de Unidades de Ictus Unificadas. San: José, Costa Rica; 2010.

[14] Gradior. Gradior [Internet]. Available from: https://www.gradior.es/; 2021.

[15] Tecnológico de Costa Rica, University of Pennsylvania. Convenio marco de colaboración entre el Instituto Tecnológico de Costa Rica y la Universidad de Pensilvania [Internet]. San José, Costa Rica. Available from: https://www.tec.ac.cr/sites/default/files/media/doc/agreements/universidad_de_pensilvania.pdf; 2018. p. 1−4.

[16] WHO, UNICEF. Global report on assistive technology. 2022. 108 p.

[17] Coto-Solano B. Measuring Access to Assitive Technology in the Public Rehabilitation Outpatient Setting in Costa Rica Using the WHO Rapid Assistive Technology Assessment (rATA) Questionnaire. In: ICCHP-AAATE 2022 Open Access Compendium "Assistive Technology, Accessibility and (e) Inclusion." 2022. p. 292.

[18] Esquivel YA, Batista AC. Wireless control system for a robotic vehicle to verify the compliance of the law 7600. In: 2014 International Work Conference on Bio-Inspired Intelligence: Intelligent Systems for Biodiversity Conservation, IWOBI 2014 - Proceedings. 2014;162−7.

[19] Carrasquilla-Batista A, Quiros-Espinoza K, Gomez-Carrasquilla C. An Internet of Things (IoT) application to control a wheelchair through EEG signal processing. In: 2017 International Symposium on Wearable Robotics and Rehabilitation (WeRob) [Internet]. IEEE; 2017. p. 1–1. Available from: https://ieeexplore.ieee.org/document/8383877/.

[20] Valverde-Arredondo V, Carrasquilla-Batista A. Haptic system for upper limb rehabilitation with hand grip strength measurements and Internet of Things capabilities. In: 2020 IEEE 11th Latin American Symposium on Circuits & Systems (LASCAS) [Internet]. IEEE; 2020. p. 1–4. Available from: https://ieeexplore.ieee.org/document/9069030/.

[21] Aguilar-Cubillo M, Calvo-Herra A, Monge-Navarro A, Vega-Araya A. Judicialización del derecho a la salud en Costa Rica desde una perspectiva de salud pública. Universidad de Costa Rica; 2013.

Chapter 16

North America and Caribbean region: Colombia

Carlos A. Cifuentes[1,2], Angie Pino[3], Andrea Garzón[4] and Marcela Múnera[1,3]

[1]Bristol Robotics Laboratory, University of the West of England, Bristol, United Kingdom,
[2]School of Engineering, Science and Technology, Universidad del Rosario, Bogotá, Colombia,
[3]Biomedical Engineering Department, Colombian School of Engineering Julio Garavito,
Bogotá, Colombia, [4]School of Medicine and Health Sciences, Universidad del Rosario,
Bogotá, Colombia

Learning objectives

At the end of the chapter, the reader will be able to:

- Understand Colombia's political, socioeconomic, and health context and status.
- Identify the current rehabilitation approaches and barriers following a stroke event.
- Recognize the country's advances in the use of technological and robotic solutions for effective rehabilitation.

16.1 Introduction: Colombia overview

Colombia is a constitutional republic located in northwestern South America, where it shares land borders with Venezuela, Brazil, Peru, Ecuador, and Panama. The country is the fourth-largest continent and the most populous Spanish-speaking nation, with 51 million inhabitants. Nearly 80.8% of the population live in urban areas, of which more than one-third of its people (16.2% of the total) live in Bogotá, the national capital, while 19.2% live in rural areas [1]. Colombia is the only South American nation with access to the Pacific Ocean and the Atlantic Ocean through the Caribbean Sea. Moreover, it has significant natural resources, biodiversity, and oil reserves and is a significant producer of gold, silver, emeralds, platinum, and coal.

Rehabilitation Robots for Neurorehabilitation in High-, Low-, and Middle-Income Countries.
DOI: https://doi.org/10.1016/B978-0-323-91931-9.00016-5

For this reason, and the attractiveness it offers to foreign investment, the country has led an economy based on merchandise exports, where coffee and oil stand out. In this way, following Brazil, Mexico, and Argentina, Colombia has the fourth-largest economy in Latin America and is currently one of the most stable economies. However, the nation's political instability has been historically tied to the unequal distribution of wealth and decades of prolonged violent conflict [2].

Corruption has been a practice that, over the years, has prevented the correct and complete investment of money by the state. This social and economic inequity has led it to be currently considered the second most unequal country in Latin America and the seventh in the world, with increased monetary poverty. Specifically, since 2018 Colombia's poverty has reached 28% of the population, where Chocó remains the poorest province in the country, covering 68.4% of the people with scarce resources [2,3].

In terms of health, by 2021, the resources allocated by the Colombian government were $36.05 billion, with an increase of 16.6% compared to 2020, due to the COVID-19 public health and medical emergency [4]. The financing of these resources is centralized, providing Colombians access to health services through the contributory and subsidized health insurance scheme regulated by the *Sistema de Seguridad Social en Salud* (SGSSS). The contributory regime (CR) is financed based on earmarked payroll taxes (i.e., 12.5% of workers' income). It includes workers earning the equivalent of one minimum wage or more, with the capacity to pay. The subsidized regime (SR) benefits poor residents, the unemployed, or those who cannot pay. In this case, government funding covers the primary benefits package's cost. Finally, private insurance has population coverage for those who can afford it [5]. Thus, as projected by National Administrative Department of Statistics (DANE) until September 2021, 47.57% of the population are affiliated with the CR, 46.97% with the SR, and 4.4% to the special or private regime, for a total coverage of 98.94% [6]. On this basis, the country has been generally lauded for providing a near-universal coverage system.

With so broad a scope, this chapter provides essential fundamentals to understand the Colombian health and disability situation, stroke statistics, and the standard of care for people with disabilities. Moreover, it is also intended to describe the background in neurorehabilitation, current attitudes and barriers to effective rehabilitation, and robotics penetration in rehabilitation systems into the country as solutions and advances for stroke therapy.

16.2 Status of public health and disability in Colombia

According to the Health Situation Analysis Report, until 2018, the crude mortality rate in the Colombian territory has ranged from 4.25 to 4.75 deaths per 1000 persons. Cardiovascular diseases, followed by stroke and respiratory diseases, are the leading causes of death, with a higher frequency in men than women. Heart attacks have grown by 11.5% between 2019 and 2020, and stroke remains the leading cause of disability in the country, with an occurrence of 45,000 Colombians per year (i.e., 2.6% of the population) [7].

By August 2020, about 1.3 million people in Colombia had some kind of disability, of which 70.3% belonged to the SR and 29.6% to the CR. Moreover, most people with disabilities are older adults (39%), with a reported stroke prevalence between 1.4 and 19.9x100,000 inhabitants in Colombia [8]. After a stroke, functional loss of the patient is one of the significant consequences affecting independence and quality of life (QoL). Therefore early care with rehabilitation has been considered one of the most critical factors in minimizing mortality and the incidence of complications and improving functionality in the in-hospital and out-of-hospital management of the disease. Information on this disease's direct and indirect costs is scarce and is not updated. However, data from 2008 estimate that the total expenses for a stroke in Colombia would be 116,000 dollars per year [9]. The high cost of treating the disease limits the most vulnerable from accessing quality service. However, some Colombian institutions such as the *Centro Especializado de Recepción* (CER), *Centro Nacional de Rehabilitación Teletón* (CNR), and *Asociación de Discapacitados y Parapléjicos* (ASODISPAR) support people with disabilities by promoting actions and giving subsidies that facilitate their social inclusion [10].

16.2.1   Standard care for poststroke patients

Early medical intervention for stroke includes reperfusion of brain tissue and neuroprotection treatments. Consequently, there are already care units in some hospitals or clinics with medical and support staff for the acute care of these patients. Indeed, Colombia has the third-best provision for acute stroke care in Latin America, following Brazil and Chile, with eight centers across the country with stroke care programs. Institutions such as the *Fundación Santa Fe de Bogotá* (Bogotá, Colombia), with its Stroke Clinical Care Center, are certified by the Joint Commission International for complying with protocols that respond to worldwide standards of care. Those stroke services include acute care, neuroimaging diagnosis, and treatment for more than 90 days after the stroke [11,12].

Another primary strategy after an emergency stroke episode, in terms of cost-effectiveness, is endovenous drug therapy included by the Colombian healthcare system in the SGSS benefit plan [13]. Also, early rehabilitation, mainly physiotherapy, is one of the most used strategies to treat neurological disorders like stroke. According to the World Health Organization (WHO), physical rehabilitation (PR) is an active process performed to achieve full recovery or at least reach optimal physical, intellectual, and social capabilities to integrate into society [14] appropriately. In contrast, neurorehabilitation is an educational and dynamic process based on adapting the individual and his environment to the neurological impairment, intending to minimize the impact of the disease to achieve the best QoL [15]. Conventional treatment is given by active training strategies involving mental and executive function work, systemic motor control theory approaches, task-oriented activities, stretching exercises, and ecological approaches that favor motor control and learning processes [9]. All the different therapeutic and conventional methods are guided by the physiotherapist's expertise and the intensity of repetitive exercises to improve patient performance. Low-cost technologies such as exercise bikes, ellipticals, and body weight support treadmill practices are standard rehabilitation techniques that promote functional locomotor recovery at an early stage of gait rehabilitation [16].

Conversely, nonconventional treatment includes support and high assistive technology such as virtual reality, robotic therapies, and isokinetic robotic dynamometry, among others [9]. Specifically, it is possible to find technology with mechanotherapy equipment for gait and balance training systems (e.g., Lokomat, THERA-Trainer, Contrex, Balance Trainer), upper extremity training systems (e.g., Armeo Power, Amadeo), and biofeedback strategies such as Pablo Plus for mobility and strength training of the shoulder, elbow, and wrist. In addition, there is physical media equipment with therapeutic purposes for muscle and nerve stimulation, magnetotherapy, thermotherapy, hydrotherapy, electrotherapy, laser therapy, cryotherapy, and ultrasound equipment [17]. Despite the wide variety of devices and services, there are currently few rehabilitation centers with these technologies in the country; some of them are the Mobility Group and *Clinica Universidad de La Sabana*, located in Bogotá. Due to the cost required to access their services, less than 1% of Colombians have the opportunity to participate in the benefits of its treatment [18].

16.2.2 Barriers

In Colombia, different factors act as barriers to addressing stroke and accessing rehabilitation services for mainly low-income patients, including:

- Physical accessibility: Although the Ministry of Health has established standards for physical accessibility, the infrastructure of the country's main cities is a cause for concern since most of them are not designed for people with disabilities. Short sidewalks and no ramps limit easy movement in outdoor spaces such as schools, universities, restaurants, movie theaters, sports facilities, etc. Moreover, the lack of elevators, small bathrooms not being adjusted, and inaccessible public transportation, among other aspects, reduce independence and the ability to participate in social life. The above reflects the lack of training of professionals in the areas of architecture and engineering regarding disability.

- Medical attention: It is estimated that about 38% of Colombian citizens do not recognize the symptoms of a stroke, and only 40% of cases go to an emergency hotline for prompt attention [12]. The lack of knowledge and identification of this type of event is the basis of the incidence of the disease over the years in the country. Likewise, the health system by itself has always presented delays in the authorization and assignment of services, significantly affecting the continuity of care. People with disabilities, rehabilitation professionals, and other stakeholders have expressed the negative impact of not having access to a wheelchair, an appropriate orthotic or prosthetic device that meets the user's needs, or training on using them properly [15]. In addition, a research paper demonstrated the absence of cognitive and behavioral rehabilitation services for stroke patients [19]. Thus, it is common to see people with disabilities and their caregivers using legal recourses to access timely and quality health services since those are often denied or delayed without getting reasons. It is common to find people need to pay out-of-pocket for additional health and rehabilitation services when the service is not competent.
- Cardiac history: People with heart disease or poor circulation due to narrowing arteries have the highest stroke risk. Added to this, a sedentary lifestyle, an unhealthy diet, and high blood pressure are factors that can be treated with a cardiac rehabilitation (CR) program to reduce the likelihood of a first or recurrent stroke. However, 24% and 30% of the population enrolling in CR programs drop out. The above signifies a high risk that may cause blood clots leading to stroke [20].
- Socioeconomic accessibility: Looking beyond the resources consolidated by the state for healthcare purposes, insufficient economic capacities of vulnerable groups prevent them from receiving full access to rehabilitation cycles or completing them successfully. It is estimated that of the 2.6% population with disabilities, 81% belong to the most vulnerable strata who are either unemployed (64% of the total) or at minimum wage to cover only basic needs (21% of the total). This way, about 59% do not attend any rehabilitation program [15].
- Lack of adherence: Around 42% of the patients deserted PR programs, generally due to anxiety, boredom, and low motivation during exercise. Currently, there is a lack of definition in the objectives of the rehabilitation processes, which is why many users remain in these programs without a clear goal to work toward. These findings are based on the perception of rehabilitation professionals [15].

16.3 Developments in technological solutions for stroke rehabilitation

Developing countries and middle-income countries like Colombia have few active, intelligent rehabilitation technology advances. Some of the emerging technological development has been supported by academic institutions and research groups to promote patient recovery, support clinical abilities, and address the country's direct problems of accessibility and coverage. The following are some developments made within the framework of the lower extremity (LE), upper extremity (UE), and social robotics technology.

16.3.1 Lower-extremity technology

As is well known, loss of motor control on one side of the body (hemiparesis) is one of the side effects of a stroke. As a result, people are often constrained to execute different daily life activities. Among these, gait parameters (i.e., cadence, step length, velocity, etc.) are constantly limited due to the inability to generate a voluntary effort in LE muscles. Thus, lower-limb exoskeletons emerged as technological tools that, from rehabilitation, aim to promote patients' recovery of walking activity, stimulate brain plasticity, and even support healthcare professionals in rehabilitating limbs during long trajectories. These solutions have been designed based on human biomechanics to assist basic body movements through control strategies, sensor measurements, and actuation systems [21].

The AGoRA exoskeleton is a unilateral, lower-limb, rigid structure developed to rehabilitate and assist the gait activity. The device comprises two active joints (hip and knee flexion and extension movements) and one passive joint (hip joint abduction and adduction movements) [22]. Moreover, it includes the physical human−robot interaction (pHRI) through a sensory interface measuring kinetic and kinematic parameters to identify the human motion intention with strain gauges and magnetic encoders, respectively. The actuation technique is performed with DC electric motors to generate torque in two modes: an assistive mode (i.e., generate desired gait patterns) and a transparency mode (i.e., unrestricted movements). A pilot test with a healthy participant during a straight-line walking activity demonstrated the device's feasibility in generating a range of torque for a complete gait cycle [21].

On the other hand, T-FLEX is a high-tech robotic device based on variable stiffness and bioinspired tendons to rehabilitate patients who have lost mobility in their ankles due to stroke spasticity. The device is a development of a branch of robotics that tries to emulate the human body's performance. It is a low-cost, open-design invention that works for repetitive flexion and extension of ankle movement in therapy and gait assistance. A case study with a stroke survivor showed improvement in the participant's spatiotemporal and kinematic parameters in a rehabilitation scenario using the therapy mode [23]. Now, in terms of gait assistance, the device has been electromechanically characterized in a test bench structure to determine the best setup with a tendons-alone configuration, where 10 N was established as the initial force to have the maximum performance [24]. Likewise, its applicability has improved ankle kinematics during poststroke gait assistance [25].

In particular, T-FLEX has been developed in conjunction with novel neurorehabilitation strategies in the therapy modality. For example, serious games are feedback approaches that solve demotivation problems in long-term therapies and access to motor and cognitive learning through developing specific tasks in an interactive environment. In this area, Jumping Guy: Ankle Rehabilitation Therapy with T-FLEX was used to involve patients in controlling the action of the orthosis to generate the jump of an avatar and evade enemies with dorsi-plantarflexion movements of the ankle. For this purpose, the system detects the movement intention based on a threshold algorithm using the angular velocity along the sagittal plane of an inertial sensor located in the paretic foot tip. Thus, when the movement intention is detected, the participant receives audiovisual feedback through the avatar's jump and kinesthetic feedback in terms of the device activation. The usability of this technique has been evaluated in a healthy subject, demonstrating its functionality at the level of graphics and orthosis activation. Furthermore, it was supposed to be a potentially usable aid in helping patients to focus on therapy goals, including cognition, motor recovery, and enjoyment [26].

Another example is the brain−computer interface (BCI) technology, which measures and processes neurological activity to produce control signals that translate into real-world interactions. The brain-controlled rehabilitation with T-FLEX considers the EEG-motor imagery (MI) technique over the specific detection of an increase in the beta-band frequency after the imagination of movement without actually performing the movement. The strategy was accessed in five stroke patients with visual and haptic stimuli to assist MI generation and control the ankle exoskeleton in an experimental context. The patient's performance favored the haptic stimulus in the MI accuracy rate. However, the electroencephalographic analysis during active periods did not generate differential activity over the affected hemisphere. These results are the basis for refining robotic solutions' design that promotes early recovery [27].

The assistance mode of both devices, developed by the *Escuela Colombiana de Ingeniería Julio Garavito* (Bogota, Colombia), is based on the real-time measurement of the sagittal angular velocity and linear acceleration signals of an inertial measurement unit (IMU) placed on the subject's foot instep for gait phase detection and the corresponding action of the devices. Several approaches have developed computational methods comprising rule-based algorithms and machine-learning approaches throughout the literature. A hidden Markov model (HMM) algorithm was implemented to perform gait phase detection, whose performance was evaluated in typical and pathological gait patterns. Moreover, it was contrasted with FSR-based values located in an insole (i.e., hallux, the first and fifth metatarsophalangeal, and the heel) and compared in an intra-subject procedure (i.e., subject-specific training) and an intersubject procedure (i.e., standardized parameter training). The gait state represented in four phases, heel strike (HS), flat foot

(FF), heel off (HO), and swing phase (SP), had a superior accuracy value of close to 80% in both intra-subject techniques with better precision in FF and SP gait phases. The results of this research represented the basis for evaluating gait variability in patients and controlling lower-limb robotic devices for motor rehabilitation [28]. These developments in LE technology are current opportunities to accelerate the rehabilitation process, improve adherence to therapy programs, and enhance cognitive skills. Particularly, T-FLEX, a low-cost robot, may offer future opportunities for low-income people to regain ankle function with comprehensive and quality strategies.

16.3.2 Upper-extremity technology

Upper-extremity dysfunction after a stroke is one of the most limiting consequences in developing activities of daily living (ADLs) to reach, grasp, and manipulate objects. The poststroke motor recovery aims to reduce impairment and facilitate performance and fine manipulation [29]. To address this, several robot-assisted technologies have been designed. One of them is an assist-as-needed robotic exoskeleton designed for hand rehabilitation to allow a sponsored system in terms of myoelectric and motion signals. The mechanical structure includes a single active degree of freedom (DOF) in each finger and three mechanical connections for the phalanges. The technology uses muscle activity and machine learning models to characterize muscle effort. Experimental results of the *Pontificia Universidad Javeriana* research group (Bogotá, Colombia) showed the functional control architecture to modulate the exoskeleton movements with EMG signals acquired from stroke patients [30].

On the other hand, the GIS Software Research Group from the *Universidad Pedagógica y Tecnológica de Colombia UPTC* (Tunja, Colombia) made an architectural proposal to provide adaptable solutions for UE in rural and urban environments. The device includes biosignal and motion real-time monitoring to execute planned movements and provide information to the clinician. According to the author, even though functional tests have not yet been conducted, the architectural proposal is supposed to provide different technology modules to support traditional rehabilitation processes of underprivileged communities around the various territories of the country [31].

On the other hand, the *Escuela Colombiana de Ingeniería Julio Garavito* (Bogota, Colombia) designed a hand exoskeleton based on soft robotics for various grasp taxonomies. Soft robotics is robotic actuator technology developed from flexible materials (e.g., polymers, elastomers, etc.) that seek to approach the human body's complex mechanical properties and movement. In this case, the hand exoskeleton is pneumatically actuated to perform each finger's selective flexion and extension. A preliminary evaluation of the device through the adapted Jebsen−Taylor Hand Function Test (JTHFT) showed all the objects' successful hand grasping. Still, a longer time is required than the condition in which the exoskeleton is not used. Thus, the experiment improved the device to decrease gripping times to potentially test it in poststroke patients for future rehabilitation [32].

Alternatively, Colombians have also designed a wearable sensor technology based on measuring the kinematic and muscle activity for poststroke upper-limb rehabilitation. Although the technology has not yet been tested on pathological subjects, functional tests show the device's reliability for recording limb movement [29]. This makes it a potential device that, like other strategies, allows tracking motor recovery, providing biofeedback [33], and even supporting bioinspired control for robotic rehabilitation [29]. Finally, as demonstrated, the contribution of each of these UE strategies goes beyond the exploration of the technology. It is a tool in development that seeks to enhance, in the future, the proper management of the disease and, thus, the QoL of citizens survivors of a stroke.

16.3.3   Socially assistive robotics

Socially assistive robots (SARs) emerged in therapy and healthcare scenarios to support clinical tasks and encourage patients to improve their self-performance through physical, cognitive, and social support. This human−robot interaction can develop several roles (i.e., companion, partner, coach, instructor, or assistive), offering neurorehabilitation services, social interaction, and constant monitoring. These interactive strategies combined with assistive robotics are potential approaches to provide feedback during physical assistance and recover the motor function of patients with neurological disorders [34].

A short-term study evaluated a social human−robot interaction system to support conventional therapies with Lokomat. The interaction system integrated a set of sensors to monitor patients' performance (i.e., cardiovascular parameters, spinal posture parameters, and cognitive parameters) and provided positive and mostly verbal feedback to motivate the patient through the NAO robotic platform. The experimental study setup included four patients under gait rehabilitation due to stroke, spinal cord injuries, and Guillain−Barré syndrome. Each patient performed two unique sessions throughout the test using the Lokomat device (one control and one assisted by the social robot) under the same device configurations and conditions of time, treadmill speed, body weight support percentage, and distance covered. The results showed the usability and correct interface operation. Moreover, the reduction of bad posture demonstrated the positive effect of the social robot interaction with the patients along with the session, being well received by users and therapists. Finally, the patient's perception of the robot's usefulness in the therapy was positive, and they agreed with the benefits of its integration as a complement to gait rehabilitation programs [16].

A clinical evaluation of the social robotic platform was performed with ten patients in Lokomat gait rehabilitation during 15 sessions to understand the long-term effects. The results validated the improvements in physiological progress and a positive acceptance of social interaction by maintaining patients' healthy postures and promoting full gait rehabilitation during the sessions. Additionally, clinicians showed trust in the system and benefit from the robot support by supporting neurorehabilitation, continuous patient monitoring, and maintaining the social distaining required because of the COVID-19 pandemic [35].

A perception study evaluated 88 Spanish and Colombian clinicians' and patients' perceptions of a social robot integrated into Lokomat therapy. Both groups of participants positively perceived the social robot in physical therapy scenarios [36]. Regarding CR, the incorporation of SAR systems demonstrated its usability by encouraging six patients to perform physical activity and improving their adherence during a treadmill-based exercise between 3 and 6 months. The strategy allowed continuous patient monitoring and engagement through the NAO robot's sensory system and feedback techniques [37].

Similarly, a 2.5-year study in the *Fundación Cardioinfantil-Instituto de Cardiología* clinic (Bogotá, Colombia) compared the patients' progress and adherence to the conventional cardiac rehabilitation program against rehabilitation supported by the socially assistive robot. Results reported significantly better heart rate recovery during the robot condition, improving cardiovascular functioning and recovery, and reducing the risk of suffering recurrent events. Furthermore, once again, the strategy's effectiveness in maintaining interaction with the robot and motivation and adherence to the therapy, even in such long sessions, was demonstrated from the perspective of both clinicians and patients [20]. This way, socially assistive robotic tools favor patient−therapy connection and, in terms of CR programs, decrease the risk of stroke in cardiovascular patients.

16.4 Conclusions

This chapter initially presented the political, socioeconomic, and health context of a developing country like Colombia. From there, it has been possible to understand social inequality, corruption, and poverty as social determinants of population health. Thus, despite the great coverage by affiliation to the SGSS, there are still discrepancies that affect the quality of the health and rehabilitation services provided in the country.

Colombia still does not have enough centers to provide immediate property care regarding stroke treatment. The standard solutions to treat this disease are essential and focus on conventional strategies such as intravenous treatment and therapy guided by physiotherapists. High-tech strategies are scarce, and accessing them requires a high amount of money, which only very few people can afford. Therefore physical and socioeconomic accessibility barriers, poor quality of medical care, poor adherence to rehabilitation treatments, and cardiac histories at risk for stroke may be evident.

Different universities and research groups have been working to provide solutions for the most vulnerable people in this context. From robotic assistive technology, low-cost upper- and lower-limb devices with novel technology (i.e., biofeedback, serious games, and BCI strategies) have been developed to favor patients' adherence and motor recovery. In conventional therapy, significant short- and long-term advances have also been demonstrated using social robotics to support the patient and the clinician. To conclude, most of this work is still under active development. However, the gradual rapprochement of the different technological strategies with the population has been significantly positive. In the medium-term future, these solutions are expected to positively impact society by improving the accessibility of users and decreasing the prevalence of this disabling disease.

Acknowledgments

The authors would like to thank the members of the Center for Biomechatronics for supporting this research.

Conflict of interest

The authors declare no conflict of interest.

References

[1] Guerrero R, Gallego AI, Becerril-Montekio V, Vásquez J. Sistema de salud de Colombia. Salud Publica Mex 2011;53(Suppl. 2):s144−55.

[2] Berry A. Aspectos jurídicos, políticos y económicos de la tragedia de la Colombia rural de las últimas décadas hipótesis para el análisis (Traducción). Estud Socio-Juridicos 2014;16 (1):8−24.

[3] Poverty and inequality. Colombia Reports [Internet]. Available from: https://colombiareports.com/colombia-poverty-inequality-statistics/ (accessed on 12 November 2021).

[4] $36,05 billones para la salud - Aprobado el Presupuesto General de la Nación - CONSULTORSALUD [Internet]. Available from: https://consultorsalud.com/aprobado-el-presupuesto-general-de-la-nacion/ (accessed on 12 November 2021).

[5] Kanavos P, Parkin GC, Kamphuis B, Gill J. Latin America Healthcare System Overview A Comparative Analysis of Fiscal Space in Healthcare. London School of Economics and Political Science, 2019.

[6] Páginas - Cifras de aseguramiento en salud [Internet]. Available from: https://www.minsalud.gov.co/proteccionsocial/Paginas/cifras-aseguramiento-salud.aspx (accessed on 12 November 2021).

[7] Dirección de Epidemiología y Demografía. Análisis de Situación de Salud (ASIS) Dirección de Epidemiología y Demografía. 2019.

[8] Ministerio de Salud. Instituto Nacional de Salud, Observatorio Nacional de Salud. Carga de enfermedad por enfermedades crónicas no transmisibles y discapacidad en Colombia. Obs Nac Salud. 2015;1−212.

[9] Hernández Álvarez E, Torres Narváez MR. Guía De Práctica Clínica Fisioterapéutica Para La Evaluación Y Tratamiento De Pacientes Con Enfermedades Cerebrovasculares En Los Primeros Seis Meses De La Enfermedad. Asoc Colomb Fisioter; 2021.

[10] Guillermo L, Suárez M, Yoana L, Garnica Z, Marina L, Daniel J, et al. Capítulo II Capacidad Económica De La Población En Condición De Discapacidad Motora En Colombia. Available from: https://orcid.org/0000-0001-8948-9188 (accessed on 14 November 2021).

[11] Bayona-Ortiz H, Useche JN, Yanez N, Velasco SC. Availability of stroke units in Colombia. Lancet Neurol 2019;18(11):988. Available from: http://doi.org/10.1016/S1474-4422(19)30332-1.

[12] Bogotá tiene la mayor prevalencia de ACV en Colombia [Internet]. Available from: https://www.elhospital.com/temas/Bogota-es-la-ciudad-colombiana-con-mayor-prevalen-cia-de-ataque-cerebrovascular + 127097 (accessed on 30 November 2021).

[13] Moreno E, Rodríguez J, Bayona-Ortiz H, Moreno E, Rodríguez J, Bayona-Ortiz H. Trombólisis endovenosa como tratamiento del ACV isquémico agudo en Colombia: una revisión sistemática de la literatura. Acta Neurológica Colomb 2019;35(3):156−66. Available from: http://www.scielo.org.co/scielo.php?script = sci_arttext&pid = S0120-87482019000300156&lng = en&nrm = iso&tlng = es.

[14] Bergen DC. Neurological disorders: public health challenges. Arch Neurol 2008;65 (1):154.

[15] Toro-Hernández ML, Mondragón-Barrera A, Múnera-Orozco S, Villa-Torres L, Camelo-Castillo W. Experiences with rehabilitation and impact on community participation among adults with physical disability in Colombia: perspectives from stakeholders using a community based research approach Int J Equity Health [Internet] 2019;18(1):18Jun 3 [cited 2021 Nov 30]. Available from: https://pubmed.ncbi.nlm.nih.gov/31155006/.

[16] Céspedes N, Múnera M, Gómez C, Cifuentes CA. Social human-robot interaction for gait rehabilitation. IEEE Trans Neural Syst Rehabil Eng 2020;28(6):1299−307. Available from: https://doi.org/10.1109/TNSRE.2020.2987428.

[17] Equipos. Mobility Group [Internet]. Available from: http://www.mobilitygroup.co/mobil-ity/page/equipos/ (accessed on 2 December 2021).

[18] Munera M, Marroquin A, Jimenez L, Lara JS, Gomez C, Rodriguez S, et al. Lokomat therapy in Colombia: current state and cognitive aspects IEEE Int Conf Rehabil Robot 2017;2017:394−9(Figure 1). Available from: https://doi.org/10.1109/ICORR.2017.8009279 (Figure 1).

[19] Olabarrieta-Landa L, Pugh M, Calderón Chaguala A, Perrin PB, Arango-Lasprilla JC. Trajectories of memory, language, and visuoperceptual problems in people with stroke during the first year and controls in Colombia. Disabil Rehabil 2019;43(3):324−30. Available from: https://www.tandfonline.com/doi/abs/10.1080/09638288.2019.1622799 (accessed on 2 December 2021).

[20] Céspedes N, Irfan B, Senft E, Cifuentes CA, Gutierrez LF, Rincon-Roncancio M, et al. A socially assistive robot for long-term cardiac rehabilitation in the real world. Front Neurorobot 2021;15:1−19. Available from: https://doi.org/10.3389/fnbot.2021.633248.

[21] Arciniegas MLJ, Múnera M, Cifuentes C. Human-in-the-loop control for AGoRA unilateral lower-limb exoskeleton. J Intell Robot Syst 2021;104(3). Available from: https://doi.org/10.1007/s10846-021-01487-y.

[22] Sanchez M, Gomez D, Casas D, Munera M, Cifuentes CA. Development of a Robotic Lower-Limb Exoskeleton for Gait Rehabilitation: AGoRA Exoskeleton. Colombia: IEEE ANDESCON. Santiago de Cali; 2018.

[23] Gomez D, Pinto M, Ballén F, Munera M, Cifuentes CA. Therapy with T-FLEX Ankle-exoskeleton for motor recovery: a case study with a stroke survivor, In: Proceedings 8th IEEE RAS/EMBS International Conference on Biomedical Robotics and Biomechatronics (BIOROB), 2020, NY, USA. Available from: https://doi.org/10.1109/BioRob49111.2020.9224277.

[24] Gomez-Vargas D, Ballen-Moreno F, Rodriguez-Guerrero C, Munera M, Cifuentes CA. Experimental characterization of the T-FLEX ankle exoskeleton for gait assistance. Mechatronics 2021;78:102608. Available from: https://doi.org/10.1016/j.mechatronics.2021.102608.

[25] Gomez-Vargas D, Ballen-Moreno F, Barria P, Aguilar R, Azorín J, Munera M, et al. The actuation system of the ankle exoskeleton T-FLEX: first use experimental validation in people with stroke. Brain Sci. 2021;11:412. Available from: https://doi.org/10.3390/brainsci11040412.

[26] Pino A, Gomez-Vargas D, Múnera M, Cifuentes CA. Visual feedback strategy based on serious games for therapy with T-FLEX ankle exoskeleton. Biosyst Biorobotics 2020;27:467−72. Available from: https://link.springer.com/chapter/10.1007/978-3-030-69547-7_75.

[27] Barria P, Pino A, Tovar N, Gomez-Vargas D, Baleta K, Díaz CAR, et al. BCI-based control for ankle exoskeleton T-FLEX: comparison of visual and haptic stimuli with stroke survivors. Sensors. 2021;21(19):1−18. Available from: https://doi.org/10.3390/s21196431.

[28] Manchola MDS, Bernal MJP, Munera M, Cifuentes CA. Gait phase detection for lower-limb exoskeletons using foot motion data from a single inertial measurement unit in hemiparetic individuals. Sensors 2019;19(13):2988. Available from: https://doi.org/10.3390/s19132988.

[29] Cifuentes CA, Braidot A, Rodriguez L, Frisoli M, Santiago A, Frizera A. Development of a wearable zigbee sensor system for upper limb rehabilitation robotics, In: Proceedings of 4th IEEE RAS & EMBS International Conference on Biomedical Robotics and Biomechatronics (BioRob). 2012, Rome, Italy. IEEE. Available from: https://doi.org/10.1109/BioRob.2012.6290926.

[30] Castiblanco JC, Mondragon IF, Alvarado-Rojas C, Colorado JD. Assist-as-needed exoskeleton for hand joint rehabilitation based on muscle effort detection. Sensors 2021;21(13):4372. Available from: https://www.mdpi.com/1424-8220/21/13/4372/htm.

[31] Velez-Guerrero MA, Callejas-Cuervo M. Data acquisition and control architecture for intelligent robotic exoskeletons in rehabilitation, In: 7th E-Health Bioeng Conf EHB, 2019. Available from: https://www.researchgate.net/publication/338949681_Data_Acquisition_and_Control_Architecture_for_Intelligent_Robotic_Exoskeletons_in_Rehabilitation. 2019 (accessed on 30 November 2021).

[32] Maldonado-Mejía JC, Múnera M, Diaz C, Wurdemann H, Moazen M, Pontes M, et al. A fabric-based soft hand exoskeleton for assistance: the ExHand Exoskeleton. Front Neurorobot 2023;17. Available from: https://doi.org/10.3389/fnbot.2023.1091827.

[33] Pinto-Bernal MJ, Cifuentes CA, Perdomo O, Rincón-Roncancio M, Múnera M. A Data-Driven approach to physical fatigue management using wearable sensors to classify four diagnostic fatigue states. Sensors 2021;21(19):6401. Available from: https://www.mdpi.com/1424-8220/21/19/6401/htm.

[34] Cifuentes CA, Pinto MJ, Céspedes N, Múnera M. Social robots in therapy and care. Curr Robot Rep 2020;1(3):59–74. Available from: https://doi.org/10.1007/s43154-020-00009-2.

[35] Céspedes N, Raigoso D, Múnera M, Cifuentes CA. Long-term social human-robot interaction for neurorehabilitation: robots as a tool to support gait therapy in the pandemic. Front Neurorobot 2021;15:1–12. Available from: https://doi.org/10.3389/fnbot.2021.612034.

[36] Raigoso D, Céspedes N, Cifuentes CA, Del-Ama AJ, Múnera M. A survey on socially assistive robotics: Clinicians' and patients' perception of a social robot within gait rehabilitation therapies. Brain Sci 2021;11(6):1–13. Available from: https://doi.org/10.3390/brainsci11060738.

[37] Casas J, Senft E, Gutiérrez LF, Rincón-Rocancio M, Múnera M, Belpaeme T, et al. Social assistive robots: assessing the impact of a training assistant robot in cardiac rehabilitation. Int J Soc Robot 2021;13(6):1189–203. Available from: https://doi.org/10.1007/s12369-020-00708-y.

Chapter 17

North America and Caribbean region: Ecuador

Gabriel Iturralde-Duenas[1] and Esteban Ortiz-Prado[2]
[1]University of Texas at Austin, Mechanical Engineering Department, Austin, TX, United States,
[2]Universidad de las Americas, Facultad de Medicina, De los Colimes, Quito, Ecuador

Learning objectives

At the end of this chapter, the reader will be able to:
1. Describe the neuroepidemiological profile of stroke in Ecuador, how its healthcare system is organized, the current status of rehabilitation for neurological impairments, and progress in rehabilitation engineering and robotics.
2. Understand the facilitators and barriers to implementing efficient and good-quality rehabilitation engineering techniques in Ecuador.

17.1 Neuroepidemiological analysis of stroke behavior in Ecuador

Stroke is one of the leading causes of death globally. In 2019, there were more than 12.2 million stroke cases, 143 million people with disability-adjusted life years, and at least 6.5 million deaths due to stroke globally [1]. Stroke is the second leading cause of death in most Latin American countries, with a regional rate of 41 deaths per 100,000 inhabitants, compared to 21.9 in North America (the United States and Canada), according to data from the Pan American Health Organization (PAHO). In Ecuador, this pathology has been one of the leading causes of death since 1975, when it ranked ninth; in 1990 it was number one; and by 2020, it ranked fourth in the list of the most common causes of death in men, with at least 2350 deaths associated with stroke in men and 2034 in women [2]. Despite the lack of historical data on stroke in Ecuador, recent years have brought to light some reports describing the epidemiology of both cerebral infarction and subarachnoid hemorrhage in Ecuador [3,4].

Rehabilitation Robots for Neurorehabilitation in High-, Low-, and Middle-Income Countries.
DOI: https://doi.org/10.1016/B978-0-323-91931-9.00010-4

Data from the last 10 years of available data show that 56,896 hospital admissions due to stroke were reported in Ecuador (CIE-10 I60, I61, I63, I64). Men accounted for 52.8% (n = 30,053) with an incidence rate of 33.8 per 100,000, while women accounted for 47.2% (n = 26,843) representing 29.7 per 100,000 (Fig. 17.1).

Using data from the National Institute of Statistics and Census (INEC), the overall sex-specific adjusted incidence rate for stroke was 36.73 per 100,000 for men and 30.78 per 100,000 for women from 2011 to 2020. The mean annual incidence for women was 32.47 cases per 100,000 as opposed to 37.08 cases per 100,000 for men (Table 17.1).

The annual mortality rate ranged from 12.82 to 20 per 100,000 for those who were at risk for stroke between 2011 and 2020. The overall distribution of case rates and the case fatality rates varied significantly by age group. For instance, the younger populations (0−49 years of age) accounted over 19.11% of the total number of cases registered; nevertheless, their average mortality was under 2.98%.

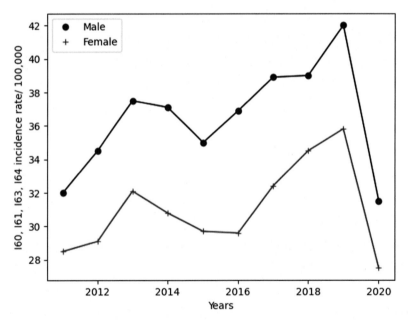

FIGURE 17.1 The adjusted incidence rate of stroke collected from 2011 to 2020 using data from the National Institute of Statistics and Census (INEC) is shown. The mean across this decade is 36.73 per 100,000 men and 30.78 per 100,000 women across all the Ecuadorian territory. The increasing trend stops in 2019 when several improvements in the healthcare system in the admission of patients and awareness campaigns started. This trend has reversed since 2019. These numbers can be affected by the COVID-19 pandemic.

TABLE 17.1 Total number of cases, incidence rates/100,000 persons, age-specific mortality rate/100,000 from 2011 to 2020 according to ranges of age (babies, young child, grown child, teenagers, young adults, adults, third age, above 80 years old).

Age	Cases— women	Incidence rate of women per 100,000	Mortality rate of women per 100,000	Cases— men	Incidence rate of men per 100,000	Mortality rate of men per 100,000
<1	75	5.71 (5.06—6.35)	2.55 (2.7—2.93)	96	6.99 (6.09—7.88)	2.19 (1.94—2.43)
1—9	181	1.25 (1.34—1.53)	0.37 (0.33—0.43)	195	1.27 (1.15—1.38)	0.40 (0.31—0.51)
10—14	138	1.73 (1.64—1.81)	0.61 (0.57—0.65)	160	1.92 (1.86—1.98)	0.38 (0.36—0.41)
15—19	254	3.35 (3.24—3.47)	1.01 (0.94—1.08)	433	5.53 (5.41—5.66)	1.34 (1.26—1.41)
20—39	1,974	8.98 (3.99—12.21)	2.27 (1.02—3.20)	2,732	11.58 (8.58—15.82)	3.26 (2.01—4.75)
40—59	5,967	42.78 (17.44—63.04)	14.78 (5.33—22.73)	6,930	57.37 (21.19—88.0)	19.72 (7.6—30.27)
60—79	11,013	179.52 (87.19—293.17)	81.43 (33.02—488.54)	13,103	226.47 (121.86—358.29)	97.95 (43.3—179.46)
>80	7,191	563.59 (547.54—579.64)	469.79 (451.05—488.54)	6,404	587.57 (571.89—603.26)	432.5 (414.03—450.98)
Total	26,843	32.56 (29.02—36.1)	16.66 (13.66—19.7)	30,053	37.16 (33.69—40.6)	17.75 (15.06—20.5)

Children under 4 years of age had a cumulative percentage of 0.329% cases for boys and 0.297% for girls; nonetheless, mortality was low for both groups (0.695%). Despite there being cases reported among the elderly (80.89%), their mortality accounted for over 120.89% of all nontraumatic subarachnoid hemorrhage (I60), nontraumatic intracerebral hemorrhage (I61), cerebral infarction (I63), and cerebrovascular accident (I64) cases, which resulted in death.

In Ecuador, the trend for each of the provinces varies in relation to the sites of presentation. For example, in our analysis, the provinces with the highest stroke rates were Galapagos with a rate of 120 cases per 100,000 inhabitants, followed by Carchi with 71.2 per 100,000 inhabitants. In relation to the provinces with the lowest incidence, we have Pichincha with 28 cases per 100,000 inhabitants followed by Cotopaxi with 30.52 cases per 100,000.

17.2 Public health policies implemented by the Ecuadorian government (including and with emphasis on the new government and how COVID-19 has influenced them)

Stroke is a multidimensional problem that requires public policy approaches to try to reduce its epidemiological impact. In a country like Ecuador, the implementation of a prevention and health promotion policy aimed at reducing stroke morbidity and mortality requires coordination among all the stakeholders of the national health system. In a country with a fragmented health system such as Ecuador, the implementation of any health plan becomes problematic. Among the main barriers are the shortage of medical specialists in the country and the lack of comparable quality care in all health subsystems in Ecuador. The country would benefit from a national project for the prevention and treatment of stroke.

For example, Brazil can provide us with the context of experience in the development of a stroke treatment unit which, in addition to being cost-effective, has been shown to decrease the risk of stroke [5]. In turn, a national program that focuses on reducing stroke risk factors and early recognition of symptoms or signs would work. For example, programs designed to reduce obesity and sedentary lifestyle, or to reduce smoking as well as the proper management and control of diseases such as atherosclerosis or hypertension, would result in a significant reduction in ischemic events in the country [6].

In 2020 and 2021, most of the projects were halted due to the pandemic, priority care for these groups was reduced, and preventive care was considerably limited. Against this background, it is necessary to strengthen the current healthcare capacities to address the critical situation.

17.3 Cardiovascular risk factors associated with stroke in the coastal, altitude, and Amazon regions

Stroke risk factors are often classified as traditional and nontraditional, and from this, they can be modifiable and not modifiable [7]. These risk factors have been classified as the most common causes of an ischemic or hemorrhagic stroke (Table 17.2).

On the other hand, several factors have been associated with increased risk of developing stroke; although their association is weaker and less studied, they need to be explored.

One of the most singular and debatable nontraditional risk factors is hypobaric hypoxia as a risk factor for those living in high-altitude environments or those who travel to mountainous regions and expose themselves to severe hypoxia [9−11]. Although very little information is available, it is presumed that exposing unacclimated humans to such conditions might trigger thrombosis, therefore stroke [12].

17.4 Rehabilitation analysis of stroke in Ecuador

The Ecuador healthcare system is strongly limited by the lack of epidemiological studies, clinical practice guidelines (CPGs), and evidence-based practice (EBP) for neurological impairments. Recent efforts have reduced the incidence of stroke by implementing protocols for patient admission and the adoption of prevention policies by the Ministry of Public Health [8,13]. From the authors knowledge, there is only one clinical practice guideline officially published by the Ministry of Public Health related to physiotherapy for the treatment of lumbar pain [14]. In addition, a study conducted in the public physiotherapy departments in Ecuador in the provinces of Chimborazo, Cotopaxi, Pastaza, and Tungurahua [15] showed the lack of awareness of EBP and inadequate training on current and state-of-the-art therapy practices and tools. There are also very few good-quality studies in the native language. This is consistent with related works on this topic [16−18].

In Ecuador, however, there is high access to rehabilitation services in terms of frequency and cost affordability. Physical, occupational, and speech therapy can be provided on average for 2 hours sessions twice a week for a range of prices between $10 and $30 per hour. Furthermore, insurance companies typically cover between 60% and 70% of all costs, and the Institute of Social Security provides free access to those services but with lower frequency. The current rate of health benefits for public medical systems established by the Ministry of Public Health of Ecuador is $23 per hour (without private or governmental insurance) for physical, occupational, or speech therapy, and this is the recommended price for private institutions [19].

TABLE 17.2 Traditional modifiable and nonmodifiable risk factors (Modified from [8]).

Traditional risk factors

		Nonmodifiable Age Sex Ethnicity Socioeconomic status Family history	Modifiable Waist circumference
Ischemic stroke	Thrombotic	Arterial dissection	Alcohol misuse
			Obesity (body mass index, waist circunference, waist-to-hip ratio)
			Diabetes
			Cigarette smoking
	Embolic	Atrial fibrillation	
		Intracardiac thrombus	
		Heart valve disease	Physical inactivity
		Trauma and fractures	Apolipoprotein B to A1
		Some types of surgeries	Hyperlipidemia
Hemorrhagic stroke	Systemic	Post-traumatic hypovolemia	
		Acute systemic hypoxia	
	ICH17[a]	Vascular malformations	Hypertension
		Bleeding diatheses	Cigarette smoking
		Trauma	Obesity Waist-to-hip ratio
	SAH[b]	Cocaine abuse	Diet
		Amphetamines	

[a]*Intracerebral hemorrhage.*
[b]*Subarachnoid hemorrhage.*

A summary of the therapy services costs for public institutions and the suggested for the private sector is shown in Table 17.3.

In addition, the rehabilitation program for a patient after stroke depends on the ease of access to high-quality healthcare services and prompt admission to a hospital. After the acute stroke is under control and the patient is assessed by a neurologist, a customized rehabilitation plan is developed by a team of multidisciplinary physicians and therapists [13]. This team commonly includes a neurologist, traumatologist, rehabilitation doctor, physiatrist, psychologist, physiotherapist, occupational therapist, and speech therapist [20]. There are however barriers and disparities to accessing these programs mainly because there is a shortage of specialists, a lack of training and understanding of the treatment of neurological pathologies, the lack of facilities in medical subsystems in Ecuador (this includes all the health centers located far from Ecuador's main cities), the ability of the patient to pay, and the enrollment in private or public insurance [21]. These obstacles could

TABLE 17.3 Prize summary of rehabilitation medicine and rehabilitation services in Ecuador according to the Ministry of Public Health updated in September 2021.

No.	Service	Relative price ($)[a]	Actual price ($)
1	Electromyography	2.12	63.60
2	Test de Lambert	2.13	63.90
3	H or F reflex	2.13	63.90
4	Evoked potentials (visual, auditive, or somatosensory)	5.76	172.80
5	Transcutaneous electric stimulation	0.44	13.20
6	Biofeedback	0.68	20.40
7	*Physical therapy session*	0.77	23.10
8	*Occupational therapy session*	0.77	23.10
9	*Speech therapy session*	0.77	23.10
10	*Early stimulation session*	0.77	23.10
11	*Respiratory therapy session*	0.77	23.10

Actual prize: This is the prize for a public hospital or clinic that has a 24/7 service; thus 30 days are considered for this cost.
[a]*Relative prize: This is the prize normalized by the number of calendar days of a month that the hospital, clinic, or health center operates.*
Modified from Minesterio de Salud Pública del Ecuador. Norma técnica sustitutiva de relacionamiento para la prestación de servicios de salud entre instituciones de la red pública integral de salud y de la red privada complementara, y su reconocimiento económico.

result in inefficient rehabilitation processes for patients with stroke that might be based only on therapist knowledge and skills without requirements of adherence to clinical criteria or guidelines.

17.5 Neurorehabilitation engineering research and development

The neurorehabilitation engineering field is in its infancy in Ecuador. There are few efforts to introduce low-cost and simple rehabilitation robotic techniques or devices for poststroke recovery. The main developments are at the research level and have not been clinically tested, or tested with limited healthy and stroke subjects. The studies performed can be categorized into:

- Myoelectric prosthesis for the upper and lower body.
- Interactive software and apps.
- Exoskeletons for the upper and lower body.

The major contributors to this research are:

- Universidad Politecnica Salesiana de Cuenca (two publications).
- Universidad Politecnica de Ambato (two publications).
- Escuela Politecnica de Chimborazo (one publication).
- Universidad de las Americas (one publication).
- Universidad Espiritu Santo (one publication).
- Universidad de las Fuerzas Armadas (one publication).
- Politecnica Nacional (one publication).

Table 17.4 shows a review of the above-mentioned literature according to its category.

As it can be observed, there are only nine articles that have been published in high-impact scientific journals (Elsevier, IEEE, Frontiers), and two of them have been tested in children with cerebral palsy. There is no study to our knowledge that has been applied in poststroke patients. However, the number of publications has increased in the last 3 years with which 67% of these papers correspond to a published date after 2018.

Furthermore, there is no patented product from Ecuador that has been successfully introduced in the neurorehabilitation industry. There are two patents that have been approved by the Ecuadorian Nacional Services of Intellectual Property that are "HandEyes" [32] and "SpeakLiz Vision" [33] that both have been awarded first and second place, respectively, in the "One Idea to Change History" contest organized by History Channel. HandEyes was developed by a group of students from Universidad de las Fuerzas Armadas (ESPE) and is intended to provide assistance for individuals with visual impairment by providing alerts via sounds and vibrations using proximity sensors and to extend mobility in a safe and confident manner to deaf people by providing sound and alerts with geo-localization. By the end of

TABLE 17.4 Review of rehabilitation engineering research prototypes in Ecuador in journals of high impact.

Category	Rehabilitation system	Reference	Type of therapy
Prosthesis	Biomimetical arm	[22]	Physical upper body
	Biomechanical right hand	[23]	Physical upper body
	Upper-limb prosthesis control through geolocation application	[24]	Physical upper limb
	Robotic low-cost knee	[25]	Physical lower limb
	EMG-controlled fore-arm bionic prosthesis	[26]	Physical upper limb
	Hip prosthesis controlled by web app	[17]	Physical hip
	Arm—hand prosthesis controlled by brain—machine interfaces	[27]	Physical arm
Interactive software and apps	Telerehabilitation web app based on Kinect	[28]	Physical whole body exercises
	Hand rehabilitation therapy system for soft robotic glove	[29]	Physical hand
Exoskeletons	Six-axis lower-limb exoskeleton controlled by neural networks	[30]	Physical lower limb
	Upper-limb robotic exo	[31]	Physical upper limb

2021, more than 1000 patients have benefited from it with a functional mobility recovery in Ecuador [32]. "SpeakLiz Vision" is a mobile app developed by *TALOV*, an Ecuadorian startup, that uses a sign language recognition algorithm based on machine learning techniques and translates this language of communication in text phrases for more than 35 idioms and has a tool that creates a voice to speak the written message. It recognizes male and female adults, kids' speech, and emergency vehicles and alarms. It is available for iOS, Android, and Windows 10, 11 platforms.

The main reasons for the small number of research articles and patents (translated to market products) are the lack of professors specialized in rehabilitation engineering, neurorehabilitation, and rehabilitation robotics in Ecuador. Furthermore, there is no graduate degree in this area offered by an

Ecuadorian university. Currently, there are just two Masters in Science programs focused on robotics in general; however, they do not include any graduate or undergraduate courses related to biomedical robotics. In the case of industry, there are strong barriers to how patents are processed, analyzed, and approved by the Ecuadorian National Services of Intellectual Property that include high cost, long waiting times, a lack of intellectual property departments in law firms, and a small number of attorneys specializing in this area.

17.6 Challenges and roadway for stroke neurorehabilitation

The engineering neurorehabilitation industry in Ecuador is strongly limited as explained in the previous section. There are efforts to introduce low-cost and simple rehabilitation robotic techniques or devices for the recovery of poststroke patients and in general for neurological impairments. They are at the research level and have not yet been clinically tested or brought to market; however, there is an increasing trend of progress in this area from the academia and industry side. We have explained in the previous section the main limitations of the Ecuadorian healthcare system. Here are some other two main obstacles that affect the effective introduction of rehabilitation engineering research and market in Ecuadorian society:

- Rehabilitation begins for many patients at the chronic stage. There is not much action done in the acute and subacute phases which is when greater functional recovery can occur.
- Illiteracy is still an issue in Ecuador that has a 92.8% of alphabetization rate index (UNESCO).

Therefore the roadway to implement an effective neurorehabilitation engineering and robotics program for the recovery of poststroke patients should focus on the implementation of a prevention and health promotion policy aimed to reduce stroke morbidity and mortality that requires coordination among all the stakeholders of the national health system. A national project for the prevention and treatment of stroke will strongly benefit the country. This plan includes campaigns for stroke prevention and awareness for the population, creating an updated epidemiological status of this disease and creating incentives for increasing the capacity of medical specialists in neurology, traumatology, rehabilitation medicine, and therapists. Another important consideration is a government investment in healthcare facilities to admit and treat patients with stroke in the country's health subsystems.

On the neurorehabilitation engineering side, there are three important players to effectively implement techniques for the treatment of neurological impairments. One of them is academia in which public and private universities must play a key role in providing more training on evidence-based practice to therapists that include a collaboration between rehabilitation therapy departments, engineering schools, and medicine schools. This joint effort will contribute to an effective implementation of rehabilitation engineering and robotics techniques in rehabilitation. Furthermore, graduate programs should be offered to build expertise in this area including neuroscience, neurorehabilitation, rehabilitation robotics, biomedical engineering, robotics, and biomechanical engineering. For this, a hiring process of professors with a master's in science, doctoral, or postdoctoral degree must be implemented to offer these degrees. Universities and the Ministry of Education should implement a strong national program to improve fluency in English and to increase the rate of literacy across the country.

Furthermore, the Ministry of Public Health is the second entity that will play a key role in the above-mentioned creation of a national plan for the prevention and treatment of stroke. Lastly, in the industry area the National Services of Intellectual Property should encourage the creation of patents for rehabilitation engineering inventions that can effectively be translated into a product in the market. Attracting international investments from developed countries is key to reaching this goal. Here, the Ministry of Economy is the key player in the private sector. Hence, the pathway is to create low-cost solutions that can create an impact on the functional recovery of poststroke patients through high healthcare access to these services.

As it can be observed, this is a complex national project for which there are three main players and include the contributions of public and private entities in which academia must play a key role. This plan might take several years, requires sound planning, and requires high public and private investment. Ecuador has two advantages over developed countries that lower the risk factors associated with stroke including the quality of daily living and a geography that makes it unique from other countries in the region. The cost and access to rehabilitation therapy sessions are ideal for the implementation of low-cost products and techniques to benefit a larger proportion of the population. Furthermore, Latin American countries have a lower risk of stroke than developed countries according to the World Health Organization, and efforts to create national plans for the prevention and treatment of stroke and similar diseases have strongly contributed to a reduction in risk factors and mortality of these diseases. Moreover, Ecuador has three regions (the coastal, mountain, and Amazon regions) in a relatively small area that can be beneficial for reducing the risk of stroke (i.e., altitude regions and Amazon) and for treatment.

There are two successful cases in which integral rehabilitation engineering studies have been implemented in Ecuador. One was implemented in an orphanage in the city of Latacunga in Ecuador with patients with cerebral palsy [16]. This was performed by a multidisciplinary group of students from Boston that included three physical therapy students, three speech therapists, and two students of engineering. The study showed that incorporating knowledge from the three areas, 3D printers to develop a low-technology communication button, and three iPad tablets and the GoTalk NOW app had positive outcomes. This study provided an effective means of communication for the children who received them. In addition, another interesting case was performed by a group of researchers from Universidad Politecnica Salesiana de Cuenca that proposed a methodological intervention that used robotic assistants, simulations environments (for practitioners and students), and uncertain reasoning controllers [18]. The preliminary outcomes of the study showed that both practitioners and students benefited from this methodology, and it contributed to their learning curve to consider new technologies in rehabilitation.

17.7 Concluding remarks

To our knowledge, this is the first work that includes an updated neuroepidemiological behavior of stroke in Ecuador, the full review of the policies implemented by the Ecuadorian government in the last 20 years to prevent and treat stroke patients, a review of cardiovascular risks associated with stroke in each of the three Ecuadorian regions, and an analysis of the current situation of rehabilitation therapy in Ecuador with an emphasis on neurological impairments. Moreover, we provided a literature review of the progress of the neurorehabilitation engineering and robotics research and industry in Ecuador, and we finally proposed a roadway for effectively introducing neurorehabilitation for stroke patients in Ecuador.

Ecuador is a promising case for providing high access and good-quality rehabilitation engineering and robotics for stroke patients. The low cost and accessibility of therapists can increase the dosage of rehabilitation in patients. Furthermore, as most of the Latin American countries, Ecuador has lower risk factors for stroke, and there is evidence that national plans to prevent and treat neurological diseases have strongly decreased the impact and morbidity of stroke. Ecuador's geography provides a unique place for stroke prevention and poststroke rehabilitation. There are three main players that in a joint effort can successfully implement this national plan: the Ecuadorian Ministry of Public Health, the Ministry of Economy in close connection to the private sector, and the academia.

Conflict of interest

There is no conflict of interest for any of the chapter's authors.

References

[1] Feigin VL. Global, regional, and national burden of stroke and its risk factors, 1990–2019: a systematic analysis for the Global Burden of Disease Study 2019. Lancet Neurol 2021;20 (10):1–26.
[2] Registro Civil Ecuador. Cifras de defunciones correspondientes a los años 2018, 2019 y 2020. Quito; 2020 April.

[3] Ortiz-Prado E, Espinosa PS, Borrero A, Cordovez SP, Vasconez JE, Barreto-Grimales A, et al. Stroke-related mortality at different altitudes: a 17-year nationwide population-based analysis from Ecuador. Front Physiol 2021;12:733928.

[4] Daniel Moreno-Zambrano C, Moreno-Zambrano D, Santamaría D, Ludeña C, Barco A, Vásquez D, et al. Artículo Original. 25, Revista Ecuatoriana de Neurología 17 Rev. Ecuat. Neurol. 2016.

[5] Martins SCO, Pontes-Neto OM, Alves CV, de Freitas GR, Filho JO, Tosta ED, et al. Past, present, and future of stroke in middle-income countries: The Brazilian experience. Int J Stroke 2013;8(100A):106−11.

[6] di Legge S, Koch G, Diomedi M, Stanzione P, Sallustio F. Stroke prevention: managing modifiable risk factors. Stroke Res Treat 2012;2012:391538.

[7] Bridgwood B, Lager KE, Mistri AK, Khunti K, Wilson AD, Modi P. Interventions for improving modifiable risk factor control in the secondary prevention of stroke Vol. Cochrane database of systematic reviews, 2018. John Wiley and Sons Ltd.; 2018.

[8] Ortiz-Prado E, Cordovez SP, Vasconez E, Viscor G, Roderick P. Chronic high-altitude exposure and the epidemiology of ischaemic stroke: a systematic review. BMJ Open 2022;12(4):e051777.

[9] Pilz S, Dobnig H, Fischer JE, Wellnitz B, Seelhorst U, Boehm BO, et al. Low vitamin D levels predict stroke in patients referred to coronary angiography. Stroke. 2008;39(9) 2611−13.

[10] Niaz A, Nayyar S. Cerebrovascular stroke at high altitude. J Coll Physicians Surg Pak 2003;13(8):446−8.

[11] Jha SK, Anand AC, Sharma V, Kumar N, Adya CM. Stroke at high altitude: Indian experience. High Alt Med Biol 2002;3(1):21−7 Mar 1.

[12] Zavanone C, Panebianco M, Yger M, Borden A, Restivo D, Angelini C, et al. Cerebral venous thrombosis at high altitude: a systematic review. Rev Neurol 2017;173(4):189−93.

[13] Matamoros et al. Retardo en llegada de pacientes con infarto esquemico hospital terciario Ecuador; 2017.

[14] Ministerio de Salud Pública del Ecuador. Guía Práctica Clínica para tratamiento de dolor lumbar; 2017.

[15] Cobo-Sevilla V, de Oliveira-Ferreira I, Moposita-Baño L, Paredes-Sánchez V, Ramos-Guevara J. Evidence-based physiotherapy clinical practice in the public health-care service in Ecuador. Physiother Res Int 2019;24(1):e1745 Jan 1.

[16] Hayward LM, Li L. Promoting and assessing cultural competence, professional identity, and advocacy in doctor of physical therapy (DPT) degree students within a community of practice. J Phys Ther Educ 2014;28(1):23−36.

[17] Rybarczyk Y, Cointe C, Gonçalves T, Minhoto V, Deters JK, Villarreal S, et al. On the use of natural user interfaces in physical rehabilitation: a web-based application for patients with hip prosthesis. J Sci Technol Arts 2018;10(2 Special Issue):15−24.

[18] Suquilanda-Cuesta P, Uguna-Uguna V, Pinos-Chuya B, Perez-Munoz A, Robles-Bykbaev V. Motor rehabilitation of children with multiple disabilities: a methodological proposal based on robotic assistants, simulation and uncertain reasoning. In: Proceedings of the 2021 IEEE Conference of Russian Young Researchers in Electrical and Electronic Engineering, ElConRus 2021. Institute of Electrical and Electronics Engineers Inc.; 2021. p. 1852−6.

[19] Ministerio de Salud Pública del Ecuador. Norma técnica sustitutiva de relacionamiento para la prestación de servicios de salud entre instituciones de la red pública integral de salud y de la red privada complementara, y su reconocimiento económico.

[20] del Brutto OH, Tettamanti D, del Brutto VJ, Zambrano M, Montalván M. Living alone and cardiovascular health status in residents of a rural village of coastal Ecuador (The Atahualpa Project). Environ Health Prev Med 2013;18(5):422−5 Sep.

[21] del Brutto OH, Peñaherrera E, Ochoa E, Santamaría M, Zambrano M, del Brutto VJ. Door-to-door survey of cardiovascular health, stroke, and ischemic heart disease in rural coastal Ecuador - the Atahualpa Project: methodology and operational definitions. Int J Stroke 2014;9(3):367−71.

[22] Calle-Siguencia J, Proaño-Guevara D. Design, simulation, and construction of a prototype transhumeral bio-mechatronic prosthesis. In: Information and Communication Technologies: 8th Conference, TICEC 2020, Guayaquil, Ecuador, November 25−27, 2020, Proceedings; 2020. p. 104.

[23] Cajamarca LF, Matute J, Calle J, Yunga F, Vargas J, Urgiles F. Design, development and implementation of a biomechanical right-hand prosthesis: second stage. In: 2017 IEEE Global Humanitarian Technology Conference (GHTC). 2017. p. 1−6.

[24] Llerena-Izquierdo J, Barberan-Vizueta M, Chela-Criollo J. Novus spem, 3D printing of upper limb prosthesis and geolocation mobile application. Rev Ibérica de Sist e Tecnologias de Informação 2020;E33:127−40.

[25] Valencia F, Ortiz D, Ojeda D. Design and testing of low-cost knee prosthesis. In: 2017 IEEE Second Ecuador Technical Chapters Meeting (ETCM); 2017. p. 1−6.

[26] Vergaray RA, del Aguila RF, Avellaneda GA, Palomares R, Cornejo J, Cornejo-Aguilar JA. Mechatronic system design and development of iROD: EMG controlled bionic prosthesis for middle-third forearm amputee. In: 2021 IEEE Fifth Ecuador Technical Chapters Meeting (ETCM); 2021. p. 1−5.

[27] Borja-Galeas C, Guevara C, Arias-Flores H, Fierro-Saltos W, Rivera R, Hidalgo-Guijarro J, et al. Control of an Arm-Hand Prosthesis by Mental Commands and Blinking. In: Human Systems Engineering and Design II: Proceedings of the 2nd International Conference on Human Systems Engineering and Design (IHSED2019): Future Trends and Applications, September 16−18, 2019, Universität der Bundeswehr München, Munich, Germany. 2019. p. 154.

[28] Palacios-Navarro G, Garcia-Magariño I, Ramos-Lorente P. A Kinect-based system for lower limb rehabilitation in Parkinson's disease patients: a pilot study. J Med Syst 2015;39(9):1−10.

[29] 2020. (Advances in Intelligent Systems and Computing; vol. 1066). Available from: Available from: http://link.springer.com/10.1007/978-3-030-32022-5.

[30] Institute of Electrical and Electronics Engineers. 2018 IEEE Third Ecuador Technical Chapters Meeting (ETCM) : Cuenca, Ecuador, October 15−19, 2018.

[31] Arteaga O, Argüello ME, Terán HC, Chacon S, Navas R, Lamingo A, et al. Design of a robotic exoskeleton force multiplier for upper limb. Int J Mech Eng Robot Res 2020;9 (1):80−6 Jan 1.

[32] Ekos R. Libro de actores del desarrollo sostenible − 20 años de la iniciativa 'Un Global Compact', Quito, Ecuador, August 18, 2019.

[33] Talov MC. SpeakLiz by Talov: Toward a Sign Language Recognition mobile application.

Chapter 18

Europe region: Serbia

Ljubica M. Konstantinovic[1,2], Andrej M. Savic[3],
Aleksandra S. Vidakovic[1,2], Olivera C. Djordjevic[1,2] and
Sindi Z. Mitrovic[1,2]

[1]*Faculty of Medicine, University of Belgrade, Belgrade, Serbia,* [2]*Clinic for Rehabilitation "Dr Miroslav Zotović", Belgrade, Serbia,* [3]*School of Electrical Engineering, Science and Research Centre, University of Belgrade, Belgrade, Serbia*

Objectives

At the end of this chapter, the reader will be able to
1. Understand the organization and capacity of the rehabilitation service in Serbia.
2. Understand the efforts to maintain the modern concept of neurorehabilitation through the development of clinical research despite barriers to implementing new technologies in everyday clinical practice.

18.1 Introduction

The Republic of Serbia is a middle-income country in Southeast Europe that includes part of the Pannonian Plain and the central part of the Balkan Peninsula. The country covers an area of 77,474 km². Serbia is a parliamentary republic with a unicameral parliament. The Republic holder of constitutional and legislative power is National Assembly with 250 numbers. Belgrade is the capital and the largest city.

18.1.1 Country history and demographics

After the long rule of the Ottoman Empire and Austro-Hungary, an independent federation was founded in 1918 known as the Kingdom of Serbs, Croats, and Slovenes, which in 1929 was formally constituted as Yugoslavia. After the disintegration of Yugoslavia, Serbia was constituted in its current form in 2006. The total population in the Republic of Serbia is 6,871,547 according to the estimates of the National Statistical Office of Serbia in 2020. It shows a trend of depopulation (-6.7%) and continuous growth of the elderly population (21%). The population's average age in the Republic of Serbia increased from 42.1 years in 2011 to 43.3 in 2019 [1].

Rehabilitation Robots for Neurorehabilitation in High-, Low-, and Middle-Income Countries.
DOI: https://doi.org/10.1016/B978-0-323-91931-9.00022-0

18.1.2 Communicable and noncommunicable disease prevalence

The main health problems of the population are cardiovascular diseases and malignant tumors that in 2019 accounted for more than two-thirds of all deaths in Serbia. In 2020, 14,933 more deaths were registered in the Republic of Serbia than the average in 2016−19, representing increased mortality of +14.7%. The overall mortality rate in 2020 was 16.9 per 1000 inhabitants. In 2020, there were 268,998 reported cases of infectious diseases on the territory of the Republic of Serbia, with an incidence rate of 3899.02 per 100,000 population. The total number of reported cases of contagious diseases in Serbia was higher than in 2019 when 46,081 cases were reported. The higher number of reported cases in 2020 compared to the previous year was due to the COVID-19 pandemic. COVID-19 is an infectious disease that has been under mandatory reporting since 2020. Despite the observed under-reporting, in 2020, COVID-19 was the most dominant disease in the group of infectious diseases and represented 78.93% of the total reported cases of communicable diseases.

Chronic noncommunicable diseases such as cardiovascular diseases, malignant tumors, diabetes, obstructive lung disease, injury and poisoning, mental health disorders, and other chronic diseases have dominated the pathologies of the nation for decades. In 2020, 55,305 people (25,617 men and 29,688 women) died from cardiovascular diseases (ICD10: I00-I199). Cardiovascular diseases account for 47.3% of all causes of death and are the leading cause of death in Serbia. Women were dying from cardiovascular diseases more often (53.7%) than men (46.3%). Ischemic heart diseases and cerebrovascular diseases are the leading causes of death in this group of diseases. The most common causes of death in 2020 belong to the following diseases groups (according to ICD10): diseases of the circulatory system 47.3% (men 42.4%, women 52.6%), neoplasms 18.3% (men 19.8%, women 16.7%), COVID-19 8.9% (men 11%, women 6.6%), and diseases of the respiratory system 5.7% (men 6.5%, women 4.9%) (Table 18.1). Apart from the leading causes, endocrine, nutritional, and metabolic diseases also contribute to the overall mortality of 3% (men 2.6%, women 3.4%) [2].

TABLE 18.1 Gender distribution of common causes of death (%).

Diseases groups (ICD10)	Men	Women	All
Diseases of the circulatory system	46.3	53.7	47.3
Neoplasms	19.8	16.7	18.3
Diseases of the respiratory system	6.5	4.9	5.7

18.1.3   Stroke incidence and prevalence

The annual incidence of stroke in Serbia is 1.72 per 1000. Mortality due to stroke per year is 1.49 per 1000 with a significantly decreasing trend. Over 100 general hospitals neurology clinics and stroke units provide acute stroke treatment for 40% of total stroke patients. Stroke has been the second leading cause of death in both sex over the last decade. Mortality rates for stroke vary significantly worldwide. The stroke mortality trend has declined substantially in the past few decades in developed countries, including the United States [3] and West European countries [4]. Contrary to this trend, in East European countries, there has been a constant increase in the mortality rate caused by stroke during the last decades of the 20th century. After a decade of growth, the mortality rate due to cardiovascular disorders (CVDs) has significantly declined in the last decade in Serbia for both sex. Trends in mortality from CVDs declined more in people younger than 70 years than those aged 70 years or older, in both men and women in Serbia [5].

18.2   Healthcare structure and resources

Healthcare for the population in the Republic of Serbia is directly provided through a network of healthcare institutions. The total number of healthcare institutions, according to the Decree on the Health Care Institution Network Plan ("Official Gazette of the Republic of Serbia," No 74/2021), was 336 in 2020 (excluding institutions from Kosovo and Metohija Province and Military Medical Institutions). This number includes primary healthcare centers and general hospitals with health centers. Hospitals are health institutions that perform inpatient and specialist consulting activities. In the Republic of Serbia, 120 healthcare institutions provide inpatient (hospital) care. These are inpatient departments in primary healthcare centers (18), general hospitals (40), special hospitals (31), institutes (14), clinics (7), clinical hospital centers (4), and university clinical centers (4).

Hospitals in the Republic of Serbia in 2020 employed 8870 doctors (of which 75.5% are specialists). The total number of beds in Serbian hospitals in 2020 was 42,089, i.e., 6.1 beds per 1000 of the population, and for rehabilitation purposes, it was 6023 (15.1%). The number of discharged patients from inpatient institutions in Serbia in 2020 was 1,623,713, and the total number of hospital days was 6,910,454. The average length of stay per patient was 6.5 days, while the average hospital bed occupancy rate was 45.0%.

18.2.1   Rehabilitation structure and rehabilitation access

The structure and strategies for physical medicine and rehabilitation services in Serbia have been established and developed over the past several decades.

The central postulate implies timely and continuous implementation of rehabilitation, which aligns with the documents adopted by the World Health Organization (WHO). In the Republic of Serbia, persons with disability make up 8% of the total population. The average age of those persons is about 67 years. Thus 71% of them fall into the age group of 65 and over. Observed by sex, the number of women with disabilities is higher (58.2%) than that of men (41.8%). The most significant percentage of persons with disabilities is in the region of Southern and Eastern Serbia (9.4%) and the smallest in the region of Belgrade city (5.9%).

During the 1990s and the first decade of the 21st century, numerous rights of persons with disabilities were established. Serbia has regulated the status and rights of persons with disabilities in the constitution and in legal acts that regulate human rights and the prohibition of discrimination. The legal acts have been harmonized with the Disability Action Plan of WHO to connect health services with users and expand the right to (re)habilitation and ancillary services [6,7]. Its goals are to connect health services to the users, extend the right to (re)habilitation and auxiliaries for activities of daily living (ADL) and ensure adequate data collection on disability.

Early acute and subacute rehabilitation is organized based on the complexity of the patient's condition at different levels of healthcare. Services are provided in the form of outpatient and inpatient rehabilitation. Healthcare is provided at three levels, primary, secondary, and tertiary. The primary level is organized territorially at the municipal level and includes general practitioners and some services such as physical medicine and rehabilitation. In addition to outpatient and home treatment, preventive examinations of children are carried out in health centers at the primary level. At the primary level, there may be special organizational units in larger centers such as the Institute for Home Treatment and Care. Rehabilitation at home is offered at the primary level of healthcare. In general, home rehabilitation is poorly developed with a small capacity to meet needs and exists only in larger centers. Patients with more complex problems are referred for outpatient or inpatient rehabilitation to institutions of secondary level of care in clinical hospital centers, general hospitals, and specialized clinics and institutes. The most complex cases are referred to tertiary-level institutions that differ from secondary-level institutions in terms of staff and equipment training and are mainly university clinical hospital centers. The State Health Fund of Serbia covers the costs of rehabilitation only in state institutions.

18.2.2   Rehabilitation capacity

Rehabilitation is carried out at all levels of healthcare. The total number of beds for rehabilitation in Serbia is 6023 (15.1%). Of the total number of hospital days in all hospitals, 13% were in the physical and rehabilitation medicine departments, of which a smaller number are in university centers. Given that rehabilitation is carried out at all levels of healthcare, the number of available services is in the hundreds of thousands.

18.2.3   Stroke rehabilitation standard of care

Stroke patients make up 15.7% of the total number of hospitalized patients in rehabilitation units. The COVID-19 pandemic has led to a reduction in the number of hospitalized and outpatients involved in rehabilitation. The Clinic for Rehabilitation "Dr. Miroslav Zotovic," Belgrade, Serbia, and Institute for Rehabilitation, Serbia, are the leading providers of neurorehabilitation services, because of the large catchment area, metropolitan location, hospital capacity, and the number of patients served.

The standard of care for stroke patients is conventional rehabilitation. Conventional rehabilitation is provided 5 days per week for 3−6 weeks, divided into two 30−60-minute occupational therapy and physiotherapy sessions. Occupational therapy for the paretic upper limb includes passive stretching within submaximal ranges of motion to inhibit spasticity, active-assisted movements, functional tasks, and activities of daily living individually which are customized for the patients progressively. Physiotherapy consists of a range of motion exercises for the upper and lower extremities, gentle stretching, splinting/casting, facilitation of active voluntary movement, and exercises to improve endurance, balance, strength, and gait. If necessary, speech therapy is provided three to five times per week.

18.3   Technology-assisted rehabilitation

Rehabilitation is generally carried out with conventional treatment, but 15% of the patients are included in clinical trials of new rehabilitation therapeutic interventions annually. The Clinic for Rehabilitation "Dr. Miroslav Zotović"[1] is the main center for clinical research in the field of neurorehabilitation (Fig. 18.1A). Collaboration with the University of Belgrade-School of Electrical Engineering, Institute for Physics, Institute for Medical Research, Tecnalia Serbia Ltd., and participation in domestic and European projects enables clinical research and development of methodologies, especially in multidisciplinary fields that introduce novel rehabilitation techniques.

Clinical research focuses on clinical efficacy testing and the development of rehabilitation systems in collaboration with engineering teams. The main topics of study were the strengthening of sensory-motor couplings by

---

1. The Rehabilitation Clinic "Dr. Miroslav Zotović" was established in 1952 based on an agreement with the UN/WHO. The contract contained all the elements from the internationally accepted documents of the concept of modern multidisciplinary rehabilitation. Preparations have lasted since 1950 when prof. Henry Kessler from the Rehabilitation Institute in New Jersey, as a UN rehabilitation expert, stayed in Belgrade, and soon after, in 1951, a team from the clinic trained at that institute.

FIGURE 18.1 (A) Clinic for Rehabilitation "Dr. Miroslav Zotovic," (B) ongoing project on BCI. Part a shows one of the buildings in the rehabilitation clinic complex of Dr. Miroslav Zotović, the building is surrounded by a small park; part b shows a subject with CVI during the performance of a learning session of voluntary control of the electrophysiological correlates of somatosensory and motor cortical activation. The patient is sitting, the therapist performs tactile stimulation of the healthy hand, and the patient imagines (initiates) the movement. The focus on tactile stimulation gives the BCI system an additional feature: it works as a tool for enabling the strengthening of the interconnection of motor and cortical sensory activation. When both the sensory and motor cortices are activated (according to the BCI protocol and calibration), this triggers sensory feedback in the form of FES-induced arm movement. When he imagined (initiated) the movement and felt the peripheral stimulus (registered in the sensory cortex), it triggered FES peripherals.

electrical stimulation for the upper and lower extremities of people with tetraplegia and hemiplegia, hybrid orthosis in walking, and the clinical efficacy of a low-cost robotic arm device. Studies are defined as small-scale experimental design studies, case series, and randomized studies. The outcomes were selected based on international classification of function (ICF) categories.

18.3.1   Assistive technologies

Clinical researchers have paid close attention to motor control in stroke in collaboration with engineering teams. Movement control problems face redundancy, nonlinear variability, and significant preference criteria based on motor tasks. Research has gone from model-based motor control to designing exoskeleton and hybrid assistive devices.

Over several decades of research, many questions have been raised, from the modeling, design, and feasibility of new technological systems to the need to predict motor response while overcoming redundancy. Functional movement is mainly a "black box" model and is an expression of sensorimotor coordination that includes three main elements: sensory information, internal coding, and motion generation. Many studies utilizing functional electrical stimulation (FES) in restorative and assistive contexts have been conducted. The mechanism of FES is, in fact, hierarchically organized as a sensorimotor closed loop and can use sensory information and correct the command motor signal with possible clinical significance. Many practice problems include unpleasant/painful sensations with higher stimulation intensities, muscle fatigue, discomfort related to electrode placement, lower selectivity of activation, determination of stimulation patterns, delay in motor response, and lack of feedback. A way to overcome these problems was to remodel the interface between stimulators and stimulated peripheral nerves by introducing a multitude of smaller electrodes with individual actuators (multiple electrodes) and multichannel stimulators instead of relatively large surface electrodes. Such hardware has been used in several studies conducted at the Clinic for Rehabilitation "Dr. Miroslav Zotović." These multi-pad electrodes allow the selective spatial-temporal targeting of motoneurons that activate synergistic muscles and produce functional movement [8].

The determination and validity of the stimulation maps for grasping were tested in a group of hemiplegic patients using electromyography. The selected set of active pads resulted in fully functional and reproducible palmar and lateral grasps like a healthy-like grasp [9]. FES protocol based on multi-pad electrodes (Fesia system, Tecnalia, Spain) to correct foot droop has shown that the system can respond to gait cycle events such as heel strike, heel and toe lowering, but also able to respond to real-time changes in mean swing time/holding and loading using algorithms that adjust the relevant stimulation parameters, without discomfort [10]. A new stimulation paradigm enables the asynchronous activation of motor neurons and could ensure controlled temporal and spatial distribution of the electric charge delivered to motoneurons by optimizing the electrode configuration [11].

A study examining peripheral sensors as a rehabilitation tool was conducted at the clinic. The hand functions assessment system (BEAGLE) for kinematic tracking of hand and finger movements is custom-designed for fast and easy placement on an impaired hand (spastic or flaccid), featuring inertial sensors integrated into simple finger caps and a hand strap. An algorithm for a range of motion (ROM) estimation was implemented to assess hand functions objectively. The results indicate that the ROM assessments can detect change with sensitivity comparable to the standardized clinical scales Fugl-Meyer and Action Research Arm Test [12].

Finally, the development and exploration of brain—computer interface (BCI) technology for neurorehabilitation is a part of the ongoing research. Namely, within a scientific project supported by the Fund for Science, Republic of Serbia, for developing a system for arm rehabilitation in stroke, research has begun on detecting motor intention or attempting to do so in real time and activating muscles by FES or controlling a robotic device. This project plans to develop a hybrid BCI platform consisting of integrated hardware for EEG, FES system, and custom software applications with innovative EEG processing algorithms within a new hybrid BCI device prototype operating based on two mental strategies: imaginary or attempted movements and selective tactile attention (Fig. 18.1B).

18.3.2 Robots

Neurorehabilitation is an essential segment of rehabilitation that is carried out by a multidisciplinary team and with a trend to introduce new technologies, such as rehabilitation robotics, to train and assess sensorimotor functions. Robotic rehabilitation devices are implemented within research studies conducted in collaboration with university partners (science and research organizations), with technology development companies, or within EU projects. The main obstacles to the broader use of robotic technology are financial reasons and the potential to reduce rehabilitation services covered by the National Health Fund.

The largest randomized clinical trial of a robotic rehabilitation device conducted in the Republic of Serbia was focused on the ArmAssist (AA) robotic system developed by Tecnalia Research and Innovation Center, Spain. AA is a low-cost robotic platform for shoulder and elbow training and assessment after stroke consisting of a portable device for providing arms support over a table and serious interactive games operating on a web-based platform. The AA system includes games for both training and assessment of motor function. The training games include tasks of different complexity levels which require variable cognitive engagement designed to motivate the user to train longer and more effectively, such as puzzles, memory, language, and card games. In contrast, the assessment games comprise shorter tasks to evaluate different aspects of upper-limb motor control, such as the range and characteristics of movements with varying degrees of freedom. This study aimed to determine the preliminary efficacy of the AA robotic device in comparison to the matched conventional arm training in subacute stroke patients undergoing rehabilitation. The results showed that the AA device's training reduced impairment and activity-related motor deficits more effectively than matched conventional arm training in the subacute phase of recovery from stroke [13]. Moreover, exploring the effects of patients' motivation for exercise using the AA device showed that training with this system is perceived as beneficial, enjoyable, and highly motivational for patients to continue and endure longer durations of training [14]. Unfortunately, AA was not used after the study because the device was not procured due to a lack of funds.

An essential line of research was focused on validating novel methods for assessing motor functions based on robotic technology since objective, reliable measures of motor function that correlate highly with human-administered clinical scales are needed for quantifying motor performance and determining the course of therapy in poststroke individuals. For this purpose, different components of the AA-based assessment of motor functions were compared against three widely used clinical tests, the Fugl-Meyer Assessment, Action Research Arm Test, and the Wolf Motor Function Test

(WMFT), showing promising results reflected in a high correlation between AA metrics and conventional clinical scales [15].

Research conducted within the Clinic for Rehabilitation "Dr. Miroslav Zotović" in collaboration with the School of Electrical Engineering, University of Belgrade, resulted in the validation of a novel method for fast and straightforward robotic assessment of shoulder/elbow function based on a computerized square-drawing test (DT) [16]. This work defined a novel set of DT-based kinematic measures and explored their relationship with WMFT scores in 47 stroke patients. This line of research validated a set of kinematic measures suitable for fast and objective motor function evaluation and functional classification, strongly correlating with WMFT scores in stroke patients.

For lower-limb robotic rehabilitation, a Walkaround device, allowing walking without hand support for individuals with limited ability to control posture, was developed and tested in Serbia. Specifically, the Walkaround system was designed to assist people with compromised posture and inability to fully support body weight and prevent falls [17,18].

18.4   Challenges and future opportunities

The COVID-19 pandemic has affected the population worldwide, creating new challenges, opportunities, and directions in rehabilitation technologies. Assistive robotics may facilitate clinical care and reduce the contagion rate. The social distancing measures and high clinical burden create barriers to rehabilitation within clinical institutions. Therefore patients may benefit from affordable and reasonable quality home-based telerehabilitation technology. Previously described studies of low-cost telerehabilitation platforms such as ArmAssist, or rapid and objective assessment robotic technology such as computerized drawing tests, or the Walkaround system which may enable appropriate patient independence and distance from a therapist during treatment, are all directly in line with emerging needs and challenges created during the COVID-19 pandemic.

Conflict of interest

All the listed authors participated in data collection, analysis, and manuscript writing.

References

[1]   Statistical Office of the Republic of Serbia. Available from: <https://www.stat.gov.rs/en-us/>.

[2]   Institute of Public Health of Serbia. Dr. Milan Jovanovic Batut. Health statistical yearbook of the Republic of Serbia 2020, 2020. Available from: <https://www.batut.org.rs/>.

[3]   Mozaffarian D, Benjamin EJ, As G, et al. American heart association statistics committee and stroke statistics subcommittee. Heart disease and stroke statistics—2015 update: a report from the American Heart Association. Circulation 2015;131:e29−e322.

[4] Levi F, Chatenoud L, Bertuccio P, Lucchini F, Negri E, La Vecchia C, et al. Mortality from cardiovascular and cerebrovascular diseases in Europe and other world areas: an update. Eur J Cardiovasc Prev Rehabil 2009;16(3):333−50.

[5] lic I, Ilic M, Sipetic Grujicic S. Trends in cerebrovascular diseases mortality in Serbia, 1997-2016: a nationwide descriptive study. BMJ Open 2019;9(2):e024417.

[6] Takáč P, Petrovičová J, Delarque A, Stibrant Sunnerhagen K, Neumann V, Vetra A, et al. Position paper on PRM and persons with long term disabilities. Eur J Phys Rehabil Med 2014;50(4):453−64.

[7] Gutenbrunner C, Negrini S, Kiekens C, Zampolini M, Nugraha B. The Global Disability Action Plan 2014−2021 of the World Health Organisation (WHO): a significant step towards better health for all people with disabilities. Chance and challenge for Physical and Rehabilitation Medicine (PRM). Eur J Phys Rehabil Med 2015;51(1):1−4.

[8] Malešević J, Štrbac M, Isaković M, Kojić V, Konstantinović L, Vidaković A, et al. Temporal and spatial variability of surface motor activation zones in hemiplegic patients during functional electrical stimulation therapy sessions. Artif Organs 2017;41(11): E166−77.

[9] Popović Maneski L, Topalović I, Jovičić N, Dedijer S, Konstantinović Lj, Popović DB. Stimulation map for control of functional grasp based on multi-channel EMG recordings. Med Eng Phys 2016;38(11):1251−9.

[10] Dedijer Dujović S, Malešević J, Malešević N, Vidaković A, Bijelić G, Keller T, et al. Novel multi-pad functional electrical stimulation in stroke patients: a single-blind randomized study. Neurorehabilitation 2017;41(4):791−800.

[11] Malešević J, Dedijer Dujovic S, Savić AM, Konstantinović L, Vidaković A, Bijelić G, et al. A decision support system for electrode shaping in multi-pad FES foot drop correction. J Neuroeng Rehabil 2017;14(1):66.

[12] Malešević J, Kostić M, Kojić V, Dordević O, Konstantinović L, Keller T, et al. BEAGLE -a kinematic sensory system for objective hand function assessment in technology-mediated rehabilitation. IEEE Trans Neural Syst Rehabil Eng 2021;29:1817−26.

[13] Tomić TJ, Savić AM, Vidaković AS, Rodić SZ, Isaković MS, Rodríguez-de-Pablo C, et al. ArmAssist robotic system versus matched conventional therapy for post-stroke upper limb rehabilitation: a randomized clinical trial. BioMed Res Int 2017;2017:7659893. Available from: https://doi.org/10.1155/2017/7659893.

[14] Rodríguez-de-Pablo C, Popović M, Savić A, Perry CJ, Belloso A, Dimkić Tomić T, et al. Post-stroke robotic upper-limb telerehabilitation using serious games to increase patient motivation: first results from ArmAssist system clinical trial. Advances in neurotechnology, electronics, and informatics. Springer International Publishing; 2016, p. 63−78.

[15] Rodriguez-de-Pablo C, Balasubramanian S, Savić A, Tomić TD, Konstantinović L, Keller T. Validating ArmAssist Assessment as an outcome measure in upper-limb post-stroke telerehabilitation. In Annu Int Conf IEEE Eng Med Biol Soc. 2015;2015:4623−6.

[16] Isaković MS, Savić AM, Konstantinović LM, Popović MB. Validation of computerized square-drawing based evaluation of motor function in patients with stroke. Med Eng Phys 2019;71:114−20.

[17] Veg A, Popović DB. Walkaround: mobile balance support for therapy of walking. IEEE Trans Neural Syst Rehabil Eng 2008;16(3):264−9.

[18] Dragin AS, Konstantinović LM, Veg A, Schwirtlich LB. Gait training of poststroke patients assisted by the Walkaround (body postural support). Int J Rehabil Res 2014;37 (1):22−8.

Chapter 19

Asia Pacific region: India

Sivakumar Balasubramanian[1], Aravind Nehrujee[1,2], Abha Agrawal[3], Guruprasad V.[4], Shovan Saha[4] and Sujatha Srinivasan[2]
[1]Department of Bioengineering, Christian Medical College, Bagayam, Vellore, Tamil Nadu, India, [2]Department of Mechanical Engineering, Indian Institute of Technology, Chennai, Tamil Nadu, India, [3]A4 Clinics, Indore, Madhya Pradesh, India, [4]Department of Occupational Therapy, Manipal College of Health Professions, MAHE, Manipal, Karnataka, India

Learning objectives

At the end of this chapter, the reader will be able to:
1. Understand neurorehabilitation in India.
2. Describe the current state of neurorehabilitation technology development and utilization in India.

19.1 Overview of India and status of neurorehabilitation in India

India, located in Southeast Asia, is a peninsula bordered by the Himalayan mountains to the North, the Indian Ocean to the South, the Bay of Bengal to the East, and the Arabian Sea to the West. It is the seventh largest country in the world and covers approximately 1,269,346 square miles. India also has a geographically diverse landscape and an extremely diverse population with thousands of ethnic groups and hundreds of languages. It is the second most populous country in the world, with roughly one-sixth of the world's population.

Though neurology (medical and surgical) has made significant progress in diagnosing and treating neurological disorders, neurorehabilitation in India is still in its early stages. Neurological disorders present a huge burden of not only physical disability but also a social stigma for patients and families in India. Approximately 1.6 million people in India experience a stroke yearly; about 500,000 people will have residual disabilities. Another 1.5 million have a disability due to brain injury from roadside accidents and trauma

Rehabilitation Robots for Neurorehabilitation in High-, Low-, and Middle-Income Countries.
DOI: https://doi.org/10.1016/B978-0-323-91931-9.00017-7

[1]. As per the statistics in India, the proportion of persons with disabilities in the total population is 1.8%−2.1%; 26%−34% are unemployed; 75% live in rural areas, and 49% are illiterate [2]. The prevalence rate of neurological disorders varies from 967 to 4070 per 100,000, significantly higher in rural than urban populations [3]. Although it is difficult to gauge the overall economic cost of neurologic disability in India, they are well-known causes of financial distress worldwide [4].

Neurorehabilitation is an essential component of neurological services. Intense, evidence-based rehabilitation improves a patient's functional ability and reduces secondary "preventable" disability. Despite increasing life expectancy following a neurological disorder, the overall quality of life remains low due to the rise in disability-adjusted life years [4]. Neurorehabilitation is a coordinated interdisciplinary care program that includes "a set of measures that assist individuals who experience (or are likely to experience) disability in achieving and maintaining optimal function in interaction with their environment" [5] for maximum independence and social reintegration [6,7]. It usually includes neurologists, rehabilitation physicians, nurses, physiotherapists, occupational therapists, speech and language pathologists, prosthetists/orthotists, neuropsychologists, and nutritionists. In a developing country like India, providing neurorehabilitation services is a considerable challenge due to financial and human resource constraints and numerous other constraints.

19.1.1 Barriers to neurological rehabilitation in India

In India, people with disabilities face many barriers to meeting their health and rehabilitation needs and accessing mainstream health services. There are several challenges to delivering neurorehabilitation in India:

1. Lack of financial resources is a significant barrier for patients to receive appropriate intensity and duration of neurological rehabilitation. The lack of insurance for physical therapy and rehabilitation services further exacerbates this problem. Less than 25% of patients in India have insurance coverage from a public or private provider. Most insurance providers have limited or no coverage for long-term outpatient or subacute rehabilitation services. Frequently, patients have already spent substantial out-of-pocket money for acute care for illnesses like stroke or spinal cord injury, which leaves them with limited ability and willingness to continue long-term rehabilitation needed for optimal recovery.
2. There is a profound lack of awareness of the role of neurorehabilitation in functional recovery. Many patients, as well as clinicians, are unaware that high-quality intense therapy can substantially reduce disability from neurological disorders. Instead, many patients and families seem "resigned" and accept disability as it is. People are much more aware of

acute care and expect rapid improvement in the patient's condition. They are often frustrated that the rehabilitation process may take months to years.

3. Unfortunately, many neurologists and neurosurgeons also do not recognize the value of neurorehabilitation in improving the patient's overall outcome. They remain primarily focused on the acute episode or pharmacotherapy. The lack of organized post-acute care infrastructure grossly exacerbates this barrier. Many patients and families receive minimal to no direction or prescription for appropriate rehabilitation services.

4. The lack of access is another substantial challenge. India has a vast and varied geography with a population of over 1.3 billion. However, almost 70% of the population resides in rural areas with limited access to primary care, let alone high-quality neurorehabilitation services. Given the financial barriers described above, acute care hospitals are prioritized over neurorehabilitation centers.

5. The country has a dearth of physical medicine and rehabilitation workforce compared to the patient population requiring such services. WHO estimates that skilled stroke rehabilitation practitioners in low- and middle-income countries number fewer than ten per one million of the population. According to the available data, the number of rehabilitation staff per million of the population in low-income and middle-income countries is rehabilitation physicians <10; physiotherapists 199; occupational therapists <50; speech and language therapists <50; and prosthetists and orthotists <1 [8]. This shortage is further worsened by the limited availability of modern therapeutic equipment.

6. Poor coordination among different health sectors (e.g., between private and public hospitals), ministries, healthcare agencies, and nongovernment organizations further fragment the patient's care and make it challenging to navigate a complex system.

7. Lack of adequate research on neurological rehabilitation and the development of devices suited to the Indian context.

19.1.1.1 Service delivery of neurorehabilitation in India

The Government of India has been establishing District Disability Rehabilitation Centers for persons with disabilities to provide comprehensive rehabilitation services at the grassroots level. These services include improving awareness, early detection and intervention, counseling, assessing and providing assistive devices, and referral to neurorehabilitation services [9]. Additionally, community-based rehabilitation programs adapted to India's local traditions have a long history [2]. There are established stroke rehabilitation centers in urban areas, but the rural population faces significant challenges accessing specialized neurorehabilitation services. Most of the rehabilitation services are provided by the private sector at a practical level. This includes hospital-based rehabilitation departments, stand-alone private clinics, and independent rehabilitation professionals who often operate on a freelance model.

19.2 Role of technology in neurorehabilitation and its current state in India

The need for technology and its importance in movement neurorehabilitation are well recognized. Like most healthcare systems worldwide, India's current rehabilitation care pathways can benefit significantly through the infusion of technological support that allows timely and effective delivery of services while efficiently using the country's healthcare (human and financial) resources. The potential advantages of using technology include:

1. Semi-automation of therapy reduces the need for one-on-one interaction to deliver movement therapy, allowing clinicians to focus on therapy planning rather than therapy delivery.
2. A single device can deliver therapy continuously without taxing clinicians' physical and mental reserves, allowing them to supervise the treatment of more patients.
3. Robotic devices can provide physical assistance to patients with limited residual motor capacity to attempt and participate in training.
4. Monitored home-based rehabilitation can reduce the need for patients to travel regularly to a clinic.
5. Biofeedback in the form of accurate, precise, and objective measurement of movement variables can supplement the assessment carried out by clinicians to understand a patient's sensorimotor state.
6. Gamification of therapy delivered by technology can make training more engaging and thus result in better compliance and outcomes for patients.

This chapter will only focus on robotic technology for upper-limb neurorehabilitation. We will summarize the current state of robotic technology for upper-limb neurorehabilitation, discuss the current clinical evidence for robot-assisted upper-limb therapy, the types of patients who will most benefit from robot-assisted therapy, and finally present the current use of rehabilitation robots in the Indian healthcare system.

19.2.1 Rehabilitation robots in India

In India, the usage and acceptance of rehabilitation robots in regular clinical practice are currently limited. Very few tertiary, private-sector hospitals in tier I cities use robotic devices to deliver neurorehabilitation. A few newly launched private clinics use upper-extremity robotic devices (along with non-invasive brain stimulation) in metropolitan and tier II cities. Almost all devices used in these hospitals are imported from manufacturers abroad (United States, Europe, China, etc.). The rehabilitation robotics industry in India is just starting to develop, and there are few commercial rehabilitation robots from Indian manufacturers. Most secondary hospitals and small private rehabilitation clinics across India currently do not use robotic devices for delivering therapy. A recent survey conducted as part of a collaborative project between CMC Vellore and IIT Madras contacted 87 hospitals and neurorehabilitation clinics across India. Only two out of the 87 centers used one or more rehabilitation robots for delivering therapy. The rest of the centers quoted the high capital cost, the poor return on investment, and inadequate support for serviceability as reasons for not incorporating robotic technology into routine clinical practice. This situation will likely change in the next 5−10 years with the increased interest in this field within India.

Over the last decade, especially over the last five years, there has been an increased focus on strengthening the state of rehabilitation devices research and development in India. Several private and government funding opportunities have been initiated to facilitate rehabilitation device research in the country. This has led to a few rehabilitation robots being developed in research labs and start-ups being spun off based on this work. Table 19.1 lists some of the upper-limb rehabilitation robots developed by academic research labs in India. Most of these are in the prototyping stage, and few have undergone usability studies. There has been only one pilot study that has evaluated the efficacy of an indigenously developed robot.

Since major technical development in technology for neurorehabilitation started only in the last decade, most robots have not been clinically tested yet, and thus, there are only three homegrown companies manufacturing and selling rehabilitation robots. Of the three companies, two sell planar upper-limb rehabilitation robots, and the third one sells a modular robot that can train the upper and lower limbs (Table 19.2).

India as a country is unique, and the rehabilitation needs are quite different from the rest of the low-middle-income countries. India is a vast nation with 29 states with different levels of socioeconomic development and public infrastructure. Thus each state has its unique challenges when it comes to the implementation of rehabilitation technologies. It also should be noted that 65% of the Indian population reside in rural areas, with 22 official languages and about 122 local languages in India. Thus it is imperative for such a technology to be valid; it should be language-agnostic and inclusive to people with minimal to no computer literacy. Additional challenges in technology-based rehabilitation, especially in rural areas, include the intermittent availability of electricity, limited transport facilities, social stigma, lack of awareness, etc. These challenges demand a multidisciplinary, codesign approach to device development with inputs and feedback from the different stakeholders (patients, caregivers, clinicians, social workers, policymakers, and engineers).

TABLE 19.1 List of rehabilitation robots published as part of academic research.

Name	Remarks	Clinical testing
Chatterjee et al. [10]	A soft exoskeleton for training finger flexion-extension	Design and prototype development stage
Singh et al. [11,12]	Finger-wrist-coupled motion with EEG control	Phase 2 trial. Pilot RCT has been completed to show feasibility and efficacy
Mathew et al. [13]	Hand exoskeleton; trains for finger flexion-extension	Design and prototype development stage
Nehrujee et al. [14]	Modular and portable robot for hand rehabilitation	Usability trail with 45 stakeholders
Balasubramanian et al. [15]	Arm rehabilitation robot: modular robot to train for arm and elbow	Design and prototype development stage
Thakur et al. [16]	Wrist rehabilitation robot	Design and prototype development stage
Seth et al. [17]	Cable-driven elbow rehabilitation robot	Design and prototype development stage

TABLE 19.2 Indigenous rehabilitation robotics companies in India.

Name of the company	Name of the robot	Remarks	Clinical testing
Beable Health	Armable robot [18]	A planar rehabilitation robot for bimanual and unimanual arm training	Preliminary usability has been reported [19]
Rymo Technologies	Mobi-L [20]	Rehabilitation robot for combined upper-limb and lower-limb training	No studies published yet
Gridbots Technologies	Physio [21]	Planar arm rehabilitation robot	No studies published yet

19.3 Factors contributing to the current lack of clinical assimilation of neurorehabilitation technology

Despite the research in rehabilitation robotics over the last 20–25 years, its uptake into routine clinical practice is limited in developed and developing countries. This section discusses some interrelated factors that we believe are significant contributors to this situation in India and worldwide.

1. **High cost-to-value of existing commercial rehabilitation robots:** The cost of commercial rehabilitation robots is high compared to the value they currently offer. The high capital cost also increases the average price of therapy for a single patient, preventing the widespread use of this approach in clinical practice. Almost all devices in India are imported, which compounds the cost of robot-assisted therapy; therefore, only large corporate Indian hospitals in tier I cities and a few private clinics presently have such technology.

2. **Lack of reimbursement for technology-aided neurorehabilitation:** Although there is limited coverage for neurorehabilitation services from selected insurance schemes in India, robot-assisted or technology-aided therapy is not covered by private insurance. Financial constraints limit the number of patients willing to undergo robot-assisted therapy either in a clinic or at home.

3. **Lack of awareness among the end users.** There is a general lack of awareness among patients and clinicians about the importance of timely and intense neurorehabilitation following neurological conditions such as stroke. This is a contributing factor for patients and caregivers, especially in rural areas, who do not aggressively pursue options following a stroke or other neurological diagnosis. Technology can deliver such therapy, but poor awareness implies low demand for such technology.

4. **Lack of sufficient work on clinical implementation.** The focus of the neurorehabilitation research community has been on the efficacy of rehabilitation robots, with little work on integrating rehabilitation robots into routine clinical practice. Several important questions need to be answered to set up guidelines for practicing clinicians to appropriately incorporate robot-assisted therapy into everyday practice. Should robot-assisted therapy be given as an add-on, or should it replace specific aspects of conventional therapy? What types of patients are best suited to undergo robot-assisted therapy? How much do rehabilitation robots decrease the workload of clinicians and improve the efficiency of clinical resource utilization? Does the use of rehabilitation robots result in added costs to patients? How much can we increase the dosage of high-quality therapy with minimally supervised home-based robot-assisted therapy? Answers to such questions in the context of the Indian healthcare system are crucial for clinically adopting technology.

5. **Too much focus on in-clinic solutions.** Most work on rehabilitation robots has focused on in-clinic solutions, which are essential. However, the poor access to rehabilitation services and care, the difficulties associated with traveling to a clinic, the lack of trained personnel to administer therapy, and the limited therapy time possible in a clinic necessitate

home-based solutions. Such solutions will address all the above problems for the patient and potentially help improve the therapy dosage delivered to patients. The lack of home-based solutions may also be a reason for the poor uptake of these technologies for delivering neurorehabilitation services. Commercial compact, robust, home-based devices that patients can rent for a prescribed duration will help them receive intense, good-quality therapy addressing poor access to therapy services.

19.4 Recommendation for taking neurorehabilitation forward in India

Addressing these problems quickly is crucial to improving the state of neurorehabilitation care pathways in India. As the Indian population ages over the next couple of decades, the incidence of conditions such as stroke will increase, further compounding the pressure on the country's healthcare system. Overhauling the country's rehabilitation healthcare infrastructure requires work at multiple levels. The following are some recommendations to this end:

1. **Amend national health policy:** The most impactful change would be to amend the national health policy to incorporate insurance coverage for rehabilitation services. The ambitious and transformative Ayushman Bharat PM-Jay scheme, launched in September 2018, provides coverage to many low-income families but has no provision for physical therapy and rehabilitation. The private and public reimbursement systems must be reformed to address the rehabilitation needs of millions of people in India.
2. **Increasing awareness among the public and the healthcare community:** Public service campaigns should be launched to advance awareness and understanding of neurorehabilitation to reduce the burden of "preventable disability." The medical profession should embrace rehabilitation as a critical component of the treatment plan to improve health outcomes for patients with neurological disorders. Similarly, rehabilitation training curricula should incorporate technology-based rehabilitation, such as robotics.
3. **Clinically relevant and accessible technology**: It is crucial for neurorehabilitation engineering research in the country to focus on developing and translating simple, compact, economical, and clinically relevant robots that will be suitable for both in-clinic and at-home use that can be delivered independent of electricity and internet connectivity. Most design and development work should not end with academic publications, but academic labs must make an effort to exploit the country's growing start-up infrastructure to push their innovations to the real world. The availability of several such technologies will:

- promote competitive pricing models for such technologies, benefitting the end user,
- stimulate research on clinical implementation of such technology for routine use,
- contribute to increasing awareness about neurorehabilitation technology among end users,
- help push policy for insurance coverage for technology-aided neurorehabilitation services, and
- eventually, improve patient outcomes and quality of life following neurorehabilitation.

Finally, the social attitudes and stigma toward any visible disability must be improved. This includes incorporating universal design principles, developing building and public transportation infrastructure to accommodate functional impairments due to disability, to changing employment and labor laws to eliminate discrimination against people with disability. India has an enormous opportunity for comprehensive, evidence-based, technology-enabled neurorehabilitation. We hope that the people in medical and rehabilitation professions, insurance providers, inventors, investors, and policymakers will collaborate to reduce the burden of unnecessary and preventable disability in India.

References

[1] Gururaj G. Epidemiology of traumatic brain injuries: Indian scenario. Neurol Res 2002;24(1). Available from: https://doi.org/10.1179/016164102101199503.

[2] Khan F, Amatya B, Mannan H, Fa R. Neurorehabilitation in developing countries: challenges and the way forward. Phys Med Rehabil Int 2015;9(2).

[3] Gourie-Devi M, Gururaj G, Satishchandra P, Subbakrishna DK. Prevalence of neurological disorders in Bangalore, India: a community-based study with a comparison between urban and rural areas. Neuroepidemiology 2004;23(6). Available from: https://doi.org/10.1159/000080090.

[4] Das A, Botticello AL, Wylie GR, Radhakrishnan K. Neurologic disability: a hidden epidemic for India. Neurology 2012;79(21):2146. Available from: https://doi.org/10.1212/WNL.0b013e3182752cdb Nov.

[5] World Health Organization. World report on disability 2011. World Health Organization; 2011.

[6] DHLCNSF Team. The national service framework for long-term conditions. *London: Department of Health*; 2005.

[7] Alliance AR. The need for a national rehabilitation strategy. Sydney, 2011.

[8] Bernhardt J, Urimubenshi G, Gandhi DBC, Eng JJ. Stroke rehabilitation in low-income and middle-income countries: a call to action. Lancet 2020;396(10260). Available from: https://doi.org/10.1016/S0140-6736(20)31313-1.

[9] Kumar Sg, Roy G, Kar S. Disability and rehabilitation services in India: issues and challenges. J Family Med Prim Care 2012;1(1). Available from: https://doi.org/10.4103/2249-4863.94458.

[10] Chatterjee S, Hore D, Arora A. A wearable soft pneumatic finger glove with antagonistic actuators for finger rehabilitation, In 2017 2nd International Conference on Communication

and Electronics Systems (ICCES), 2017, pp. 341−345. Available from: https://doi.org/ 10.1109/CESYS.2017.8321294.

[11] Singh N, Saini M, Kumar N, Srivastava MVP, Mehndiratta A. Evidence of neuroplasticity with robotic hand exoskeleton for post-stroke rehabilitation: a randomized controlled trial. J Neuroeng Rehabil 2021;18(1). Available from: https://doi.org/10.1186/s12984-021-00867-7 Dec.

[12] Singh N, Saini M, Anand S, Kumar N, Srivastava MVP, Mehndiratta A. Robotic exoskeleton for wrist and fingers joint in post-stroke neuro-rehabilitation for low-resource settings. IEEE Trans Neural Syst Rehabil Eng 2019;27(12):2369−77. Available from: https://doi.org/10.1109/TNSRE.2019.2943005.

[13] Mathew M, Arun M, Francis RN, Sudheer AP. Exoskeletal Development of a Hand Complex for Rehabilitation Activities, In 2021 International Conference on Intelligent Technologies (CONIT) June 2021. Available from: https://doi.org/10.1109/CONIT51480.2021.9498443.

[14] Nehrujee A, et al. Plug-and-train robot (PLUTO) for hand rehabilitation: design and preliminary evaluation. IEEE Access 2021;9:134957−71. Available from: https://doi.org/ 10.1109/ACCESS.2021.3115580.

[15] Balasubramanian S, Guguloth S, Mohammed JS, Sujatha S. A self-aligning end-effector robot for individual joint training of the human arm J Rehabil Assist Technol Eng 2021;820556683211019864. Available from: https://doi.org/10.1177/20556683211019866.

[16] Thakur S, Das S, Bhaumik S A. Smart-Band Operated Wrist Rehabilitation Robot, In Proceedings of 2020 IEEE Applied Signal Processing Conference, ASPCON 2020, Oct. 2020, pp. 119−122. Available from: https://doi.org/10.1109/ASPCON49795.2020.9276666.

[17] Seth D, Vardhan Varma VKH, Anirudh P, Kalyan P. Preliminary Design of Soft Exo-Suit for Arm Rehabilitation, In Lecture Notes in Computer Science (including subseries Lecture Notes in Artificial Intelligence and Lecture Notes in Bioinformatics), 2019, vol. 11582 LNCS, pp. 284−294. Available from: https://doi.org/10.1007/978-3-030-22219-2_22.

[18] Available from: https://www.beablehealth.com/armable-pro.

[19] Kumaran DS, Ali AS, Swathi GA, Nayak N, Raja RA, Arumugam A. Effects of a Novel Game-Based Arm Rehabilitation Device, the ArmAble, on Improving Upper Limb Function in People With Stroke, In Proceedings - International Conference on Developments in eSystems Engineering, DeSE, 2021, vol. 2021-December, pp. 259−264. Available from: https://doi.org/10.1109/DESE54285.2021.9719548.

[20] Available from: https://www.rymo.in/.

[21] Available from: https://www.gridbots.com/physio.html.

Chapter 20

Asia Pacific region: Malaysia

Eileen L.M. Su[1], Fazah Akhtar Hanapiah[2], Natiara Mohamad Hashim[2], Che Fai Yeong[1], Kang Xiang Khor[3] and Yvonne Y.W. Khor[4]

[1]*Faculty of Electrical Engineering, Universiti Teknologi Malaysia, Johor Bahru, Johor, Malaysia,* [2]*Faculty of Medicine, Universiti Teknologi MARA, Sungai Buloh, Selangor, Malaysia,* [3]*Techcare Innovation Sdn. Bhd., Taman Perindustrian Ringan Pulai, Skudai, Johor, Malaysia,* [4]*YK Natural Physio & Academy, Taman Austin Perdana, Johor Bahru, Johor, Malaysia*

Learning objectives

At the end of this chapter, the reader will be able to:

- Provide an overview of the methods and technology used for stroke rehabilitation in Malaysia.
- Understand the challenges in providing stroke rehabilitation in Malaysia.

20.1   Malaysia and health statistics

Malaysia had a population of approximately 32.7 million in 2021 with a growth rate of 0.2% as compared to 32.6 million and an annual growth rate of 0.4% in 2020. The growth rate of citizens remained stable at 1.0%. The older age population in Malaysia is increasing every year; the percentage of the population aged 65 and over (old age) increased from 7.0% to 7.4% from 2020 to 2021. In contrast, the percentage of young age population (0−14 years) and working age population (15−64 years) declined by 0.3% and 0.1% within the same period. In 2021, the male population outnumbered the female with 16.8 million and 15.9 million, respectively [1]. The major causes of death in Malaysia were largely attributed to diseases of the circulatory system and the respiratory system that accounted for over 22% and 21%, respectively. Hospitalizations were largely due to childbirth, diseases of the respiratory system, and infectious diseases [2,3]. Each year, a large share of the government's annual budget is allocated for health expenditures. In 2021, healthcare expenditure was 9.89% of the national budget, totaling RM31.9 billion [1,2]. Despite this, Malaysia has long experienced a lack of professionals to provide care for the population. The ratio of medical doctors to the

Rehabilitation Robots for Neurorehabilitation in High-, Low-, and Middle-Income Countries.
DOI: https://doi.org/10.1016/B978-0-323-91931-9.00020-7

population was 1:482, and the ratio of nurses to population was 1:302, while community nurses to population were 1:1386 [3].

The Malaysian healthcare system adopts a dual-tiered healthcare system comprising a heavily subsidized tax-funded public sector and a booming private sector funded by self-paying patients or health insurance, providing a binary but interdependent public—private model [4,5]. The Ministry of Health governs public sector health services through central, state, and district administrations and acts as the governing body regulating the private sectors [4]. Health-related services for persons with disabilities (PWDs) are provided by many government agencies, including health, welfare, education, human resources, housing, and the private sector and nongovernmental organizations [6]. In the early 1990s, disability preventive strategies were the mainstay of healthcare service via immunization programs, health education, antenatal and postnatal care, continuous child health assessment, nutrition, and healthy lifestyle [6]. These strategies have evolved and catapulted, currently, at the second phase (2011—20) with its Plan of Action (POA) on Health Care for PWDs to ensure a more comprehensive healthcare program [7]. The specific objectives of this POA are providing equal opportunities for health care, empowering individuals, families, and communities for self-care, and developing support services and adequate multilevel medical rehabilitation services and care [7]. The strategies to achieve these specific objectives are as follows: (1) advocate for PWDs' issues and policies, (2) increase accessibility to facilities and services, (3) empower individuals, families, and communities, (4) encourage and strengthen the multi- and inter-sectorial collaboration, (5) ensure adequate and competent workforce, (6) intensify research and development, and (7) health program development for specific disabilities [7,8].

Earlier, rehabilitation services were mostly delivered by physiotherapists (PTs) and occupational therapists (OTs) in government hospitals, focused mainly on physical rehabilitation of mobility and activities of daily living [6]. Rehabilitation practice then evolved beyond the conventional approach, integrating dynamic multidisciplinary medical approaches led by rehabilitation physicians, treating beyond physical impairments, and adopting a more holistic approach in a biopsychosocial spectrum of care [6]. There are currently approximately 106 rehabilitation physicians practicing in Malaysia. As of May 2020, 75%, that is, 79 practitioners are serving in Ministry of Health (MOH) tertiary hospitals, 18 in university-based settings (with or without a university hospital) (UM, UKM, UiTM, USM, UMS, UIA), and nine in the private practice [6].

Currently, active rehabilitation care is provided by governmental organizations, private institutions, and NGOs [6]. The government initiatives as a service provider are generally operated by general hospitals, university hospitals, primary health, community-based rehabilitation centers, special schools, and welfare organizations with approximately 1502 rehabilitation facilities [6]. The Ministry of Health provides specialized multidisciplinary rehabilitation services via its tertiary and secondary government hospitals; Ministry of Education via university hospital; and Ministry of Human Resources via rehabilitation and vocational center owned and funded by Social Security Organization (SOCSO) [6]. Ministry of Social Welfare and other NGO sectors also supplement this rehabilitation service continuum beyond government-funded medical facilities via their community-based rehabilitation program estimating a total of 511 centers [6,9]. However, these institutions and organizations do not always individually provide the complete spectrum of comprehensive rehabilitation services, but a few endeavors have taken place to ensure these facilities are acceptably equipped [6].

The private sectors have actively participated in providing rehabilitation services by filling the gaps of the governmental sectors whose facilities are very much stretched and limited [6]. The services span from rehabilitation and medical equipment provider companies, for example, orthotics and prosthetics, specialized wheelchairs, rehabilitation mobility services, private hospital services, private stand-alone specialized rehabilitation hospitals, nursing homes, and other few alternative therapies. However, many of these private practices are mainly managed by allied health personnel and lack comprehensive multidisciplinary approaches [6]. The number of private centers that provide some rehabilitation services is approximately around 686 centers [6].

Malaysia has seen a dynamic, proactive change in how rehabilitation medicine is practiced from adopting physician-led multidisciplinary rehabilitation strategies, integrating technological advancement in advance tertiary centers with the establishment of Hospital Rehabilitasi Cheras (HRC) serving as the national referral center, and increased government and private partnerships.

20.2   Stroke and neurorehabilitation

20.2.1   Stroke statistics and trends in neurorehabilitation

The National Neurology Registry (NNEUR) of Malaysia states that Malaysia is witnessing an escalating incidence of stroke, with it being the third most common cause of mortality and topping the nation's disability rate [10]. The Malaysia Stroke Council recorded 92 daily admissions in 2016 with 32 deaths daily from stroke. From the statistics in 2016, out of the 11,284 cases of stroke in Malaysia, 55% were men and 60% were over 60 years.

An 8-year registry of stroke patients in Malaysia between 2009 and 2016 noted that the mean age of stroke patients was 62.5 years old. The average stroke age in Malaysia is comparable to countries in Southeast Asia. However, this age is relatively younger compared to statistics obtained from the United Kingdom (74.2 years) and the United States (69.2 years) [10].

Awareness of neurorehabilitation is making its birth as part of stroke treatment in Malaysia. However, the outcome of stroke rehabilitation is still poorly documented. The disability outcome from stroke captures that of the acute discharges; 35% of patients are independent (Modified Rankin Scale 1−2), while the remaining 54% have various degrees of cognitive and physical impairment that will benefit from rehabilitation [10].

Stroke-related disability is best addressed by a multidisciplinary team of professionals [11]. As training for rehabilitation medicine in Malaysia is still in its infancy, the development of curricula in both the Ministry of Health and universities is now being consolidated to address subspecialty training and formation of post-basic courses. Neurorehabilitation is delivered in specialized centers, hospitals with appointed rehabilitation physicians (Ministry of Health general and state hospitals in Malaysia), major public teaching university hospitals, and some private hospitals [12].

In Malaysia, 39% of stroke patients are 59 years and below. This cohort of patients is considered within the productive age group, as the retirement age in Malaysia is 60 years old. Hence, the stroke neurorehabilitation program must also address return to work (RTW). The Social Security Organization (SOCSO) of Malaysia established its RTW program in 2007 and helps insured persons to overcome disability (including those as a result of stroke) and assist them in returning to meaningful employment. This program works closely with healthcare providers in hospitals and rehabilitation centers [13]. The call for infrastructure development and human resources to address the needs of stroke survivors has seen positive changes in the trend of the neurorehabilitation service provided [14].

20.2.2 Standard of care for stroke survivors

It is not uncommon for patients in Malaysia to receive acute stroke care and poststroke rehabilitation in a hospital that does not have stroke units or stroke rehabilitation units [12]. The acute care is managed by a team led by a general physician or a neurologist. On discharge, patients are referred to an outpatient rehabilitation program. If the hospital is equipped with a multidisciplinary rehabilitation team led by a rehabilitation physician, the inpatient care may be extended to allow a few days of intensive inpatient neurorehabilitation.

The standard stroke care model begins early, from acute, to subacute to long-term and sustained rehabilitation [15]. It is also common practice that an inpatient rehabilitation facility delivers 3 hours of therapy per day [16]. Early intensive multidisciplinary rehabilitation that is started earlier (within the first 9 days) has been shown to have the most improvement [17]. A local study also states that rehabilitation may play a role in reducing the 28-day readmission poststroke [18].

Typical recovery is dependent on individual patients and the severity of the stroke. Expected prognosticated and realistic goals are set, and the inpatient intensive rehabilitation duration can be between 2 weeks to 3 months. It is not uncommon that patients do not receive immediate rehabilitation poststroke due to limited resources and manpower. Coverage of insurance also lags in preparedness to provide funding for rehabilitation poststroke in a private setting.

There are 47 "stroke-ready" listed hospitals in Malaysia [19]. These hospitals are equipped with at least physiotherapists to accommodate the acute phase of rehabilitation. However, most fall short, except for a few that can provide intensive multidisciplinary inpatient rehabilitation. The Ministry of Health Malaysia has taken the initiative to build a stand-alone rehabilitation hospital in Cheras, Malaysia. HRC is a 166-bedded hospital that has a dedicated stroke/neurorehabilitation team. HRC serves as a specialist referral hospital for neurorehabilitation for the country. There are also two private stand-alone rehabilitation hospitals in Malaysia: Daehan Rehabilitation Hospital in Putrajaya and ReGen Hospital in Petaling Jaya.

20.3   Technological solutions for stroke therapy

20.3.1   Available technologies

The use of technological solutions for stroke therapy in Malaysia is still at an early stage. Most of these devices are imported into the country. The commonly used equipment for stroke therapy is electrical stimulation, EMG, TENS, shockwave, and ultrasound. Several bigger rehabilitation centers have imported robotic solutions, as shown in Table 20.1 [20−22]. From the list in Table 20.1, the only locally developed systems are the CR2-Motion and CR2-Haptic that are used in NASAM. The Malaysian government and many local organizations have expressed the need to drive technological innovations and solutions for stroke rehabilitation because most institutions cannot afford imported technologies. Several nationwide programs have strived to increase the awareness of research-driven innovations from universities. Some of these programs include The Great Lab by CREST, Innovate Malaysia, and Microsoft Imagine Cup, among others. Hackathons and innovation competitions also helped to kickstart market viability and prototype development.

20.3.2   Case studies of innovation within the country

One of our success stories is the development of Compact Rehabilitation Robot (CR2), which is currently commercialized through a Malaysian technological company (http://www.techcareinnovation.com). The CR2-Haptic (Fig. 20.1) was a robot designed to help stroke patients to train the upper limb, especially for pronation/supination, with haptics and virtual reality. The training was also gamified to increase the motivation and engagement of the users. The CR2-Haptic design was modular and reconfigurable, whereby

TABLE 20.1 Centers with robotic stroke therapy in Malaysia (as of August 2023) [20−22].

Types of robot	Centers with robotic service	Rehabilitation indications
Body weight support exoskeletal treadmill robotic gait trainer		
Lokomat Hocoma exoskeletal robotic	• Sunway Medical Centre • Prince Court Medical Centre • SOSCO Neuro-Robotics Rehabilitation & Cybernics Centre (PERKESO) • Therapedic Brain and Spine Robotic Centre • Cheras Rehabilitation Hospital • (Hospital Rehabilitasi Cheras) • iREHAB Rehabilitation Center	• Provides body weight support for repetitive high-intensity treadmill gait training exercise • Provides motor relearning biofeedback for gait training • Improves overall aerobic endurance performance • Improves joint range of motion by reducing the hypertonic condition of the limb • Improves overall balance and lower-limb strength by partial active-assisted therapy • Provides spatiotemporal and biomechanical parameters to guide the therapy delivery
Exoskeletal over-ground robotic suit gait trainer		
Hybrid Assistive Limb (HAL) robotic suit	• PERKESO • Hospital Universiti Sains Malaysia • Hospital Pengajar Universiti Putra Malaysia (HPUPM)	• Provides an exoskeletal over-ground gait training • Provides partial active-assisted gait training by detecting electromyography (EMG) signals by surface EMG to activate or assist the desired movement • Improves overall balance, endurance, and reciprocal gait training
Angel Robotics Lower Limb Exoskeleton	• Daehan Rehabilitation Hospital • Hospital Universiti Teknologi MARA (HUiTM)	• Provides repetitively an over-ground gait training for patients who has residual muscle power to the bilateral lower limb • Suitable for patients who can initiate walking • Provides assistive force control by actuator precision control technology that is

(Continued)

TABLE 20.1 (Continued)

Types of robot	Centers with robotic service	Rehabilitation indications
		AI-driven to provide customized assistive force that improves walking motor control and posture • Provides balance, strength, and coordination trainin
Freewalk Exoskeletal Robotic Gait trainer	Regen Rehabilitation Hospital	• Provides an exoskeletal over-ground gait training • Provides balance, strength, and coordination training that are essential for walking
End-effector robotic gait trainer		
REHA technology—GEO robotic gait-assisted trainer	• University Malaya Medical Centre (UMMC)	• Provides body weight support end-effector treadmill gait training • Provides repetitive passive, partial-assisted, or active gait training • An excellent gait training for patients who is not fit for exoskeleton robotic gait training due to hypertonic condition of the proximal lower-limb joint (hamstring spasticity) • Provides endurance, balance, coordination, and motor control training that is essential for walking • Provides biofeedback to enhance engagement and encourage motivation
REHA technology—NexStep robotic assisted trainer	• Daehan Rehabilitation Hospital	• Provides body weight support end-effector treadmill gait training • Provides repetitive passive, partial-assisted, or active gait training • An excellent gait training for patients who is not fit for exoskeleton robotic gait training due to hypertonic condition of the proximal lower-limb joint (hamstring spasticity)

(Continued)

TABLE 20.1 (Continued)

Types of robot	Centers with robotic service	Rehabilitation indications
		• Provides endurance, balance, coordination, and motor control training that is essential for walking • Provides biofeedback to enhance engagement and encourage motivation
Over-ground body weight support robotic gait trainer		
Andago Robotic Self-Directed Body weight Support (Hocoma)	• PERKESO	• Provides guided body weight support over-ground gait training • Provides real experience with partially assisted gait training • Improves overall balance, coordination, and motor control required for walking
Robotic joint mobilizer and trainer		
Erigo Standing (Hocoma)	• PERKESO	• Provides and assists very early mobilization and verticalization of immobile patients with little or no capacity for interaction or therapy engagement • Provides the treatment of orthostatic hypotension by gradual verticalization and end-effector robotic footplate providing repetitive passive-assisted motion of the ankle • Provides a preventive method of thrombus formation in the lower limb • Prevents the occurrence of joint stiffness or contracture formation • Prevents other complications of prolonged immobility such as pressure injuries, constipation, or orthostatic pneumonia.

(Continued)

TABLE 20.1 (Continued)

Types of robot	Centers with robotic service	Rehabilitation indications
		• Enhances cognitive performance, awareness, and general wellbeing by assisting the patient to be in an upright position
Robert Robotic Exoskeletal Mobilizer	• PERKESO	• An adjunctive approach to provide highly repetitive passive or resistive joint mobilization training, especially for ankle • Provides preventive intervention for thrombus formation in the lower limb (blood clot formation) • Prevention of ankle joint contracture formation secondary to hypertonic condition that is prevalent in stroke population
Cyberdyne single-Joint exoskeletal	• PERKESO • HPUPM • Hospital Universiti Sains Malaysia	• Provides specific partial-assisted joint motion to improve strength and promote motor relearning by detecting EMG signals of the muscles • Provides flexibility exercises toward the joint to maintain a good range of motion, reduces tone, and prevents joint stiffness or contractures
Upper-limb robotic trainer		
• CR2-Motion • CR2-Haptic	• National Stroke Association of Malaysia centers	• Provides active, assistive, and passive training for a range of motion exercise and strength training for wrist and forearm • Provides performance feedback and gamification to enhance therapy engagement and promote motivation

(Continued)

TABLE 20.1 (Continued)

Types of robot	Centers with robotic service	Rehabilitation indications
Techcare Hand Robot (HR-30)	• PERKESO • Tung Shin Hospital • Queen Elizabeth Hospital	• Provides repetitive passive and partial-assisted hand and functional training • Provides customizable finger exercise to increase finger flexibility, strength and range of motion • Provides intermittent air compression therapy to reduce hand spasticity and edema
• Armeo Spring • Armeo Boom	• PERKESO • UMMC • Hospital Rehabilitasi Cheras	• Provides passive or partial active-assisted exoskeleton hand therapy, especially for proximal hand weakness • Provides customization of required therapy according to the stage of stroke (early or late) or the severity of the hand weakness • Provides biofeedback for upper-limb training via gamification concept to enhance therapy engagement and promote motivation
• Armeo Senso • Gamified finger rehabilitation device	Therapedic Brain and Spine Robotic Centre	• Provides advanced hand training to improve proximal upper-limb coordination, motor control, and strength • Provides self-initiated, active, and repetitive hand training for selected patients with good cognitive function and availability of residual upper-limb muscle power • Provides self-guided repetitive training by augmented performance feedback to enhance appropriate upper-limb motor control by reducing the compensatory mechanism

(Continued)

TABLE 20.1 (Continued)

Types of robot	Centers with robotic service	Rehabilitation indications
Fourier Arm Motus M2	• HUiTM • Daehan Rehabilitation Hospital • ReGen Rehabilitation Hospital • UMMC	• Provides repetitive passive partial-assisted and active proximal hand training • Provides haptic sensory feedback to enhance the neurorecovery • Provides flexibility exercise to maintain proximal upper-limb joint range of motion • Provides performance feedback by gamification concept to enhance engagement and promote motivation
Neofect Rapael Smartboard	• Daehan Rehabilitation Hospital	• Provides repetitive passive partial-assisted and active proximal hand training • Provides flexibility exercise to maintain proximal upper-limb joint range of motion • Provides performance feedback by gamification concept to enhance engagement and promote motivation

the cost and size were reduced through the adoption of different therapeutic end effects for different training movements on a single robot [23–27].

The robot was initially developed by a group of undergraduates from Universiti Teknologi Malaysia (UTM) for their final-year project. This robot went through a series of design iterations and considered inputs from rehabilitation centers, physiotherapists, and patients. The product was submitted to various competitions, including MyInovasi Malaysia, Innovate Malaysia, Microsoft Imagine Cup, and British Invention Show, among others. The work received much recognition and accolades from these competitions that helped popularize the rehabilitation robot among local communities. The project was then continued as PhD research and went through trials with stroke survivors, to further investigate and improve the design, viability, and effectiveness [23–27].

Another project was a balance rehabilitation device called FIBOD (Fig. 20.2), which was an interactive balancing device with biofeedback

FIGURE 20.1 The earlier prototype (left) and the commercialized version of CR2-Haptic (right). The left picture shows an earlier prototype of the CR2-Haptic, with plastic covers at the top and bottom, held in place with a few rods. The wires, circuit board, and motor within the prototype were visible. The motor was connected to a round, white colored knob, and a hand was seen holding the knob with five fingers. On top of the prototype cover was a red color emergency stop button.

The picture on the right showed a person sitting on a wheelchair, trying out the recent design of the CR2-Haptic. The CR2-Haptic system was placed on table. The user was looking at a monitor with animated games, while the right hand was placed on the table; the forearm was strapped to the device, and the hand was gripping a CR2-Haptic handle, trying to rotate the handle. The handle is connected to a black color metal box in front of the hand, which housed the CR2-Haptic main controller. This box has a small part colored in orange and a red emergency button on top of the box.

[28−30]. This product was developed by another UTM undergraduate in the final year of study. Similarly, the product was submitted to several national-level competitions, namely Innovate Malaysia, Inatex, and Microsoft Imagine Cup, to obtain exposure and validation for the design. The invention was finally commercialized as a consumer product FIBOD and has been exported to a few countries. FIBOD has been used at Universiti Sains Malaysia Hospital (HUSM), Malaysian Research Institute on Ageing (MyAgeing), Sultanah Nur Zahirah Hospital and PERKESO Rehabilitation Center.

The CR2 and FIBOD present two cases where innovations from university research could be commercialized, providing a reference for researchers in this field who are also striving to offer affordable technology for the community.

20.4 Barriers or opportunities to rehabilitation

20.4.1 Barriers in stroke neurorehabilitation

Despite being recognized as an upper-middle-income country, the need for rehabilitation in Malaysia is still largely unmet, and this gap is becoming more extensive due to disrupted healthcare services during the COVID-19 pandemic. The facilities available in some smaller government hospitals are basic, unlike those in tertiary hospitals equipped with more advanced

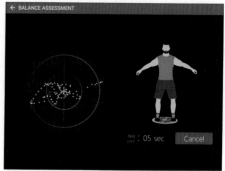

FIGURE 20.2 Fibod (left) is a smart fitness balance board that allows users to train and access their balance skills with virtual reality games and an objective assessment program. An AI program with machine learning feature (right) in predicting risk of injury. The left image showed an illustration of a human standing on a round fitness balancing board with both the arms outstretched. The board was wirelessly connected to a monitor. The large monitor screen showed an animated game, where a vehicle controlled by the balancing board was chasing after a figure running on a jungle trail, with tall trees along the trail.

The image on the right showed a program interface titled Balance Assessment. The interface was divided into two main parts, with one part on the left illustrating a top view of the balance board movement. The board movement was depicted in white dots on two circles: an bigger diameter outer circle and a smaller diameter inner circle. A red dot which formed the target was located within the inner circle. The other side of the interface showed an avatar of a man standing on the balancing board with both the arms outstretched. The time left for the trial was shown at the bottom of the avatar. A cancel button is available at the lower right corner of the interface. Source: *Available from: http://www.techcareinnovation.com.*

technologies and therapists [6]. However, not all tertiary hospitals have a complete and comprehensive rehabilitation service team led by rehabilitation medicine physicians [6]. There are only about 106 rehabilitation medicine specialists for an estimated 3 million disabled in the population. The ratio of rehabilitation physicians to PWDs is still significantly low in Malaysia [6].

The national health service provides rehabilitation free of charge for those who have registered with the Social Welfare Department and obtained their PWD card. However, the demand for the service exceeds the supply in the government setting. Private healthcare facilities do provide rehabilitation services to address this gap. The provision of private services is costly, especially when insurance coverage for rehabilitation is limited. A recent study looking at the costs of receiving poststroke outpatient care in a government university hospital concluded that costs were primarily to cover attendant care and traveling, as service provision was free of charge for PWD card holders [31]. Another study looking at case-mix costing analysis for acute ischemic stroke and follow-up in another government university hospital quoted a cost of RM9000 (USD2140) for acute major stroke hospitalization and RM3353 (USD797) for a minor stroke. Outpatient setting specialist visit was RM103 (USD25), and the cost of rehabilitation therapy was RM43 (USD10) per patient per session [32].

Neurorehabilitation takes time and effort before the results may be evident; hence with escalating costs of living, attitude toward investing in neurorehabilitation may not take precedence as compared to paying for care attendants as an immediate need of care. The healthcare system, strained by a lack of resources and trained staff, often does not prioritize neurorehabilitation. This is evident by the lack of multidisciplinary teams to provide neurorehabilitation in tertiary hospitals and community health facilities. As reported in 2016, 144 government hospitals throughout the country are equipped to offer rehabilitation services, though only 16 of them have rehabilitation physicians. Hence, only some would have the privilege to receive the intensive neurorehabilitation of 3 hours per day as outlined above. The remaining hospitals can provide rehabilitation, though not specialized services such as neurorehabilitation [33].

The lack of standardized discharge protocol adds to the barrier of neurorehabilitation on discharge from acute hospitalization poststroke [34]. A discharge pathway to improve accessibility to rehabilitation along the spectrum will reduce the possibility of patients slipping through the system.

Lack of knowledge on the availability of the services also appears to be a barrier [12]. This is not only among the general public but also among healthcare providers. The lack of knowledge is amplified more in the rural area of Malaysia. Understanding stroke, or the lack of it, has the potential negative impact on the health outcome poststroke [35]. Nonetheless, as knowledge increases and the demand and expectation from the public grow on receiving neurorehabilitation, the government and private sectors have to be prepared to do so.

20.4.2 Barriers to the implementation of robotic rehabilitation

Robotic training solutions have been recommended by American Stroke Association [36]; however, there are many challenges in implementing robotic solutions widely in the country. For any technology or rehabilitation robot to be implemented in a medical facility in Malaysia, the product must be registered under the Medical Device Authority (MDA) [37]. MDA is a statutory body under the Ministry of Health Malaysia, which was established to control and regulate the medical device industry and its activities, as well as to enforce the Medical Device Act 2012.

Additionally, hospitals would have both the engineering and clinical teams evaluate a product prior to the adoption of any rehabilitation technology. Clinical teams include therapists or rehabilitation physicians who test the product and ensure it can deliver its intended usage in the actual setting. The engineering team's role is to check the product specification and the maintenance of the product to ensure smooth product operations throughout the usage and to consider its maintenance requirements.

From the perspective of local researchers, the high development cost and the high costs of lengthy clinical trials to fulfill requirements by local regulatory authorities result in many small teams or companies being unable to hold out till commercialization. In the end, products from larger international corporations that can afford the developmental costs and the medical device certifications are those left standing. The stringent rules set by the government are indeed needed to protect the welfare of the community, but it inadvertently also forms a high barrier of entry for newer researchers with little capital.

Most of the available robotic solutions require the need of high investment costs and large space to operate, both of which are scarce resources in Malaysian hospitals, particularly in government-run facilities. This equipment also needs well-trained robotic therapists, which Malaysia still lacks [20]. Typically, robotic devices designed for rehabilitation are limited to only certain components of recovery; for complete training, multiple devices are needed for different components in a complete rehabilitation plan. Thus this further increases the overall cost of using technology for the full continuum of recovery.

Many robotic solutions are designed for clinical settings and cannot be installed in patients' homes easily. Home-based devices should be more user-friendly, easy to use, portable, and affordable [38−40]. The limited opportunity to use the robotic device may also hamper patient recovery when they could only use the robot for a few hours a month. The seemingly slow recovery rate may reduce the confidence of patients and practitioners in the potential of robotic rehabilitation. In actual fact, robotic systems should only be considered as vehicles that enable the delivery of evidence-based, impairment-oriented treatment, providing highly repetitive, intensive, and interactive treatment that is not possible in usual care. Further improvements in function might come from combining treatments that target different impairments [41].

20.4.3   Policies for implementing effective rehabilitation

These limitations of unmet rehabilitation need span from lack of awareness among the community of the importance of rehabilitation, limited funding, inadequate integration of rehabilitation pathway in every tier of the healthcare system, shortage of trained rehabilitation professionals to form an effective multidisciplinary team, and also constrained resources which include facilities, equipment, assistive technology, and consumables [42,43]. However, these gaps provide an avenue to encourage research and development activities and partnerships among the government and nongovernmental organizations, academics, and other stakeholders in searching for effective, affordable strategies to overcome these problems. A few strategies that can be adopted in fulfilling the criteria of an optimum rehabilitation healthcare provision are as follows:

- Encourage research, development, and innovation of homegrown affordable, sophisticated, technologically advanced devices underpinning the fundamental principles of rehabilitation interventions.
- Provide training and upgrading skills to all healthcare professionals involved as a rehabilitation service provider.
- Optimize high-quality, specialized rehabilitation institutions and improve their accessibility.
- Establish policy and implementation of health insurance coverage toward rehabilitation health services and their needs.
- Expand sustainable frameworks pertaining to financial support toward lower socioeconomic PWDs.
- Develop and establish an effective method of remote- or home-based rehabilitation services via telerehabilitation.
- Encourage a transdisciplinary discussion and collaboration of essential stakeholders to establish a cost-effective and time-effective rehabilitation services framework that includes policymakers, healthcare professionals, technologists, scientists, health economists, etc.

20.5   How COVID-19 has changed rehabilitation

Stroke rehabilitation, particularly neurorehabilitation training, is a labor-intensive process that requires thousands of training repetitions, according to the concept of neuroplasticity. Most of the time, training relies on a lot of human assistance from caregivers to enable patients to make a required movement. Getting patients to have sufficient training is extremely difficult even during normal conditions due to the steep ratio between therapists and PWDs. The difficulty was amplified even further during the COVID-19 outbreaks where many rehabilitation centers and hospitals limited their rehabilitation services to reduce the spread of the virus. Besides, many patients also voluntarily ceased their rehabilitation training amid the COVID-19 situation due to concerns about their safety. Thus the rise of COVID-19 cases called for the reshaping of rehabilitation services to ensure that patients continue to receive quality and personalized rehabilitation care [44].

Because of movement limitations and physical interaction restrictions, technology and robotic solutions have become critical in providing access to affordable and good quality care through home-based rehabilitation [44]. During the pandemic, some service providers quickly launched their virtual rehabilitation training for stroke rehabilitation. National Stroke Association of Malaysia (NASAM) introduced telehealth program for stroke with group sessions of aerobic exercises to improve strength and balance as well as "qi gong" exercises for body and mind relaxation via the teleconferencing app Zoom [45]. Techcare Innovation also implemented home-based robotic hand telerehabilitation, with remote supervision by therapists [46]. When telerehabilitation was made available, private rehabilitation centers were able to provide rehabilitation sessions to their patients with remote guidance, maintained patient progression, and also maintained their business operations. Online courses or webinars for physiotherapists and patients increased in the social media sphere to build awareness about exercises and musculoskeletal issues. In a way, this also helped patients gain knowledge about their conditions and encouraged the community to practice a healthier lifestyle.

COVID-19 has accelerated the rise of digital health and wellness technologies. The digital revolution, along with COVID-19 and its restrictions, has significantly shifted customer demands and expectations for digital health in Malaysia [47].

20.6   Future of rehabilitation and opportunities

The area in which evidence-based medicine studies on the use of rehabilitation robotics is a growing area of interest among clinicians and engineers alike [48–54]. Rehabilitation robots primarily are not to replace therapists but are used to enhance and augment conventional therapy by adding value toward the quality and repetition of movements. In Malaysia, a recent review article by Nik Ramli et al. explored the effectiveness of robot-assisted physical therapy poststroke. The complexity and heterogeneity of the population studied and differences in robotic device used made it difficult to have a single conclusion. However, the authors remarked that robot-assisted therapy as an adjunct is potentially beneficial for stroke rehabilitation [20]. During the COVID-19 lockdown, digital technology and robotic solutions have been used to provide telerehabilitation, overcoming the barriers in rendering care to rural communities. With the emergence of more local research in rehabilitation robotics, in both R&D and application, the prospect of its adoption in the standard of care for neurorehabilitation in Malaysia looks promising.

Acknowledgments

The contributors would like to thank the Ministry of Health, Malaysia, and the Ministry of Higher Education, Malaysia, for supporting the work published here.

Conflict of interest

The contributors declare no conflict of interest.

References

[1] Department of Statistics Malaysia Official Portal. Population & Demography, Available from: https://www.dosm.gov.my/; 2021 (accessed 16 December 2021).

[2] Ministry of Health Malaysia. Health Facts 2020, Available from: https://www.moh.gov.my/moh/resources/Penerbitan/Penerbitan%20Utama/HEALTH%20FACTS/Health%20Facts%202020.pdf; 2020 (accessed 16 December 2021).

[3] International Trade Administration. Malaysia Healthcare, Available from: https://www.trade.gov/market-intelligence/malaysia-healthcare; 2021 (accessed 2 October 2021).

[4] World Health Organisation. Malaysia health system review. Manila: WHO Regional Office for the Western Pacific; 2012.

[5] Quek D. The Malaysian healthcare system: a review. Intensive Workshop on Health Systems in Transition; 2009.

[6] Naicker AS, Yatim SM, Engkasan JP, Mazlan M, Yusof YM, Yuliawiratman BS, et al. Rehabilitation in Malaysia. Phys Med Rehabil Clin 2019;30(4):807−16.

[7] Public Health Department,Ministry Health Malaysia. Health care program for person with disabilities, plan of action 2011−2020. Book Publication by Family Health Development Division, Public Health Department,Ministry Health Malaysia; 2011.

[8] Global Disability Plan of Action, Family Health Development Division, Public Health Department, Ministry of Health Malaysia.

[9] Official Portal Department of Social Welfare. Available from: https://www.jkm.gov.my/jkm/index.php; 2021 (accessed 10. October 2021).

[10] Zariah AA, Norsima NS, editors. Annual Report of the Malaysian Stroke Registry 2009−2016. Available from: https://www.neuro.org.my/assets/guideline/Stroke%20registry%20report%202009-2016.pdf; 2017 (accessed 8 July 2022).

[11] Leonardi M, Fheodoroff K. Goal setting with ICF (International Classification of Functioning, Disability and Health) and multidisciplinary team approach in stroke rehabilitation. Clinical pathways in stroke rehabilitation: evidence-based clinical practice recommendations. Thomas Platz. Springer; 2021. p. 35−56.

[12] Ministry of Health Malaysia. Stroke Rehabilitation Health Technology Assessment Report. MOH/PAK/45.02(TR).

[13] Return to Work Program. Available from: https://www.perkeso.gov.my/en/our-services/rehabilitation/return-to-work-program.html; 2021 (accessed 5 November 2021).

[14] Kurubaran G, editor. Malaysia's Stroke Care Revolution. Discover! Special Edition. 2019; 2(1) [Special Issue: ISSN 2600−9110].

[15] Wasti SA, Surya N, Stephan KM, Owolabi M. Healthcare settings for rehabilitation after stroke. Clinical pathways in stroke rehabilitation: evidence-based clinical practice recommendations. Thomas Platz, editor. Springer; 2021. p. 261−82.

[16] Inpatient Rehabilitation Therapy Services: Complying with Documentation Requirements. Department of Health and Human Services. Centers for Medicare & Medicaid Services. 16 July 2012 ICN 905643.

[17] Serce A, Karaca Umay E, Cakci FA. Early intensive multi-faceted rehabilitation in stroke patients: what is the best effective rehabilitation time? J Phys Med Rehabilitation Sci 2021;24(3):267−76.

[18] Ang SH, Hwong WY, Bots ML, Sivasampu S, Abdul Aziz AF, Hoo FK, et al. Risk of 28-day readmissions among stroke patients in Malaysia (2008−2015): trends, causes and its associated factors. PLoS One 2021;16(1):e0245448. Available from: https://doi.org/10.1371/journal.pone.0245448.

[19] Mystroke Hospital. Available from: https://mystrokehospital.my/hospital/; 2021 (accessed 5 November 2021).

[20] Nik Ramli NN, Asokan A, Mayakrishnan D, Annamalai H. Exploring stroke rehabilitation in Malaysia: are robots better than humans for stroke recuperation? Malaysian J Med Sci 2021;28(4):14−23. Available from: https://doi.org/10.21315/mjms2021.28.4.3. Available from: http://www.mjms.usm.my/MJMS28042021/MJMS28042021_03.pdf.

[21] Bulletin Mutiara. First Therapedic Brain and Spine Robotic Centre launched in Penang. Available from: https://www.buletinmutiara.com/first-therapedic-brain-and-spine-robotic-centre-launched-in-penang/; 2020 (accessed 08 July 2022).

[22] News Straits Times. Futuristic Rehab Centre, Available from: https://www.rehabmalaysia.com/my_asset/Futuristic%20Rehab%20Centre%20-NST%20(1%20Oktober%202019).pdf; 2019 (accessed 08 July 2022).

[23] Khor KX, Rahman HA, Fu SK, Sim LS, Yeong CF, Su ELM. A novel hybrid rehabilitation robot for upper and lower limbs rehabilitation training. Procedia Computer Sci 2014;42:293−300 [ISSN 1877-0509].

[24] Khor KX, Chin PJH, Rahman HA, Yeong HA, Yeong CF, Su ELM, et al. A novel haptic interface and control algorithm for robotic rehabilitation of stoke patient. In: Proc. 2014 IEEE Haptics Symposium (HAPTICS); 2014; 421−426. Available from https://doi.org/10.1109/HAPTICS.2014.6775492.

[25] Khor KX, Chin PJH, Rahman HA, Yeong HA, Yeong CF, Narayanan ALT, et al. Development of CR2-Haptic: a compact and portable rehabilitation robot for wrist and forearm training. In: 2014 IEEE Conference on Biomedical Engineering and Sciences (IECBES), 2014; 424−429. Available from: https://doi.org/10.1109/IECBES.2014.7047535.

[26] Khor KX, Chin PJH, Rahman HA, Yeong HA, Yeong CF, Narayanan ALT, et al. Development of reconfigurable rehabilitation robot for post-stroke forearm and wrist training. J Teknol 2015;72(2). Available from: https://doi.org/10.11113/jt.v72.3888. Available from.

[27] Khor KX, et al. Portable and reconfigurable wrist robot improves hand function for post-stroke subjects. IEEE Trans Neural Syst Rehabil Eng 2017;25(10):1864−73. Available from: https://doi.org/10.1109/TNSRE.2017.2692520. Available from.

[28] Khor KX, Mustar MF, Abdullah N, Yeong CF, Darsim MN, Su ELM. Development of InnovaBoard: an interactive balance board for balancing training and ankle rehabilitation. In: 2016 IEEE International Symposium on Robotics and Intelligent Sensors (IRIS). 2016; p. 128−133. Available from: https://doi.org/10.1109/IRIS.2016.8066078.

[29] Khor KX, et al. Smart balance board to improve balance and reduce fall risk: pilot study. In: Ibrahim F, Usman J, Ahmad MY, Hamzah N, editor. Proceedings of 2nd International Conference for Innovation in Biomedical Engineering and Life Sciences. 2018; (67) p. 35−39. Available from: https://doi.org/10.1007/978-981-10-7554-4_6.

[30] Khor KX, et al. Balance assessment for double and single leg stance using FIBOD balance system. In: Ibrahim F, Usman J, Ahmad MY, Hamzah N, editor. Proceedings of 3rd International Conference for Innovation in Biomedical Engineering and Life Sciences. ICIBEL 2019. vol 81. Springer, Cham; 2021. p.103−110. Available from: https://doi.org/10.1007/978-3-030-65092-6_12.

[31] Hejazi SMA, Mazlan M, Abdullah SJF, Patrick Engkasan J. Cost of post-stroke outpatient care in Malaysia. Singap Med J 2015;56(2):116−19.

[32] Aznida FAA, Nor MNA, Amrizal MN, Saperi S, Aljunid SM. The cost of treating an acute ischaemic stroke event and follow-up at a teaching hospital in Malaysia: a casemix costing analysis. BMC Health Serv Res 2012;12(Suppl 1):p6.

[33] The Sunday Mail. Malaysia faces chronic shortage of rehab specialists. Available from: https://www.thesundaily.my/archive/1819729-YSARCH370512; 2016 (accessed 5 November 2021).

[34] Abdul Aziz AF, Mohd Nordin NA, Ali MF, Abd Aziz NA, Sulong S, Aljunid SM. The integrated care pathway for post stroke patients (iCaPPS): a shared care approach between stakeholders in areas with limited access to specialist stroke care services. BMC Health Serv Res 2017;17(35):1−14.

[35] Yap KH, Warren N, Allotey P, Reidpath DD. Understandings of stroke in rural Malaysia: ethnographic insights. Disability Rehabil 2021;;43(3):345−53.

[36] Winstein CJ, et al. Guidelines for adult stroke rehabilitation and recovery: a guideline for healthcare professionals from the American Heart Association/American Stroke Association. Stroke 2016;47(6):e98−e169.

[37] Ministry of Health Malaysia. Medical Devices: Guidance document − Guidance on the rules of classification for general medical devices. MDA/GD/0009 First Edition. 2014.

[38] Van Norman GA. Drugs and devices: comparison of European and U.S. approval processes. JACC Basic Transl Sci 2016;1(5):399−412. Available from: https://doi.org/10.1016/j.jacbts.2016.06.003.

[39] Weber LM, Stein J. The use of robots in stroke rehabilitation: a narrative review. NeuroRehabilitation 2018;43(1):99−110. Available from: https://doi.org/10.3233/NRE-172408.

[40] Li L, Fu Q, Tyson S, Preston N, Weightman A. A scoping review of design requirements for a home-based upper limb rehabilitation robot for stroke. Top Stroke Rehabil 2021;29 (6):449−63. Available from: https://doi.org/10.1080/10749357.2021.1943797.

[41] Mazzoleni S, Duret C, Grosmaire AG, Battini E. Combining upper limb robotic rehabilitation with other therapeutic approaches after stroke: current status, rationale, and challenges. Biomed Res Int 2017;2017:p1−p11. Available from: https://doi.org/10.1155/2017/8905637.

[42] World Health Organization. Rehabilitation in health systems. Available from: https://apps.who.int/iris/handle/10665/254506; 2017. Licence: CC BY-NC-SA.

[43] Hughes AM, et al. Translation of evidence-based assistive technologies into stroke rehabilitation: users' perceptions of the barriers and opportunities. BMC Health Serv Res 2014;14(124):p1−p12. Available from: https://doi.org/10.1186/1472-6963-14-124.

[44] Manjunatha H, Pareek S, Jujjavarapu SS, Ghobadi M, Kesavadas T, Esfahani ET. Upper limb home-based robotic rehabilitation during COVID-19 outbreak. Front Robot AI 2021;8:1−14. Available from: https://doi.org/10.3389/frobt.2021.612834.

[45] The Star. NASAM Helping stroke survivors in a pandemic, Available from: https://www.thestar.com.my/news/focus/2020/10/25/helping-stroke-survivors-in-a-pandemic; 2020 (accessed 8 July 2022).

[46] Techcare Innovation. Techcare Tele-Rehab, Available from: https://techcareinnovation.com/tele-rehab/; 2021 (accessed 8 July 2022).

[47] Shukor A.A.. Digital Health & Wellness: Then, Now and Tomorrow Foreword by MaGIC. Available from: https://www.mymagic.my/media-centre; 2021 (accessed 8 July 2022).

[48] Duret C, Grosmaire AG, Krebs HI. Robot-assisted therapy in upper extremity hemiparesis: overview of an evidence-based approach. Front Neurol 2019;10:412.

[49] Hamzah N, Giban NI, Mazlan M. Robotic upper limb rehabilitation using Armeo®Spring for chronic stroke patients at University Malaya Medical Centre (UMMC). In: Ibrahim F, Usman J, Ahmad M, Hamzah N, Teh S, editors. Proceedings of 2nd International

Conference for Innovation in Biomedical Engineering and Life Sciences (ICIBEL 2017). 2018; 67:p225−230. Springer, Singapore.

[50] Hau CT, Gouwanda D, Gopalai AA, Low CY, Hanapiah FA. Gamification and control of nitinol based ankle rehabilitation robot. Biomimetics. 2021;6(3):53. Available from: https://doi.org/10.3390/biomimetics6030053.

[51] Baronchelli F, Zucchella C, Serrao M, Intiso D, Bartolo M. The effect of robotic assisted gait training with Lokomat® on balance control after stroke: systematic review and meta-analysis. Front Neurol 2021;12:661815.

[52] Porciuncula F, Baker TC, Arumukhom Revi D, Bae J, Sloutsky R, Ellis TD, et al. Targeting paretic propulsion and walking speed with a soft robotic exosuit: a consideration-of-concept trial. Front Neurorobot 2021;15:689577.

[53] Shamsuddin S, Abdul Malik N, Hashim H, Yussof H, Hanapiah FA, Mohamed S. Robots as adjunct therapy: reflections and suggestions in rehabilitation for people with cognitive impairments. In: Khairuddin O, et al., editors. Intelligent robotics systems: inspiring the NEXT. FIRA 2013. Communications in computer and information science, 376. Berlin, Heidelberg: Springer; 2013. p. 390−404. Available from: https://doi.org/10.1007/978-3-642-40409-2_33.

[54] Flynn N, Kuys S, Froude E, Cooke D. Introducing robotic upper limb training into routine clinical practice for stroke survivors: perceptions of occupational therapists and physiotherapists. Australian Occup Ther J 2019;66(4):530−8. Available from: https://doi.org/10.1111/1440-1630.12594.

Chapter 21

Asia Pacific region: China

Ning Cao[1], Koh Teck Hong (Zen)[2], Zhen Chen[3,4,5] and Zikai Hua[6]

[1]Department of Physical Medicine and Rehabilitation, Johns Hopkins University School of Medicine, Baltimore, MD, United States, [2]MotusAcademy, Zürich, Switzerland, [3]Neurorehabilitation Center, The First Rehabilitation Hospital of Shanghai, Shanghai, P.R. China, [4]Tongji University School of Medicine, Shanghai, P.R. China, [5]Tenth People's Hospital of Tongji University, Shanghai, P.R. China, [6]Shanghai University, Shanghai, P.R. China

Learning objectives

At the end of the chapter, the reader will be able to:

1. Understand the status of neurorehabilitation and rehabilitation robotics in China.
2. Describe how the rehabilitation process was handled during the pandemic.

21.1 Current status of neurorehabilitation in China

21.1.1 Overview

A fifth of the world's population resides in China, with a population amounting to 1.4 billion people [1,2]. With over 85 million disabled people and an aging population over 60 years that numbers a staggering 260 million, the rehabilitation demands have steadily increased in the past years [3].

Moreover, the global burden of stroke is huge, and China bears the biggest burden in the world with an estimated 2.4 million new stroke cases and 1.1 million stroke-related deaths annually, with 11.1 million stroke survivors at any given time [2,4,5]. Over the past decade, there has been an enormous growth of research in stroke leading to an increase in evidence-based stroke care and improvements in in-hospital outcomes [5]. The government has also launched the China National Stroke Registry to improve the quality of stroke care throughout the country [4]. A notable recent nationwide study such as the National Epidemiological Survey of Stroke in China (NESS-China) included 480,687 individuals over 20 years old from 31 provinces between 2012 and 2013. This reports China with the highest stroke prevalence and mortality in the world with a prevalence of 1115 cases per 100,000 people and mortality of 115 per 100,000 cases [5].

Over half of the population lives in rural China, and the overall stroke incidence is higher in rural (298 cases per 100,000 person-years) than in urban areas (204 cases per 100,000 person-years) [5], and the trend tends to be increasing [2]. A comparison of the Chinese geographical regions in stroke incidence and mortality shows apparent variations with higher numbers in the northeast compared to the south of China. However, based on the NESS-China study, the estimated mortality-to-incidence ratio (MIR) of stroke shows the highest MIR in the southwest and the lowest along the eastern and southern coasts. These regional differences prove the disparities in access and quality of stroke care across the country and are consistent with regional Healthcare Access and Quality (HAQ) index patterns. The distribution of MIR and HAQ might be associated with the variation in the proportion of registered medical doctors per 1000 people, with higher numbers in the north and east of China and the lowest in the southwest. There are variations in the proportion of secondary and tertiary hospitals with certified stroke centers with higher numbers in the east and south of China and lower numbers in the northeast and western regions [5].

21.1.2 Healthcare situation

To cope with the healthcare needs of a burgeoning population, the Healthy China 2030 blueprint was released by the Chinese government in 2016. It stated that "all people can enjoy the required quality and affordable health services such as prevention, treatment, rehabilitation, and health promotion." As an important link in the modern health service continuum, rehabilitation is of special significance to promoting public health. The five major goals of this national strategy were to improve the level of health, reduce major risk factors, increase the capacity of the health service provision, scale up the health industry, and perfect the health service system [2]. This blueprint is based on four core principles, that is, health priority, innovation, scientific development, and fairness and justice. The framework is shown in Fig. 21.1.

Medical rehabilitation services have been drastically promoted since 2008, after the Wen Chuan earthquake incident [1]. In 2009, the Chinese government announced the National Health Care Strategy which is made up of three major elements—prevention, treatment, and rehabilitation whereas the Healthy China 2030 strategy has six key principles related to rehabilitation medicine [1,2]. Health integration has become a priority in all government departments to ensure health in every aspect across the lifespan of citizens. Most provinces are adopting and investing in health promotion projects [1]. In 2012, the Ministry of Health issued a Guidance on Rehabilitation Medicine during the 12th Five-Year Plan for Health Development, which proposed to establish tiered and staged rehabilitative

Goal				
Put health on the priority list of development to a strategic position; promote the concept of health in the whole process of public policy implementation;enable everyone to be involved health and everyone to share health care services;focus on the health of all the people all their life in China.				
Principles				
Health Priority	Reform and Innovation	Scientific Development		Justice and Equity
HC 2030: China's vision for health care				
1. Health Level	2. Healthy life	3. Health Services and Health Security	4. Environmental Health	5. Health Industry
The 13 Core Indicators				
A. The average life expectancy B. The mortality rate of infants C. The mortality rate of children below 5 years of age D. The mortality rate of pregnant women and mortality E. The proportion of those meeting the national physique determination standard among urban and rural residents	A. The level of health literacy among residents B. The number of people taking part in physical exercise	A. Premature mortality as a result of major non–communicable diseases B. The number of registered doctors per 1000 residents and registered nurses per 1000 residents C. The proportion of personal health spending in the total health expenses	A. Good air quality rate of all cities at prefecture level or above B. The rate of surface water quality better than III	A. The total investment scale of health services

FIGURE 21.1   The framework of Healthy China 2030 vision [6]. Source: *Adapted from Tan X, Liu X, Shao H. Healthy China 2030: a vision for health care. Value Health Reg Issues. 2017;12: 113.*

service networks and to build two-way referral systems between general hospitals, rehabilitation centers, and community health centers. The policy shared details for the service network including:

- enhancing development and management of rehabilitative services,
- developing rehabilitation professionals,
- increasing capacity to deliver rehabilitation services,
- establishing tiers and stages of rehabilitation care, and
- pooling and coordinating the use of rehabilitative resources.

Fig. 21.2 illustrates the framework of the rehabilitative service network in China [3].

The current models of rehabilitation services in China include rehabilitation departments in general hospitals, specialized rehabilitation hospitals, community-based rehabilitation, and home-based rehabilitation which can be either private or government hospitals [1]. Since the early 2000s, patients with stroke have received routine therapy in three-stage rehabilitation networks. In 2012, the Ministry of Health endorsed rehabilitation assessment as part of a national standard for acute stroke care [4]. However, 30%−60% of patients still do not have access to in-hospital rehabilitation and those who did receive it in the form of acupuncture, passive massage, or isolated physical therapy with less than 10% receiving any form of physiotherapy, occupational therapy, speech-language therapy, or psychotherapy [5]. This is due to the uneven distribution of medical resources between regions, low bed counts, low talent acquisitions, and lack of proper rehabilitation care management, leading to unmet potential needs for rehabilitation [3]. Looking to the latest data in 2018, there are 38,260 physicians and 15,514 nurses in rehabilitation hospitals with an estimated 40,000 therapists in practice [1].

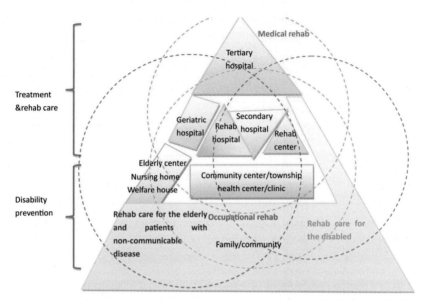

FIGURE 21.2 Current rehabilitation care system in China. *Source: Adapted from Xiao Y, Zhao K, Ma Z-X, Li X, Qiu Y-P. Integrated medical rehabilitation delivery in China. Chronic Dis Transl Med. 2017;3(02):77.*

The projected need for physicians, nurses and therapists is estimated to rise to 60,000, 60,000, and 150,000, respectively [1]. In June 2021, eight departments of the State Council, including the National Health Commission, jointly formulated the notice on accelerating the development of rehabilitation medical work, proposing to improve the rehabilitation medical service system, strengthen the construction of rehabilitation medical professional team, promote the reform and innovation in the field of rehabilitation medical service, and promote the high-quality development of rehabilitation medical service. It will certainly promote the further development of rehabilitation medicine in China.

21.1.3 Technology-assisted rehabilitation

In late 2020, the Intelligent Rehabilitation Professional Committee of the Chinese Association of Rehabilitation Medicine was established in China [7]. The goal of this exclusive committee is to assemble the top talent in rehabilitation medicine and biomedical engineering in China and work toward building an integrated exchange platform that involves intelligent technologies and rehabilitation medicine. The plan to extensively integrate intelligent technologies consisting of artificial intelligence, cloud networking, virtual reality, and wearable sensors into rehabilitation will significantly enhance the process of rehabilitation assessment and therapy that is still considered to be inefficient in the current stage of rehabilitation medical services.

21.2 Current status of robotics therapy in neurorehabilitation within China

21.2.1 Overview

In China, a "state−province−city−county" type of vertical management system is used in the national public health system. There are notable discrepancies observed across 34 provinces, municipalities, and autonomous regions in China [8]. The southeast region had the highest GDP in 2020 where most provinces exceeded or reached almost 1 trillion USD [9]. Most of the central and northern provinces had a GDP of 500−1000 billion USD in the same year [9]. On the other hand, most southwest provinces had the lowest GDP below 500 billion USD [9]. In a similar fashion and as stated previously, the healthcare and rehabilitation services/resources vary notably across these regions in China. The rehabilitation services capacity in Southwest China is lower than that of northern, southeast, and central China [10]. Nonetheless, the average number of rehabilitation services has tripled in recent decades [1].

21.2.2 ICF Framework in China according to WHO

According to the outline of Healthy China 2030 approved by the Chinese government, one of the key principles related to rehabilitation therapy was to advocate for the health industry in adopting and developing rehabilitation robotics, smart assistive devices, virtual reality, prosthesis and orthosis, wearable devices, meta-data collection and analysis, and telerehabilitation [1]. As discussed above, rehabilitation services in China had significant improvement once the ICF was adopted. In a recent multicenter clinical study that evaluated the effectiveness of the ICF in stroke rehabilitation assessment in China, the ICF checklist was found to be a rehabilitation assessment instrument that is compatible with typically used clinical assessment scales for stroke [11]. It can also be used in combination with the commonly used scales. Therefore clinical implementation of ICF that is in line with the key principle related to rehabilitation technology mentioned above has been widely initiated in hospitals and other rehabilitation facilities and clinics [1].

21.2.3 Robotics therapy

In China, conventional rehabilitation therapy is generally used in clinics [12,13]. Such a method consumes substantial resources in the form of manpower, time, material, and finances, which is currently inadequate to meet the rehabilitation needs of most patients. Therefore an alternative approach such as rehabilitation robotics is urgently needed to aid rehabilitation therapists in providing more efficient and effective treatment to patients using fewer resources. Presently, rehabilitation robot technology between China and other developed countries still has a substantial gap because the broad adoption of the technology in China has only happened in recent years [14]. As stated above, with the prevalence of stroke, China has the largest market potential in the world for rehabilitation technologies. Therefore rehabilitative robotics are urgently needed in China, and moving forward into the 2020s, the industrial scale is expected to exceed 700 billion.

21.2.4 Upper-limb rehabilitation robots

Conventional upper-limb rehabilitation training involves one-to-one manually assisted motor training in China, which has several limitations. An optimal therapeutic effect is hindered due to therapy shortages where training duration, intensity, and work efficiency are affected. It is also difficult to measure the quantitative changes that occur throughout rehabilitation. Therefore robot-assisted upper-limb rehabilitation robotics have been proposed to overcome the shortcomings of conventional therapy [13]. Development and research into rehabilitation robotics are still in their infant stage.

21.2.4.1 Technological development

Structurally, upper-limb rehabilitation robots can be classified into end-effector-based robots and exoskeleton robotics. Their driving modes can also be divided into electric, pneumatic, hydraulic, and non-driven forms. With links to the rehabilitation control strategy, the robots may adopt many control strategies such as proportional−integral−derivative, fuzzy, neural network, and adaptive control [15].

End-effector-based upper-limb rehabilitation robots are mainly composed of tandem robots or a linkage mechanism. The traction force is connected to the patient's limb and the robot, and the movement of the upper limb is assisted by the robot to achieve the purpose of rehabilitation [15,16]. The advantage of an end-effector-based robot is that structurally it is simple and portable with a simpler control method and is relatively independent of the patient. This makes trajectory planning easy [15,16].

Tsinghua University is one of the first universities to successfully develop a rehabilitation robot with complex movement that meets the training requirements such as needs of the shoulder, elbow, and hand. The university has developed a 2-DOF traction-type upper-limb robot with various functions for passive training, auxiliary active training, and constrained impedance training [17].

Southeast University Song Aisong has developed a compound 3-DOF end-effector robot based on the remote control. This system allows both upper- and lower-limb rehabilitation training to be carried out. Harbin Industrial University, Harbin Engineering University, Beijing University of Technology, and Taiwan University have similar studies [17].

The development of upper-limb rehabilitation robotics by multiple research institutes has paved the way for many domestic rehabilitation robotics and technology startup companies to develop and improve their upper-limb rehabilitation robots. Upper-limb rehabilitation robots such as ArmMotus and HandyRehab (Fourier Intelligence), Buffalo-Robot Upper Limb Exoskeleton (Buffalo Robotics Technology), and Arm Rehabilitation Robotics A2 and Hand Rehabilitation Robotics A5 (Guangzhou Yeecon) are already commercially available in China market.

21.2.4.2   Clinical application

Some barriers to adopting robotics for rehabilitation include the fear of robotics and perceived risks and benefits. Exposure remains sparse especially within the rural areas of China due to therapy imbalances and the cost that is associated with robotic devices. There is also the perceived ease of use from the therapists. Safety remains a big factor, and it may be wise to consider the safety strategies within the software and hardware. Security strategies such as compliance control, back-drivable mechanisms, emergency stop buttons, and force/speed limits need to be considered. Common hazards with robot-aided training include pinch injuries and collision-type injuries.

21.2.5   Lower-limb rehabilitation robots

Presently, lower-limb rehabilitation robots from overseas, such as Lokomat by Hocoma (Switzerland) and HAL by Cyberdyne (Japan), are adopted by some hospitals in China [12]. These robots are sophisticated yet expensive, making them unaffordable for small-scale clinics and hospitals in rural areas. To solve this issue, domestic universities have been actively researching lower-limb rehabilitation robots. The typical design requirements of a lower-limb rehabilitation robot consist of structural and design safety, clinical and functional effectiveness, and universal design to meet various patients' needs. Due to the limited supply of lower-limb rehabilitation robots in the domestic market, the price of the robot also needs to be reasonable and affordable, and at the same time, it needs to be compact and refined [12].

21.2.5.1 Technological development

There are two types of lower-limb rehabilitation robot design that universities and research institutes in China focus on, that is, exoskeleton and foot pedal structure design.

Harbin Engineering University has been focusing on the horizontal foot pedal structure robot since the early 2010s [18]. The robot can provide additional active−passive training mode when compared to the exoskeleton counterpart. It uses impedance and fuzzy controls to account for any possible abnormal movement and spasm to minimize the risk of secondary injury to the lower limb. Nonetheless, the monitoring of the real-time motion of the lower limb cannot be done effectively due to the placement of the force and position sensors at the foot pedal.

Tsinghua University is one of the earliest universities in China to explore the lower-limb rehabilitation robot. They started with a horizontal foot pedal design that follows a bicycle scheme but expanded further on the design into a vertical exoskeleton with a weight-reducing balance frame [19]. The robot works on a similar principle as the Lokomat and provides positive gait recovery training for a patient. However, position control is only activated during the swing phase to counter the problem of the position control algorithm lagging behind the actual position. This approach caused slippage between the foot sole and the conveyor belt of the medical treadmill. In addition, the constant need to switch the motor causes vibration.

Yanshan University developed a sitting horizontal type of lower-limb exoskeleton that has a form factor comparable to a conventional massage chair with the control system embedded under the seat [20]. The prototype also delivers a good platform for human−computer interaction. The main advantage of this robot is the incorporation of EMG signal acquisition hardware that can monitor the lower-limb muscles' intention to move in real time and with great accuracy.

Shanghai Jiao Tong University has also been actively involved in lower-limb rehabilitation using the vertical exoskeleton. They developed their exoskeleton by adapting and improving upon the legs design of Berkeley Bionics Tech Company. To solve the issue of treatment being restricted to one site only, the robot utilizes crutches to support the patient's body weight [21]. Such an approach allows the training of various gaits and movements in diverse environments and terrain types. However, the exoskeleton can only act as an assistive robot with substandard rehabilitation effects due to the absence of a sensing and measuring system.

Harbin Institute of Technology also developed a vertical exoskeleton which was an adaptation of Lokomat. By replacing the conventional proportional−derivative control with fuzzy adaptive control, the flexibility was improved with two degrees of freedom also added to the sole to achieve heel landing first and toe landing later [22]. Nonetheless, the absence of conveyor belts and the difficulty to replicate normal gait can be disadvantageous [22].

Similar to the upper limb, the development of lower-limb rehabilitation robotics by multiple research institutes also kickstarted many startup companies that developed their own lower-limb rehabilitation robots. Lower-limb rehabilitation robots such as ExoMotus (Fourier Intelligence), Buffalo-Robot (Buffalo Robotics Technology), AiLeg and AiWalker (Beijing Ai-Robotics Technology), and UGO (Hangzhou RoboCT Technology) are already commercially available in the China market.

21.2.5.2   Clinical application

As mentioned above, many lower-limb rehabilitation robots by startup companies are commercialized, and some of these products are already certified by China's National Medical Products Administration (NMPA) for clinical application.

Recently, China conducted their first multicenter clinical trial using their domestically designed lower-limb-powered exoskeleton as orthosis for paraplegic patients [23]. The clinical trial aims to substantiate the safety and effectiveness of their exoskeletons in rehabilitation training. Forty subjects with T6 to L2 spinal cord lesions and post-injury time of 67.8 ± 112.8 months were recruited for the study. All patients underwent a program consisting of walking and standing using the isocentric reciprocating gait orthosis (IRGO) and the powered gait orthosis (PGO). The outcome measures such as the walking speed, distance covered, time to put on the orthosis, physiological cost index, and rating of perceived exertion were determined. The outcome measures for PGO showed improvement compared to IRGO. Patients with paraplegia also achieved efficient overground ambulation and thus highlight the potential use of PGO for functional gain and improved fitness. However, a few patients had calcaneal compression fractures and minor skin pressure ulcers, thus highlighting some of the main risks of using PGO in clinical application.

21.3   Adaptation of rehabilitation delivery during the pandemic

Ever since the outbreak of the COVID-19 virus in 2019, healthcare services worldwide have been massively overwhelmed with confirmed cases of the virus, and its priority shifted to providing care for the surge of critically ill patients. In China, the impact of COVID-19 caused a drastic change in the essential healthcare services framework and delivery nationwide. This change was mainly due to the healthcare resources from provinces other than Hubei being diverted to Wuhan (the epicenter of the COVID-19 virus outbreak) and its neighboring region to aid the shortage of healthcare professionals in the area. As a result, there was a significant reduction in the capacity of rehabilitation services resulting in limited unnecessary inpatient admissions and shorter inpatient stays. In a general survey conducted in China, a decrease in activities of daily living (ADLs) and quality of life (QoL) among patients with continued care needs was highlighted [24]. Therefore affected patients were subjected to a high risk of disability and morbidity due to the disrupted rehabilitation services and inevitable shutdown.

Since late February 2020, the fight against COVID-19 in China saw a significant decline in new confirmed cases and mortality. However, without the support of evidence, the procedure for reopening and operating rehabilitation facilities presents a multitude of challenges to decision-makers, nurses, clinicians, and therapists. Even though global experts had developed several recommendations on the delivery of rehabilitation services [25], there was no solid evidence or a comprehensive framework for healthcare professionals to follow on operating rehabilitation services during the pandemic. Nonetheless, through the guidance of an expert group summoned by health commission authorities, the mentioned general recommendations were adapted in China to prevent the spread of COVID-19 and rapidly bring back suspended rehabilitation services. Correspondingly, a standardized procedure was established among the society of physical and rehabilitation medicine followed by remote education via portals or smartphone apps that were provided to all healthcare professionals [26]. China's strategic preparedness and response plan for COVID-19 in rehabilitation centers involve two key components, that is, to prevent the introduction of cases and enforce strict social distancing measures to prevent transmission [27].

From the plan, five measures of safety management were taken into effect:

a. "First-contact responsibility" and management system

Before rehabilitation assessment and treatment, the therapist or clinician will carefully perform contact tracing and assess the patient's COVID-19 symptoms and prescribe a test when necessary. Recently admitted patients will be placed in a temporary inpatient quarantine ward (equipped with at least ten separate care beds) until they are tested negative for COVID-19. It is the responsibility of the first-contact doctor to report the case to the authority and help transfer the case to a designated nearby hospital if a patient tested positive three consecutive times. Consequently, the first-contact doctor and all close contacts will need to be quarantined at home or in a designated accommodation for 14 days.

b. Full-coverage system for nosocomial infection control

To prevent COVID-19 cross-infection within rehabilitation centers, staff and patients must wear a mask all the time and perform proper hand hygiene routines. In addition, the air within the wards was circulated 24/7, and designated staff will carry out the disinfection process in the rehabilitation facilities and space twice a day.

c. "Closed-off management" system

At each entry point of the rehabilitation centers, a gatekeeper will perform a temperature scan and epidemiological investigation on each patient. Following the procedure, a permit to enter a particular area of the rehabilitation center will be issued that is valid for 3–7 days only. If an issued permit needs further updates on its accessibility, a public health officer will further review it. The outpatient and inpatient sections are also strictly separated. Upon assessing the independence and ADL, inpatients are allowed to be escorted by a caretaker. Visitors are prohibited in the center but encouraged to do remote communication.

d. Full-coverage system for body temperature monitoring

Temperature monitoring and symptom checking are conducted in the rehabilitation center three times a day for all patients, and if the temperature is higher than the normal range, a COVID-19 test will be administered. This protocol also applies to all staff in the center.

e. Telerehabilitation

Any intervention/drug prescriptions, video guidance rehabilitation, and follow-up treatment were operated remotely through a telerehabilitation platform with 5G internet capabilities, set up to combat the COVID-19 outbreak. This network platform also allows real-time minor needs from patients to be addressed by clinicians effectively using instant messaging functions or communication alternatives such as Tencent virtual clinic, WeChat, etc., to provide quick feedback.

In February 2020, the number of patient visits in outpatient clinics and new admission in inpatient wards reported a tremendous decrease due to the epidemic compared to the previous 2 months. However, the rehabilitation center encountered a surge of patient visits between March and April 2020 once the outbreak was successfully contained due to good safety management. Such an increase in patient visits can also be attributed to the increased demand for rehabilitation among post-COVID-19 patients who require extensive and individualized rehabilitation treatment [28]. Consequently, the highly affected province of Hubei also opened their COVID-19 rehab outpatient clinics in the early March of 2020 once the cases were under control.

The future direction of rehabilitation robotics in China shall not only focus on hospital therapy, but more emphasis should be placed on home therapy. With the experience of lockdown in China during the pandemic, home therapy was of high importance, and even with the implementation of telerehabilitation, the lack of physical therapy with clinician can hinder a patient's recovery. Rehabilitation robotics can be an alternative solution to this issue with proper therapy protocol and training prescribed to clinicians. Outside of a pandemic, rehabilitation robots can also help patients with limited access to rehabilitation center with expert clinicians, especially in rural areas.

With respect to the design of the rehabilitation robots, a more lightweight material and structural design is needed to improve the safety and comfort of using both upper- and lower-limb robots, so that the patient can do self-training with less supervision from clinicians. Research collaborations on rehabilitation robotics with research institution outside of China should also be advocated since the rehabilitation robot system is a multidisciplinary and complex system that requires various experts in the field to develop cost-effective robots that are suitable for home therapy.

21.4 Future direction of robotics therapy in China

The role and future of rehabilitation robotics in China are to expand from a hospital setting to home-based therapy. The recent pandemic lockdown has created an increased demand for home-based robotics therapy, which usually involves a complex network of equipment and personnel like therapists, nurses, and physical therapists. With the implementation of remote patient management, rehabilitation centers can provide patients with a personalized treatment plan from the comfort of their homes, possibly allowing them to continue physical therapy after a hospitalization. Outside of a pandemic, with proper training, prescription of exercises, and remote monitoring by an expert clinician, rehabilitation robots can also help housebound patients due to disability or limited access to therapy, especially in rural areas.

The rehabilitation robot system is a multidisciplinary and complex system that requires various experts in the field to develop safe, cost-effective robots suitable for home therapy. This is especially so in China, where rehabilitation robots must meet stringent safety requirements and be affordable for home-based treatment. Research collaborations on rehabilitation robotics with research institutions outside China should be advocated to expedite the development and adoption. The future direction of rehabilitation robotics research in China is to use existing technologies better, integrate them with innovative technologies, and improve the design of rehabilitation robots. The aim is to improve the safety and comfort of home-based rehabilitation and telerehabilitation robots, giving patients more therapy hours through self-training with less clinician supervision.

Acknowledgments

A special thanks to Chen-Onn Leong, Ph.D., from MotusAcademy, Sarah Lim, PT from Fourier Intelligence, and Jiaxi Tang from Ruijin Hospital for helping with the successful completion of this chapter.

Conflict of interest

There is no conflict of interest for all contributors.

References

[1] Li J, Li LS. Development of rehabilitation in China. Phys Med Rehabil Clin. 2019;30 (4):769−73.

[2] Wu S, Wu B, Liu M, Chen Z, Wang W, Anderson CS, et al. Stroke in China: advances and challenges in epidemiology, prevention, and management. Lancet Neurol 2019;18 (4):394−405.

[3] Xiao Y, Zhao K, Ma Z-X, Li X, Qiu Y-P. Integrated medical rehabilitation delivery in China. Chronic Dis Transl Med 2017;3(02):75−81.

[4] Liu L, Liu J, Wang Y, Wang D, Wang Y. Substantial improvement of stroke care in China. Stroke. 2018;49(12):3085−91.

[5] Wang W, Jiang B, Sun H, Ru X, Sun D, Wang L, et al. Prevalence, incidence, and mortality of stroke in China: results from a nationwide population-based survey of 480 687 adults. Circulation. 2017;135(8):759−71.

[6] Tan X, Liu X, Shao H. Healthy China 2030: a vision for health care. Value Health Reg Issues 2017;12:112−14.

[7] Huang G. The implementation path of intelligent rehabilitation under the background of healthy China construction. Wearable Technol 2022;2(1):41−50.

[8] Chen F, Yang Y, Liu G. Social change and socioeconomic disparities in health over the life course in China: a cohort analysis. Am Sociol Rev 2010;75(1):126−50.

[9] GDP. GDP-2020 is a preliminary data "Home - Regional - Quarterly by Province" [press release]. China NBS, March 1 2021.

[10] Zhang R, Chen Y, Liu S, Liang S, Wang G, Li L, et al. Progress of equalizing basic public health services in Southwest China−health education delivery in primary healthcare sectors. BMC Health Serv Res 2020;20(1):1−13.

[11] Zhang T, Liu L, Xie R, Peng Y, Wang H, Chen Z, et al. Value of using the international classification of functioning, disability, and health for stroke rehabilitation assessment: a multicenter clinical study. Medicine 2018;97(42).

[12] Liu Z, Lu D. Research progress of lower limb rehabilitation robots in Mainland China. Open J Ther Rehabil 2019;7(03):92.

[13] Cheng X, Zhou Y, Zuo C, Fan X. Design of an upper limb rehabilitation robot based on medical theory. Procedia Eng 2011;15:688−92.

[14] Shi D, Zhang W, Zhang W, Ding X. A review on lower limb rehabilitation exoskeleton robots. Chin J Mech Eng 2019;32(1):1−11.

[15] Meng Q., Xie Q., Yu H., editors. Upper-limb rehabilitation robot: State of the art and existing problems. In: Proceedings of the 12th International Convention on Rehabilitation Engineering and Assistive Technology; 2018.

[16] Zhang K, Chen X, Liu F, Tang H, Wang J, Wen W. System framework of robotics in upper limb rehabilitation on poststroke motor recovery. Behavioural Neurol 2018;2018.

[17] A review of upper and lower limb rehabilitation training robot. In: Hu W, Li G, Sun Y, Jiang G, Kong J, Ju Z, et al., editors. International conference on intelligent robotics and applications. Springer; 2017.

[18] Li-Xun Z, Hongying S, Zhenmei Q. Kinematics analysis and simulation of horizontal lower limbs rehabilitative robot. J Syst Simul 2010;22(8):2001−5.

[19] Chen K, Liu Q, Wang R. Development of a body weight-support gait training robot. Chin J Rehabil Med 2011;26(9):847−51.

[20] Shi X, Ren L, Liao Z, Zhu J, Wang H. Design & analysis of the mechanical system for a spacial 4-DoF series-parallel hybrid lower limb rehabilitation robot. J Mech Eng 2017;53:48−54.

[21] Rao L-J, Xie L, Zhu X-B. Research and design on lower exoskeleton rehabilitation robot. Ji Xie She Ji Yu Yan Jiu(Machine Des Res) 2012;28(3):24−6.

[22] Li F, Wu ZZ, Qian J. Trajectory adaptation control for lower extremity rehabilitation robot. Chin J Sci Instrum 2014;35(9):2027−36.

[23] Chen S, Li J, Shuai M, Wang Z, Jia Z, Huang X, et al. First multicenter clinical trial of China's domestically designed powered exoskeleton-assisted walking in patients with paraplegia. Ann Phys Rehabil Med 2018;61:e495.

[24] Zhu S, Gao Q, Yang L, Yang Y, Xia W, Cai X, et al. Prevalence and risk factors of disability and anxiety in a retrospective cohort of 432 survivors of Coronavirus Disease-2019 (Covid-19) from China. PLoS One 2020;15(12):e0243883.

[25] Thomas P, Baldwin C, Bissett B, Boden I, Gosselink R, Granger CL, et al. Physiotherapy management for COVID-19 in the acute hospital setting: clinical practice recommendations. J Physiother 2020;66(2):73−82.

[26] Hong Z, Li N, Li D, Li J, Li B, Xiong W, et al. Telemedicine during the COVID-19 pandemic: experiences from Western China. J Med Internet Res 2020;22(5):e19577.

[27] Zhu S, Zhang L, Xie S, He H, Wei Q, Du C, et al. Reconfigure rehabilitation services during the Covid-19 pandemic: best practices from Southwest China. Disabil Rehabil 2021;43(1):126−32.

[28] Li Z, Zheng C, Duan C, Zhang Y, Li Q, Dou Z, et al. Rehabilitation needs of the first cohort of post-acute COVID-19 patients in Hubei, China. Eur J Phys Rehabil Med 2020;56(3):339−44.

Chapter 22

Middle East region: Iran

Narges Rahimi[1], Saeed Behzadipour[2,3] and Shafagh Keyvanian[4]

[1]*Department of Neurology, Jefferson-Einstein Medical Center, Philadelphia, PA, United States,*
[2]*Department of Mechanical Engineering, Sharif University of Technology, Tehran, Iran,*
[3]*Djavad Mowafaghian Research Center in Neuro-rehabilitation Technologies, Sharif University of Technology, Tehran, Iran,* [4]*Department of Mechanical Engineering and Applied Mechanics, University of Pennsylvania, Philadelphia, PA, United States*

Learning objectives

At the end of this chapter, the reader will be able to:

1. Understand the strategy of Iran's healthcare system to manage stroke rehabilitation.
2. Identify the current capacity of high-tech and robotics approaches in stroke rehabilitation based on the research being conducted in academic institutions and laboratories and a survey through rehabilitation units.

22.1 Healthcare in Iran

Based on the UN data released in 2021, Iran is an upper-middle-income country located in Southern Asia and is the 16th largest country in the world with an area of over 1.6 million km^2 [1,2]. Iran has a population of 85 million, which are mostly concentrated in urban areas due to migration from rural areas to cities (with 9 million people living in the capital, Tehran), with life expectancy of 76.3%, 10.6%, and 6.56% of the population aged 60 and 65 years old or over, respectively [3,4]. The economy of the country depends on petroleum and natural gas exports and is among the 20 largest economies in the world [5]. The GDP at current prices reported by the UN is 603,780 USD, and the health expenditure on health is 8.7% of the GDP [6,7].

The Iranian constitution entitles Iranians to basic healthcare, and the Ministry of Health and Medical Education (MOHME) has executive responsibility for health and medical education within the Iranian government to provide and monitor healthcare delivery [8,9]. In each province, there is at least one medical science university, which is the official representative of

Rehabilitation Robots for Neurorehabilitation in High-, Low-, and Middle-Income Countries.
DOI: https://doi.org/10.1016/B978-0-323-91931-9.00028-1
339

the MOHME in that province. Generally, Iran provides healthcare services through the public and private sectors. Public hospitals are usually affiliated with medical schools and are the primary provider of specialty and higher levels of care. The private sector and NGOs also play a significant role in healthcare provision in Iran. Additionally, in order to provide healthcare services in rural and remote areas, there is a national healthcare network and also "health houses," which are small health centers located in villages [8,10]. Over the last couple of decades, there has been a major improvement in education and health services. According to the UN data, the number of physicians and R&D personnel in the country are around 130 thousand and 184 thousand, respectively [11,12].

22.2   Stroke rehabilitation

Stroke is the leading cause of long-term disability worldwide. It has become a major health concern in Iran since it was the second most common cause of death in 2019 from the sixth in 1990 [13]. A recent study estimated the stroke burden from 1990 to 2019 using the results of the Global Burden of Disease (GBD) study 2019. The results show that the stroke cases increased from 48,274 (42,265−55,623) in 1990 to 102,778 (90,115−117,821) in 2019, revealing a 2.1-fold increase, whereas the age-standardized rate of stroke decreased from 166.6 (146.6−190.9) in 1990 to 138.8 (121.8−159.6) per 100,000 population in 2019 [14]. Stroke affects physical, social, and emotional aspects of life for a patient and their family. Rehabilitation plays a crucial role in helping patients relearn skills and gives them the opportunity to participate and engage in the community. The most common impairment after a stroke is a motor disability which can have negative consequences on almost every part of a patient's life [15]. Iran has a national rehabilitation standard for stroke that includes preventative and therapeutic approaches for the consequences of stroke. Preventative approaches include the prevention of comorbidities such as deep vein thrombosis, limb edema, falls, contractures, pain, and decreased function. In therapeutic and assistive approaches, motor and mobility deficits are targeted most often. Stroke results in mobility limitations in a majority of patients. Around 80% of stroke survivors have movement problems within the first 3 weeks after stroke that results in dependence on ADLs [16]. Treatment goals such as increasing social participation require focusing on motor recovery by complex interventions often involving a combination of therapeutic approaches with several interacting components. Physical intervention is a mainstay of treatment that includes the improvement of motor skills, mobility training, constraint-induced therapy, and range-of-motion therapy [17].

Three main world-class methods (hospital rehabilitation, rehabilitation in stroke units, and home-based rehabilitation) are being used in Iran, as well as in different countries worldwide [18]. The rehabilitation process starts in hospitals and is then being followed by providing services in stroke units and mostly by home-based programs for stroke patients since home-based rehabilitation is known to be less expensive and more effective than stroke unit and hospital-based rehabilitation in Iran [19]. There are different stroke units that are operating in the public and the private sector. Stroke rehabilitation services are done by traditional rehabilitation methods (all over the country), as well as technology-assisted physical activities (mostly in large cities) [17].

22.3 Robotic rehabilitation—technical perspective

Robot-aided rehabilitation is an example of a technology-based approach, using robotics to help patients regain their function and strength. The Iranian rehabilitation clinics have shown interest in utilizing this technology according to their local capacity and availability of technical services.

The application of high-tech imported devices such as robotic manipulators in the clinics is mostly limited due to economic considerations and exchange rates. The hourly cost of physio/occupational therapy according to the public health tariff is 2−3 USD [20]. The purchasing price of imported rehabilitation robotic devices begins in the tens of thousands of dollars. Adding the maintenance cost to the capital investment makes it an unattractive line of service for private clinics. As a result, on the one hand, the application of the popular robotic systems has become limited to a few public health service centers funded by the government. On the other hand, several local companies have started filling the gap by developing domestic robotic systems at more affordable prices. Such systems are mostly inspired by international rival products but customized according to local needs and preferences.

The available technologies, mostly provided by the local manufacturers, can be categorized as: (1) gait trainers, (2) exoskeletons (for upper and lower extremities), (3) robotic devices for a particular joint, and (4) virtual reality (VR) and biofeedback. In the last category, although there is no mechanical component to apply force or motion, the basis of operation and the underlying mechanism to boost neuroplasticity are very similar to robotic devices and hence will be reviewed here.

Robotic gait training services are provided in a few clinics in major cities. The earliest utilization of this technology dates back to 2010 when a hospital in Tehran installed a Lokomat. Since then, several other public and private centers have started using similar devices developed by local companies. The most inclusive version provides a weight balance system, an exoskeleton for passive and/or assist-as-needed movement of the lower extremities, and VR component (Fig. 22.1A and B). More affordable versions have been also developed and are in service in which the exoskeleton is removed, and the VR component is optional (Fig. 22.1C).

The development of robotic exoskeletons has been pursued by several university research groups and private companies for their clinical and personal applications. As a rehabilitation device, these robots are able to provide assistive torque to particular targeted joints. They can be integrated with EMG feedback to detect the intention of the user and FES for electrical stimulation in parallel with mechanical actuation to maximize neuroplasticity [21]. Exoped is an Iranian lower-extremity exoskeleton that has been installed in a few clinics (Fig. 22.2). It has four active joints (hip and knee flexion extension). The legs of the robot are on the ground, and therefore the weight of the robot (29 kg) is not carried by the patient. This facilitates gait training and weight-bearing exercises. The gait parameters (i.e., step size, height, and pace) are determined by the clinician, and the robot executes a constant gait cycle. The energy consumption of the robot is reported as an indication of the amount of assistance provided to the patient. In order to maintain the balance, the patient should use his hands; therefore for hemiplegics or those with weakness in hands, ceiling-suspended monorails are used.

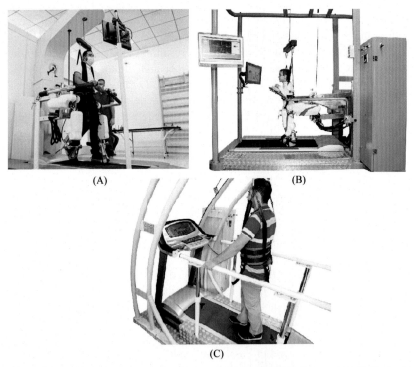

FIGURE 22.1 Gait training robots: (A, B) a domestic version of Lokomat, AmbuReR, developed for adults and children; (C) an affordable version of gait trainer systems developed by local companies which provide body weight support but lack lower-extremity exoskeletons.

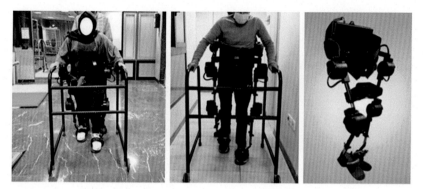

FIGURE 22.2 A clinical lower-extremity exoskeleton (Exoped by Pedasys Co.) developed by a local company and used for gait rehabilitation.

Besides the clinical exoskeletons, many research centers are working on the development of this technology as an assistive device. An exoskeleton to assist with flexion/extension of the hip joint is developed to run an assist-as-needed control to help the users perform normal gait [22]. There are many attempts to optimize either the controller design of such robotic devices or the kinematic synthesis [23–26].

Although control of such robots has a major impact on their feasibility, the clinical and biomechanical aspects should not be overlooked. Most exoskeleton robots minimize the number of active joints to reduce the price. This increases the design sensitivity of the passive joints. Proper selection of the passive joints facilitates the compliance of the robot with the complex human body kinematics [27]. Also, the mechanics of the connections between such robots and the body have been investigated and shown to have a major impact on the safety and/or comfort of the user [28,29]. Note that the exoskeleton robots have relatively simple kinematics formed by standard kinematic joints. In the human body, however, the kinematics of the joints are usually complex. For instance, the knee joint is a simple revolute joint in robots, while biomechanics of the actual joint suggest five or six degrees of freedom [30]. This kinematic mismatch between the robot and the body should be compensated at the robot–body connections by properly designed compliance, otherwise internal tension may rise and compromise the safety of the user [28,29].

The application of robotic exoskeletons for upper extremities is limited to a few clinics in the capital, Tehran. These devices are mostly continuous passive motion (CPM) machines with extended options such as programmability of the applied effort, range of motion, and amount of assistance (Fig. 22.3). For stroke patients, they are usually utilized in the early stages of motor rehabilitation to address spasticity and contractures of joints. The research on such devices is, however, more extensively conducted in technical

FIGURE 22.3 Sharif Exo, an arm exoskeleton for passive and assistive motion for hand rehab (elbow and wrist joint).

universities. Many prototypes have been developed for rehabilitation purposes with a focus on the kinematic design and the interaction force measurements [31]. The complex kinematics of the shoulder requires innovating a combination of passive translational motions along with actuated rotational joints [32]. Measurement of the kinematic parameters and adaptation of the robot accordingly are other problems pursued to improve their acceptability for clinical applications [33]. There have been also some attempts to utilize such exoskeletons for performance measurement in patients with stroke [34] rather than training (Fig. 22.4). The exoskeleton, in this application, provides a weight support for the arm and accurate measurement of the hand/arm movement to calculate the performance indices such as smoothness, accuracy, and speed while performing controlled tasks through a VR/video gaming interface.

Robotic devices that are developed for particular joints are more commonly used in some rehabilitation centers. Robohab is developed for wrist rehabilitation (Fig. 22.5A). It can be arranged for pronation/supination or flexion/extension motions. Several operation modes are implemented: (1) passive motion in which it applies a predefined movement with a limited torque or impedance to avoid injuries, (2) limited assistance to support the voluntary movement of the patient, and (3) resistive mode in which the robot applies opposing torques to the joints. There is an option of gamified exercise for more intensive involvement of the patient. A similar device is introduced with the ability to be configured for elbow, shoulder, knee, and ankle joints (Fig. 22.5B). It can provide several motion functions such as isometric, isokinetic, and isotonic to the specified joint. Another joint-specific robot is shown in Fig. 22.5C which is designed for mobilizing the glenohumeral joint. The mobilization is a delicate maneuver usually performed by an expert PT/OT. The robotic device utilizes a custom-designed brace and a one-degree-of-freedom motion. The robot has two control modes. The first mode is a force control to determine the arthrokinematics range of motion. Then, a position control is applied to perform the glide motions to mobilize

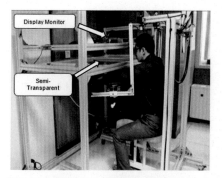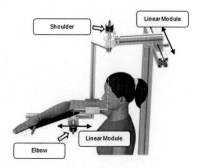

FIGURE 22.4 An upper-extremity exoskeleton for performance measurement in stroke patients. The device provides weight support for the arm and shoulder and accurate kinematic measurement while performing reaching tasks in a VR environment. *VR*, virtual reality.

the joint. The process can be repeated as many times as required by the therapist.

Joint-specific robots have been extensively researched in universities and research centers. This might be due to the simplicity and affordability of such robots when it comes to the proof of concept or development of new treatment modalities. The research activities on this subject in the Iranian centers can be categorized as: (1) design and fabrication of novel robots, (2) development of novel control techniques for better performance, and (3) integration of biosignals such as surface electromyogram (sEMG) and/or electrical stimulation with the rehabilitation robots. The design and fabrication of several robots have been reported in the literature to provide controlled passive/assistive/resistive rehabilitative exercises for hip [35], ankle [36], wrist and arm [37–42], and fingers [43,44]. In wrist and ankle robots, the kinematic complexity has been the focal point of the innovation. In finger robots, which are very relevant for stroke patients with grasp impairments, the compactness of the design and application of tendon-driven mechanisms were pursued. The complexity of the human body dynamics and the central nervous system (CNS) interferences, on the one hand, and high safety standards, on the other hand, make the control of the robot a rather challenging problem. Several attempts have been made as reported in the literature to address these issues such as using fuzzy and intelligent approaches [45–47], novel impedance/admittance control approach [48,49], and methods to handle high variability of the body dynamics [50,51]. To improve the efficacy of rehabilitation robots, there are many attempts to detect and integrate the patient intention through sEMG signal [52], apply FES along with the mechanical force and motion [53], and even combine the robotic maneuvers with serious gaming [54].

VR, biofeedback, and exergaming are considered as offspring of robotic technologies with high potential for motor and cognitive rehabilitation.

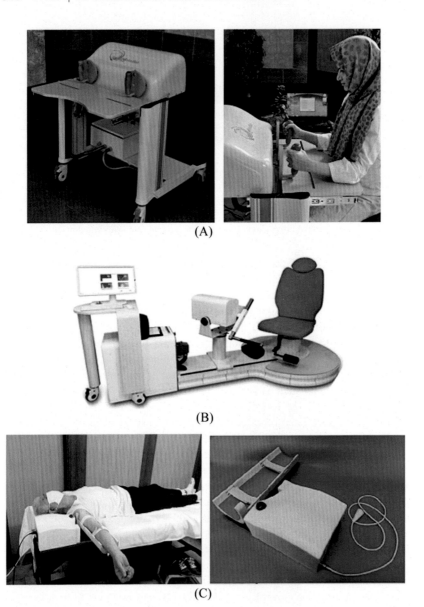

FIGURE 22.5 Robotic systems with particular anatomy target: (A) Robohab, a robotic device for wrist rehabilitation, (B) a joint-specific rehab robot with isokinetic, isotonic, and isometric function modes, (C) a robotic device for the mobilization of the glenohumeral joint.

Although they do not apply external forces to the body, they share some of the same technologies such as motion and force sensing, human body kinematics and dynamics models, machine learning, and vision. Mass production of the hardware components of these systems such as the sensors and AV elements for other areas such as video gaming has brought down their prices significantly. As a result, they are becoming more attractive for Iranian rehabilitation centers that desire to utilize advanced technologies. Several local companies have started developments in this line of products. Sana is a VR-based exercise/assessment system developed for in-clinic and at-home neurorehab (Fig. 22.6A). Taking advantage of a skeleton tracker camera (Kinect version 1) and a custom-built force plate, it provides several training programs such as: (1) upper-extremity gross motor training including reaching, tracking moving targets, drag and drop of the targets, (2) dynamics and static balance training using biofeedback from the center of pressure, and (3) dual tasking (motor and cognitive functions). All details of the exercises can be determined and tuned by the therapist including load, accuracy, speed, range of motions, and cognitive functions (e.g., working memory and attention). The system can also provide detailed reports on the performance of the patient for one rehabilitation session or throughout the entire course of the therapy program. About 20 rehabilitation centers in the country are using the system mostly for neurorehabilitation according to the manufacturer. The same concept has been pursued by some other companies using different sensing devices and biofeedback. Surface EMG sensors and/or

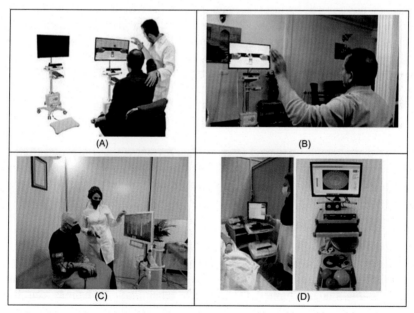

FIGURE 22.6 Virtual reality, exergaming, and biofeedback devices used for motor/cognitive neurorehabilitation: (A, B) Sana is a virtual reality exergaming system with potential for motor/cognitive training and testing used both in clinics and at home; (C, D) EMG and pressure biofeedback for upper-extremity muscle functions and cervical traction.

pressure gauges, for instance, are integrated with gamified training tasks to provide intensive and efficient neurorehabilitation therapies (Fig. 22.6B and C). According to the manufacturer, more than 80 rehab clinics have installed the device and are providing this service. The higher popularity of these devices in comparison with the rehab robots is believed to be mostly due to their lower purchase and maintenance costs.

22.4 Robotic rehabilitation—survey

A questionnaire was developed and sent to rehabilitation facilities equipped with high-technology devices. The survey included facility general information, patients' demographics, financial affairs, robots and devices, and capacities and barriers.

Based on questions about finances, public rehabilitation hospitals and facilities are funded by either Ministry of Health and Wellness, or donor support (e.g., charity foundations, nongovernmental organization, etc.). On the other hand, private rehabilitation hospitals and facilities are mostly self-funded and/or private-funded. Assessments on insurance coverage for patients show that stroke rehabilitation in the form of traditional services is mostly covered by insurance companies, while robotics and high-tech services are not covered by insurance and patients have to pay out of pocket.

Based on the results, high-tech rehabilitation is mostly taking place in the capital (Tehran). Based on the survey and to the best of the authors' knowledge, available technologies are gait trainers, exoskeletons (for upper and lower extremities), robotic devices for particular joints, VR, and biofeedback, as mentioned earlier. The disorders under treatment in the facilities equipped with high-tech devices are as follows: stroke, muscle spasticity, motion limitation, CVA, ataxia, Parkinson's disease, TBI, CP, vestibular, proprioceptive, Ms, and UMNS.

The demographics gathered from this survey show that on average one-fifth of the patients in need of rehabilitation were stroke survivors. The mean values of male/female patients' percentage for general and stroke cases were 48/52 and 45/55, respectively.

An overall assessment of robotics rehabilitation from a clinical perspective is presented in Tables 22.1 and 22.2, including the advantages/disadvantages and barriers/solutions, respectively.

In conclusion, the capacity of adopting robotics in rehabilitation is promising as long as some major barriers such as the cost of robotics equipment and the low budget allotted to high-tech rehabilitation methods can be addressed. Increasing the public knowledge of rehabilitation technologies was also addressed as vital.

It should be noted that in gathering the data, only a handful of facilities were surveyed, and they were all facilities that had high-tech devices.

TABLE 22.1 Advantages and disadvantages of rehabilitation robotics.

Advantages	Disadvantages
Attractive/entertaining experience for the patients	Higher costs for the patients
Real-time feedback for patients (visual and auditory)	Need for maintenance
Higher effectiveness and faster progress	Financially not justifiable
Higher accuracy	Hard to understand for senior patients (in some cases)
Less injury probability during rehabilitation	

TABLE 22.2 Barriers and solutions of rehabilitation robotics.

Barriers	Solutions
Patients' distrust of the methods	Creative commercial strategies
Shortage in technology	Updating the technologies
Import issues	-
Affordability for the patients	Insurance coverage
Lack of insurance coverage	

Acknowledgments

We would like to express our deepest appreciation to Dr. Mansour Rayegani, for his contribution to this chapter.

We would like to show our gratitude to the following research centers and companies for sharing their research and/or rehabilitation products with us during the course of this research: Djavad Mowafaghian Research Center in Neuro-rehabilitation Technologies, Taharrok Fannavar Robotic Co., Tavan Danesh Mehrsam Co., Danesh Salar Iranian Co., Pedasys Co., and Physical Medicine and Rehabilitation Department of Shahid Beheshti Medical University.

We are also grateful to the following rehabilitation centers for their contribution to the survey mentioned in the manuscript, which helped us to have a better vision of the rehabilitation robotics situation in the active centers in the country: Firoozgar Hospital, Rasoul Hospital, Rofeideh Hospital, Sabaye Salamat Rehabilitation Center, Noora Physiotherapy Rehabilitation Center, Omid Rehabilitation Clinic, and Khaneye Omid Rehabilitation Center.

Conflict of interest

There is no conflict of interest for all authors.

References

[1] Iran's Third National Communication to UNFCCC. Tehran. Tehran: Department of Environment; December 2017. [Online]. Available from: https://unfccc.int/sites/default/files/resource/Third%20National%20communication%20IRAN.pdf.

[2] United Nations Department of Economic and Social Affairs. World Economic Situation and Prospects 2021; 2021, pp. 173–204.

[3] United Nations. SYB64_1_202110_Population, Surface Area and Density. 2021. [Online]. Available from: http://data.un.org/.

[4] The Global Economy. Population ages 65 and above - Country rankings. 2020. [Online]. Available from: https://www.theglobaleconomy.com/rankings/elderly_population/.

[5] "The World Bank Data. Iran, Islamic Republic of. The World Bank," 2020. [Online]. Available from: https://data.worldbank.org/country/iran-islamic-rep.

[6] United Nations. "SYB64_230_202110_GDP and GDP Per Capita," [Online]. Available from: https://data.un.org/.

[7] United Nations. "SYB64_325_202110_Expenditure on health," 2021. [Online]. Available from: https://data.un.org/.

[8] Mehrdad R. Health system in Iran. Jpn Med Assoc J 2009;52(1):69–73.

[9] MOHME. Ministry of Health and Medical Education, Official website [Online]. Available from: https://ird.behdasht.gov.ir/MOHME.

[10] Shahjouei S, et al. Acute management of stroke in Iran: obstacles and solutions. Iran J Neurol 2017;16(2):62–71.

[11] United Nations. "SYB64_154_202110_Health Personnel," [Online]. Available from: https://data.un.org/.

[12] United Nations. "SYB64_285_202110_Research Development Staff," [Online]. Available from: https://data.un.org/.

[13] Institute for Health Metrics and Evaluation. [Online]. Available from: https://vizhub.healthdata.org/gbdcompare; 2019 (Accessed on October 2021).

[14] Fallahzadeh F, et al. National and subnational burden of stroke in Iran from 1990 to 2019. Ann Clin Transl Neurol 2022;5(5):669–83.

[15] Dalvandi A, et al. Post stroke life in Iranian people: used and recommended strategies. Iranian Rehabilitation J 2009;7(9):17–24.

[16] Shah S, et al. Efficiency, effectiveness, and duration of stroke rehabilitation. Stroke 1990;21:241–6.

[17] Standards for Rehabilitation in Stroke by Ministry of Health and Medical Education; 2020.

[18] World Health Organization - Iran (Islamic Republic of). WHO statistical profile. [Online]. Available from http://www.who.int/gho/countries/irn.pdf; 2015 (Accessed on February 2016).

[19] Farzaneh Miri RG, et al., Evaluating the cost-effectiveness analysis of rehabilitation methods for patients with stroke; 2021.

[20] Rehabilitation. The National Program of Rehabilitation (in Farsi). Ministry of Health and Medical Education, Iran; 2020.

[21] del-Ama AJ, Gil-Agudo Á, Pons JL, Moreno JC. Hybrid FES-robot cooperative control of ambulatory gait rehabilitation exoskeleton. Neuroeng Rehabil 2014;11:27.

[22] Naghavi N, Akbarzadeh A, Tahamipour SM, Kardan I. Assist-As-needed control of a hip exoskeleton based on a novel strength index. Robot Autonomous Syst 2020;134:103667.

[23] Zarandi SMT, Sani KH, Tootoonchi MRA, Tootoonchi AA, Farajzadeh-D M-G. Design and implementation of a real-time nonlinear model predictive controller for a lower limb exoskeleton with input saturation. Iranian J Sci Technology, Trans Electr Eng 2021;45:309−20.

[24] Ehsani-Seresht A, Moghaddam MM, Hadian MR. Joint function control method for robotic gait training of stroke patients. Int J Robot Autom 2020;35(3).

[25] Salarieh H. Adaptive impedance control of exoskeleton robot. Modares Mech Eng 2013;13(7):26−111.

[26] MahdiehB., Ghanbari A., NooraniS.M.R.. Mechanical design, simulation and nonlinear control of a new exoskeleton robot for use in upper-limb rehabilitation after stroke. In *Iranian Conference on Biomedical Engineering (ICBME)*, 2013.

[27] Taherifar A, Shariat A, Khezrian R, Zibafar A, Rashidi AR, Rostami H, et al. Design and control of an assistive exoskeleton with passive toe joint. Int J Mechatron Autom 2018;6(3).

[28] Shafiei M, Behzadipour S. The effects of the connection stiffness of robotic exoskeletons on the gait quality and comfort. J Mechanisms Robot 2020;12(1):011007.

[29] Shafiei M, Behzadipour S. Adding backlash to the connection elements can improve the performance of a robotic exoskeleton. Mechanism Mach Theory 2020;152:103937.

[30] Li G, Gil J, Kanamori A, Woo SL. A validated three-dimensional computational model of a human knee joint. ASME J Biomech Eng 1999;121(6):657−62.

[31] MahdavianM., Yousefi-Koma A., ToudeshkiA. Design and fabrication of a 3DoF upper limb exoskeleton. In *RSI International Conference on Robotics and Mechatronics*, Tehran, 2015.

[32] Babaiasl M, Ghanbari A, Noorani SMR. Mechanical design, simulation and nonlinear control of a new exoskeleton robot for use in upper-limb rehabilitation after stroke. In *Iranian Conference on Biomedical Engineering (ICBME 2013)*, Tehran, 2013.

[33] Beigzadeh B, Ilami M, Najafian S. Design and development of one degree of freedom upper limb exoskeleton. In *RSI International Conference on Robotics and Mechatronics*, Tehran, 2015.

[34] Nikzad A, Behzadipour S, Hajihosseinali M. Design and construction of a planar robotic exoskeleton for assessment of upper limb movements. In *7th International Conference on Robotics and Mechatronics (ICRoM)*, 2019.

[35] Hadipour M, Ghobadpour Z, Najafi F. Hip rehabilitation mechanism optimization. In *International Conference on Roboitcs and Mechatronics (IcRoM 2017)*, Tehran, 2017.

[36] Alipour A, Mahjoob M. A rehabilitation robot for continuous passive motion of foot inversion-eversion. In *International Conference on Robotics and Mechatronics*, Tehran, 2016.

[37] Baniasad FFNNAM. Wrist-RoboHab: A robot for treatment and evaluation of brain injury patients. In *IEEE International Conference on Rehabilitation Robotics*, Zurich, 2011.

[38] Nikafrouz N, Mahjoob M, Tofigh M. Design, modeling, and fabrication of a 3-DOF wrist rehabilitation robot. In *International Conference on Robotics and Mechatronics (IcRoM 2018)*, 2018.

[39] Mohammadia E, Zohoor H, Khadema S. Design and prototype of an active assistive exoskeletal robot for rehabilitation of elbow and wrist. Sci Iran 2016;23(3):998−1005.

[40] Sajadi MR, Nasr A, Moosavian SAA, Zohoor H. Mechanical design, fabrication, kinematics and dynamics modeling, multiple impedance control of a wrist rehabilitation robot. In *RSI International Conference on Robotics and Mechatronics*, Tehran, 2015.

[41] Sepahi S, Hashemi A, Jafari M, Sharifi M. A novel upper-limb rehabilitation robot with 4 DOFs: Design and prototype. In *International Conference on Robotics and Mechatronics (IcRoM 2018)*, 2018.

[42] Faghihi A, Haghpanahi S, Farahmand F, Jafari M. Design and fabrication of a robot for neurorehabilitation; smart robowrist. In *International Conference on Knowledge-based Engineering and Innovation*, Tehran, 2015.

[43] Kermanshahani AH, Samavati FC. Design, analysis and control of the 4 fingers rehabilitation robot. J Rehabil Sci Res 2019;160−8.

[44] Norouzi M, Karimpour M, Mahjoob M. A finger rehabilitation exoskeleton: design, control, and performance evaluation. In *RSI International Conference on Robotics and Mechatronics IcRoM*, Tehran, 2021.

[45] Azar WA, Nazar PS. An optimized and chaotic intelligent system for a 3DOF rehabilitation robot for lower limbs based on neural network and genetic algorithm. Biomed Signal Process Control 2021;69:102864.

[46] Baniasad MA, Akbar M, Alasty A, Farahmand F. Fuzzy control of a hand rehabilitation robot to optimize the exercise speed in passive working mode. In *Medicine Meets Virtual Reality*, https://doi.org/10.3233/978-1-60750-706-2-39.

[47] Abbasimoshaei A, Mohammadimoghaddam M, Kern TA. Adaptive fuzzy sliding mode controller design for a new hand rehabilitation robot. In *Haptics: Science, Technology, Applications. EuroHaptics 2020. Lecture Notes in Computer Science*, https://doi.org/10.1007/978-3-030-58147-3_56, 2020.

[48] Ozgoli S. A new impedance control structure for leg rehabilitation robot. In International Conference on Control, Instrumentation and Automation (ICCIA), 2011.

[49] Yousefi F, Alipour K, Tarvirdizadeh B, Hadi A. *Artificial Intelligence and Robotics (IRANOPEN)*, 2017.

[50] Delavari H, Jokar R. Intelligent fractional-order active fault-tolerant sliding mode controller for a knee joint orthosis. J Intell & Robotic Syst 2021;102:39.

[51] Rakhtala S. Adaptive gain super twisting algorithm to control a knee exoskeleton disturbed by unknown bounds. Int J Dyn Control 2021;9:711−26.

[52] Shabani A, Mahjoob M. Bio-signal interface for knee rehabilitation robot utilizing EMG signals of thigh muscles. In *International Conference on Robotics and Mechatronics*, 2016.

[53] Rastegar M, Kobravi HR. A hybrid-FES based control system for knee joint movement control. Basic Clin Neurosci 2021;12:441−52.

[54] Ardakani PS, Moradi H, Bahrami F, Akbarfahimi M. A web-based gamification of upper extremity robotic rehabilitation. In *International Serious Games Symposium (ISGS)*, 2021.

Chapter 23

Middle East region: Turkey

Duygun Erol Barkana[1], Ismail Uzun[2], Devrim Tarakci[3], Ela Tarakci[4], Ayse Betul Oktay[5] and Yusuf Sinan Akgul[6]

[1]*Department of Electrical and Electronics Engineering, Yeditepe University, İstanbul, Turkey,* [2]*İNOSENS, Kocaeli, Turkey,* [3]*Department of Occupational Therapy, Istanbul Medipol University, İstanbul, Turkey,* [4]*Department of Physiotherapy and Rehabilitation, Istanbul University-Cerrahpaşa, İstanbul, Turkey,* [5]*Department of Computer Engineering, Yildiz Technical University, İstanbul, Turkey,* [6]*Department of Computer Engineering, Gebze Technical University, Kocaeli, Turkey*

Learning objectives

At the end of this chapter, the reader will be able to:

1. Describe the current developments and future directions in rehabilitation robotics and technologies in Turkey.
2. Understand existing barriers to effective rehabilitation robotics and give an overview of the policies and legal and ethical issues around rehabilitation robotics in Turkey.

23.1 Neurorehabilitation process and therapies in Turkey

Turkey consists of seven different geographical regions. The socioeconomic conditions of each region are different from each other. The population of Turkey was 83,614,362 in December 2020. According to Organization for Economic Co-operation and Development—European Union (OECD-EU) and Turkey data, approximately 15% of the world's population comprise individuals with disabilities. While the rate of people living in the city and district centers in Turkey is 93%, the rate of those living in towns and villages is 7%. Although there is not enough data on the epidemiology of stroke in the country, it is the second most common cause of death in the country, according to the National Disease Burden and Cost-Effectiveness Project Report [1]. In addition, stroke is the first most common condition hospitalized in neurology clinics.

Rehabilitation Robots for Neurorehabilitation in High-, Low-, and Middle-Income Countries.
DOI: https://doi.org/10.1016/B978-0-323-91931-9.00014-1

Three institutions are important in the provision of health services in Turkey. These institutions are the Ministry of Health, social insurance, and universities. The presence of various institutions in the health service delivery process causes fragmented and complex situations in terms of services. The biggest role in the delivery of health services belongs to the Ministry of Health. For example, only the Ministry of Health provides preventive health services. Health centers, tuberculosis dispensaries, and health houses can be examples of institutions where preventive health services are provided [2].

About 1.9 million stroke cases were diagnosed in Turkey, according to the Turkish Social Security Institution [3,4]. Current rehabilitation includes physiotherapy using a motor control perspective for stroke rehabilitation [5]. The rehabilitation therapists create environmental conditions where the patients can functionally use their affected extremities to improve motor control. Various approaches are currently being used to rehabilitate individuals with stroke in Turkey. Some of these approaches are neurophysiologically based Bobath therapy, Brunnstrom method using reflexes, mirror method using mirror neurons, restraint induction movement therapy, and biofeedback. Although speculations continue about the superiority of these approaches over one another, it is known that behavior modification techniques (target setting, movement planning, self-observation, etc.) added to the exercise or training improve the performance of patients. Therefore the rehabilitation approaches in Turkey can be divided into two groups: muscle-focused and task-oriented.

Muscle weakness is the most common and notable consequence of stroke. Patients with muscle weakness experience severe immobilization and decreased physical activity levels. Therefore muscle-strengthening techniques and approaches are used, especially in functional rehabilitation activities for walking [6,7]. Progressive resistance exercises are the most effective method for muscle strengthening in individuals with paralysis. In addition, strengthening exercises have been shown to increase the strength of the lower-extremity (LE) muscles when intensively planned according to the patient's needs [8]. The other muscle-focused approach used in Turkey is functional electrical stimulation (FES). FES is used as a treatment modality for gait training after stroke or as an alternative to mechanical orthotic devices [9].

In task-oriented approaches, the patient's active participation in the treatment is ensured in contrast to muscle-focused approaches. The rehabilitation program is built on theories of motor learning and motor control. Treatment aims to prevent the progression of disease symptoms, incorporate task-oriented strategies, and adapt target-oriented tasks to ever-changing environmental conditions. Because of repetitive task training improves LE function (walking distance, functional ambulation, sit-stand, and standing balance) in individuals with stroke [10]. Constraint-induced movement therapy is one of the commonly used task-oriented approaches that support the repetitive use of the affected upper extremity (UE) in people with neurological motor deficits in Turkey [11]. The biofeedback technique, another task-oriented approach, aims to regulate the person's body functions. Variables of the patient's impaired physiological activities are extracted and measurably shown to the patient with the appropriate devices selected [12]. Thus patients learn to regulate these impaired physiological activities. Virtual reality (VR)-supported interventions are another commonly used task-oriented approach in Turkey. Evidence for the effectiveness of VR applications in stroke rehabilitation shows positive outcomes for different motor performance parameters such as gait, balance, and upper-limb functionality [13–15].

23.2   Robotic devices in rehabilitation centers in Turkey

The number of LE/UE rehabilitation robots has recently increased in hospitals and rehabilitation centers in Turkey. Thus the number of patients with neurological diseases (multiple sclerosis, Parkinson's), stroke, traumatic brain injury, and spinal cord injury who use these robotic devices has also increased. The most commonly used ones are Lokomat [16], LokoHelp [17], ArmeoPower [18], Amadeo [19], Erigo [20] and RoboGait [21].

Lokomat [14] is commonly used in neurological diseases with gait disturbance, in patients with loss of balance, and in any disease that causes loss of gait functions. LokoHelp [17] is another rehabilitation robot used to help patients who have partially or completely lost their walking ability regain their ability to walk. ArmeoPower [18] is a robotic technology used in the early stages of rehabilitation in patients who experience loss of shoulder and arm functions due to neurological or orthopedic reasons. Amadeo [19] is commonly used to allow each finger, including the thumb, to work independently or together. Erigo [20] is intensively used in patients who cannot move and are bedridden, especially in the early stages of their treatment. RoboGait [21] device is also used for balance and walking.

Lokomat [16], LokoHelp [17], ArmeoPower [18], Amadeo [19], Erigo [20], and RoboGait [21] robots are used in private rehabilitation centers and physical therapy and rehabilitation departments in private and public hospitals in Turkey. Generally, no additional fee is charged to the patient for robotic rehabilitation treatment.

23.3   Upper/lower-extremity rehabilitation robots developed in university laboratories in Turkey

Various upper- and lower-extremity robots have been developed in university laboratories in Turkey. An exoskeleton-type rehabilitation robotic system, called RehabRoby, is developed for rehabilitation in the Robotics and Research Laboratory at Yeditepe University (Fig. 23.1) [22–24]. A control architecture containing a high-level and low-level controller is designed so RehabRoby can complete the given rehabilitation task in a desired and

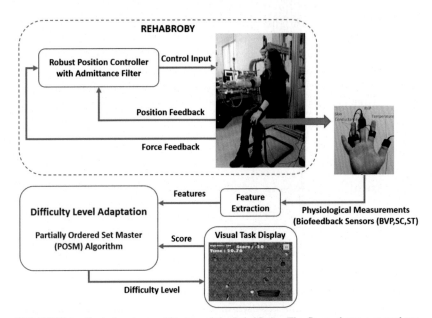

FIGURE 23.1 Control system architecture of the RehabRoby. The figure shows a control system architecture of the robot-assisted rehabilitation system called RehabRoby. The robust position controller with an admittance filter is seen at the top left. The control input from the robust position controller with admittance filter is provided to the RehabRoby robot, where position and force feedbacks are then feedback to the controller again. The biofeedback sensors used to collect blood volume pulse (BVP), skin conductance (SC), and skin temperature (ST) are seen at the right top. The features are extracted from these physiological signals. The visual task display is seen in the middle bottom, where scores are recorded. The difficulty level adaptation model, which uses extracted features and scores to find the best difficulty level of the task, is seen on the bottom left. Partially ordered set master (POSM) algorithm is used for difficulty level adaptation.

safe manner. A hybrid system modeling technique is used for the high-level controller. An admittance control with an inner robust position control loop is used for the low-level control of the RehabRoby. RehabRoby is potentially more effective because it recognizes the patients' emotional states and modifies the task's difficulty level considering these emotions to increase the patient's engagement. Different feedback strategies are used to determine the difficulty level of motivating subjects during RehabRoby. First, the physiological signals of the subjects are measured using biofeedback sensors. Second, the excitement levels of the subjects were explored while dynamically adapting the task difficulty level of RehabRoby for each subject using score feedback, physiological signal feedback, or their combination [25–27].

AssistOn-Arm has been shown to increase the range of motion (ROM) for UE exercises and to enable the delivery of glenohumeral mobilization and scapular stabilization exercises. AssistOn-Gait enables its users to freely execute pelvic movements [28]. AssistOn-Walk is designed to provide active support to hip and knee movements in the sagittal plane while allowing passive hip rotations to enable natural gait patterns. AssistOn-Wrist is a self-aligning bimanual rehabilitation robot designed to assist forearm and wrist rotations [29]. AssistOn-Finger is an under-actuated active exoskeleton for the therapy of human fingers [30]. AssistOn-Knee can assist flexion/extension movements of the knee joint and accommodate its translational movements in the sagittal plane [31]. Finally, AssistOn-Ankle is a reconfigurable, powered exoskeleton for ankle rehabilitation that features reconfigurable kinematics to deliver ROM/strengthening and balance/proprioception exercises [32].

A self-balancing lower-body exoskeleton prototype called Compliant Exoskeleton for Human-Robot Co-Existence (Co-Ex) has been developed [33−36]. Co-Ex possesses two major features that distinguish it from most state-of-the-art exoskeleton systems: (1) self-balancing capability and (2) a leg structure with no directly attached actuators. Each leg consists of a two degrees of freedom (DoF) hip joint that allows motions along with flexion/extension and adduction/abduction axes. Furthermore, it has one DoF knee and ankle joints that allow motions along with flexion/extension and dorsi plantar/flexion axes [33−37].

Physiotherabot, which assists the hip and leg, has been developed [38]. In addition, Physiotherabot/wf [39] is developed to perform therapeutic exercises for the wrist and forearm. Physiotherabot/wf has been tested on patients, and an improvement has been observed in flexion and extension isometric force values. DIAGNOBOT is another UE rehabilitation robot developed for treatment [40].

A lower-extremity exoskeleton robot similar to human joint biomechanics, whose stiffness and damping can be changed in each joint, has been developed. It has been shown that using equipment with variable damping in the joints of LE robots provides advantages such as high stability and effective oscillation control [41]. Furthermore, a variable stiffness ankle joint design named VS-AnkleExo is developed in which the stiffness adjustment mechanism is embedded at the calf level below the knee [42]. Moreover, a flexible LE exoskeleton robot named BioComEx, with variable stiffness in the wrist joint and constant stiffness in the knee and hip joints, has been designed, and the preliminary tests are carried out [43]. Furthermore, the first variable impedance (stiffness and damping) LE exoskeleton robot is developed [44].

23.4 Barriers to effective rehabilitation robotics in Turkey

The advantage of rehabilitation robots is that their movements are similar to normal activity and constantly stimulate the brain's centers. As a result, the recovery process of patients is accelerated, and they develop movement patterns close to normal. Rapid progress is achieved compared to manual treatments because of the longer and more intensive functional training sessions [45,46]. The patient's movement activities are easily supervised and evaluated with rehabilitation robots. Additionally, the movement patterns and supporting forces can be adjusted individually to the patient's needs to accommodate functional training. As a result, it is possible to obtain easy and reproducible measurements of patient improvement. Although the benefits of rehabilitation robots for the LE/UE have been observed, more research and development are needed to reduce their limitations and obstacles. The most obvious limitations are their high cost and the lack of a standardized treatment protocol. Other limitations are their large size and lack of internal power supply time in mobile units. New studies focus on low power consumption, integration with new communication technologies, data collection, data analysis, easy transfer and relocation capabilities that need industrial design approaches, and flexible size and person body characteristics that may need new material techniques.

23.5 Case studies with rehabilitation robots in Turkey

Various LE/UE rehabilitation robotic devices are used in Turkey for treatment [47], and their effects are investigated in different studies. The efficiency of robotic gait training with the Lokomat [16] device was evaluated during the subacute period for 34 patients with spinal cord injuries [48]. After ten robotic treatment training sessions, improvements were observed in functional status, gait, and daily living activities. In another study [49], the effects of VR-based robotic rehabilitation on weight transfer and mobility levels were evaluated with a Lokomat [16] device on 45 patients who were divided into early (within 6 months after stroke) and late stroke (greater than 6 months after stroke) groups for 4 weeks, and no significant differences were found between them. An additional study investigated training with robotic rehabilitation with the RoboGait [21] device on balance and walking [50]. The results showed that conventional physiotherapy and robotic rehabilitation are useful for functional independence, balance, and walking.

The satisfaction and motivation of patients trained with the Lokomat [16] device were also evaluated [51]. About 30 patients were trained for 30 minutes, 5 days a week for 1 month. Barthel index (BI) for activities of daily living (ADLs), the SF-36 quality of life scale, the internal motivation inventory (IMI), and the robotic rehabilitation patient motivation and satisfaction questionnaire were used as outcomes. The BI and SF-36 scores increased during treatment, and 73.3% of the patients recommended robotic rehabilitation. In addition, BI scores and SF-36 quality of life scores of patients with central nervous system disease increased significantly and were positively correlated. Furthermore, the effects of robotic treadmill training on LE functions and walking were evaluated on 30 male patients with traumatic incomplete spinal cord injuries [52]. The results showed that conventional and additional robotic training improved LE motor function and functional independence.

The satisfaction and motivation of 30 neurologically impaired patients trained with an ArmeoPower [18] 5 days a week for a month were investigated [53]. Statistically, significant changes were detected in scores of physical functions, energy, buoyancy, vitality, and pain according to SF-36 scores, but no significant changes were detected in the BI results. A UE rehabilitation robot for hand functions and activity has been used in 39 patients with chronic stroke [54]. A statistically significant improvement has been observed in performance and satisfaction parameters after 3 weeks of robotic rehabilitation with the Amadeo [19]. The effects of ArmeoSpring [55] rehabilitation robot on hand function and quality of life in 49 stroke patients were also investigated [56]. The patients were trained five times weekly for 6 months with conventional rehabilitation. Improvements were found in both hand function and ADLs.

The most common approach followed in the case studies is evaluating the effects of robotic rehabilitation on diagnostic and control groups with or instead of conventional therapy. Although some studies did not find significant differences between these groups [49], most studies reported improvements gained with rehabilitation robots [51,52,54]. Thus additional research is needed to be performed with a larger number of participants for different types of illnesses to comprehensively evaluate the use of rehabilitation robots.

23.6 The policies, legal and ethical studies conducted during the use of rehabilitation robotics in Turkey

The only certification requirement in the European Community (EC) is to comply with EC directives. Applying International Organization for Standardization (ISO) standards is not mandatory, but it is a strong argument when requesting certification. It is also sometimes a mandatory requirement expressed by customers.

A Social Security Institution Health Implementation Legislation has been prepared for robotic rehabilitation applications in Turkey. According to this legislation, physical medicine and rehabilitation specialists must prepare the board of health report [57]. The Social Security Institution covers the costs when the rehabilitation robot is used for diagnoses, but the robotic rehabilitation sessions cannot be billed together with other therapy and rehabilitation procedures. A maximum of 30 sessions per year is allowed; therefore 15 patients can be billed daily for each rehabilitation robot. As technology will be a more important part of rehabilitation, Turkey may need to adopt regulations to benefit patients who need rehabilitation.

23.7 Current state of rehabilitation robots under COVID-19 conditions and future directions

The COVID-19 pandemic prevented many patients from going to rehabilitation centers or hospitals for treatment. Therefore rehabilitation robots in the centers often could not be used during the pandemic. Furthermore, even if it was possible to use them for short periods, it was necessary to assure the patients about the system's safety against the COVID-19 virus, and the whole system had to be thoroughly inspected for virus contamination.

Patients recovering from the COVID-19 infection, especially those who required intensive care treatments, will probably require rehabilitation [58]. Since only very little is known about the long-term effects of this disease, the exact nature of rehabilitation processes is not yet clear. Rehabilitation robotics interventions with these patients are a possibility that needs to be investigated.

Many therapists consider the current telerehabilitation solutions under COVID-19 conditions temporarily [59]. Furthermore, most therapists have no plans to continue with digital remote therapy after the pandemic, making it difficult to make current practitioners accept remote robotic-based rehabilitation therapy. For example, in stroke rehabilitation, the therapist must guide the exercises by giving tactile stimulation to increase the patient's motor and cognitive abilities and facilitate correct movements. The physical size and transportation difficulties of rehabilitation robots currently on the market must be addressed for the wider incorporation of these systems. Several design features and development paths need to be incorporated into the next generation of rehabilitation robots for effective rehabilitation. First, more portable systems with specifically developed building materials and technologies are needed for interactive applications at home with remote expert intervention. Second, systems with VR games that direct the patients for purposeful exercises and can increase patient motivation must be installed and operated at home by nonprofessional caregivers under the guidance of remote experts. Finally, personalized data for automatic follow-up of patient-specific rehabilitation programs must be considered while keeping the patient data private and safe during communication with remote rehabilitation sites.

23.8 Future of neurorehabilitation in rehabilitation robotics

Rehabilitation robotics is a growing field that seeks to incorporate advances in robotics combined with neuroscience and rehabilitation to define new methods for treating problems related to neurological diseases. According to the findings obtained by the Global Initiative on Neurology and Public Health carried out by the World Health Organization (WHO), many neurological disorders are chronic and progressive, constitute a global public health problem, and affect especially older people.

eHealth, which uses information and communication technologies (ICTs) for health, can improve prevention, diagnosis, treatment, monitoring, and management. Furthermore, it can benefit the entire community by improving access to care and quality of care and making the health sector more efficient. It includes information and data sharing between patients and health service providers, hospitals, health professionals, and health information networks; electronic health records; telemedicine services; portable patient-monitoring devices; operating room scheduling software; rehabilitation robots; robotized surgery; and blue-sky research on a virtual physiological human.

The adoption of eHealth, along with organizational changes and other technical innovations, can improve the quality of life for patients and ensure the sustainability of care. Despite the worldwide economic crisis, the market potential of eHealth is strong. The well-being market enabled by digital technologies is rapidly growing. Thus eHealth's 5G-enabled telerehabilitation robots and technological rehabilitation will be important application areas. The convergence between wireless communication technologies, healthcare devices, and health and social care creates new businesses. Redesigning the delivery of care and the silver economy are highly promising markets. Turkey is preparing telecommunication infrastructure for 5 G, and some companies are working on European Union-supported Research and Development (R&D) projects which use 5 G and eHealth [60].

More than half of the WHO Member States now have an eHealth strategy, and 90% of eHealth strategies reference the objectives of Universal Health Coverage or its key elements. In addition, many countries report at least one eHealth initiative. However, despite the rapid growth, very few WHO Member States reported evaluations of government-sponsored eHealth programs, limiting knowledge of what works well and what mistakes to avoid.

Healthcare data sharing is complex as it touches on privacy requirements, legal requirements, security issues, and the more obvious structural and infrastructural problems. Telehealth is only implemented to a minor extent and is mostly available for consultations with other healthcare practitioners. However, when telehealth capabilities are implemented, they are mostly used (usage in around 90% of the surveyed hospitals on average), demonstrating that they offer genuine utility for healthcare professionals. European regulators have published high-level cybersecurity recommendations for industries, including medical devices involved in the Internet of Things (IoT) paradigm. As a part of telehealth, rehabilitation robot manufacturers and technological rehabilitation solution providers should meet European data privacy requirements under the General Data Protection Regulation (GDPR).

Artificial intelligence (AI) for rehabilitation robots can also produce further options. It is not enough if robots are simply therapist extenders and save human efforts when executing repetitive and exhausting exercises.

AI can be an additional advantage of robots: continuously evaluating the patients' status and choosing the next exercise accordingly.

AI-based clinical decision support systems will be one of the main features of eHealth services. AI-based eHealth systems can utilize data from various sources to support decision-making inside clinics and during rehabilitation. At each stage, data are provided/collected relevant to diagnosis and better support for health. The collected health data may include exercise data, data from the environment, and vital data from a patient. AI methods can analyze the collected data, make clinical decisions more efficient, and make data quickly available outside the clinics.

With 5 G technology that will allow remote and secure communication, rehabilitation robots will enter a new era like other eHealth applications. 5G technology allows high security, reliability, coverage, micro-operator possibilities, transferring large amounts of data, low (or no) latency, collection of large amounts of patient data, and edge computing for faster data analysis for smarter medication.

AI and AI-based decision support systems will be among eHealth's most important support systems, especially neurorobotics and rehabilitation. As a result, we will see more smart clinics, rehabilitation centers, and smart homes, including robotic systems [61,62].

Acknowledgments

This work is supported by the Scientific and Technological Research Council of Turkey Technology and Innovation Funding Programs Directorate (TÜBİTAK-TEYDEB) under grant numbers 3210036 (1501) and 9180068 (1509−CELTIC-NEXT Program).

References

[1]   TÜİK Kurumsal. Adrese Dayalı Nüfus Kayıt Sistemi Sonuçları 2020, Available from:<https://data.tuik.gov.tr/Bulten/Index?p = Adrese-Dayali-Nufus-Kayit-Sistemi-Sonuclari-2020-37210>; 2021.

[2]   Diger H, Bilgin A. OECD Ülkelerinde sağlık sistemlerinin sağlık göstergeleri açısından değerlendirilmesi. Int J Innov Approaches Soc Sci 2021;5(4):212−35.

[3]   Ince B, Necioglu D. Organization of stroke care in Turkey. Int J Stroke 2017;12(1):105−7.

[4]   Ozturk Y, Demir C, Gursoy K, Koselerli R. Analysis of stroke statistics in turkey. Value Health 2015;18(7):A402.

[5]   Díaz-Arribas MJ, Martín-Casas P, Cano-de-la-Cuerda R, Plaza-Manzano G. Effectiveness of the Bobath concept in the treatment of stroke: a systematic review. Disabil Rehabil 2020;42(12):1636−49.

[6]   Kara B, Aytekin E, Sayiner Caglar N, Pekin Dogan Y, Caglar S, Aydemir K, et al. Neuromuscular electrical stimulation therapy effects on the functional and motor recovery of the upper extremity in patients after stroke: a randomized controlled trial. Istanb Med J 2021;22(3):202−7.

[7] Aydogan Arslan S, Ugurlu K, Sakizli Erdal E, Keskin ED, Demirguc A. Effects of inspiratory muscle training on respiratory muscle strength, trunk control, balance and functional capacity in stroke patients: a single-blinded randomized controlled study. Top Stroke Rehabil 2021;7:1–9.

[8] Mustafaoglu R, Erhan B, Yeldan I, Ersoz Huseyinsinoglu B, Gunduz B, Razak Ozdincler A. The effects of body weight-supported treadmill training on static and dynamic balance in stroke patients: a pilot, single-blind, randomized trial. Turk J Phys Med Rehabil 2018;64(4):344–52.

[9] Karaahmet OZ, Gurcay E, Unal ZK, Cankurtaran D, Cakci A. Effects of functional electrical stimulation-cycling on shoulder pain and subluxation in patients with acute-subacute stroke: a pilot study. Int J Rehabil Res 2019;42(1):36–40.

[10] Kayabinar B, Alemdaroglu-Gurbuz I, Yilmaz O. The effects of virtual reality augmented robot-assisted gait training on dual-task performance and functional measures in chronic stroke: a randomized controlled single-blind trial. Eur J Phys Rehabil Med 2021;57 (2):227–37.

[11] Huseyinsinoglu BE, Ozdincler AR, Krespi Y. Bobath concept versus constraint-induced movement therapy to improve arm functional recovery in stroke patients: a randomized controlled trial. Clin Rehabil 2012;26(8):705–15.

[12] Dost Surucu G, Tezen O. The effect of EMG biofeedback on lower extremity functions in hemiplegic patients. Acta Neurol Belg 2021;121(1):113–18.

[13] Ikbali Afsar S, Mirzayev I, Umit Yemisci O, Cosar Saracgil SN. Virtual reality in upper extremity rehabilitation of stroke patients: a randomized controlled trial. J Stroke Cerebrovasc Dis 2018;27(12):3473–8.

[14] Simsek TT, Cekok K. The effects of Nintendo Wii(TM)-based balance and upper extremity training on activities of daily living and quality of life in patients with sub-acute stroke: a randomized controlled study. Int J Neurosci 2016;126(12):1061–70.

[15] Ogun MN, Kurul R, Yasar MF, Turkoglu SA, Avci S, Yildiz N. Effect of leap motion-based 3D immersive virtual reality usage on upper extremity function in ischemic stroke patients. Arq Neuropsiquiatr 2019;77(10):681–8.

[16] Hocoma. Lokomat®, Available from: <https://www.hocoma.com/solutions/lokomat/>; 2022.

[17] Woodway. Loko Help the Way to Walk, Available from: <https://www.woodway.com/products/loko-help/>; 2020.

[18] Hocoma. Armeo®Power, Available from: <https://www.hocoma.com/solutions/armeo-power/>; 2022.

[19] Tyromotion. Amadeo® Finger-Hand-Rehabilitation, Available from: <https://tyromotion.com/en/products/amadeo/>.

[20] Hocoma. Erigo®, Available from: <https://www.hocoma.com/solutions/erigo/>;2022.

[21] Bamateknoloji, Robogait, Available from: http://www.bamateknoloji.com/products/robotic-rehabilitation-3-en/robogait/?lang = en >.

[22] Ozkul F, Erol Barkana D. Upper-extremity rehabilitation robot RehabRoby: methodology, design, usability and validation. Int J Advaced Robotic Syst 2013;10(12):401.

[23] Ozkul F, Erol Barkana D, Badıllı Demirbaş Ş, Inal S. Evaluation of elbow joint proprioception with RehabRoby: a pilot study. Acta Orthop Traumatol Turc 2012;46(5):332–8.

[24] Ozkul F, Erol Barkana D. A Robot-assisted rehabilitation system – RehabRoby. Interdisciplinary Mechatronics. Wiley; 2013. p. 145–62.

[25] Ozkul F, Palaska Y, Masazade E, Erol Barkana D. Exploring dynamic difficulty adjust-ment mechanism for rehabilitation tasks using physiological measures and subjective rat-ings. IET Signal Process 2018;13(3):378−86.

[26] Ozkul F, Erol Barkana D, Masazade E. Dynamic difficulty level adjustment based on score and physiological signal feedback in the robot-assisted rehabilitation system. Rehabroby IEEE Robot Autom Lett 2021;6(2):447−54.

[27] Gumuslu E, Erol Barkana D, Kose H. Emotion recognition using EEG and physiological data for robot-assisted rehabilitation systems. In Companion Publication of the 2020 International Conference on Multimodal Interaction, 2020;379−87.

[28] Munawar H, Yalcin M, Patoglu V. AssistOn-Gait: an overground gait trainer with an active pelvis-hip exoskeleton. In 2015 IEEE International Conference on Rehabilitation Robotics (ICORR), 2015;594−9.

[29] Erdogan A, Satici AC, Patoglu V. Passive velocity field control of a forearm-wrist reha-bilitation robot. In 2011 IEEE International Conference on Rehabilitation Robotics, 2011;1−8.

[30] Ertas IH, Patoglu V. AssistOn-Finger: an under-actuated finger exoskeleton for robot-assisted tendon therapy. Robotica 2014;32(8):1363−82.

[31] Celebi B, Yalcin M, Patoglu V. AssistOn-Knee: a self-aligning knee exoskeleton. In 2013 IEEE/RSJ International Conference on Intelligent Robots and Systems, 2013;996−1002.

[32] Erdogan A, Celebi B, Satici AC. AssistOn-Ankle: a reconfigurable ankle exoskeleton with series-elastic actuation. Autonomous Robot 2017;41:743−58.

[33] Coruk S, Soliman AF, Dalgic O, Yildirim MC, Ugur D, Ugurlu B. Towards crutch-free 3-D walking support with the lower body exoskeleton Co-Ex: self-balancing squatting experiments. In: Moreno JC, Masood J, Schneider U, Maufroy C, Pons JL, editors. Wearable robotics: challenges and trends. WeRob 2020. Biosystems & Biorobotics, vol. 27. Cham: Springer; 2022.

[34] Yildirim MC, Kansizoglu AT, Emre S, Derman M, Coruk S, Soliman AF, et al. Co-Ex: a torque-controllable lower body exoskeleton for dependable human-robot co-existence. IEEE Int Conf Rehabil Robot 2019;2019:605−10.

[35] Yildirim MC, Sendur P, Kansizoglu AT, Uras U, Bilgin O, Emre S, et al. Design and development of a durable series elastic actuator with an optimized spring topology. In Proceedings of the Institution of Mechanical Engineers, Part C: Journal of Mechanical Engineering Science, 2021.

[36] Ugurlu B, Sariyildiz E, Kansizoglu AT, Ozcinar EC, Coruk S. Benchmarking torque con-trol strategies for a torsion based series elastic actuator. In IEEE Robotics & Automation Magazine, 2021.

[37] Soliman AF, Ugurlu B. Robust Locomotion Control of a self-balancing and underactuated bipedal exoskeleton: task prioritization and feedback control. IEEE Robot Autom Lett 2021;6(3):5626−33.

[38] Akdogan E, Adli MA. The design and control of a therapeutic exercise robot for lower limb rehabilitation: physiotherabot. Mechatronics 2011;21(3):509−22.

[39] Akdogan E, Koru AT, Aktan ME, Arslan MS, Atlihan M, Kuran KB. Hybrid impedance control of a robot manipulator for wrist and forearm rehabilitation: performance analysis and clinical results. Mechatronics 2018;49:77−91.

[40] Aktan ME, Akdogan E. Design and control of a diagnosis and treatment aimed robotic platform for wrist and forearm rehabilitation: DIAGNOBOT. Adv Mech Eng 2018;10 (1):1−13.

[41] Baser O, Demiray MA. Selection and implementation of optimal magnetorheological brake design for a variable impedance exoskeleton robot joint. In Proceedings of the Institution of Mechanical Engineers, Part C: Journal of Mechanical Engineering Science, 2017;231(5):941−60.

[42] Baser O, Kizilhan H. Mechanical design and preliminary tests of VS-AnkleExo. J Braz Soc Mech Sci Eng 2018;40(442):1−16.

[43] Baser O, Kizilhan H, Kilic E. Biomimetic compliant lower limb exoskeleton (BioComEx) and its experimental evaluation. J Braz Soc Mech Sci Eng 2019;41(226):1−5.

[44] Baser O, Kizilhan H, Kilic E. Employing variable impedance (stiffness/damping) hybrid actuators on lower limb exoskeleton robots for stable and safe walking trajectory tracking. J Mech Sci Technol 2020;34(6):2597−607.

[45] Masiero S, Poli P, Rosati G, Zanotto D, Iosa M, Paolucci S, et al. The value of robotic systems in stroke rehabilitation. Expert Rev Med Devices 2014;11(2):187−98.

[46] Takahashi CD, Der-Yeghiaian L, Le V, Motiwala RR, Cramer SC. Robot-based hand motor therapy after stroke. Brain. 2008;131(Pt 2):425−37.

[47] Bilir Kaya B, Ayaz O. Demographic characteristics of robotic rehabilitation patients in a public rehabilitation hospital according to age groups. Med J Istanb Kanuni Sultan Suleyman 2019;11(3):118−24.

[48] Cinar C, Oneş K, Yildirim MA, Goksenoglu G. Comparison of the patients with complete and incomplete spinal cord injury administered robotic-assisted gait training treatment. Turkish J Phys Med Rehabilitation 2020;23(1):12−19.

[49] Koc S. The comparison of the effects of the virtual reality based robotic rehabilitation program on weight transmission and movement participation [MS Thesis]. Inonu University, 2020.

[50] Bulbul IE. Effects of robotic rehabilitation application on functional walking, balance and functional independence in children with spastic type cerebral palsy [MS Thesis]. Medipol University, 2021.

[51] Bener G. Evaluation of patient motivation and satisfaction perception of robotic walking systems in central nervous system diseases [MS Thesis]. Medipol University, 2018.

[52] Midik M, Paker N, Bugdayci D, Midik AC. Effects of robot-assisted gait training on lower extremity strength, functional independence, and walking function in men with incomplete traumatic spinal cord injury. Turk J Phys Med Rehabil 2020;66(1):54−9.

[53] Tekeci E. The effect of hand arm robot used in neuro impaired patients for patients' satisfaction and motivation for robotic rehabilitation [MS Thesis]. Istanbul Medipol University, 2018.

[54] Cekmece C, Sade SI. Efficacy of robotic rehabilitation on hand function and activity of daily living in stroke patients. J Health Sci Kocaeli Univ 2020;7(1):35−8.

[55] Hocoma. Armeo®Spring, Available from: <https://www.hocoma.com/solutions/armeo-spring/>.

[56] Mustafaoglu R, Yildiz A, Kesiktas N. The effect of robot-assisted upper extremity training on hand function and quality of life in stroke patients. Osmangazi J Med 2021;43(3):224−33.

[57] T.C. Cumhurbaşkanlığı Mevzuat Bilgi Sistemi. Sosyal Güvenlik Kurumu sağlık uygulama tebliği, Available from: <https://www.mevzuat.gov.tr/mevzuat?MevzuatNo = 17229& MevzuatTur = 9&MevzuatTertip = 5>; 2013.

[58] Salawu A, Green A, Crooks MG, Brixey N, Ross DH, Sivan M. A proposal for multidisciplinary telerehabilitation in the assessment and rehabilitation of COVID-19 survivors. Int J Env Res Public Health 2020;17(13):4890.

[59] Rausch AK, Baur H, Reicherzer L, Wirz M, Keller F, Opsommer E, et al. Physiotherapists' use and perceptions of digital remote physiotherapy during COVID-19 lockdown in Switzerland: an online cross-sectional survey. Arch Physiother 2021;11 (1):18.

[60] Health5G. HEALTH5G is a project labelled by CELTIC-PLUS, Available from: <https://health5g.eu/>; 2022.

[61] European Commission. eHealth Action Plan 2012−2020 - Innovative healthcare for the 21st century, Available from: <https://ec.europa.eu/health/sites/default/files/ehealth/docs/com_2012_736_en.pdf>; 2012.

[62] World Health Organization. Global diffusion of eHealth: Making universal health coverage achievable, Available from: <http://apps.who.int/iris/bitstream/10665/252529/1/9789241511780-eng.pdf?ua = 1>; 2016.

Chapter 24

Africa region: Nigeria

Morenikeji A. Komolafe[1], Kayode P. Ayodele[2],
Matthew O.B. Olaogun[3], Philip O. Ogunbona[4], Michael B. Fawale[1],
Abiola O. Ogundele[5], Akintunde Adebowale[1],
Oluwasegun T. Akinniyi[2], Sunday O. Ayenowowan[5],
Abimbola M. Jubril[2], Ahmed O. Idowu[1], Ahmad A. Sanusi[1],
Abiodun H. Bello[6] and Kolawole S. Ogunba[2]

[1]Department of Medicine, Obafemi Awolowo University, Ile-Ife, Osun, Nigeria, [2]Department of
Electronic and Electrical Engineering, Obafemi Awolowo University, Ile-Ife, Osun, Nigeria,
[3]Faculty of Medical Rehabilitation, University of Medical Sciences, Ondo, Ondo, Nigeria,
[4]School of Computing and Information Technology, University of Wollongong, Wollongong,
NSW, Australia, [5]Department of Medical Rehabilitation, Obafemi Awolowo University Teaching
Hospitals Complex, Ile Ife, Osun, Nigeria, [6]Department of Medicine, University of Ilorin
Teaching Hospital, Ilorin, Kwara, Nigeria

Learning objectives

At the end of this chapter, the reader will:
1. Understand the challenges and opportunities of neurorehabilitation in Nigeria;
2. Understand why robot-mediated therapy is an option for improving neurorehabilitative care in Nigeria.

24.1 Introduction

Nigeria is a West African country adjoining the Gulf of Guinea. With an estimated population of 206 million, Nigeria is the most populous country in Africa, and the seventh most populous in the world. The country has a young population, with a median age of 17.9 years, and a life expectancy of 53 years. Due to its young population and population growth rate, Nigeria is projected to become the world's third most populous country by 2050.

Nigeria is classified by the World Bank as a lower-middle-income economy with a gross national income per capita of $2030. Severe poverty and income inequalities are pervasive nationwide, with Nigeria's inability-adjusted human development index of 0.349 placing it 128 out of 150 countries ranked in 2019 [1].

Rehabilitation Robots for Neurorehabilitation in High-, Low-, and Middle-Income Countries.
DOI: https://doi.org/10.1016/B978-0-323-91931-9.00008-6
Copyright © 2024 Elsevier Inc. All rights reserved.

Nigeria adopts a multitiered healthcare system, with healthcare centers of varying funding, size, and specialization mapping approximately to the three tiers of the government. Primary health services are provided through dispensaries, health posts, and clinics largely controlled by local government units. Secondary healthcare centers are operated mainly by the 36 state governments and include district hospitals, comprehensive health centers, general hospitals, and specialist hospitals. Tertiary healthcare centers form the highest tier and are classified as either teaching hospitals or federal medical centers.

Healthcare is primarily funded by the government, with the lion's share going to secondary and tertiary centers. Partly due to the neglect of primary centers, most patients bypass them and take their cases, no matter how seemingly mild, to higher tiers. This increases the pressure on secondary and tertiary centers.

The establishment in 1995 of a National Health Insurance Scheme (NHIS) was an important step toward more equitable financing mechanisms and universal health coverage. Coverage started from the formal sector, with a mandatory enrollment of all federal employees. Other categories of the formal sector were captured in distinct subsequent phases. Contributions are earnings-related, with 10% of an enrollee's basic salary contributed by the employer and 5% by the employee. Enrollees' out-of-pocket payments for applicable health services are thereafter capped at 10% of the total service cost. The scheme expanded to informal sectors of the economy by 2009. Progress toward universal coverage has been limited, with approximately 5% of the total population now covered [2].

24.2   Relevant medical statistics and trends in Nigeria

The most common medical conditions necessitating neurorehabilitation in Nigeria include Parkinson's disease, spinal cord injury, traumatic brain injury, and stroke.

In the first detailed report on parkinsonism and Parkinson's disease (PD) in Nigeria, Osuntokun [3] described extrapyramidal disorders in 1.9% of 9600 patients with neurological conditions seen at the University College Hospital Ibadan, Southwestern Nigeria, over 12 years (1957−69). The estimated crude prevalence of PD in Nigeria was lower (10−249/100,000) compared to studies published in Europe (65.6−12,500/100,000) which has been attributed to underdiagnosis or misdiagnosis [4]. The estimated crude prevalence of PD from neurological hospital admission studies was 75/100,000 in Ibadan [5], 63/100,000 in Kano [6], with the lowest estimate from the Niger Delta area (32/100,000) [7], and the highest from Enugu (165/100,000) in southeast Nigeria [8].

The Nigeria Parkinson Disease Registry (NPDR) revealed that the mean age at onset for PD was in the seventh decade of life (60.3 ± 10.7 years), and the median disease duration (interquartile range) was 36 months (18−60.5 months) with a male-to-female ratio of 2.64−1 [9]. A systematic review and meta-analysis found increased mortality in PD, with a pooled mortality ratio of 1.5 [10]. Okubadejo et al. revealed the case fatality rate of PD in Nigeria to be 25% compared to 7.1% in controls [11].

A study done in Southeast Nigeria found the prevalence of spinal cord injury (SCI) to be 31.8% [12]. A retrospective study done in a Nigerian university hospital over 15 years in 468 patients showed a male to female ratio of 2.3:1, similar to other local studies with the majority in the mean age of 30−40 years [13]. The leading causes of SCI were motor vehicle accidents, falls, and gunshot injuries.

An incidence rate of 2710/100,000 per year has been reported for traumatic brain injury (TBI) in Nigeria [14]. The majority of studies on TBI across Nigeria revealed that more men than women were affected, and the mean incidence ages were in the second to third decade of life [14,15]. The most common etiological factors were road traffic accidents involving motorcycles and automobiles [14,15]. A hospital-based cross-sectional study in Nigeria that covered all ages for patients treated for trauma in the emergency room found that 31% of all trauma deaths were due to severe TBI [15]. In another hospital-based retrospective study, the case fatality rate for TBI was 22.6% [15]. Mortality varies by severity but is high in those with severe injury and in the elderly. The risk factors for TBI identified in Nigeria range from male gender, motorcycle riding, illiteracy and ignorance of traffic laws, trading, and extremes of age [14].

A community-based study in 2007 put the prevalence of stroke in Nigeria at 1.14 per 1000 with a higher prevalence of 1.51 per 1000 in men compared with 0.69 per 1000 in women [16,17]. Enwereji et al. [18] reported a prevalence rate of 1.63 per 1000 in Southeast Nigeria. A recent systematic review and meta-analysis found the crude prevalence of stroke in Nigeria to be 6.7 (5.8−7.7)/1000 population; incidence is higher among men at 6.4 (5.1−7.6) per 1000, compared to women at 4.4 (3.4−5.5) per 1000. The prevalence of stroke survivors increased minimally from 6.0 (95% CI: 4.6−7.5) per 1000 to 7.5 (95% CI: 5.8−9.1) per 1000 over the period 2000−09 [19].

The 30-day case fatality rate from stroke ranges from 28% to 40%; the outcome for those with complications is worse [20]. The estimated mortality from stroke per annum in Nigeria was 1.26 per 1000 [21]. Hence, there is an immense burden of stroke as it may leave survivors with a permanent disability, suboptimal health-related quality of life, and death.

Data from the first INTERSTROKE study showed that the proportions of ischemic and hemorrhagic stroke in Africa were about 66% and 34%, respectively [22]. Recent data from the Stroke Investigative Research and Educational Network (SIREN) study in Nigeria and Ghana reported 68% of ischemic stroke and 32% of hemorrhagic stroke, which partly confirm the proportions of stroke subtypes in Africa reported by the INTERSTROKE study [23].

24.3   Standard of neurorehabilitation care in Nigeria

Neurorehabilitation in Nigeria is multidisciplinary and interdisciplinary, involving the neurologist, physiotherapist, occupational therapist, nutritionist, speech and language therapist, psychiatrist, and neuropsychologist. Rehabilitation begins immediately after diagnosis, and life-threatening problems are controlled during the acute hospitalization. CT or MRI scans are essential for diagnosis and treatment, but in some cases, patients' financial constraints may preclude their use. As they are not covered by health insurance, payment is out-of-pocket, and most patients struggle to raise the funds. The cost of MRI scans in Nigeria typically ranges between $125 and $250, which is not affordable for the majority of the population. This has a domino effect on rehabilitation, as many patients' funds are depleted by the scans, leading them to treat rehabilitation as a luxury they can ill afford. There are approximately 100 CT and 60 MRI equipment in the country, and they are distributed across 20 cities. This is grossly inadequate for the population of Nigeria, and the lopsided distribution often implies long-distance travel for most patients requiring imaging service. There is also an attending delay in diagnosis.

Nigeria has scant resources for neurorehabilitation. There are approximately 200 neurologists, 4587 physiotherapists, and 137 occupational therapists in Nigeria [24], making the number of health professionals approximately 1, 23, and 0.67 per million people, respectively. This reveals the marked inadequacy of professionals involved in caring for and rehabilitation of survivors of nervous system insults. The already bad situation is further exacerbated by a recent economic downturn that has resulted in a significant increase in emigration of medical professionals. The total number of tertiary hospitals in Nigeria is 85, of which only about 60 possess physiotherapy facilities. A total of seven hospitals have facilities for both physiotherapy and occupational therapy, and these facilities are unevenly distributed.

Most guidelines advocate a minimum of 45 minutes of relevant daily therapy for at least 5 days a week [25]. This is difficult in Nigeria. While the typical number of sessions per week depends on factors such as proximity of patients' homes to the care center, inadequate personnel, and overcrowded clinics, patients in Nigeria typically have one rehabilitation session per week [26].

Stroke rehabilitation typically takes place during acute in-hospital care in general medical wards, usually within the premises of the hospitals where patients are admitted. Rehabilitation plans and services depend on the type of accompanying functional impairment. The physiotherapist and occupational therapist are invited immediately after admission; however, the review will depend on the patient paying for five sessions of treatment. The dietetic unit and speech therapist are also invited to review if their services are needed. In healthcare centers with competence to manage stroke cases, there are weekly or biweekly stroke unit meetings in which medical status, complications, and rehabilitation needs are discussed and decisions implemented.

Due to severe resource constraints, post-discharge rehabilitation either takes place in patients' homes (for patients who have money to engage physiotherapists on a private basis) or outpatient physiotherapy facilities. Typically, the patient continues outpatient rehabilitation at the clinic for 30 min to 1 h sessions. The large patient load increases waiting time and reduces the therapy session time. There is absence of community-based physiotherapy services or structured home rehabilitation services. The challenges of home care physiotherapy services include poor working environment, transportation costs, and lack of regulation of care [27].

Motor impairment is the most common deficit after a stroke. Other deficits include sensory loss (e.g., hemianesthesia), dysphagia, aphasia or dysarthria, impaired vision and partial blindness (hemianopia), and spatiotemporal neglect for the right-sided stroke. Stroke reduces mobility in more than half of stroke survivors aged 65 and over [28]. Physiotherapists engage in physical activities that may improve motor function in terms of movement, balance, stretching, and strengthening of muscles. These physical activities include progressive active or isokinetic muscle strengthening, muscle stretching, mobility training, motor skill learning, constraint-induced movement therapy (CIMT), focused-use therapy, and range of motion therapy.

24.4 Rehabilitation robotics in Nigeria: motivation, status, and future prospects

24.4.1 The case for rehabilitation robotics in Nigeria

Healthcare funding, personnel, and general socioeconomic data suggest that current healthcare delivery models in use in Nigeria cannot adequately address emerging epidemiological trends in the country. Viable alternatives must be sought out and tested. Toward this, Ekechukwu et al. [29] identified 25 therapies that can be considered for adoption by developing countries. Robot-mediated therapy (RMT) was viewed as a potentially feasible approach, although the authors opined that it was still too expensive for LIMCs. This observation is corroborated by studies that supported the effectiveness of robots in stroke rehabilitation equally as high-intensity physical therapy [30].

An essential component of most rehabilitation therapies is the provision of augmentative or substitutional force or motion in an assistive or therapeutic capacity. Traditionally, this augmentation is provided by a human. RMT provides a system in which the augmentation can be provided through a robot. RMT is therefore not so much a single class of therapy as it is a paradigm in which the provision of augmentative force or motion and some associated decision-making are transferred from a human agent to a robot. Thus, RMT versions of numerous rehabilitation therapies have been developed, including body weight-supported treadmill training [31], cognitive therapy [32], speech therapy [33], CIMT [34], electrical stimulation [34], mirror therapy [34], proprioceptive neuromuscular facilitation [34], gait therapy [21], telerehabilitation [35], home-based rehabilitation [36], task-oriented therapy [37], and muscle strength training [38], among others.

By allowing some elements of force or motion (along with the associated decision-making) normally provided by a therapist to be transferred to a machine, robots can hypothethetically reduce the average time spent on patients by rehabilitation professionals to a fraction of what they are now, greatly increasing their effective capacity. There is evidence that this is possible [30].

Obvious barriers to the adoption of RMT in Nigeria include cost (including the cost of technical expertise) and nonexistent or unreliable infrastructure. Both of these are surmountable. Although RMTs are still relatively expensive, cautious optimism can arise for two reasons. First, unit costs of robots have tended to drop over time in every industry where they have been widely adopted. Second, there is evidence that even simpler and cheaper robots can be quite effective [39]. The key to circumventing the effects of unreliable infrastructure may be the development of home-grown robotic technology, leading to robots suited to a Nigerian context characterized by limited resources, harsh weather, intermittent electrical power, and shortage of skilled technicians. Preliminary observations from the work of the Ife Rehabilitation Robotics Research Group (I3RG) suggest that this can be done.

Given the available evidence of the efficacy of robots, their force multiplier effect, and the possibility of circumventing the factors militating against their adoption, there is reason to be confident that the adoption of rehabilitation robotics can lead to significant improvements in the quality of neurorehabilitation in Nigeria.

24.4.2 The state of rehabilitation robotics in Nigeria

There have been no developments of note in the clinical application of medical robots in Nigeria. There is no public record of rehabilitation robots being used in the country, nor any mention of rehabilitation robotics or indeed any other type of medical robot in government policy documents. In the area of rehabilitation robotics research, however, there have been some promising recent developments, mainly centering on the work of the I3RG.

The I3RG is the first rehabilitation robotics research group in Nigeria. The group consists primarily of academics, engineers, and technicians from three departments (the Departments of Electronic and Electrical Engineering, Medical Rehabilitation, and Medicine) in Obafemi Awolowo University

(OAU), Ile-Ife, and clinicians from the neighboring OAU Teaching Hospital Complex (OAUTHC). A long-term goal of the I3RG is to foster local expertise in rehabilitation robotics as a way of developing nationwide research and development capability as well as accelerating clinical adoption.

One of the IR3G's first research projects is the Ife Robotic Glove (IRG) [40,41]. The IRG is an exoskeleton device through which repetitive active training of finger flexion and extension can be carried out to aid rehabilitation (Fig. 24.1). This device is composed primarily of a 6061 aluminum alloy base layer padded with skin-friendly leather material. The upper base layer is connected to 3D-printed rigid joint layers through a sliding stainless steel strip for force transmission between the base layer and the joint layers. The robotic glove design was adapted from designs in several past studies into a simple hand brace-like object which could be strapped to each finger, all connected to a simple, lightweight, and high-power linear actuator for fine control of the patient's fingers. The actuation system is constructed in such a way that the simple linear motion of the actuator performs three rotational motions of the finger joints to achieve complete finger flexion and extension with the use of a single actuator. The use of a single linear actuator and the monolithic structure of the compliant elements enables the device to be compact and lightweight.

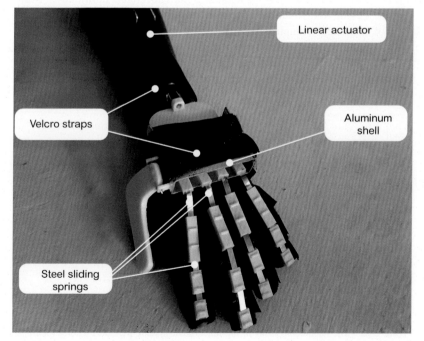

FIGURE 24.1 The "Ife Robotic Glove."

The group is also working on a device called the Platform for Upper Limb Stroke Rehabilitation (PULSR). PULSR can train the upper limb of stroke survivors, facilitating extension and flexion movements. The device is a hybrid end-effector rehabilitation device constrained to a single plane of movement. It consists of a five-link parallelogram robotic arm powered via two brushless DC motors, a functional electrical stimulation unit, an electroencephalogram (EEG) unit, and a control unit built around an embedded computer (Fig. 24.2). The next iterations will also include an electromyogram (EMG) and goniometer for real-time monitoring of patients' arm muscle activity and improvement in the arm range of motion. PULSR can be used to train patients' impaired arm by repetitively moving the arm through a diagonal trajectory similar to those followed by patients' arm during activities of daily living, while the bicep brachii muscles of the arm are electrically stimulated. The training time, speed, and movement trajectory are all predetermined by a computer program. The EEG and EMG feedback mechanisms are to allow the rehabilitation of stroke survivors during the acute phase, by attempting to detect the intention of the user to move their limbs, which can then be integrated into the control loop.

PULSR-2 underwent usability testing for 6 weeks. Ten healthy participants and 6 patients were recruited from the neurology and medical rehabilitation departments at OAUTHC, after meeting the inclusion criteria of the study. The test was centered on obtaining patient-related information while using the device, the structural/operational capabilities of the device, and the safety of users and clinicians during use. Before a rehabilitation session, a clinician measured the vitals of a participant to determine whether they were fit for the session (10 minutes). During a session, the user's hand was strapped to the device's end effector (as shown in Fig. 24.3), while the

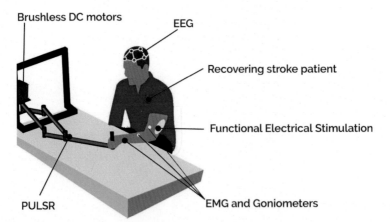

FIGURE 24.2 The main components of the full PULSR system. *PULSR*, Platform for Upper Limb Stroke Rehabilitation.

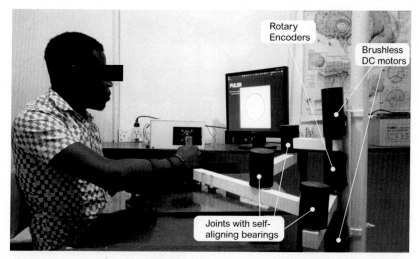

FIGURE 24.3 PULSR-2 shown here during usability testing with a normal subject. *PULSR*, Platform for Upper Limb Stroke Rehabilitation.

device repeatedly moved the user's arm at constant speed around a trajectory. Testimonials from all the users were positive; they were optimistic about the impact of the device because they believe it would grant them more training time and improved therapy.

The development of both IRG and PULSR has followed an adapt-and-evolve strategy. The original design of IRG is similar to a device previously described in reference [42]. PULSR uses a variant of the parallelogram arm manipulator structure first used as far back as 1991 [43]. IRG and PULSR designs have both gone through three iterations, with fourth and fifth generation designs currently under development, with each generation incrementally introducing new innovations. For example, PULSR-3 now uses a novel technique called sensing via arm-integrated load cell (SAIL). Multi-axis force/torque sensors are some of the more expensive components in end-effector rehabilitation devices. Their costs increase nonlinearly with complexity so that a three-axis sensor is usually much more than thrice as expensive as a single-axis sensor. The SAIL design (Fig. 24.4) dispenses with the multi-axis sensor in the end effector and instead introduces low-cost single-axis load cells into the links. Although the design increases errors in the quantification of user effort, preliminary data show that the impact of such errors can be mitigated through appropriate control laws. Even without the economics of scale kicking in, locally developed rehabilitation robots cost less than a quarter of roughly comparable robots in the United States.

A key takeaway from I3RG's approach is the adoption of a pragmatic adapt-and-evolve strategy. The reality for most researchers in developing countries is that, on the one hand, infrastructural constraints and a dearth of

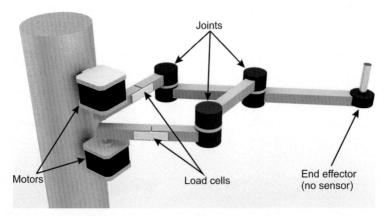

FIGURE 24.4 Parallelogram arm robot with sensing via arm-integrated load cells (SAIL).

expertise make cutting-edge rehabilitation robotics research more challenging, while on the other hand, the medical robotics industries in their countries may be nonexistent, limiting the appetite of local or international funders for funding work in the area. In essence, prospective rehabilitation robotics researchers may find themselves needing to bootstrap the entire specialization area for their respective countries. The adapt-and-evolve strategy has helped the IR3G to successfully navigate this bootstrapping, and growing expertise has coincided with increasingly original contributions in the area of rehabilitation robotics.

24.4.3 Toward sustainable rehabilitation robotics in Nigeria

According to [30], sustainable adoption of RMT in developing countries must be preceded by strategies for ensuring economical total costs of ownership and improved local expertise through strengthened educational programs. A key strategy toward the former is to emphasize locally developed solutions rather than importation. Even if imported devices can guarantee higher overall quality initially, local development is essential for RMT to take root in resource-constrained areas. For one, the higher manpower and certification costs in developed countries, factored into the cost of imported devices, can be eliminated by local development. Second, imported robots are not attuned to developing country realities, and the minuscule market sizes of such countries are insufficient incentives for global manufacturers to address this.

One tactic for stimulating sustainable local development is to replace high-tech components with low-tech or off-the-shelf alternatives [30]. This motivated the replacement of the switched-mode power supply units (PSUs) originally used by PULSR with off-the-shelf batteries. Most PULSR power consumption is by a pair of 24 V (48 V in PULSR-3) brushless DC motors. When constant power outages and low power supply quality led to the failure of imported PSUs, the repair was rejected as a solution since they would be liable to fail again without additional expense on protective equipment. The solution was to replace them with a bank of 40 Ampere-hour (Ah) deep-cycle absorbent glass mat batteries. This provided two advantages, both attuned to Nigerian realities. First, unlike the switched-mode PSUs (which are hard to obtain and difficult to maintain in Nigeria) the batteries represent a well-known technology and are available in even remote villages. In addition, the batteries serve as an uninterruptible power supply system, decoupling PULSR from the effects of intermittent electrical power supply. The SAIL technique for user effort quantification is another one in which a cheaper, low-tech solution has been introduced. While multi-axis force/torque sensors are expensive and must be imported, single-axis load cells are cheap and quite common.

Emphasizing software-centric approaches can also be an effective strategy for developing countries. Due to lower infrastructural requirements, software alternatives (where possible or appropriate) tend to be cheaper than hardware solutions. Software is a relative area of strength for developing countries like Nigeria. Such countries can therefore adopt strategies in which complexities are moved to software. For example, in the SAIL design for PULSR-3, the higher cost of multi-axis sensors is being converted into the cheaper software problem of mitigating errors caused using low-cost single-axis sensors.

Strengthening related educational programs is also important [30]. More work needs to be done in this regard in Nigeria. The interdisciplinary synergy required for rehabilitation robotics research and development is still inadequate. There are 32 accredited medical schools in the country and a total of 57 engineering faculties, but only 28 universities have both. Only 12 of these also have programs in physiotherapy. Biomedical engineering is offered by less than a dozen universities, but only one of them is currently accredited by the Council for the Regulation of Engineering in Nigeria (COREN). There is still no accredited graduate program in biomedical engineering in the country.

Important distinctions must be made between developing and developed countries in terms of the value proposition of RMT. Developed countries have highly functional healthcare systems, and unless treatment outcomes for comparatively priced RMTs can be established to be superior to conventional therapies, the value proposition of RMTs to such countries remains questionable. The trade-offs are totally different for a developing country like Nigeria. The healthcare system is so overwhelmed and understaffed that RMTs do not need to be superior to conventional therapies. If they have somewhat comparable outcomes to conventional therapies, they emerge as one of only a few options to improve rehabilitation services in the country—as long as the consequential matter of their economic and technical viability can be addressed.

The viability barrier appears insurmountable at first glance. Indeed, the prevailing wisdom is still that developing countries are not ready for RMT. Hopefully, this chapter provides some evidence to question that wisdom. Perhaps, what is needed is for research and development agendas that factor into the unique contexts of developing countries like Nigeria. Nobody is better placed to do this than Nigerian researchers and developers. International corporations have little motivation to do so since the country's share of the global medical robotics market is infinitesimal. Furthermore, only local researchers can fully grasp socioeconomic nuances.

Considering this, I3RG's work to date should provide guarded optimism that developing countries like Nigeria are capable of adapting and evolving RMT technologies to make them economically and technically viable for in-country use. If indigenous robotic technologies become mature and cost-effective enough, developing countries like Nigeria may emerge as leaders in RMT adoption since their need is far more pressing, and the value proposition of RMTs to them is much greater than it is to developed countries.

24.5 Conclusion

Although robots are traditionally regarded as expensive high-tech devices and Nigeria is a resource-constrained developing country, RMT can provide a viable path toward addressing some of the challenges of neurorehabilitation in the country. A lot remains to be done before this can happen, most importantly by developing curricula and new programs in, and fostering collaborations across relevant medical and engineering subdisciplines. However, the work of IR3G shows that as long as they are based on pragmatic strategies, local research and development can lead to the introduction of low-cost but effective robotic devices. Hopefully, this will foreshadow increasing clinical adoption.

Acknowledgments

The IRG and PULSR projects are both funded by the Tertiary Education Trust Fund (TETFUND). Some of the studies described in this chapter were supported by grants from the NVIDIA Corporation and utilized an NVIDIA Titan Xp GPU and a Quadro 8000. EEG bio-amplifiers and headsets were kindly donated by OpenBCI. The Opey Akinlolu Foundation supported the first generation of IRG.

Conflict of interest

The authors of this chapter certify that they have NO affiliations with or involvement in any organization or entity with any financial interest (such as honoraria; educational grants; participation in speakers' bureaus; membership, employment, consultancies, stock ownership, or other equity interest; and expert testimony or patent licensing arrangements), or nonfinancial interest (such as personal or professional relationships, affiliations, knowledge or beliefs) in the subject matter or materials discussed in this manuscript.

References

[1] United Nations Development Programme. UNDP Human Development Reports - Nigeria Human Development Indicators. Available from: http://hdr.undp.org/en/countries/profiles/NGA (Accessed on 2 December 2021).

[2] Punch Newspapers 'Less than 5% of Nigerians covered by NHIS' – Punch Newspapers. Available from: https://punchng.com/less-than-5-of-nigerians-covered-by-nhis/ n.d. (accessed on 04 October 2020).

[3] Osuntokun BO. The pattern of neurological illness in tropical Africa: experience at Ibadan, Nigeria. J Neurol Sci 1971;12(4):417−42.

[4] Oluwole OG, Kuivaniemi H, Carr JA, Ross OA, Olaogun MO, Bardien S, et al. Parkinson's disease in Nigeria: a review of published studies and recommendations for future research. Park Relat Disord 2019;62:36−43.

[5] Talabi OA. A 3−year review of neurologic admissions in University College Hospital Ibadan, Nigeria. West Afr J Med 2003;22(2):150−1.

[6] Owolabi LF, Shehu MY, Shehu MN, Fadare J. Pattern of neurological admissions in the tropics: experience at Kano, Northwestern Nigeria. Ann Indian Acad Neurol 2010;13 (3):167.

[7] Chapp-Jumbo E. Neurologic admissions in the Niger Delta area of Nigeria: a ten year review. Afr J Neurol Sci 2004;23:14−20.

[8] Ekenze OS, Onwuekwe IO, Ezeala BA. Profile of neurological admissions at the University of Nigeria Teaching Hospital Enugu. Nigerian J Med 2010;19(4):419−22.

[9] Ojo OO, Abubakar SA, Iwuozo EU, Nwazor EO, Ekenze OS, Farombi TH, et al. The Nigeria Parkinson disease registry: process, profile, and prospects of a collaborative project. Mov Disord 2020;35(8):1315−22.

[10] Macleod AD, Taylor KS, Counsell CE. Mortality in Parkinson's disease: a systematic review and meta - analysis. Mov Disord 2014;29(13):1615−22.

[11] Okubadejo NU, Ojini FI, Danesi MA. Longitudinal study of mortality predictors in Parkinson's disease in Nigerians. Afr J Med Med Sci 2005;34(4):365−9.

[12] Ibikunle PO, Ekuma OO. A fifteen years retrospective study of spinal cord injury in South-Eastern Nigeria. EC Orthop 2018;9(1,120):21−30.

[13] Obalum DC, Giwa SO, Adekoya-Cole TO, Enweluzo GO. Profile of spinal injuries in Lagos, Nigeria. Spinal Cord 2009;47(2):134−7.

[14] Emejulu JK, Isiguzo CM, Agbasoga CE, Ogbuagu CN. Traumatic brain injury in the accident and emergency department of a tertiary hospital in Nigeria. East Cent Afr J Surg 2010;15(2):28−38.

[15] Onwuchekwa RC, Echem RC. An epidemiologic study of traumatic head injuries in the emergency department of a tertiary health institution. J Med Tropics 2018;20(1):24.

[16] Danesi M, Okubadejo N, Ojini F. Prevalence of stroke in an urban, mixed-income community in Lagos, Nigeria. Neuroepidemiology. 2007;28(4):216−23.

[17] Wahab KW. The burden of stroke in Nigeria. Int J Stroke 2008;3(4):290−2.

[18] Enwereji KO, Nwosu MC, Ogunniyi A, Nwani PO, Asomugha AL, Enwereji EE. Epidemiology of stroke in a rural community in Southeastern Nigeria. Vasc Health Risk Manag 2014;10:375.

[19] Adeloye D, Ezejimofor M, Auta A, Mpazanje RG, Ezeigwe N, Ngige EN, et al. Estimating morbidity due to stroke in Nigeria: a systematic review and meta-analysis. J Neurol Sci 2019;402:136−44.

[20] Ogun SA, Ojini FI, Ogungbo B, Kolapo KO, Danesi MA. Stroke in south west Nigeria: a 10-year review. Stroke. 2005;36(6):1120−2.

[21] Strong K, Mathers C, Bonita R. Preventing stroke: saving lives around the world. Lancet Neurol 2007;6(2):182−7.

[22] O'Donnell MJ, Chin SL, Rangarajan S, Xavier D, Liu L, Zhang H, et al. Global and regional effects of potentially modifiable risk factors associated with acute stroke in 32 countries (INTERSTROKE): a case-control study. Lancet 2016;388(10046):761−75.

[23] Sarfo FS, Ovbiagele B, Gebregziabher M, Wahab K, Akinyemi R, Akpalu A, et al. Stroke among young west Africans: evidence from the SIREN (Stroke Investigative Research and Educational Network) large multisite case−control study. Stroke. 2018;49 (5):1116−22.

[24] Ogbole GI, Adeyomoye AO, Badu-Peprah A, Mensah Y, Nzeh DA. Survey of magnetic resonance imaging availability in West Africa. Pan Afr Med J 2018;30(1):240.

[25] Hebert D, Lindsay MP, McIntyre A, Kirton A, Rumney PG, Bagg S, et al. Canadian stroke best practice recommendations: stroke rehabilitation practice guidelines, update 2015. Int J Stroke 2016;11(4):459−84.

[26] Olaleye OA, Lawal ZI. Utilization of physiotherapy in the continuum of stroke care at a tertiary hospital in Ibadan. Niger Afr Health Sci 2017;17(1):79−87.

[27] Onyeso OK, Umunnah JO, Ezema CI, Anyachukwu CC, Nwankwo MJ, Odole AC, et al. Profile of practitioners, and factors influencing home care physiotherapy model of practice in Nigeria. Home Health Care Serv Q 2020;39(3):168−83.

[28] Benjamin EJ, Virani SS, Callaway CW, Chamberlain AM, Chang AR, Cheng S, et al. Heart disease and stroke statistics—2018 update: a report from the American Heart Association. Circulation. 2018;137(12):e67 −492.

[29] Ekechukwu EN, Olowoyo P, Nwankwo KO, Olaleye OA, Ogbodo VE, Hamzat TK, et al. Pragmatic solutions for stroke recovery and improved quality of life in low-and middle-income countries—a systematic review. Front Neurol 2020;11:337.

[30] Demofonti A, Carpino G, Zollo L, Johnson MJ. Affordable robotics for upper limb stroke rehabilitation in developing countries: a systematic review. IEEE Trans Med Robot Bionics 2021;3(1):11−20.

[31] Hornby TG, Zemon DH. Campbell D. Robotic-assisted, body-weight−supported treadmill training in individuals following motor incomplete spinal cord injury. Phys Ther 2005;85 (1):52−66.

[32] Dino F, Zandie R, Abdollahi H, Schoeder S, Mahoor MH. Delivering cognitive behavioral therapy using a conversational social robot. In 2019 IEEE/RSJ International Conference on Intelligent Robots and Systems (IROS); 2019 Nov 3 (pp. 2089−2095).

[33] Estévez D, Terrón-López MJ, Velasco-Quintana PJ, Rodríguez-Jiménez RM, Álvarez-Manzano V. A case study of a robot-assisted speech therapy for children with language disorders. Sustainability. 2021;13(5):2771.

[34] Kim J, Gu GM, Heo P. Robotics for healthcare. Biomedical engineering: frontier research and converging technologies. Cham: Springer; 2016. p. 489−509.

[35] Song A, Wu C, Ni D, Li H, Qin H. One-therapist to three-patient telerehabilitation robot system for the upper limb after stroke. Int J Soc Robot 2016;8(2):319−29.

[36] Li L, Fu Q, Tyson S, Preston N, Weightman A. A scoping review of design requirements for a home-based upper limb rehabilitation robot for stroke. Top Stroke Rehabil 2021;29 (6):449−63.

[37] Chisholm KJ, Klumper K, Mullins A, Ahmadi M. A task oriented haptic gait rehabilitation robot. Mechatronics. 2014;24(8):1083−91.

[38] Mehrholz J, Hädrich A, Platz T, Kugler J, Pohl M. Electromechanical and robot-assisted arm training for improving generic activities of daily living, arm function, and arm muscle strength after stroke. Cochrane Database Syst Rev 2012;(6):CD006876.

[39] Johnson MJ, Rai R, Barathi S, Mendonca R, Bustamante-Valles K. Affordable stroke therapy in high-, low-and middle-income countries: from theradrive to rehab CARES, a compact robot gym. J Rehabil Assist Technol Eng 2017;4 2055668317708732.

[40] Osayande E, Ayodele K, Komolafe M. Development of a robotic hand orthosis for stroke patient rehabilitation. Int J Online Biomed Eng 2020;16(13).

[41] Akinniyi OT, Ayodele KP, Komolafe MA, Olaogun MO, Owolabi DG, Fajimi MA, et al. A Low-Cost Orthosis Using a Compliant Spring Mechanism for Post-Stroke Hand Rehabilitation. In 2022 IEEE Nigeria 4th International Conference on Disruptive Technologies for Sustainable Development (NIGERCON) 2022 Apr 5 (pp. 1−4). IEEE.

[42] Nycz CJ, Bützer T, Lambercy O, Arata J, Fischer GS, Gassert R. Design and characterization of a lightweight and fully portable remote actuation system for use with a hand exoskeleton. IEEE Robot Autom Lett 2016;1(2):976−83.

[43] Hogan N, Krebs HI, Charnnarong J, Srikrishna P, Sharon A. MIT-MANUS: a workstation for manual therapy and training. I. In 1992 Proceedings IEEE International Workshop on Robot and Human Communication 1992 (pp. 161−165). IEEE.

Chapter 25

Africa region: Botswana

Maikutlo Kebaetse[1], Michelle J. Johnson[2], Billy Tsima[1],
Cassandra Ocampo[1], Justus Mackenzie Nthitu[3], Ntsatsi Mogorosi[4],
Lingani Mbakile-Mahlanza[5], Kagiso Ndlovu[6], Venkata P. Kommula[7],
Rodrigo S. Jamisola, Jr[8] and Timothy Dillingham[2]

[1]*Faculty of Medicine, University of Botswana, Gaborone, Botswana,* [2]*Departments of Physical Medicine and Rehabilitation, Bioengineering, and Mechanical Engineering and Applied Mechanics, University of Pennsylvania, Philadelphia, PA, United States,* [3]*Department of Occupational Therapy, Mahalapye District Hospital, Mahalapye, Botswana,* [4]*Botswana-UPenn Partnership, Gaborone, Botswana,* [5]*Faculty of Social Sciences, University of Botswana, Gaborone, Botswana,* [6]*Faculty of Computer Science, University of Botswana, Gaborone, Botswana,* [7]*Faculty of Engineering & Technology, University of Botswana, Gaborone, Botswana,* [8]*Faculty of Engineering & Technology, Botswana International University of Science & Technology, Palapye, Botswana*

Learning objectives

At the end of this chapter, the reader will be able to:
1. Understand the state of rehabilitation technology in Botswana.
2. Describe limitations to the development and use of rehabilitation robotics in Botswana.
3. Outline opportunities available for the development of rehabilitation technologies in Botswana.

25.1 Introduction

25.1.1 Country and healthcare overview

Botswana, which got its independence in 1966, is a landlocked country in Southern Africa, sharing its borders with South Africa to the south, Namibia to the west, Zimbabwe to the northeast, and Zambia to the north. The country is approximately 581,000 km^2 in size and has a population of approximately 2.4 million people, 51.6% of which was women as of 2020 [1]. Most of the country is sparsely populated, with most people concentrated along the eastern corridor that follows the main railroad and north–south highway. Botswana is a member of the African

Rehabilitation Robots for Neurorehabilitation in High-, Low-, and Middle-Income Countries.
DOI: https://doi.org/10.1016/B978-0-323-91931-9.00018-9

Union, the Southern African Development Community, the Commonwealth of Nations, and the United Nations. The relatively young sub-Saharan nation is known for its stable democracy and economic growth from being poor at independence to being an upper middle-income country since 2005 [2].

The Government of Botswana has placed oversight of the nation's healthcare and its delivery under the Ministry of Health and Wellness (MoHW), whose mission is to "promote and provide integrated, holistic, and sustainable preventive, curative, and rehabilitative quality health services" [3,4]. The ministry is subdivided into various departments, including *corporate; clinical; public health; HIV and AIDS prevention and care; regulatory; health policy development, monitoring, and evaluation;* and *health hub* services. Through these departments and several ministerial facilities, the overall functions of the ministry are (1) formulation of the nation's healthcare policy and guidelines, (2) provision of public healthcare services, which occurs primarily under strategically placed regional District Health Management Teams (DHMTs), and (3) training of several healthcare professions through the Institutes of Health Sciences [4].

The provision of healthcare has historically been a joint venture between the Ministry of Health and the Ministry of Local Government. Botswana has a five-tier health system comprising health posts, clinics, primary hospitals, district hospitals, and referral hospitals. The Ministry of Local Government provides the bulk of primary healthcare services that are coordinated or administered on a day-to-day basis by DHMTs. The DHMTs have a network of three national referral hospitals, 15 district hospitals, 17 primary hospitals, 318 clinics, 347 health posts, and 973 mobile stops (Master Health Facility List 2016, Health Statistics Unit), while the Ministry of Health directly administers two referral hospitals, six district hospitals, 17 primary hospitals, and 1 mental hospital. MoHW retains the portfolio responsibility for health policy development, professional/technical guidance, and supervision of healthcare, irrespective of the provider or institution [4]. People who are on medical aid schemes (health insurance) mostly use the private healthcare system. There are two types of medical aid schemes: one for people who are employed in the private sector and another that is for public civil servants. Unemployed people have the option to join the medical aid schemes, but the majority of Batswana uses the public healthcare system because patients pay nominal fees, and a patient is never turned away for lack of funds. Currently, consultations in public health facilities are subsidized, adult citizens are expected to pay a minimum of P8.00 (approximately US$0.70) which covers consultation and treatment (inclusive of drugs), and noncitizens pay P50 upon registration.

In 2019, the national healthcare budget was increased from 8% to 16.2% (0.763 billion USD) of the annual budget [5]. However, a number of challenges are yet to be overcome or addressed. For example, (1) it has been suggested that the organization and governance of health services and inadequate resources limit service delivery at the hospital level [6]; (2) the country is still battling high rates of HIV/AIDS and other infectious diseases [7]; (3) the country is battling increased rates of noncommunicable diseases, such as diabetes and hypertension [8]; (4) there is a severe shortage of rehabilitation services and professionals, especially in the public health sector; and (5) like most countries around the world, Botswana was ill-prepared for the COVID-19 pandemic as available resources (financial, medical supplies and facilities, and manpower) were overstretched for a prolonged period.

As a worldwide trend, people are generally living longer, and consequently, there is an increase in both the number of people living with disabilities (PWDs) and years lived in disability. Botswana is a relatively young country, where 62.9% of the country are aged less than 24 years and those over 65 years make up only 5.1% of the population. A recent report suggests that PWD represent 4.2% ($\sim$100,800) of the population, but this may be underreported due to inadequate monitoring of disability [9]. Hanass-Hancock and Carpenter [10] showed that the greatest contributors toward disability-adjusted life years (DALYs) in Botswana are mental disorders (13.7%), HIV and other sexually transmitted diseases (13%), musculoskeletal disorders (9.9%), neurological disorders (8.0%), and sense organ diseases (6.2%) [10]. Given the nation's rise in disability prevalence from communicable and noncommunicable diseases, the authors stressed the need for better data on rehabilitation to prepare Botswana's health systems to accommodate the increasing need for disability support and rehabilitation services [10].

In Botswana, the main reference document on the rights and provisions for PWD is the 1996 National Policy on the Care of People with Disabilities, which recognizes the importance of disability rights and dignity for all individuals [11]. This policy assigns responsibilities for the care of PWD to different government ministries and departments. Despite the government's good policy intentions, one study suggested that the disability movement was weak, and efforts to advocate for the rights of PWD were not strong, especially in that the constitution of Botswana did not include disability-specific legislation [12]. The introduction of the Office of the Coordinator for People with Disabilities at the Office of the President in 2010 has, however, added an impetus to the call for equalization of opportunities for PWD. Similarly, the signing and ratification of the United Nations Convention on the Rights of Persons with Disabilities in 2021 signaled major progress in the recognition of their rights. Despite these improvements, PWDs continue to experience discriminatory health service delivery with rural/urban disparities existing in the quality and supply of healthcare [9].

25.1.2   Stroke and rehabilitation access

Stroke is growing in sub-Saharan Africa and the rest of the continent. The annual incidence rate of stroke in Africa is up to 316 per 100,000 individuals, which is among the highest incidence rates in the world [13]. There are no prevalence data on stroke in Botswana; however, in 2018, the World Health Organization (WHO) reported that the stroke death in Botswana reached 933 or 6.42% of total deaths, with an age-adjusted death rate of 91.56 per 100,000 of population that placed the nation at a rank of number 64 in the world [13].

During the year 2007, a population-based STEPS survey was conducted in eight selected districts in collaboration with the Ministry of Local Government, and 4003 adults aged 25−64 participated in the study [14]. The survey identified the major reasons were smoking, alcohol consumption, limited physical activity (especially in women), and less consumption of fruits and vegetables in the general population. Approximately, 38.6% of the population were found to be overweight, and 33.1% had high blood pressure, with women much more disproportionately affected when compared to men. The survey thus unveiled the high prevalence of risk factors for developing noncommunicable diseases in the population.

The country has been adversely affected by the HIV/AIDS epidemic. Despite the success of programs making treatments available and educating the populace about prevention, the prevalence of HIV/AIDS was 18.5% among those aged 18 months to 49 years, with the average life expectancy being around 69 years in 2018 [9]. Bain et al. [15] reported that although combination antiretroviral therapy is helping the people of Botswana live longer, the success is being challenged by increased incidences of cardiovascular and metabolic diseases. The study also reported that HIV infection can result in stroke via several mechanisms, including opportunistic infection, vasculopathy, cardio-embolism, and coagulopathy [15]. HIV-related neurocognitive impairment typically affects motor skills, such as fine grasping and manipulation, and cognitive domains, such as attention, executive function, information processing, and working memory [16]. Up to 40% of people living with HIV (PLHIV) develop acute neurologic complications from either central nervous system infections or noncommunicable central nervous system disorders, such as stroke, leading to associated motor and cognitive dysfunction which requires rehabilitation [17,18].

The prevalence of HIV has clearly impacted the face of stroke rehabilitation. Evidence now reveals that HIV is associated with an increased risk for stroke [19,20], especially in individuals aged 45 years or younger [21]. Bearden et al. compared risk factors and outcomes among Botswana adults with and without HIV admitted for acute stroke [22]. Hypertension was the most common risk factor identified in both groups, and when compared to non-HIV patients, those with HIV were significantly more likely to have a small vessel lacunar syndrome. In a younger HIV-associated stroke cohort, long-term disability is more debilitating, thus compromising the ability to work and quality of life [23]. Stroke, especially, is known to cause major motor and cognitive impairments that affect both the upper and lower limbs, leading to functional mobility issues and difficulties with activities of daily living and instrumental activities of daily living. Nonetheless, stroke patients were less likely to have access to rehabilitation at the rates recommended by the WHO, had poor functional outcomes, and were typically discharged without inpatient rehabilitation [22].

Hospitals in Botswana lack the rehabilitation staff to provide an assessment of disability and regular follow-up care [2]. The number of rehabilitation professionals available in Botswana is disproportional to the population like many other African countries. During the year 2022, Princess Marina Hospital, the largest tertiary care facility with 530-bed capacity in the capital city Gaborone, had 0 physiatrists, five occupational therapists (OTs), and seven physical therapists (PTs), making it impossible to meet recommended rehabilitation standards for every patient. This has resulted in scaling down the services to the neediest patients, with a greatly reduced frequency of therapy sessions. As a developing country, Botswana's healthcare system is mainly focused on the curative model of service delivery, leaving marginal resources for rehabilitation services. Stroke survivors' needs for multiple therapies span over long periods, and limited specialized human resources and low or inadequate frequency of therapy sessions lead to slow progress and, often, secondary physical complications.

Healthcare services in Botswana are to a large extent free and offered through public facilities. Rehabilitation services are mainly available only at the district and the referral hospitals. The service is, however, fraught with barriers including poor referral processes, inadequate resources, and the lack of appropriate transport. The standard of care for stroke survivors with motor impairment involves mobility training, provision of braces and walking aids, environmental adaptation, and training in independent living. The neurological management of complications such as abnormal muscle tone varies according to the knowledge and experience of the therapist. These services often lack a multidisciplinary approach due to human resource constraints. There are, however, several private facilities in urban areas and major villages mainly serving citizens and residents who have private health insurance coverage. Having health insurance and funds to pay for rehabilitation services out of pocket can ensure that some get adequate rehabilitation services at a frequency that is recommended by clinical practice guidelines.

It is important to note that in Botswana, from January 3, 2020, to the present, there have been 263,950 confirmed cases of COVID-19 with a little over 2500 deaths, reported to WHO. Active immunization has been done with 1.4 million vaccine doses administered. The country has adopted preventive and control strategy measures since 2020. COVID-19 has impacted access and uptake of essential health services with the decline in the use of motor rehabilitation across the country.

To meet rehabilitation access needs creatively, we must first understand the current system/situation in Botswana and its rehabilitation capacity for

PWD, especially due to HIV and stroke. Section 25.2 provides a summary of a rehab capacity survey conducted from March 1, 2021, to March 1, 2022.

25.2 Rehab capacity in Botswana

Primary data on the locations and capacities of rehabilitation services in Botswana were obtained from an Assessment of Rehabilitation Capacity survey adapted from a situational analysis questionnaire specific to rehabilitation medicine originally implemented in Ghana by Asare et al. [24]. The survey targeted four main sectors (infrastructure, training resources/technology, human resources, and patient characteristics). Rehabilitation services were defined as occupational therapy, speech therapy, physiotherapy, prosthetics and orthotics, psychological counseling/social work, and others. Facilities were excluded from the analysis if they did not provide any form of rehabilitation service.

Survey participants were asked to identify three major successes and three major challenges for rehabilitation in Botswana and their facilities as well as asked to identify recommendations to address challenges facing rehabilitation services. Responses to the question of success were mixed. Many agreed that the number of therapists has increased over the last 10−15 years, but the numbers still lag, which prevents many in remote areas from getting consistent rehabilitation services. There is a growing recognition of the importance of rehabilitation, and there are plans to improve services in the public sector. The private sector is also growing robustly. Despite the limitations, when patients get access, their conditions do improve.

Most respondents agree that more work is needed and the challenges for their facilities were in the following areas: (1) shortage of qualified rehabilitation staff; (2) poor or inadequate insurance coverage for patients; (3) low reimbursement for rehabilitation, with delayed payments; (4) lack of local rehabilitation manufacturers and distributors that limit the ability to expeditiously source assistive technology equipment of all types; (5) lack of academic programs in institutions to develop rehabilitation professionals and specialists; (6) lack of capital funding and investment sources to expand services; and (7) lack of affordable and adequate gym space and equipment.

There are many consequences for these challenges, such as (1) the lack of rehabilitation specialists limits the country's ability to respond to rehabilitation needs and expand services to those in need at the frequency required; (2) poor insurance and lack of staff lead to patients having to wait long to obtain service, which leads to poor functional outcomes and medical complications; (3) poor reimbursements limit patients' ability to get the care and private therapists' ability to provide the care; and (4) the lack of local manufacturers leads to lack of equipment and forces facilities to import essentially all of the equipment, resulting in exorbitant prices to the government, facilities, and patients and taxpayers.

When asked about the major challenges of rehabilitation in Botswana, responders echoed many of the issues already described for their facilities but added more information about the government's role and its impact on patients. There is a need for the government prioritization of rehabilitation that should be manifested in increased funding for personnel training, the building of a better support structure and rehab framework that protects the rehab profession and patients, and public education about rehabilitation to help reduce issues of neglect of children with cerebral palsy and other patient populations, which is sometimes extreme in rural areas. Another facility pointed out that the government needed to implement "proper policies to facilitate participation of clients in their communities post-rehabilitation." Some of the respondents suggested that rehabilitation in Botswana needs to become more important to MoHW than at present and stated, "we [rehab] are not valued at the same level as other services."

One clear takeaway is that there needs to be an infusion of investment in the local rehabilitation sector at all levels, from local training and development of rehab staff—therapists, doctors, nurses, prosthetists, and others—to the development of local rehabilitation business and infrastructure to spur innovation around the development and manufacturing of rehabilitation products. MoHW's increasing investments in the rehabilitation space can increase access to assistive technologies including high-tech and robotic ones. For example, hospitals should be facilitated to have dedicated rehabilitation beds, general rehabilitation services overseen by interdisciplinary rehabilitation teams, and well-equipped rehabilitation spaces/gyms with assistive devices and therapy technologies.

Many facilities surveyed report having a physiotherapy gym, but no other specialized area. Most gyms were relatively small and lacked space for large equipment and housed some low-tech assistive technologies such as exercise bikes. Many assistive technologies recognized by WHO as essential [25,26] were not readily available, and due to lack of local manufacturing, the cost to import them was high. Most reported there is no diagnostic equipment such as MRIs, even at government facilities. Many facilities mentioned that they did not use high-tech or robotic rehabilitation equipment for the following reasons—(1) they were too costly; (2) they had not heard of them and felt they do not have the skill to use them; (3) the equipment may not be needed; or (4) no space. When asked specifically about the use of telehealth/telerehab or phone calls or video conferencing to consult patients who are unable to physically attend, there was some use of web-based technology with increases occurring mainly due to COVID-19. One primary reason for the lack of telehealth use among private rehab clinics was that mobile data time was expensive and the government had no mechanism to reimburse their use.

Given the limited number of rehabilitation technology-driven activities in Botswana, Section 25.3 presents rehabilitation technology-based projects currently occurring in Botswana that can revolutionize health and rehab care for stroke and other patients with motor and cognitive impairments.

25.3 Rehabilitation technology and robotics

25.3.1 Upper-limb therapy robots

A project entitled "Robot-Assisted Rehabilitation After HIV-Associated Stroke" is an NIH-sponsored project via the Center for AIDS Research (CFAR) at the University of Pennsylvania (UPenn). The PIs are Dr. Maikutlo Kebaetse of the University of Botswana (UB) and Dr. Michelle Johnson at UPenn. The research has direct sponsorship from the Botswana UPenn Partnership (BUP)—local center that coordinates collaborations between UB and UPenn. The project leverages the pioneering work of Dr. Johnson in Mexico and the United States which posits that an affordable robot therapy gym with multiple one-degree-of-freedom end-effector robots can augment the standard of care and potentially allow one therapist to concurrently treat up to four patients [27]. The project leverages the development of an affordable intervention based on simple haptic robots, which is not only highly cost-effective but also affordable, feasible, and culturally acceptable.

The Haptic Theradrive provides adaptive forces to the user to provide assistance or resistance in completing movement tasks. Fig. 25.1 shows a person using the robot to complete an assessment task. Assistance or resistance is created by controlling the torque motor, and for safety, the maximum forces are limited by a torque limiter. The user operates Theradrive by manipulating a vertically mounted crank handle equipped with force sensors

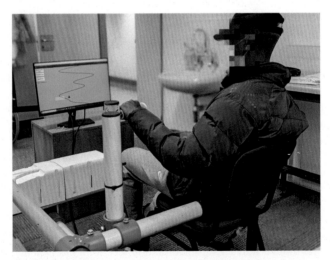

FIGURE 25.1 Rehab CARES robot gym with Haptic Theradrive and a patient.

and an optical encoder. For assessment purposes, it is run in zero-impedance mode. In this mode, the forces applied by the user's arm onto the end effector in the handle are used to calculate the necessary response by the motor to give the sensation that there is no resistance when the user pushes or pulls on the handle.

The goal of this project is to examine the feasibility of using simple haptic robots to (1) assess motor and cognitive impairment in people with HIV only, stroke only, and both HIV and stroke and (2) provide augmented therapy to the upper limb by providing force assistance or resistance as needed during game-based therapy. If proven useful, these systems could potentially address gaps between the supply of rehab staff and the patient demand for rehabilitation access. Since 2021, five Haptic Theradrive robots have been deployed in Botswana. There are two robot gym systems, each containing two Haptic Theradrives at about $2500 per gym. One robot gym is set up in the new Rehabilitation Prosthetics and Robotics Lab at the Sir Ketumile Masire Teaching Hospital (SKMTH) (Fig. 25.2). T is a parastatal hospital, meaning both private- and government-owned facility, located in the south-east district of Botswana in the capital, Gaborone. The other robot gym system is deployed inside the occupational therapy department at the Mahalapye District Hospital (MDH)—a public referral hospital in the central district of Botswana in the village of Mahalapye (Fig. 25.3). One robot is deployed in a private outpatient clinic in the capital city called Stroke Rehabilitation Center. These three locations serve as testing sites for the research, which is ongoing.

The success of deployment would not have been possible without the collaboration with the Department of Mechanical Engineering at UB. With the guidance of the designs from the Rehab Robotics Lab at UPenn, the

FIGURE 25.2   Rehab CARES robot gym at SKMTH in Gaborone.

FIGURE 25.3 Rehab CARES robot gym at MDH in Mahalapye.

Department of Mechanical Engineering translated the designs into products, and colleagues such as Dr. Venkata P. Kommula, Professor in Mechanical Engineering at UB, are playing a vital role in this process. This work raises the need for capacity building within the engineering community to enable Botswana to equip the next generation of engineers with the ability to support local medical device manufacturers in terms of development and innovation. Capacity building is currently being practiced through student attachment to the rehab care lab during their industrial training attachment period, subsequently providing them support to produce components locally with available material. The 3D printing technology and computer numerical control machines are used while producing the critical and precise components of the robotic system. The Rehab CARES robot gym platform which is shown in Fig. 25.1 was manufactured at the Faculty of Engineering Technology workshops. To sustain long-term human resource capacity building, it is necessary for UB to develop multidisciplinary study programs such as mechatronics and biomedical engineering.

25.3.2 Lower-limb wearable robots

Dr. Rodrigo S. Jamisola, an Associate Professor at Botswana International University of Science and Technology (BUIST), is developing wearable systems for rehabilitation. Powered robotic exoskeletons are a good way to intervene in rehabilitation as they offer the possibility of many repetitions of the same motion inside and outside clinics. This allows faster gait rehabilitation for poststroke patients. His lab has built and controlled a lower-limb exoskeleton robot to perform leg rehabilitation where the user commands the leg to move using brain signals from the Emotiv Epoc Flex headset [28]. This exoskeleton leg robot is a simple two-degrees-of-freedom electromechanical system and is fabricated using mainly steel. The system is designed for individuals whose stroke affected their left leg, but the right leg is functional. The notable design feature is the adjustable hip joint shown in Fig. 25.4. It has a slot in which the fittings can be slid and tightened at the subject's hip joint. To move each joint, measured human gait was used as the desired joint displacements for the lower limb.

Control commands to the robot were generated from a trained machine learning model that was trained using EEG data acquired from the motor imagery experiment and captured using an Emotiv Epoc system. The experimental setup for motor imagery was such that the subject is set on a comfortable chair and shown a clip of the stimuli with instructions on it while recording their brain signals with the Emotiv Epoc device. A picture of the

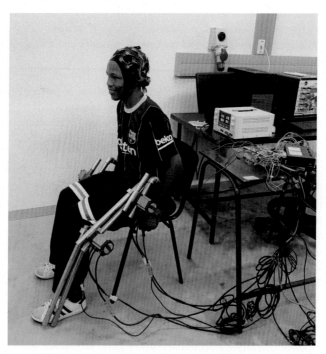

FIGURE 25.4 Subject wearing affordable exoskeleton with Emotiv device on head.

left hand, right hand, and feet would appear on the video clip as motor imagery instructions. In between these instructions, there is a 4 seconds delay to allow the subject to perform motor imagery tasks. Data cleaning, artifact removal, feature extraction and selection, and classification were done on the acquired raw EEG signals to allow the user to choose from three actions such as stop, walk slow, and walk fast only using brain signals (Fig. 25.4).

25.3.3 Mobile health technologies initiatives

High-impact, low-cost sustainable mHealth initiatives are critical for strengthening the healthcare system in Botswana and other resource-limited settings. According to the recently launched Botswana eHealth Strategy (2020–24), the perceived impact of mHealth interventions includes contributing to health system strengthening, ensuring equity, affordability, sustainability, the discovery of new knowledge, and improvement of health outcomes and clinical decision-making [29]. Since 2010, BUP has collaborated with MOHW, UB, and other local partners to implement and research mHealth interventions in the areas of telemedicine to increase access to specialty care, information retrieval, public health data collection, and medical education. After pilot research is complete, if local partners are interested in scaling up the mHealth project nationally, BUP supports the initiative as a technical advisor, facilitating public–private partnerships and working closely with partners to identify and pursue operational and financial sustainability milestones.

BUP mHealth Sustainable Scale: Kgonafalo Telemedicine: Kgonafalo Telemedicine is BUP's cornerstone mHealth initiative, born from four pilot studies that researched the application of store-and-forward telemedicine using mobile phones in the specialties of dermatology, radiology, women's health (cervical cancer screening), and oral medicine [29,30]. In 2012, BUP began working with the MoHW to sustainably scale up the project with local technical support and a public–private partnership with the telecommunications company, Orange Botswana (a mobile telephone company), to support the national rollout for the first 3 years. Some technical and social challenges were encountered during the initial studies, such as malfunctioning mobile devices, accidental damage of devices, and cultural misalignment between information technology and healthcare providers. Lessons learned from these challenges include a strong need for unwavering senior management support, the establishment of solid local public–private partnerships, and efficient project sustainability plans.

Television Whitespace (TVWS): TVWS is unused television channels in the ultrahigh frequency (UHF) band between 470 and 698 MHz which can be used opportunistically by secondary users [31]. Botswana leveraged this unused television frequency spectrum to augment the limited connectivity in support of synchronous telemedicine services at participating healthcare facilities. The project was supported by multiple stakeholders including BUP, Botswana Innovation Hub, Microsoft, Vista Life Sciences, and Global Broadband Solutions. It facilitated teledermatology, cervical cancer screening, and family medicine for HIV/AIDS, tuberculosis, and general adult and pediatric medicine across connected healthcare facilities in Botswana [32]. The technology was relatively inexpensive and faster to set up. It offered a farther-reaching Internet connection (up to 10 km radius) that is a promising option for linking the previously unconnected populations of remote and underserved areas to specialist care [32].

Although there presently are no established telerehabilitation use cases in Botswana, knowledge and experience gained from the above mHealth initiatives could be leveraged and tailored to telerehabilitation. For example, given the lack of human, structural, and resources in Botswana, telerehabilitation could be used for improved public education on rehabilitation, remote access to rehabilitation services, and staff training.

25.3.4 Tablet-based assessment and treatment of cognition

While the persistence of motor deficits is most common, cognitive impairment is also prevalent in 46%−61% of stroke survivors even 10 years after their stroke [33]. This is further compounded by the presence of HIV. Neuropsychologists are rare in Botswana, despite the high prevalence of neurological disorders. Dr. Lingani Mbakile-Mahlanza, one of the few practitioners, is attempting to build capacity for culturally appropriate cognitive assessments. In Botswana, there are currently no validated neuropsychological tests to assess cognitive functions, and there is an immense need to improve service provision for people with neurological disorders and to enhance neuropsychological resources. Cognition is typically assessed using extensive clinical neuropsychological batteries which require time and trained professionals, so there is a recognized need to innovate using technology. Bui et al. developed novel robot-based metrics for the assessment of motor and cognition [34], and these assessments are being tested in Botswana under the CFAR initiative described above. Cognitive rehabilitation, either through compensational or restitution-based strategies, aims to improve impaired cognitive function. One affordable approach to restitution-based intervention is the use of game-based computer-assisted cognitive rehabilitation (CACR), which has been shown to improve executive and visual-spatial abilities, speech, attention, and memory skills in neurological patients [35,36]. CACR can be delivered at a relatively low cost using tablets and computers, is commercially available, and can be tailored to patients [35,36]. Many of these CACR tasks often involve both motor and cognitive actions, but conventional CACR focuses solely on the cognitive actions. One study recently awarded by the Alzheimer's Association is focused on examining the use of tablet-based cognitive assessment to detect dementia and cognitive impairment in the elderly population. This study is in the initial stages.

25.3.5 Immediate-fit prosthetics systems

There are currently 25 million persons lacking a prosthetic device globally. The lack of trained prosthetists, high device cost, and inaccessibility of prosthetic services leave many patients in low-resource countries without prostheses. To address this problem, an immediate-fit, adjustable, modular, prosthetic system was developed by iFIT Prosthetics, LLC. The iFIT transtibial prosthesis is a modular, immediate-fit, fully adjustable, transtibial prosthetic system made of high-strength aerospace-grade polymer materials—as strong as aluminum, but lighter [37]. In contrast to conventional fabrication with casting, molding, and shaping the socket that can take weeks to complete, the iFIT prosthesis is fitted and aligned directly on the residual limb in a single setting. The prosthesis features a soft inner liner that can be customized to the patient through added padding. The patient is immediately able to walk away with a comfortable and adjustable prosthesis. A buckle system with safety locks allows the patient to adjust the fit whenever they wish. The iFIT prosthetic system can be used with a variety of commercially available prosthetic feet and silicone locking liners. The iFIT transfemoral prosthesis is similar to the transtibial version, except that it is fitted above the knee and requires a commercial knee. Fig. 25.5 shows a transfemoral and a transtibial iFIT prostheses on two amputees.

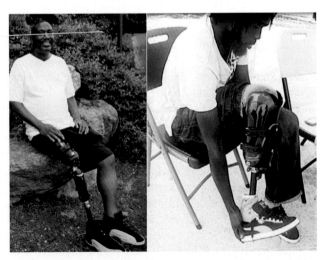

FIGURE 25.5 Above-knee immediate-fit prosthetic and a below-knee immediate prosthetics.

In 2021, the founder of iFIT Prosthetics visited Botswana to determine how to create capacity and support for amputees in Botswana. Seventeen persons in total, six transfemoral and 11 transtibial with limb loss, were fitted in seven days by iFIT founder, his son, and local physiotherapist and academic, Dr. Maikutlo Kebaetse. Seven of the persons fit did not have a prosthesis previously. The prosthetic kits were all transported to various locations via four large bags and fitted with hand tools contained in a backpack. The devices were modified by adding padding and cutting or heating the sockets to accommodate each person. Only one person with a transtibial amputation could not be comfortably fitted. The prosthetic fittings took under 2 hours, and most persons were able to ambulate with basic gait training. Prosthetists at each location were trained and assisted with the fittings. The immediate-fit prostheses were easily transported to several locations, patients were rapidly fitted, and some fittings occurred outdoors. This makes them ideal for areas lacking prosthetic clinics and labs. Persons unable to travel distances for fittings and those with compromised immune systems needing to be fitted in their home, nursing home, or hospital during the COVID-19 pandemic could benefit from this easy-to-fit feature. The prostheses were well received by participants as the ability to adjust them on-the-fly enabled a rapid and comfortable fit. The iFIT prosthetic system and this care delivery model (training prosthetists to fit and align the devices in their clinics) could prove useful in meeting the current and future prosthetic demands and costs in Botswana and other low-resource countries.

25.4 Future directions

In Botswana, there is a distinct gap between rehabilitation service needs and resources; there is limited local capacity to run specialized rehabilitation programs, a lack of assistive technologies, and a lack of local capacity to develop and maintain these technologies. Access to rehabilitation services is limited on three major fronts by: (1) economics: rehabilitation services and associated technologies often are quite limited outside the major urban areas, and many times are not affordable for low-income patients; (2) training: skilled therapists or physiatrists are often not available in large numbers inside or outside of cities, especially in the public health system; and (3) technology: access to state-of-the-art assistive or rehabilitation technologies may be limited, and gaining access to them may be too costly and require long waiting times. Finding effective ways to cascade services below district healthcare facilities will ensure better coverage to the deserving populations, the majority of whom reside in rural areas.

Developing affordable rehabilitation technologies including robotic ones is a viable direction for Botswana to help bridge care gaps [38]. In high-income countries (HICs) such as the United States, evidence supports the use of smart mechatronic systems and rehabilitation robots after brain injuries to bridge healthcare gaps and provide treatment equivalent to conventional and high-intensity therapy [39]. One review study indicates that, generally, some innovative rehabilitation technologies such as exoskeleton and end-effector robots available in HICs are usually too expensive to be directly mapped to rehabilitation environments in LMICs, yet the technology's ability to augment care is remarkable [40].

Partnerships with institutions in HICs such as the United States and the establishment of memoranda of understanding are proving to be critical to building capacity for healthcare staff and implementation of technology-assisted rehabilitation, including the use of robotics. In order to be successful in bridging healthcare gaps, more investment is needed in rehabilitation on multiple levels. Investments must focus on infrastructure and funds to equip and train engineers in developing innovative technology, and forming local manufacturing capacity, and for nurses, therapists, and doctors to become specialists in rehabilitation and medicine and experts in utilizing practice methods that incorporate technologies in the expansion of service.

Acknowledgments

We would like to thank the following sponsors and funders: US Fulbright Scholarship, IIE, National Institutes of Health, Center for Aids Research at the University of Pennsylvania, Botswana Upenn Partnership, University of Pennsylvania: Department of Physical Medicine and Rehabilitation, Department of Bioengineering, Department of Mechanical Engineering and Applied Mechanics, the Rehabilitation Robotics Lab— GRASP Lab, University of Botswana: Department of Biomedical Sciences, Department of Mechanical Engineering, Department of Cognitive Sciences, and the e-Health Unit.

We would like to thank our clinical partners in Botswana: Jessica Makonyago from Chesire Foundation; Lavonah Gure, David Damba from Stroke Rehab Center; Clifford Seleka from Sir Ketimule Masire Teaching Hospital; and Tauya Simbanegavi from Merribah Occupational Therapy Solutions for tirelessly working to support this effort. We would also like to thank the students at UB and BIUST for their contribution to gathering the data and conducting the experiments for this work: Tumelo Mafokate, Aryaan Lambe, Ngadzi Goitsema, Olebogeng Mbedzi, Lucky Odirile Mohutsiwa, and Kaloso Mpho Tlotleng.

Potential conflict of interests

Dr. Dillingham is the founder of iFIT Prosthetics LLC. Dr. Michelle Johnson is co-founder of Recupero Robotics LLC and inventor of the Haptic Theradrive and the Rehab CARES gym.

References

[1]   World Bank. Population, female (% of total population) [Internet]. Available from: https:// data.worldbank.org/indicator/SP.POP.TOTL.FE.ZS?locations = BW. World Bank. http:// Population, female (% of total population) 2021 (accessed on 10 December 2021).

[2] World Bank. Data for Botswana, Upper middle income [Internet]. World Bank. Available from: http://Data for Botswana, Upper middle income. 2021 (accessed on 10 December 2021).

[3] Ministry of Health. Strategic Foundations [Internet]. Botswana Government. Available from: https://www.moh.gov.bw/strategic_foundations.html. 2022 (accessed on 10 December 2021)

[4] Ministry of Health Departments & Programs [Internet]. Botswana Government. Available from: https://www.moh.gov.bw/about_us_departments.html. 2021 (accessed on 8 December 2021).

[5] Mathambo O. Budget speech [Internet]. Gaborone: Government Printing and Publishing Services; 2019. Available from: <https://www.tralac.org/documents/resources/by-country/botswana/2614-botswana-2019-budget-speech/file.html> (accessed on 8 December 2021).

[6] Seitio-Kgokgwe O, Gauld RD, Hill PC, Barnett P. Assessing performance of Botswana's public hospital system: the use of the World Health Organization Health System Performance Assessment Framework. Int J Health Policy Manag 2014;3(4):179.

[7] Solomon M, Furuya-Kanamori L, Wangdi K. Spatial analysis of HIV infection and associated risk factors in Botswana. Int J Environ Res Public Health 2021;18(7):3424.

[8] Silva R. Non communicable diseases: the payoffs of investing early in prevention efforts [Internet]. Available from: https://www.healthpolicyproject.com/pubs/858_FINALBotswana NCDBriefA.pdf; 2015 (accessed on 9 March 2022).

[9] United Nations. Un botswana annual results report. United Nations; 2020.

[10] Hanass-Hancock J, Carpenter B. Trends in health and disability in Botswana. An analysis of the global burden of disease study. Disabil Rehabil 2021;43(25):3606−12.

[11] Mukhopadhyay S, Moswela E. Situation analysis of disability rights in the context of Botswana. Gaborone: The University of Botswana. Open Society Initiative for Southern Africa; August 2016.

[12] Mukhopadhyay S, Moswela E. Situation analysis of disability rights in the context of Botswana. Gaborone: The University of Botswana. Unpublished Report; August 2016.

[13] Akinyemi RO, Ovbiagele B, Adeniji OA, Sarfo FS, Abd-Allah F, Adoukonou T, et al. Stroke in Africa: profile, progress, prospects and priorities. Nat Rev Neurol 2021;17 (10):634−56.

[14] World Health Organisation. Chronic disease risk factor surveillance [Internet]. Available from: https://www.who.int/ncds/surveillance/steps/2007_STEPS_Report_Botswana.pdf; 2007 (accessed on 6 December 2021).

[15] Bain LE, Kum AP, Ekukwe NC, Clovis NC, Enowbeyang TE. HIV, cardiovascular disease, and stroke in sub-Saharan Africa. Lancet HIV 2016;3(8):e341−2.

[16] Elliott R. Executive functions and their disorders: imaging in clinical neuroscience. Br Med Bull 2003;65(1):49−59.

[17] Heaton RK, Franklin Jr DR, Deutsch R, Letendre S, Ellis RJ, Casaletto K, et al. Neurocognitive change in the era of HIV combination antiretroviral therapy: the longitudinal CHARTER study. Clin Infect Dis 2015;60(3):473−80.

[18] O'Brien KK, Ibáñez-Carrasco F, Solomon P, Harding R, Cattaneo J, Chegwidden W, et al. Advancing research and practice in HIV and rehabilitation: a framework of research priorities in HIV, disability and rehabilitation. BMC Infect Dis 2014;14(1):1−2.

[19] Ovbiagele B, Nath A. Increasing incidence of ischemic stroke in patients with HIV infection. Neurology. 2011;76(5):444−50.

[20] Abdallah A, Chang JL, O'Carroll CB, Musubire A, Chow FC, Wilson AL, et al. Stroke in human immunodeficiency virus-infected individuals in Sub-Saharan Africa (SSA): a systematic review. J Stroke Cerebrovasc Dis 2018;27(7):1828−36.

[21] Benjamin LA, Corbett EL, Connor MD, Mzinganjira H, Kampondeni S, Choko A, et al. HIV, antiretroviral treatment, hypertension, and stroke in Malawian adults: a case-control study. Neurology. 2016;86(4):324−33.

[22] Bearden DR, Omech B, Rulaganyang I, Sesay SO, Kolson DL, Kasner SE, et al. Stroke and HIV in Botswana: a prospective study of risk factors and outcomes. J Neurol Sci 2020;413:116806 Jun 15.

[23] Onwuchekwa AC, Onwuchekwa RC, Asekomeh EG. Stroke in young Nigerian adults. J Vasc Nurs 2009;27(4):98−102.

[24] Geberemichael SG, Tannor AY, Asegahegn TB, Christian AB, Vergara-Diaz G, Haig AJ. Rehabilitation in Africa. Phys Med Rehabil Clin N Am. 2019;30(4):757−68.

[25] Smith RO, Scherer MJ, Cooper R, Bell D, Hobbs DA, Pettersson C, et al. Assistive technology products: a position paper from the first global research, innovation, and education on assistive technology (GREAT) summit. Disabil Rehabil Assist Technol 2018;13(5): 473−85.

[26] Cooper RA. Commentary on WHO GATE Initiative. J Spinal Cord Med 2017;40(1):2−4.

[27] Johnson MJ, Rai R, Barathi S, Mendonca R, Bustamante-Valles K. Affordable stroke therapy in high-, low-and middle-income countries: From Theradrive to Rehab CARES, a compact robot gym. J Rehabil Assist Technol Eng 2017;4 2055668317708732.

[28] Mbedzi O, Jamisola Jr. RS, Mohutsiwa L and Tlotleng K. Lower-limb Robot Exoskeleton for Leg Rehabilitation with Commands from a 32-Channel Emotiv Epoc Flex Headset. Unpublished.

[29] Ndlovu K, Littman-Quinn R, Park E, Dikai Z, Kovarik CL. Scaling up a mobile telemedicine solution in Botswana: keys to sustainability. Front Public Health 2014;2:275 Dec 11.

[30] Littman-Quinn R, Mibenge C, Antwi C, Chandra A, Kovarik CL. Implementation of m-health applications in Botswana: telemedicine and education on mobile devices in a low resource setting. J Telemed Telecare 2013;19(2):120−5.

[31] Ndlovu K, Mbero ZA, Kovarik CL, Patel A. Network performance analysis of the television white space (TVWS) connectivity for telemedicine: a case for Botswana. 2017 IEEE AFRICON 2017, September (542−547).

[32] Chavez A, Littman-Quinn R, Ndlovu K, Kovarik CL. Using TV white space spectrum to practise telemedicine: a promising technology to enhance broadband internet connectivity within healthcare facilities in rural regions of developing countries. J Telemed Telecare 2016;22(4):260−3.

[33] Delavaran H, Jönsson AC, Lövkvist H, Iwarsson S, Elmståhl S, Norrving B, et al. Cognitive function in stroke survivors: a 10-year follow-up study. Acta Neurol Scand 2017;136(3):187−94.

[34] Bui KD, Wamsley CA, Shofer FS, Kolson DL, Johnson MJ. Robot-based assessment of HIV-related motor and cognitive impairment for neurorehabilitation. IEEE Trans Neural Syst Rehabil Eng 2021;29:576−86.

[35] Bangirana P, Giordani B, John CC, Page C, Opoka RO, Boivin MJ. Immediate neuropsychological and behavioral benefits of computerized cognitive rehabilitation in Ugandan pediatric cerebral malaria survivors. J Dev Behav Pediatr 2009;30(4):310.

[36] Boivin MJ, Busman RA, Parikh SM, Bangirana P, Page CF, Opoka RO, et al. A pilot study of the neuropsychological benefits of computerized cognitive rehabilitation in Ugandan children with HIV. Neuropsychology. 2010;24(5):667.

[37] Dillingham T, Kenia J, Shofer F, Marschalek J. A prospective assessment of an adjustable, immediate fit, transtibial prosthesis. PM R 2019;11(11):1210−17.

[38] Demofonti A, Carpino G, Zollo L, Johnson MJ. Affordable robotics for upper limb stroke rehabilitation in developing countries: a systematic review. IEEE Trans Med Robot Bionics 2021;3(1):11.

[39] Duret C, Grosmaire AG, Krebs HI. Robot-assisted therapy in upper extremity hemiparesis: overview of an evidence-based approach. Front Neurol 2019;10:412 Apr 24.

[40] Ekechukwu EN, Olowoyo P, Nwankwo KO, Olaleye OA, Ogbodo VE, Hamzat TK, et al. Pragmatic solutions for stroke recovery and improved quality of life in low-and middle-income countries—a systematic review. Front Neurol 2020;11:337.

Chapter 26

Africa region: Ghana

Abena Yeboaa Tannor[1], Frank Kwabena Afriyie Nyarko[2],
Benedict Okoe Quao[3,4] and Ebenezer Ad Adams[5]
[1]Department of Health Promotion and Disability, School of Public Health, College of Health
Sciences, Kwame Nkrumah University of Science and Technology, Kumasi, Ghana, [2]Department
of Mechanical Engineering, Kwame Nkrumah University of Science and Technology, Kumasi,
Ghana, [3]Ankaful Leprosy & General Hospital, Ankaful, Ghana, [4]National Leprosy Control
Programme, Disease Control & Preventive Department, Ghana Health Service Public Health
Division, Korle-Bu, Accra, Ghana, [5]National Stroke Support Services, Accra, Ghana

Learning objective

At the end of this chapter, the reader will be able to:
1. Understand the incidence and prevalence of stroke in Ghana.
2. Gain a deeper knowledge of stroke rehabilitation in Ghana.
3. Appreciate the challenges facing the introduction of stroke rehabilitation robots in Ghana.

26.1 Introduction

26.1.1 Ghana's history and demographic

The Republic of Ghana is one of 16 coastal countries located along the Gulf of Guinea, in the western part of Africa, with a shoreline of about 560 km. Away from its coast with the Atlantic Ocean, Ghana is bordered by Burkina Faso to the north, Togo to the east, and Cote d'Ivoire to the west. The country has a total land area of 238,537 sq. km. The average annual temperature is about 26°C (79°F).

Ghana became the first Black nation in sub-Saharan Africa to gain independence from British colonial rule on the March 6, 1957. The country became an independent republic on July 1, 1960, and Accra is its administrative and political capital. Ghana's population of 30,792,608 is divided into 75 ethnic groups [1]. The median age and life expectancy are 21.5 and 64.42 years, respectively. Persons with disabilities comprise 3% of Ghana's

Rehabilitation Robots for Neurorehabilitation in High-, Low-, and Middle-Income Countries.
DOI: https://doi.org/10.1016/B978-0-323-91931-9.00003-7

population with more being located in the rural areas compared to the urban centers [2].

26.1.2 Communicable disease and noncommunicable disease prevalence

Ghana has its fair share of communicable diseases resulting from infection. Major communicable diseases recorded across hospitals in Ghana include lower respiratory infections, HIV/AIDS, diarrheal diseases, malaria, tuberculosis, and neglected tropical diseases [3]. The leading cause of death due to communicable diseases according to WHO is lower respiratory tract infections with a death rate of 53.49 per 100,000 population [4]. Ghana had made numerous efforts to manage communicable diseases in collaboration with WHO. These include malaria treatment guidelines, sputum sample referral system to improve tuberculosis case findings, and mass drug administration to control tropical diseases such as onchocerciasis [5].

The major noncommunicable diseases (NCDs) in Ghana are cardiovascular diseases, cancers, diabetes, chronic respiratory diseases, and sickle cell disease. The first four share common risk factors, namely use of tobacco and alcohol, unhealthy diet, and physical inactivity. Up to 48% of Ghanaian adults have hypertension, and 9% have diabetes [6].

Among the top 10 diseases in Ghana, NCDs cause disability-adjusted life years (DALYs), the number of years lost due to ill health, disability, or early death, ranging from 1165.97 to 1462.52 per 100,000 population for ischemic heart diseases and stroke, respectively [4]. NCDs account for 43% of all deaths in Ghana with 94,400 total deaths in 2020 and a 21% probability of premature mortality [7].

The burden of NCDs is projected to increase due to aging, rapid urbanization, and unhealthy lifestyles. Steps taken and fully achieved to curtail this include banning tobacco advertisements and developing an NCD policy as well as guidelines for the management of cancer, cardiovascular diseases, and diabetes.

26.1.3 Stroke incidence and prevalence

Stroke affects 12,800–16,000 persons in Ghana annually with 67.8% being below the age of 60 years. Stroke in Ghana results in 1462.52 DALYs per 100,000 population leaving many Ghanaians with disabilities during their productive paid working years, with grave implications for the individual, family, and society in terms of mental capital, productivity, and socioeconomic progress [4]. Stroke is the leading cause of death from NCDs with 49.88 deaths per 100,000 population and ranks third in the overall top 10 causes of death [4]. In a few decades, stroke incidence in Ghana is likely to surge significantly with hypertension being the leading cause of stroke in

Ghana. Adopting and implementing innovative programs such as the community-based life after stroke rehabilitation program developed by SASNET-Ghana and partners are one of the many strategies for improving the quality of life of persons with stroke. It is also very crucial for the government of Ghana to implement effective and efficient evidence-based interventions to improve stroke outcomes.

26.2   Rehabilitation structure

In Ghana, although all the teaching hospitals and many of the regional hospitals in the country can boast of multispecialist medical teams capable of providing advanced medical care for stroke, the Korle-Bu stroke unit set up in 2014 through the Wessex Ghana Stroke Partnership that has been in existence since 2009 [8] remains the only specialized stroke unit in the country. Located in the heart of the capital, it boasts of seasoned stroke professionals including neurologists, physiotherapists, occupational therapists, psychologists, rehabilitation nurses, dieticians, and speech-language therapists all providing care under one roof in a multidisciplinary approach, but sadly, this assortment of expertise remains largely inaccessible by most stroke patients who reside outside Accra.

26.2.1   Rehabilitation access for stroke

Access to poststroke rehabilitation following discharge from acute care is also inadequate, a situation made more complex by the limited coordinated interdisciplinary rehabilitation care available, particularly outside major hospitals located in urban areas. One study even showed that at one teaching hospital in Ghana, less than half of the stroke survivors were either assessed for or received rehabilitation [9]. Additionally, with most stroke survivors having to usually continue their follow-up care at the centers where they were treated acutely, the distance from their areas of residence becomes another barrier and contributes to many being lost to follow-up, especially if they have little family support to assist with their commute.

Even when patients followed up, their rehabilitation needs are usually not the focus of their healthcare providers. With the relatively high prevalence of comorbidities such as hypertension and diabetes among stroke survivors in Ghana [10], it is not surprising that secondary prevention of stroke dominates the objectives of most poststroke care. Many stroke survivors and their families are therefore left to manage the impairments of stroke, physical, psychoemotional, and cognitive that affect their quality of life with little knowledge and support [10].

The bulk of poststroke rehabilitation is provided by physiotherapists in Ghana, and many stroke survivors would usually access physiotherapy services in parallel to their regular reviews after discharge from acute stroke care. The distribution of centers providing physiotherapy services is however not equitable (Fig. 26.1). While about half of the available physiotherapy units are located in only two regions, Ashanti and Greater Accra, some of the country's 16 regions have none or only one based on data from the Ghana Physiotherapy Association [11], with the distance a typical patient travels to receive physiotherapy services averaging 63 km [12]. With half of the six new regions created in 2018 to help improve access to government and public services [13] having no established physiotherapy units, it can be expected that in the near future access to such services will be extended to some of these underserved areas in the country.

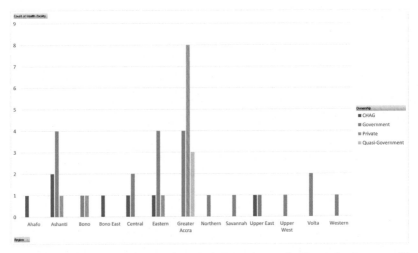

FIGURE 26.1   Regional distribution of 43 physiotherapy centers in Ghana, showing ownership status, based on data sourced from the Ghana Physiotherapy Association (GPA) website [11]. This chart shows how the 43 available physiotherapy centers in Ghana are distributed in the 16 administrative regions. The Greater Accra region has the most, with 15 physiotherapy centers majority of which (8) are privately owned. It is followed in descending order by the Ashanti, Eastern, and Central regions, with seven, six, and three physiotherapy centers, respectively. Bono, Upper East, and Volta regions all follow with two physiotherapy centers each, while Ahafo, Bono East, Northern, Savannah, Upper West, and Western regions have a single physiotherapy center. The three remaining regions—Oti, Northeast, and Savannah—have no such physiotherapy center. Concerning ownership, aside from the eight privately owned physiotherapy centers in the Greater Accra region, there is one privately owned physiotherapy center in each of the following three regions—Ashanti, Bono, and Eastern. Greater Accra is the only region with quasi-government-owned physiotherapies, having three in number. The Christian Health Association of Ghana (CHAG) owns seven of the 43 physiotherapy centers, with two located in the Ashanti region and one each in the Ahafo, Bono East, Central, Eastern, and Upper East regions. The remaining 22 physiotherapy centers are all government-owned.

26.2.2   Rehabilitation capacity for stroke

Although the country boasts of a multiplicity of poststroke medical professionals, there is still a huge deficit in rehabilitation professions excluding physiotherapy [12,14]. There has however been some improvement over the last 15 years, with the establishment of the School of Allied Health Sciences at the University of Ghana in 2012 [15] and the commencement of subspecialty training of family physicians in sports, exercise, and rehabilitation medicine since 2018 through the Ghana College of Physicians and Surgeons [16], being two notable milestones in the country's quest to bridge the rehabilitation human resource gap.

Christian et al. [12] carried out the first major cross-center study in Ghana to critically assess the capacity of the country for providing rehabilitation care, with findings suggesting persisting deficiencies in humans resource, similar to that observed almost a decade earlier [14]. Significantly in the Christian et al. study [12], there were no physiatrists, rehab psychologists, or recreational therapists and only one prosthetist, one orthotist, and one rehab nurse across the nine sampled centers serving a population of about 6.6 million people. Additionally, Christian et al. discovered that while poststroke rehabilitation was being provided by 100% of sampled centers, other forms of rehabilitation such as musculoskeletal and medical rehabilitation were not being offered by all the centers surveyed. The study findings highlight the need to increase the available human capacity for providing rehabilitation care in the country and improve the diversity of rehabilitation care being provided.

The most recently published Labour Market Survey (LMS) report by the Ghana Statistical Service (GSS) and Fair Wages and Salaries Commission (FWSC) [17] is quite revealing on the availability of rehabilitation skills within public service institutions, with no listed position determined to have 100% of vacancies filled. Looking at Table 26.1 which attempts to summarize the data from the LMS report, some important rehabilitation skills such as "speech-language pathologists (SLPs)" and "prosthetic and orthotic (P&O) technicians" are not listed at all, while for many of the other listed skills, if even 100% vacancies were filled, the population-to-staff ratio will be so large—ranging from above 100,000 population per physiotherapist to above 3,000,000 per prosthesis nurse—that these professionals will not be able to provide any significant impact at the population level.

26.2.3   Stroke rehabilitation standard of care

There is currently no consensus on a national stroke rehabilitation standard of care. In many hospitals providing stroke services, poststroke rehabilitation is rarely provided by an interdisciplinary team, especially after discharge from acute care. In what may be the best multidisciplinary approach being

TABLE 26.1 Rehabilitation human resource needs in public sectors from 2019 Labour Market Survey [17].

Type of rehabilitation skills	Number required	Number available	Number of vacancies	Percent of vacant skill	Percent of staff at post	Population-to-staff ratio
Orthopedic physical therapist	20	5	15	0.1	25.0	1,539,630
Physiotherapist (PT)	297	116	181	1.0	39.1	103,679
PT assistant	86	80	6	0.0	93.0	358,054
Prosthesis nurse	10	0	10	0.1	0.0	3,079,261
Occupational therapist (OT)	13	5	8	0.0	38.5	2,368,662
OT assistant	16	10	6	0	62.5	1,924,538
Speech-language pathologist	Not listed					
Technician—P&O	Not listed					

implemented in the country at the Korle-Bu Stroke Unit, each team of rehab professionals has equal access to the patients undergoing acute care, and combined multidisciplinary ward rounds are organized weekly. This approach allows for the early initiation of rehabilitation care even before discharge, in line with the standard of care and utilization of the early window of opportunity for maximum recovery [18].

The standard approach of establishing rehabilitation goals at the start of care and reviewing the progress of various multidisciplinary approaches is however not being consistently employed. Additionally, in Ghana, where most poststroke interventions are being provided by physiotherapists, less than half of the practicing physiotherapists are using any standardized stroke rehabilitation outcome measure routinely [19]. Nonetheless, among those working to some standard, the most frequently used outcome measure scale is the Barthel index (BI) [19] which has been adapted in one study as the modified Barthel scale of activities of daily living (m-DALY) [20].

26.3 Technology-assisted rehabilitation

WHO estimates that by 2030, 2 billion people will require assistive technology (AT) [21]. This estimation includes persons recovering from stroke living in Ghana. AT for stroke can range from extremely low-level technology such as communication boards all the way to very sophisticated technology such as wheelchairs, visual aids, and specialized computer software and hardware, including robots, that increase mobility [22]. The goal of AT in stroke rehabilitation is to improve patients' function, independence, and quality of life.

26.3.1 Robots in stroke rehabilitation

In Ghana, the use of robots in stroke rehabilitation is virtually nonexistent in health facilities. A survey [12] revealed that very few hospitals have stroke recovery equipment such as therapeutic ultrasound, functional electrical stimulation, and transcutaneous electrical nerve stimulation machines. However, there have been a few related case studies piloted by the College of Engineering students at the Kwame Nkrumah University of Science and Technology, Ghana. This report presents two case studies of locally developed ATs, namely "**The Arm Recovery Device**" and the **Inflated Body Strap**.

26.3.1.1 The arm recovery device

The arm-strengthening device (shown in Fig. 26.2) is a device for therapeutic loading of an arm that has lost strength due to stroke or a fracture, to regain arm strength in order to restore normal functionality of the arm. The device was designed by a group of mechanical engineering students of KNUST,

FIGURE 26.2   Model of Arm Recovery Device at KNUST, Ghana. This photograph shows an arm recovery device which is made up of a basic chair structure with pulleys, weights, and footrests attached for arm-strengthening exercises. The weights are attached to the weight rods which are hooked to the ropes; when the ropes are pulled, the weights slide along the vertical length on the backside of the chair. There are two pulleys in this design which position to facilitate the kind of upward motion exercise the patient is required to do. There is also a footrest to help patients relax their feet during exercises. The footrest is spring-loaded to serve as a means for feet-strengthening exercises.

Ghana, in 2020. The load-bearing device allows for the affected bones or joints to be pressed together to their correct positions, and as per physical medicine, any part of the body subjected to weight-bearing develops strength.

The device is made up of a basic chair structure with pulleys, weights, and footrests attached for arm-strengthening exercises. The weights are attached to the weight rods which are hooked to the ropes; when the ropes are pulled, the weights slide along the vertical length on the backside of the chair. There are two pulleys in this design which position to facilitate the kind of upward motion exercise the patient is required to do. There is also a footrest to help patients relax their feet during exercises. The footrest is spring-loaded to serve as a means feet-strengthening exercises.

The tension in the ropes provides resistance for the exercise. The pulley has grooves inside to provide easy rotational motion. The physiotherapist selects the suitable tension in the ropes after the affected arm of the patient has been strapped. The patient is then made to grab on to the handle of the pulley rope. Subsequently, the handle of the rope is unclipped for the patient to bear the given load. This process is repeated 5−10 times in a session with 5 minutes break between sessions (depending on the physical condition of the patient).

26.3.1.2   Inflated body strap

This device (shown in Fig. 26.3) has an air sack strapped around the waist. The air sack has a suction pump that fills by sucking air from the atmosphere and is fastened around the waist. Attached to the air sack is a strap that is worn around the lower arm to hold it in position. A shoulder strap that is independent of the inactive part of the stroke patient is also attached to it. Thus, if the right part of the body is inactive, the shoulder strap goes around the left shoulder. Pressurized air from the air sack (generated from compression of the inflation bulb) is channeled into a bulb on which the palm is placed. The inflation bulb can be compressed by the unimpaired hand of the patient or by the physiotherapist.

The mechanism for the rehabilitation training is based on a simple method: automatic inflation and deflation of the bulb. As the bulb gets inflated, the fingers are forced to expand and contract as the bulb gets deflated. This trains the fingers to gain strength to move.

With time, the rate of inflation and deflation is minimized. This will allow the patient to control the movement of the fingers voluntarily. The rate of inflation and deflation of the bulb is controlled by a pressure control

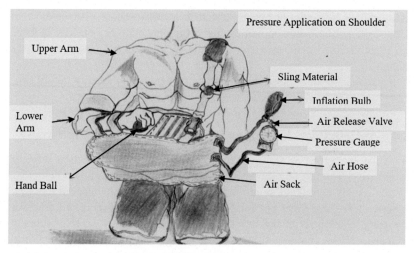

FIGURE 26.3 Schematic diagram of Inflated Body Strap developed at KNUST, Ghana. This picture shows a schematic diagram of Inflated Body Strap device developed at KNUST, Ghana. The device consists of an air sack strapped around the waist. The air sack has a suction pump that fills by sucking air from the atmosphere and is fastened around the waist. Attached to the air sack is a strap that is worn around the lower arm to hold it in position. A shoulder strap that is independent of the inactive part of the stroke patient is also attached to it. Thus, if the right part of the body is inactive, the shoulder strap goes around the left shoulder. Pressurized air from the air sack (generated from compression of the inflation bulb) is channeled into a bulb on which the palm is placed. The inflation bulb can be compressed by the unimpaired hand of the patient or by the physiotherapist.

valve. As the rehabilitation process progresses successfully, the patient is tasked to compress the already air-filled bulb with no aid from the device. The bulb in this state sucks air directly from the air sack via the air ducts.

26.3.2 Challenges to adoption of assistive technologies in Africa

The Convention on the Rights of Persons with Disabilities (CRPD) requires governments to meet the AT needs of the citizens. However, access to AT in Africa is severely limited. According to WHO, only 5%−15% of the people needing AT have access to it [23]. Most hospitals in Ghana lack the services needed to effectively and efficiently utilize services that AT gives, and the ATs are sometimes supplied without considering the need for associated services. These associated services include individual assessment, selection, fitting, training, and follow-up and have a great impact on the outcome of AT in a region. Incompatibility with the environment mostly results in the abandonment of the technology provided.

Skilled labor or personnel is one of the associated services to effectively utilize AT. In most hospitals in Ghana, the capacity of skilled personnel to provide AT services and products is very low. The situation is worsened by the lack of training in the workforce to improve their knowledge, skills, and quality. Some hospitals in Ghana also lack a well-defined plan on how to handle certain emergency situations due to the unavailability of ATs. If there was a well-defined plan, provisions for the ATs would be made in a timely manner.

Inadequate service provision, insufficient regulatory capacity, and a limited procurement system are challenges to the use of AT in Ghana.

26.4 Anticipated challenges to adoption of robots in Ghana

Ghana has numerous existing challenges in access to rehabilitation services and capacity. The use of robot systems for stroke has been shown in studies to be extremely beneficial in surmounting most of the challenges. Robot systems therefore have great potential to be used in Ghana. There are however anticipated challenges to the introduction of robots for stroke rehabilitation in Ghana. These challenges have been grouped using four of the World Health Organization's (WHO) building blocks of health systems, namely health service delivery, health workforce, health financing, and leadership and governance.

26.4.1 Health service delivery

With Ghana having only one dedicated stroke unit, the health service delivery system may not be adequately set up to provide rehabilitation robotics. A study on stroke care in Ghana revealed that essential and basic medical equipment for stroke care including rehabilitation is not available in majority of hospitals [24]. Providing robots for rehabilitation when basic equipment is not available may therefore be a challenge.

If a service facility can provide robots for rehabilitation, it is important to consider the healthcare user's perception and beliefs about robots. People in Ghana are more familiar with a hands-on approach, so are less willing to accept and use robots. On the flip side, Lutong Li et al. highlighted the potential for users to be over-reliant on robots and refuse other types of rehabilitation [25]. Finally, the common belief in the spiritual cause of stroke is a major challenge to users' attendance in outpatient care. This may influence the frequency of use of rehabilitation robots [26].

26.4.2   Health workforce

There is a dearth of rehabilitation professionals in Ghana. Although rehabilitation robots have the potential to improve service delivery, the lack of training and time spent in setting up the robot and monitoring its use can be challenging. In addition, the fear of losing one's job is a key factor in rehabilitation professionals' uptake or acceptance of robots. Several studies have shown that there is a huge concern among healthcare workers that robots, who do not have ethical principles, will replace them and lower the standards of healthcare provided [27,28]. Research conducted among healthcare professionals in Finland however revealed that among healthcare workers, physiotherapists and other rehabilitation workers had the second-highest positive reviews of robots mainly due to having personal experiences [29].

26.4.3   Health financing

One major challenge anticipated in the use of rehabilitation robots is cost. Research has shown that previously developed rehabilitation robots cost between $75,000 and $350,000 US dollars exclusive of shipment, taxes, installation, and maintenance costs [30]. WHO's "Choosing Interventions that are Cost-Effective" project revealed that a health intervention costing less than three times the national annual gross domestic product (GDP) per capita is considered cost-effective [31]. Ghana's GDP per capita for the year 2020 was 2328.53 USD. Any rehabilitation robotic device costing more than $6985.59 is therefore not cost-effective. This means that without price adjustments, or the development of cheaper but effective rehabilitation robots, it will be difficult to make a case for policymakers and Ministries of Health to purchase them. A lack of technicians trained in maintaining and repairing the rehabilitation robots will also result in the robots becoming nonfunctional. This might discourage investment in purchasing the equipment.

26.4.4   Leadership and governance

Ghana currently does not have a rehabilitation policy or a national stroke rehabilitation standard of care. The national health policy [32] revised in the year 2020 does not specifically mention robotic devices for stroke although it states that the government will "promote the availability and use of high-quality assistive devices and technologies (including prostheses, orthoses, etc.) at an affordable cost." Without clear knowledge of the benefits of rehabilitation robots, this type of AT may not be promoted by the government. Healthcare managers may also not be motivated to provide robot services for rehabilitation when there is no specific policy governing it.

26.5   Future opportunities

Despite the numerous challenges anticipated in the introduction of robotic devices for stroke rehabilitation, their use in this setting has great potential and benefits in increasing the amount of rehabilitation sessions available to patients and making up for the dearth of human resources for stroke rehabilitation. Future opportunities exist in all four areas of the health system blocks are discussed above.

26.5.1   Health service delivery

With many Ghanaian health institutions lacking dedicated stroke units to provide care including rehabilitation, dedicated beds for patients with stroke can be created on general medical wards with partitions. The introduction of rehabilitation robots can begin in university hospitals with a collaboration formed between the hospital and the university's rehabilitation and engineering departments. Students can be enrolled in courses aimed at designing and building low-cost robots for stroke rehabilitation as demonstrated by Ethan Danahy et al. [33]. A few nongovernmental organizations in Ghana are already training students in both second- and tertiary-cycle institutions on how to build and use robotics, using systems such as the LEGO Mindstorms NXT 2.0 [34]. This can be formalized and scaled up countrywide. An added benefit should be the software providing a dosage of rehabilitation interventions, tracking the rehabilitation process, and providing feedback that is easily transferrable to clinic notes while keeping the data secure.

26.5.2   Health workforce

Continuous professional development workshops can be used to sensitize rehabilitation professionals on the importance of robots in stroke rehabilitation and address concerns surrounding job replacement by robots using case studies. Adequate training through short courses is important to ensure safety, identify the indications for the use of robots in patients with stroke, manage complications that may arise, and provide basic maintenance. In addition to initial training, there is a need for ongoing training to build the capacity of rehabilitation professionals.

26.5.3   Health financing

The main argument against the introduction of rehabilitation robots for stroke is cost. Calabro et al., however, after analyzing this, note that the introduction of rehabilitation robots must not be seen only from an economic standpoint [35]. There is a good financial return on investment which may not be evident in the short term as shown in Carpino et al.'s study [36] which showed that robot-mediated therapy for lower-extremity therapy for stroke is more effective allowing more patients to walk compared to conventional therapy. There is however still the need to develop affordable rehabilitation robots similar to the $5500 Rehab CARES robot gym platform piloted by Johnson et al. [37] in Mexico which is less than WHO's $6985.59 cut-off recommendation for cost-effective health interventions in Ghana [31].

In 2021, Ghana launched its first strategy for Health Technology Assessment (HTA). This can be leveraged to provide the evidence behind the selection and procurement of robots for stroke rehabilitation. Using evidence from similar low-to-middle-income countries, the government can invest in purchasing robots for a few institutions as part of a pilot project while monitoring and analyzing data on its cost-effectiveness and return on investment.

26.5.4   Leadership and governance

A policy dedicated to rehabilitation robots in the Ghanaian health system needs to be advocated for, highlighting the importance of procuring the most cost-effective robots in the right quantities. Monitoring the bidding of these robots is important to ensure suppliers' open and competitive bidding with transparency in their distribution. Rehabilitation advocacy groups can play a major role in ensuring good governance and leadership in this process.

Based on WHO's recommendation on good leadership and governance in healthcare [38], patient satisfaction and utilization of rehabilitation robot surveys will be useful tools for obtaining information on the quality and responsiveness of services.

References

[1]   Ghana Statistical Service. Population & Housing Census Preliminary Report [Internet]. Ghana: Ghana Statistical Service. Available from: https://census2021.statsghana.gov.gh/. 2021 (updated 2021 Sept; cited 2021 Dec 15).

[2]   Ghana Statistical Service. Population & Housing Census Disability in Ghana[Internet]. Ghana: Ghana Statistical Service. Available from: https://www2.statsghana.gov.gh/docfiles/2010phc/Mono/Disability%20in%20Ghana.pdf. 2010 (updated 2014 Oct; cited 2021 Dec 15).

[3]   The Institute for Health Metric and Evaluation. Ghana [Internet]. Seattle: IHME. Available from: https://www.healthdata.org/ghana n.d. (updated 2022; cited 2021 Dec 15).

[4]   World Health Organization. The Global Health Observatory-Ghana [Internet]. Geneva: World Health Organization. Available from: https://www.who.int/data/gho/data/themes/mortality-and-global-health-estimates/ghe-leading-causes-of-death. 2019 (updated 2021; cited 2021 Dec15).

[5]   World Health Organization. Ghana Annual Report. [Internet]. Ghana: World Health Organization. Available from: https://www.afro.who.int/publications/who-ghana-2020-annual-report. 2020 (updated 2020; cited 2021 Dec16).

[6]   Ministry of Health Ghana, National Policy for the Prevention and Control of Chronic Non Communicable Diseases in Ghana [Internet]. Ghana: Government of Ghana. Available from: https://www.iccp-portal.org/sites/default/files/plans/national_policy_for_the_prevention_and_control_of_chronic_non-communicable_diseases_in_ghana(1).pdf. n.d. (updated 2012 Aug; cited 2021 Dec16).

[7]   World Health Organization. NCD Progress Monitor [Internet]. Ghana: World Health Organization. Available from: https://www.who.int/publications/i/item/9789240000490. n.d. (updated 2020; cited 2021 Dec16).

[8]   Wessex Ghana Stroke Partnership. Korle Bu Stroke Unit [Internet]. Dorset: Wessex Ghana Stroke Partnership. Available from: http://www.wgstroke.org/korle-bu-stroke-unit/. n.d. (updated 2016; cited 2022 Jan 17).

[9]   Kumi F, Bugri AA, Adjei S, Duorinaa E, Aidoo M. Quality of acute ischemic stroke care at a tertiary Hospital in Ghana. BMC Neurol 2022;22(1):1.

[10] Donkor ES, Owolabi MO, Bampoh PO, Amoo PK, Aspelund T, Gudnason V. Profile and health-related QOL of Ghanaian stroke survivors. Clin Interv Aging 2014;9:1701.

[11] Ghana Physiotherapy Association. Physiotherapy Units in Ghana [Internet]. Ghana: Ghana Physiotherapy Association. Available from: http://physioghana.org/physiotherapy-in-ghana/physiotherapy-units-in-ghana/. n.d. (updated 2021; cited 2022 Jan 16).

[12] Christian A, Bentley J, Aryeetey R, Ackuaku D, Mayer RS, Wegener S. Assessment of rehabilitation capacity in Ghana. CBR and Inclusive Development. 2016;27:33−60.

[13] Commission of Inquiry into the Creation of New Regions. Report of the Commission of Inquiry into the Creation of New Regions: Equitable Distribution of National Resources for Balanced Development [Internet]. Ghana: Government of Ghana. Available from: https://media.peacefmonline.com/docs/201812/762942613_800218.pdf; n.d. (updated 2018 Aug; cited 2022 Jan 18).

[14] Tinney MJ, Chiodo A, Haig A, Wiredu E. Medical rehabilitation in Ghana. Disabil Rehabil 2007;29:921−7. Available from: https://doi.org/10.1080/09638280701240482.

[15] Hample SK, Sundsted MJ. Occupational therapy's role: a foundational occupational therapy education resource for Ghana. Occup Ther Capstones 2015;77.

[16] Ghana College of Physicians and Surgeons. Family Medicine [Internet]. Ghana: Ghana College of Physicians and Surgeons. Available from: https://gcps.edu.gh/family-medicine/. n.d. (updated 2021 Aug; cited 2022 Jan 25).

[17] Ghana Statistical Service. Labour Force Report [Internet]. Ghana: Ghana Statistical Service. Available from: https://www.statsghana.gov.gh/gssmain/fileUpload/Demography/LFS%20REPORT_fianl_21-3-17.pdf. 2015 (updated 2016; cited 2022 Jan 22).

[18] Zeiler SR. Should we care about early post-stroke rehabilitation? Not yet, but soon. Curr Neurol Neurosci Rep 2019;19(3):1−9.

[19] Agyenkwa SK, Yarfi C, Banson AN, Kofi-Bediako WA, Abonie US, Angmorterh SK, et al. Assessing the use of standardized outcome measures for stroke rehabilitation among physiotherapists in Ghana. Stroke Res Treat 2020;2020:9259017.

[20] Sarfo FS, Acheampong JW, Oparebea E, Akpalu A, Bedu-Addo G. The profile of risk factors and in-patient outcomes of stroke in Kumasi. Ghana Ghana Med J 2014;48 (3):127−34.

[21] World Health Organization. Assistive Technology [Internet]. Geneva: World Health Organization. Available from: https://www.who.int/news-room/fact-sheets/detail/assistive-technology. n.d. (updated 2018 Feb 18; cited 2021 Oct 28).

[22] Physiopedia. Assistive devices. [Internet]. United Kingdom: Physiopedia. Available from https://www.physio-pedia.com/Assistive_Devices. (updated 2022; cited 2021 Oct 28).

[23] Borg J, Lindström A, Larsson S. Assistive technology in developing countries: a review from the perspective of the Convention on the Rights of Persons with Disabilities. Prosthet Orthot Int 2011;35(1):20−9.

[24] Baatiema L, Otim M, Mnatzaganian G, Aikins AD, Coombes J, Somerset S. Towards best practice in acute stroke care in Ghana: a survey of hospital services. BMC health Serv Res 2017;17(1):1.

[25] Li L, Tyson S, Weightman A. Professionals' views and experiences of using rehabilitation robotics with stroke survivors: a mixed methods survey. Front Med Technol 2021;66:780090.

[26] Baatiema L, Aikins AD, Sav A, Mnatzaganian G, Chan CK, Somerset S. Barriers to evidence-based acute stroke care in Ghana: a qualitative study on the perspectives of stroke care professionals. BMJ Open 2017;7(4):e015385.

[27] Hofmann B. Ethical challenges with welfare technology: a review of the literature. Sci Eng Ethics 2013;19(2):389−406.

[28] Sharkey A, Sharkey N. Granny and the robots: ethical issues in robot care for the elderly. Ethics Inf Technol 2012;14(1):27−40.

[29] Turja T, Van Aerschot L, Särkikoski T, Oksanen A. Finnish healthcare professionals' attitudes towards robots: reflections on a population sample. Nurs Open 2018;5(3):300−9.

[30] Duret C, Grosmaire AG, Krebs HI. Robot-assisted therapy in upper extremity hemiparesis: overview of an evidence-based approach. Front Neurol 2019;10:412.

[31] Marseille E, Larson B, Kazi DS, Kahn JG, Rosen S. Thresholds for the cost−effectiveness of interventions: alternative approaches. Bull World Health Organ 2014;93:118−24.

[32] Ministry of Health. National health policy [Internet]. Ghana: Government of Ghana. Available from https://www.moh.gov.gh/wp-content/uploads/2020/07/NHP_12.07.2020. pdf-13072020-FINAL.pdf. n.d. (updated 2020; cited 2022 Jan 10).

[33] Danahy E, Wang E, Brockman J, Carberry A, Shapiro B, Rogers CB. Lego-based robotics in higher education: 15 years of student creativity. Int J Adv Robotic Syst 2014;11(2):27.

[34] Ghana Robotics Academy Foundation [Internet]. Ghana: Ghana Robotics Academy Foundation. Available from: https://foundation.ghanarobotics.org/about-us/know-more-about-us; n.d. (updated 2022; cited 2022 Jan 17).

[35] Calabrò RS, Müller-Eising C, Diliberti ML, Manuli A, Parrinello F, Rao G, et al. Who will pay for robotic rehabilitation? The growing need for a cost-effectiveness analysis. Innov Clin Neurosci 2020;17(10−12):14.

[36] Carpino G, Pezzola A, Urbano M, Guglielmelli E. Assessing effectiveness and costs in robot-mediated lower limbs rehabilitation: a meta-analysis and state of the art. J Healthc Eng 2018;2018:7492024.

[37] Johnson MJ, Rai R, Barathi S, Mendonca R, Bustamante-Valles K. Affordable stroke therapy in high-, low-and middle-income countries: from theradrive to Rehab CARES, a compact robot gym. J Rehabil Assist Technol Eng 2017;4 2055668317708732.

[38] World Health Organization. Monitoring the Building Blocks of Health Systems: A Handbook of Indicators and their Measurement Strategies [Internet]. Geneva: World Health Organization. Available from: https://www.who.int/healthinfo/systems/WHO_MBHSS_2010_full_web.pdf. n.d. (updated 2010; cited 2022 Jan 05).

Chapter 27

Africa region: Cameroon

Sinforian Kambou[1,2], Valery Labou Tsinda[3], Asongafac[2,4],
Philomène Synthia Tonye[5], Eric Gueumekane Bila Lamou[6] and
Michael Temgoua[1]

[1]Institute of Applied Neurosciences and Functional Rehabilitation, Yaoundé, Cameroon,
[2]Center for Promotion of Rehabilitation Medicine and Disability Research, Yaoundé, Cameroon,
[3]Neuro-rehabilitation and Movement Disorders Society, Yaoundé, Cameroon, [4]Physiotherapy
Unit, Buéa Regional Hospital, Buéa, Cameroon, [5]Physiotherapy Unit, Nkongsamba Regional
Hospital, Nkongsamba, Cameroon, [6]Department of Internal Medicine, Gynaeco-Obstetric and
Paediatric Hospital, Douala, Cameroon

Learning objectives

After completing this chapter, the reader will be able to:
- Demonstrate basic understanding of Cameroon's healthcare system.
- Identify the different facets of stroke demographics and stroke rehabilitation in Cameroon.
- Describe facilitators and barriers to the use of robotics in stroke rehabilitation in Cameroon.

27.1 Introduction to Cameroon context and healthcare system

27.1.1 Introduction to country context

Cameroon is a Central African country located at the bottom of the Gulf of Guinea. It is bordered by Nigeria to the west, Chad to the northeast, the Central African Republic to the east, and Congo, Gabon, and Equatorial Guinea to the south. Cameroon covers an area of 475,650 km^2. In 2020, its population was approximately 24,910,305 inhabitants according to the Central Bureau of the Census and Population Studies. Administratively, the country is divided into 10 regions, 58 departments, and 268 districts. The two official languages are English and French. However, there are over 200 national languages.

Rehabilitation Robots for Neurorehabilitation in High-, Low-, and Middle-Income Countries.
DOI: https://doi.org/10.1016/B978-0-323-91931-9.00006-2

This population is extremely young with an average age of 22.1 years. The under-15 age group accounts for 43.6% of the total population while those aged below 25 years represent 64.2%. The majority of the population lives in urban areas (52%). There is a high population density in big cities: Douala (2,717,695 inhabitants in 2015) and Yaounde (2,785,637 inhabitants in 2015).

27.1.2 National health system

In Cameroon, the health system is organized into three levels: central, intermediate, and peripheral. The system has a pyramid organization with central level at the summit, intermediate level at the middle, and peripheral level at the base. Central services and intermediate-level structures ensure, respectively, strategic organization of the system and support to health districts. Peripheral levels ensure operational implementation. The health system comprises three sublevels: public subsector (includes all public health structures at the three levels of the health pyramid as well as health structures under the supervision of other ministerial departments); private subsector (including nonprofit and for-profit); and traditional subsector (includes structures working on traditional medicine). Central, intermediate, and peripheral levels of the health pyramid have administrative, health, and dialog structures. At the central level, administrative structures comprise the Minister's Office, Secretary General, Secretary of State to MOH Technical departments, and others. Intermediate-level administrative structures involve the 10 regional delegations in the country, while the peripheral level is made of 189 health districts. Dialog structures in the health pyramid have the crucial role of bringing together all actors involved in health activities for coordinated planning, implementation, and evaluation of policies. At the central level, there is the National Council of Health, Hygiene, and Social Affairs. The Regional Fund for Health Promotion is the dialog structure at the intermediate level in each region, and at peripheral level, we have the management committee and health committee.

The health sector is also divided into five components: health promotion; disease prevention; case management; health system strengthening and governance; and lastly strategic steering. Several stakeholders are involved in the development of the health sector in Cameroon. These include the state, households/communities, the private sector, and external partners [1,2].

The epidemiological profile of Cameroon highlights a predominance of communicable diseases with a rising prevalence of noncommunicable diseases notably cardiovascular diseases, cancers, and mental diseases. The five major causes of death are HIV/AIDS (14.24%), cancers and lower respiratory tract infections (10.52%), malaria (8.78%), neonatal diseases (8.47%), and diarrheal disease (5.01%) [1]. According to the Demographic and Health Survey in Cameroon (EDSC-V), infant mortality rate in Cameroon is 48 deaths per 1000 live births. This survey highlights that from 2004 to 2018, the mortality rate was 467 per 100,000 live births. To improve access to affordable and quality healthcare for all Cameroonians, the government is preparing the official launching of the pilot phase of Universal Health Coverage. In Cameroon, researchers estimate disability prevalence using the Washington Group tools which capture self-reported activity limitations as described by the International Classification of Functioning, Disability, and Health. Using the Washington Group tools Short Set (six items focusing on seeing, hearing, mobility, memory/concentration, self-care, and communication), disability prevalence ranged from 6.1% with the standard threshold to 66.3% with the wide threshold.

27.2   Rehabilitation in Cameroon

The World Health Organization (WHO) defines rehabilitation as "a set of interventions designed to optimize functioning and reduce disability in individuals with health conditions in interaction with their environment" [3]. WHO's Rehabilitation Needs Estimator reports that in 2019 in Cameroon, 5.3 million people experienced conditions that could benefit from rehabilitation, and 670k years, across the population, were lived with disability, corresponding to an increase of 195.5% of years lived with a disability between 1990 and 2019.

In the country, the field of rehabilitation developed with the occurrence of poliomyelitis cases. The main interventions consisted in the treatment of motor impairments caused by this pathology. Since that time, rehabilitation has remained poorly structured at the national level. Despite this situation, the Ministry of Social Affairs ensures coordination in this sector with the contribution of other ministries and organizations.

At the legal level, a set of legal instruments aimed at protecting the rights of people with disabilities have been put in place. These include Presidential Decree No. 2021/751 of December 28, 2021, ratifying the United Nations Convention on the Rights of Persons with Disabilities and Law No. 2010/ 002 of April 13, 2010, on the Protection and Promotion of Persons with Disabilities [4,5]. This law defines rehabilitation of the disabled person and its three areas of intervention: medical rehabilitation, psychosocial support, and special education.

In terms of health policies, rehabilitation is underrepresented in various documents published such as the National Health Development Plan (2016−2020) and the Health Sector Strategy (2016−2027). Financial resources allocated to this sector are almost nonexistent. Similarly, there are no specific indicators of population functioning in the National Health Information System [1,2].

The country has various rehabilitation professionals. The main ones are physiotherapists, occupational therapists, prosthetists and orthotists, psychomotor therapists, speech therapists, psychologists, and social workers. However, the distribution of these professionals in public or private health facilities remains very disparate. This underscores the crucial issue to train and employ these professionals, especially physical and rehabilitation specialists, in order to respond efficiently to the growing rehabilitation needs in Cameroon. Regarding assistive technology and products, there is no harmonized provision policy at the national level. It is a challenge for many disabled persons to access quality and affordable devices.

The National Center for Rehabilitation of Persons with Disabilities Cardinal Paul Emile LEGER of Yaoundé is the biggest rehabilitation center in the country. This center is under the authority of the Ministry of Socials Affairs. Physiotherapy units are available in many hospitals depending on the Ministry of Health. These units offer inpatient and outpatient rehabilitation while private settings deliver outpatient rehabilitation.

Regarding the private nonprofit health sector, Baptist Convention and Catholic Church are involved in the fields of rehabilitation with many rehabilitation centers and community-based rehabilitation programs in the country. Cameroon Baptist Convention Health Services are involved in various projects for the rehabilitation of persons living with disabilities. These include the Cameroon Clubfoot Care project in partnership with the Ministry of Health and the "Rehabilitation Compass for Inclusion Project" with the support of the Ministry of Social Affairs for the professional training of physiotherapists, occupational therapists, and CBR field workers in Cameroon [6].

27.3 Stroke situation in Cameroon

Globally, the burden of neurological disorders has increased in the past decades. In this way, stroke represents a leading cause of mortality and long-term disability worldwide [7].

In Cameroon, there is a paucity of statistics on stroke at the national level. However, some hospital studies show the prevalence of this pathology. A 3-year cross-sectional study in a Teaching Hospital in Cameroon showed that in the 325 patients enrolled, hypertension was the main cerebrovascular risk factor (81.15%). Ischemic stroke accounted for 52% of cases while 48% were hemorrhagic. The mean duration of hospitalization was 8.58 ± 6.35 days with a case fatality rate of 26.8% [8].

A review of 1277 medical charts of cerebrovascular accidents from urban and rural centers in Cameroon showed that the main symptoms observed in stroke patients were paresis (83.1%) and speech problems (25%) [9]. Another study of medical records of inpatients and outpatients from two urban public hospitals in Douala and two rural healthcare centers highlighted also that the most frequent neurological complaint was paresis/weakness

(25.2%), and the most common concurrent medical history was hypertension (40.0%). The most common neurological diagnosis was cerebrovascular disease (51.5%) [10].

Spastic paresis is a major cause of poststroke motor impairment observed in rehabilitation services in the country. Cognitive impairments are also a significant cause of disability after stroke. In a series of 114 stroke patients seen at the neurology outpatient department, the prevalence of cognitive impairment was 41.2%. These impairments were mild in 26.3% and moderate in 8.8% of participants. Dementia was found in 6.1% of patients [11].

As in many developing countries, the financial cost of stroke care remains high and is borne mostly by patients and their families. A study evaluating the cost of stroke management in a reference hospital found that physiotherapy accounted for 48% of inpatient and outpatient expenses [12]. Considering that, population screening and treatment of cerebrovascular risk factors remain an indispensable strategy to fight against in a limited resources context.

At this moment, there is no stroke unit in Cameroon. A patient presenting a sudden neurological deficit is taken to the nearest health facility to consult. If the technical capability of this center is insufficient, the center transfers the patient to a more specialized health facility where cerebral imagery, neurology, and intensive care units could be available. During hospitalization, stroke patients receive physiotherapy sessions, and after discharge, a follow-up program is developed with the neurologist and physiotherapist.

27.4   Stroke rehabilitation

Stroke remains across the world, the second leading cause of death and disability and one of the leading causes of depression and dementia globally [13,14]. It is well known that stroke leads to many impairments. In this line, it is crucial to promote multidisciplinary rehabilitation approaches to ensure the best functioning for stroke survivors.

27.4.1   Stroke rehabilitation policy

In Cameroon, there are no national guidelines for stroke rehabilitation. At the local level, the "North West Best Practices in Stroke Rehabilitation Group" has developed guidelines for stroke management and rehabilitation in the North-West region of Cameroon [15]. This publication provides best practice recommendations for inpatient and outpatient rehabilitation and community-based rehabilitation for stroke in this region.

27.4.2   Management of cases

Rehabilitation is part of the continuum of care for stroke survivors. Globally, rehabilitation interventions are passive and active exercises, posture control and training, splinting, assistive devices provision, speech therapy, and psychosocial support. The creation of structured neurorehabilitation unit could be an important step to improve the standard of rehabilitation care for stroke patients in the country.

The COVID-19 pandemic heavily affected health systems and healthcare providers worldwide. In Cameroon, this crisis highlighted vulnerabilities of the health system, and maintaining essential healthcare provision was a crucial challenge. The pandemic has led to profound psychological distress among patients and health personnel. In Cameroon, a survey at the Gyneco-Obstetric and Pediatric Hospital of Douala reported a prevalence of anxiety and depression of about 57.14% and 31.43%, respectively, in the nursing staff. Another detrimental effect of this health emergency was the decrease in in-person healthcare visits by the population. This could be explained by the stay-at-home orders or fear to be infected by the virus in health facilities. The number of patients attending outpatient rehabilitation decreases significantly. During this period, outpatient rehabilitation services were cut down or unavailable for many patients with functioning limitations despite all the barriers and measures against COVID-19. Inpatient rehabilitation was still operational, and care was offered as long as needed. This outbreak highlighted the need for rehabilitation services (physiotherapy, speech therapy, prosthetics and orthotics, psychomotor therapy, etc.) to be better prepared to deliver rehabilitation care during public health emergencies by organizing online rehabilitation sessions (telerehabilitation) to avoid cessation or declines in rehabilitation services delivery.

27.5   Robotics in stroke rehabilitation and other technologies

Robotic devices for neurorehabilitation have developed considerably with a greater focus on the treatment of poststroke paretic upper limbs. This breakthrough in rehabilitation engineering is well noticed in developed countries where innovative technologies are used to improve poststroke functioning.

In Cameroon, very few robots for rehabilitation are available in health facilities. Robotics rehabilitation within health care could significantly decrease the burden of physiotherapy staff and improve functional outcomes. Using robotics devices could enable therapists to overcome issues of traditional rehabilitation approaches by improving the number of repetitions, providing safe positioning for exercises, and improving quality patient assessment. This could be facilitated by training rehabilitation professionals on the use of these technological devices and managing some barriers as the initial cost of the devices and ergonomic issues related to the installation of the machines in rehabilitation services.

Regarding poststroke rehabilitation, only continuous passive mobilization machines (arthromotor) for the upper or lower limb are available.

Additionally, assistive devices such as wheelchairs, prosthetics and orthotics, crutches, and braces are also available. Although the literature has demonstrated the benefits of body weight-supported treadmill in the management of poststroke gait impairment, these technologies are not available in our context. To compensate for this lack of equipment, most rehabilitation units use sling suspension therapy, which allows active and passive mobilization. We acknowledge that the intensity of rehabilitation with suspension therapy is lower than with robotic rehabilitation. It would therefore be important to provide affordable robotics devices for hospitals and train rehabilitation professionals in their optimal use to improve functional outcomes of stroke patients in Cameroon. Research on robotics for rehabilitation is also lacking in our context. There is a need to advocate for more investment in research and funding in this domain to improve the quality of care and outcome of patients who are in need of these technologies.

27.6   Conclusion

Stroke is a major concern in Cameroon health system. In the country, using robotics devices for stroke rehabilitation is recent and not well developed. To date, passive continuous movement devices are available. Suggestions are made to invest in robotics devices such as robotic-assisted body weight-supported treadmill training to improve the functional ability of stroke survivors. Creating appropriate robotics units in rehabilitation facilities and training human resources are crucial for the successful development of robotics in stroke rehabilitation.

Conflict of interest

The authors declare that they have no conflicts of interest.

References

[1] Cameroon Minstry of Health. National Health Development Plan 2016–2020 [Internet]. Available from: https://www.minsante.cm/site/?q = en/content/national-health-development-plan-nhdp-2016-2020. 2016 (cited 2021 Feb 21).
[2] Cameroon Minstry of Health. Health Sector Strategy 2016–2027 [Internet]. Available from: https://www.minsante.cm/site/?q = en/content/health-sector-strategy-2016-2027-0. 2016 (cited 2021 Feb 21).
[3] Mills T, Marks E, Reynolds T, et al. Rehabilitation: Essential along the Continuum of Care. In: Jamison DT, Gelband H, Horton S, et al., editors. Disease Control Priorities: Improving Health and Reducing Poverty. 3rd edition. Washington (DC): The International Bank for Reconstruction and Development/The World Bank; 2017 Nov 27. Chapter 15. Available from: https://www.ncbi.nlm.nih.gov/books/NBK525298/, http://doi.org/10.1596/978-1-4648-0527-1_ch15.

[4] Biya P. Decree No. 2021/751 of 28 December 2021 to ratify the United Nations Convention on the Rights of Persons with Disabilities [Internet]. Available from: https://www.prc.cm/en/news/the-acts/decrees/5585-decree-no-2021-751-of-28-december-2021-to-ratify-the-united-nations-convention-on-the-rights-of-persons-with-disabilities. n.d. (cited 2022 Feb 21).

[5] Biya P. Law [Internet]. N°2010/OO2 Apr 13, 2010. Available from: https://www.un.org/development/desa/disabilities/wp-content/uploads/sites/15/2019/11/Cameroon_-protection-and-promotion-of-persons-with-disabilities.pdf.

[6] Events — Cameroon Baptist Convention Health Services [Internet]. Available from: https://cbchealthservices.org/afas-rci-project/events/. n.d. (cited 2022 Feb 28).

[7] Murray CJL, Lopez ADWHOrganization, World Bank & Harvard School of Public Health. The Global burden of disease : a comprehensive assessment of mortality and disability from diseases, injuries, and risk factors in 1990 and projected to 2020 : summary [Internet]. World Health Organization; 1996. Available from: <https://apps.who.int/iris/handle/10665/41864> (cited 2022 Feb 28).

[8] Mapoure YN, Kuate C, Tchaleu CB, Ngahane HBM, Mounjouopou GN, Ba H, et al. Stroke epidemiology in Douala: three years prospective study in a teaching hospital in Cameroon. World J Neurosci 2014;4(5):406−14.

[9] Doumbe J, Mapoure Y, Nyinyikua T, Kompoliti K, Shah H, Delgado EC. Burden of cerebro-vascular disease in Cameroon. J Neurol Sci 2015;357:e372.

[10] Doumbe J, Mapoure YN, Nyinyikua T, Kuate C, Kompoliti K, Shal H, et al. Neurological disease surveillance in Cameroon, a rural and urban-based inout patient population study. Clin Neurol Neurosci 2019;3(1):24.

[11] Mapoure Y, Essoga MM, Ba H, Cyrille N, Gams MD, Luma H. Post stroke cognitive impairment in Douala (Cameroon). Afr J Neurol Sci 2017;36.

[12] Mapoure Y, Kuate C, Pe B, Luma H, As M, Njamnshi A. Cost of stroke at the Douala General Hospital. Health Sci Dis 2014;15:3.

[13] GBD 2015 Neurological Disorders Collaborator Group. Global, regional, and national burden of neurological disorders during 1990−2015: a systematic analysis for the Global Burden of Disease Study 2015. Lancet Neurol 2017;16(11):877−97.

[14] Owolabi MO, Sarfo F, Akinyemi R, Gebregziabher M, Akpa O, Akpalu A, et al. Dominant modifiable risk factors for stroke in Ghana and Nigeria (SIREN): a case-control study. The Lancet Glob Health 2018;6(4):e436−46.

[15] Cockburn L, Ngole M. Best practice guidelines for stroke in Cameroon: an innovative and participatory knowledge translation project. Afr J Disabil 2014;3::92.

Chapter 28

Africa region: Morocco

Said Nafai[1], Amin Zammouri[2] and Abderrazak Hajjioui[3,4]

[1]*School of Health Sciences, Department of Occupational Therapy, American International College, Springfield, MA, United States,* [2]*Engineering Department, EPF Graduate School of Engineering, Cachan, France,* [3]*Clinical Neuroscience Laboratory, Faculty of Medicine, Pharmacy and Dentistry, University Sidi Mohamed Ben Abdellah, Fez, Morocco,* [4]*Department of Physical & Rehabilitation Medicine, University Hospital Hassan II, Fez, Morocco*

Learning objectives

At the end of this chapter, the reader will be able to:

- Learn about stroke in Morocco.
- Learn about the current status of stroke rehabilitation in Morocco.
- Learn about the development and use of robotic and assistive technology in Morocco.

28.1 Introduction and healthcare statistics

The Kingdom of Morocco is a country located in North Africa across the Straits of Gibraltar from Spain. Morocco has a fast-growing economy and a rich history that spans centuries. Morocco has a population of approximately 37 million. The population is mainly comprised of Arabs and Amazigh people, known as Berbers. Arabic is the first language in Morocco with French as the second language. Tamazight, the language of the Amazigh people, is now taught in schools, and Moroccans' use of the English language is on the rise [1].

Morocco's human development index value for 2019 was 0.686, which increased 50.1% from 0.457 in 1990. This puts Morocco in the "medium human development" category of 121 out of 189 countries and territories [2]. Between 1990 and 2019, Morocco's life expectancy at birth increased by 11.9 years to the current 76.6 years of age. Expected years of schooling increased by 7.2 years. Morocco's gross national income per capita increased by 98.6% between 1990 and 2019, and it is currently at $7368. [2]. The areas of healthcare and education still challenge Morocco with mean years of schooling at 5.6 in 2019 with expected years of schooling at 13.7 [2]. The Moroccan government has given emphasis to preventive medicine by increasing the number of health centers. However, rural areas account for 37.5% of the population, and more than half of these rural areas still lack access to preventive healthcare facilities [1]. In addition, access to rehabilitation is limited especially for older adults who suffer from neurological disorders such as stroke.

Stroke and ischemic heart disease are the top two conditions responsible for the most death and disabilities combined in Morocco. These were the largest contributors to the total number of disability-adjusted life years (DALYs) in 2019, and their major risk factors are metabolic (high blood pressure) and behavioral [3]. According to the results of the national survey on common risk factors for noncommunicable diseases (2017−18), 33.6% of Moroccans suffer from high blood pressure, 21% from sedentary behavior, 20% from being overweight and obesity, 17.2% from smoking, and 10.6% from diabetes [4].

28.1.1 Most common health conditions responsible for morbidity, mortality, and disability in Morocco

Morocco is going through an epidemiological transition characterized by a shift in the overall burden of morbidity and mortality from infectious diseases to noncommunicable diseases (NCDs) and injuries. The country faces the growing impact of NCDs which account now for 75% of all deaths [5]. In terms of the number of years of life lost due to premature death in Morocco, ischemic heart disease, stroke, and hypertensive heart disease were the top three leading causes in 2019. According to the results of the national survey on common risk factors for NCDs in Morocco published in 2018, the leading risk factors in Morocco are arterial hypertension with a prevalence of 33.6%, followed by physical inactivity (21%), high body mass index (20%), tobacco (17.2%), and diabetes (10.6%) [6].

28.1.2 Current organization of healthcare system—levels of care including rehabilitation

The Moroccan healthcare system combines different levels of health services organized into two main sectors: the public sector that is mainly led by the Ministry of Health, and the private sector that is more developed in larger cities, which only a small proportion of the general population can afford [7]. The public healthcare sector offers medical services that need to be either paid directly by the patients (self-pay out-of-pocket) or indirectly through insurance and some free basic medical services (e.g., infant vaccinations, emergency first aid, pregnancy checkups, assisted delivery). It is important to highlight that only 62% of the Moroccan population are granted health insurance, covering employees with a formal work contract, low-income individuals, and students in higher education [7,8]. The basic insurance that is granted to low-income individuals only gives access to free hospital services and is limited to what is available in these hospitals such as surgical tools and medication, meaning that patients usually need to additionally pay out-of-pocket what is needed for their medical and/or surgical interventions. Hence, people sustaining spinal cord injury, even with insurance coverage, still rely on private pay to get access to quality and comprehensive acute care and rehabilitation services [9].

28.1.2.1 Multidisciplinary team

Although there are well-trained PRM physicians, with 25 working in the public health system, the lack of inpatient rehabilitation wards has a negative effect on the quality of care. Furthermore, the rehabilitation services provided by the general rehabilitation centers are not reliably assessed. On the other hand, occupational therapy is not yet available in all rehabilitation facilities, and physiotherapists are not trained to give functional rehabilitation care. Psychologists do not exist in public health facilities and are only available in the private sector for the patients who can pay for it. Hence, we believe that stroke survivors in Morocco have very limited access to proper medical, rehabilitation care, and assistive technologies.

28.1.2.2 Rehabilitation facilities

The Moroccan health system does not have specialized rehabilitation care for stroke patients. However, the provision of general inpatient rehabilitation care remains very limited with a total number of hospital beds not exceeding 200 beds for a population of over 37.90 million (2022).

The university hospital centers in Casablanca and Rabat have about 10 general rehabilitation beds each; in the private sector, there is the Noor Center in Casablanca with 100 beds and the Salmia Center with 20 beds; and in Rabat, there is rehabilitation and neurosciences with about 10 beds.

To the scarcity of general rehabilitation facilities located mainly in Casablanca and Rabat, is added an unequal geographical distribution of these rehabilitation structures in the 12 regions of Morocco.

28.1.3 Rehabilitation needs in general and disparities in access

Functional limitations and disability are very common among Moroccan patients, and there is a great need for inpatient rehabilitation facilities in the Moroccan health system [10]. The most needed rehabilitation interventions among adult hospitalized patients were physiotherapy (36.2%), occupational therapy (20.4%), and prosthetics (15.8%), and only 24.5% of patients were eligible for inpatient rehabilitation support [11].

28.1.4 Insurance and funding systems for healthcare needs

Even if Morocco has taken giant steps to generalize universal health coverage, the lack of general and specialized rehabilitation facilities and the shortage of rehabilitation workforces strongly limit access to appropriate rehabilitation services and long-term care and worsen health inequalities in the Moroccan health system [12]. Morocco's health insurance system is a mixture of government-run and privately owned insurance businesses. Most in Morocco have coverage through the primary source of health insurance. This is the mandatory health insurance, L'Assurance Maladie Obligatoire (AMO) [13]. The second insurance policy that Morocco implemented is the Regime Assistance Medical (RAMED). RAMED is a public, government-financed program to fund insurance for those living in poverty and without the income needed to access the AMO. The private insurance sector, which people often choose simply due to availability, is a system based on a fee-for-service policy. For whatever the service may be, private insurance requires the individual to pay a minimum of 20% of the fees due. However, fees sometimes range as high as 50%. Morocco's health insurance system guarantees free care to anyone. However, it is specifically free for anyone living in poverty at any clinic that Morocco's government runs. Thankfully, the poverty rate in Morocco is low at 3.6%. However, healthcare remains concentrated in the cities leaving the rural population without easy access. The rural population often remains uncovered and without the funds to be a part of private insurance operations. The impending health insurance expansion promises to cover rural workers. This will ease the economic burden of health insurance from their income. The expansion to cover more workers is not the first one the government has made since 2019. In 2020, the Moroccan government expanded its health insurance system to cover all costs, for every citizen, for COVID-19 treatment. The treatment coverage is available through the AMO. Morocco's health insurance system will expand pending the implementation of six drafted policy proposals. The overarching plan for Morocco's health insurance system is to generalize all health insurance for uncovered workers. The first step in this plan is the creation of coverage beginning with the farmers in the outlying reaches of Morocco, the taxi drivers in the cities, and the artisans spread around the country.

28.2 Stroke statistics and management

A cerebrovascular accident, also known as a stroke, is defined by the World Health Organization (WHO) as "an acute neurological dysfunction of vascular origin with symptoms and signs corresponding to the involvement of focal areas of the brain" [14]. There are two types of stroke: ischemic and hemorrhagic (refer to Chapter 1 for more details). Stroke is a major public health concern and a real growing burden in the Moroccan health system, especially its cost on the social, psychological, and economic levels [15]. Even if there is a lack of Moroccan data concerning the temporal patterns of incidence or the long-term evolution of this debilitating disease, the incidence of stroke continues to increase in Morocco, with an estimated prevalence of 283.85/100,000 inhabitants in Morocco and an incidence, which increases with age [16]. According to the latest WHO data published in 2018, stroke deaths in Morocco reached 15,675 or 9.42% of total deaths [17]. The age-adjusted death rate is 58.96 per 100,000 population ranking Morocco #114 out of 189 in the world. Stroke is the second cause of death in Morocco, all ages combined since 2009 with a variation of 22.5% between 2009 and 2019 [4].

Ischemic stroke affects the younger population especially men, with long prehospital delays influencing the access to care for Moroccan patients [18]. Hemorrhagic stroke is the rupture of a weakened cerebral blood vessel where blood accumulates outside of the vascular space and pressures surrounding brain tissue. There are two types of hemorrhagic stroke, intracerebral (bleeding into the brain itself) and subarachnoid (bleeding into an area surrounding the brain). Aneurysms and arteriovenous malformations are the most common types of blood vessels that are weak causing hemorrhagic strokes [19]. Hemorrhagic strokes are less common than ischemic strokes and account for 13% of all strokes. However, they have a higher mortality rate than ischemic strokes [20]. Inadequate care in the acute phase and the lack of appropriate care and rehabilitation structures are believed to be the causes of poststroke disability reaching 50%. The incidence is higher in the elderly, with a death rate of 25% in the first year.

The management of stroke has been addressed and has clearly improved over the past 2 years, but the improvements are still insufficient given the high number of patients annually. "We estimate that there are 5000 cases of stroke per year in the Rabat region alone, but we only manage to take care of some 1000 patients" according to the neurology department at the Center Hospital University (CHU) Ibn Sina, Rabat, Morocco. In Morocco, neurovascular units exist only at the level of the CHU. Interventional neuroradiology is only practiced in the CHUs of Rabat and Fez, which are dense urban locations. These two facilities are especially dedicated to stroke care and are equipped with the technical devices necessary to reduce mortality and improve the chances of recovery. Intravenous thrombolytic therapy with recombinant tissue plasminogen activator (rTPA) is an approved method to treat acute ischemic stroke (AIS). Its efficacy and safety have been proven by numerous studies [21,22]. Unfortunately, rTPA remains unavailable in the majority of the healthcare facilities in Morocco, and only 1.8%−2.9% of patients with AIS benefit from this treatment [18]. According to the 2020 health map, published by the Ministry of Health, there are 152 neurologists in the public sector, while there are 77 in the private sector. This number is too small in relation to the needs of the population for neurological services [23]. This severe lack of neurologists and their poor distribution with a concentration in certain cities such as Rabat and Casablanca constitute a great challenge to the appropriate management of all neurovascular diseases, in particular stroke.

In the top 10 leading causes of death and disability (DALYs) in Morocco in 2019, compared to other countries with equivalent sociodemographic and economic indicators (age-standardized DALY rate per 100,000), stroke was ranked after ischemic heart disease as the second cause of death and disability (DALY) in Morocco in 2019 [3]. According to the study on the outcome and health-related quality of life after stroke in Morocco, conducted by Hajjioui A. et al., 78% of patients were sent home without any initial rehabilitation. Only 18% fully regained independent ambulation, 56% walked with assisted devices (tripod cane, walker), and 26% were limited to using a wheelchair. About 48% of patients had a Barthel index greater than 60 and 24% had a Barthel index less than 40 [15]. Several domains of quality of life were compromised. Quality of life was significantly negatively correlated with the Barthel index values, indicating the lower quality of life in people with worse functional status and greater clinical severity of stroke [24].

In terms of governance at the Ministry of Health, there is no specific plan or strategy for stroke management. Recently, the Minister of Health started working on a strategic plan for neurovascular emergencies. There are no good practice recommendations or guidelines for stroke management by scientific societies in the country at this time. There is a need for health authorities to set up local neurovascular units in the different regions of Morocco to initiate treatment in the acute phase. There is also a need to strengthen the role of the general practitioner and the family doctor, to better control risk factors and allow the reduction of number of stroke cases to improve care. Finally, it is imperative to create rehabilitation centers to allow better integration of stroke survivors.

28.3 Rehabilitation for stroke

Morocco signed the Convention on the Rights of Persons with Disabilities in 2007 and formally ratified it in 2009 [25]. Despite Morocco signing the treaty, progress for people with disabilities has been slow. However, many associations and organizations advocate for and serve people with disabilities including those with stroke [11]. The Moroccan Constitution of 2011 is one of the few constitutions in the world that devotes Article 34 to the right of persons with disabilities to rehabilitation. In this positive legislative context, the Moroccan government adopted the National Health and Disability Action Plan (2015−2021) and the national action plan of the integrated public policy for the promotion rights of people with disabilities (2017−2021), promulgated by *Dhahir* 1−16−52 (King's decree) of April 27, 2016, a framework law number 97−13 of April 27, 2016, relating to the protection and promotion of people in disability status (Dhahir 1−16−52 of April 27, 2016) [10].

In Morocco, general physicians and neurologists refer people with stroke for rehabilitation. Rehabilitation typically consists of occupational therapy, physical therapy, and speech therapy. In 2017, the Ministry of Health, with the support of the WHO office in Morocco, developed a guide to good practice, advice, and home rehabilitation care for people after a stroke, intended for health professionals and associations. Apart from the fact that this guide has not been widely disseminated at the community level, there are no protocols or recommendations for rehabilitation after a stroke. In the same year, the Ministry of Health validated, for the first time, the norms and standards of rehabilitation facilities which had not been implemented in the Moroccan health system due to a lack of understanding of decision-makers and professionals of the importance of rehabilitation for the well-being and quality of life of the population.

Rehabilitation is offered in some public hospitals at a free or discounted cost under the support of the Ministry of Health. Private clinics attempt to fill the gap in rehabilitation in Morocco. However, access to these services is limited to only those who can afford it. The major providers of outpatient rehabilitation for people with stroke and other disabilities are organizations and charity associations that include occupational therapists, physical therapists, and speech-language pathology therapists. These associations offer rehabilitation services for a discounted price or free to people. Often a doctor's referral and a medical diagnosis are required for these types of therapy. Some orthopedics doctors also have outpatient rehabilitation services in their private clinics, which however can be very costly to the public [10,15].

Rehabilitation in Morocco focuses more on motor recovery than cognition recovery. In addition, due to the Moroccan culture and societal norms, often family members provide a great deal of caretaking assistance to their loved one affected by stroke. This can inadvertently lead to a "learned nonuse phenomenon" and overreliance on caregivers to perform self-care tasks such as self-feeding, bathing, and dressing despite their ability to perform these tasks. Managing the rehabilitation process for people with stroke is often complicated due to other comorbidities they suffer from. Many people die within the first few years poststroke or experience a long-term disability in Morocco due to lack of or limited access to rehabilitation during the acute phase. During the chronic phase of stroke rehabilitation often, people who did not receive adequate rehabilitation services during the acute phase eventually lead to long-term disability and reliance on others for assistance even for basic self-care tasks. Due to this lack, stroke survivors live with limitations of motor and cognitive abilities, which can make rehabilitation during the chronic phase especially difficult.

COVID-19 further limited the ability of people with stroke and their families to access rehabilitation. All rehabilitation centers were closed from March 2020 to June 2020 [26]. In developed countries, virtual rehabilitation sessions were frequently used due to COVID-19 restrictions. However, this was not the case in Morocco. A lack of preexisting infrastructure to support confidential virtual sessions, a nationwide lack of consistent internet access, limited access to smartphones, and a general lack of awareness or expectation of a virtual rehabilitation session hindered the process despite the potential benefits of these telehealth sessions [27]. Essentially, stroke rehabilitation infrastructure is barely intact for in-person rehabilitation, and virtual rehabilitation was simply too much of a burden on an already overextended system during the acute pandemic crisis.

The lack of access to appropriate rehabilitation in Morocco for people with stroke presents the largest challenge for this population and adds to the overall level of disability in Morocco. However, with the increased awareness of the benefits of rehabilitation in stroke recovery and the increase in educational programs to produce adequately trained therapists, these challenges may start to decline. Morocco needs to ensure that healthcare workers are well trained in stroke recovery to facilitate access to rehabilitation and assistive devices and minimize the challenges of people living with stroke.

28.4 Rehabilitation robotics in Morocco

The introduction of technological tools in rehabilitation in Morocco has grown through research projects that are at the crossroads of robotics, artificial intelligence, neurosciences, and other disciplines. Nowadays, there is a variety of devices that support the activity of patients and rehabilitation specialists. These devices include treadmills, motorized verticalizers, and connected exoskeletons, to name a few. Equipped with artificial intelligence, robotic devices used in rehabilitation are increasingly becoming an integral part of the care of patients in Moroccan rehabilitation centers and practices. Generally, the design of these devices aims to support and improve patient mobility as well as to assist the practitioner's work. Robotic devices used in rehabilitation are considered mechanical machines, which act to improve the mobility of the upper and/or lower limbs. The designs of such machines involve the use of sensory sensors as well as a variety of actuators. These sensors and actuators are the key elements that allow continuous adaptability of the rehabilitation device to the patient's capacities [28]. These basic elements in a rehabilitation device are linked to an electronic system, generally called an "embedded system," which represents the treatment unit and allows the practitioner to adapt the degree of intensity and the level of performance to the rehabilitation program.

The use of robotic devices is aimed at improving the quality of therapeutic care and providing assistance to practitioners. Robotic devices make it possible to provide longer and more adapted training sessions. In Morocco, rehabilitation centers and practices predominantly use robotic devices for the rehabilitation of patients with cerebral palsy, multiple sclerosis, and degenerative diseases. For example, there is the use of an exoskeleton to initiate movement of the arms or fingers or legs in patients with cerebral palsy or a neurodegenerative muscle disease such as myopathy.

The current designs of dedicated robotic devices for rehabilitation aim to extend use outside of dedicated centers and practices and to make them tools for assisting those with disabilities. The objective behind the extension of the use of robotic devices in everyday life is to restore and/or increase the autonomy of patients. This could be achieved through the treatment of sensorimotor impairment offered by these devices. For example, in the event of an accident resulting in motor disturbances (e.g., cerebral vascular accident), muscular capacities decrease and may even be lost in the affected limb(s). Frequently, there is a decrease in physical strength and loss of muscle tone. Faced with these symptoms, the patient is forced to produce significant effort, which in turn can lead to loss of motivation and withdrawal from a rehabilitation program. In this context, robotic rehabilitation devices make it possible to ensure the continuity of the rehabilitation program. This is achieved through programs that run on embedded systems, which allow the practitioner to prepare longer and more intense exercises that are not subject to the patient's endurance capacity. Indeed, the majority of robotic-based rehabilitation devices are used as assistance tools. An intensive rehabilitation program combining shoulder/elbow training assisted by the InMotion 2.0 robot and conventional occupational therapy led to an improvement in shoulder and elbow movements [29]. Also of note, repeated movements by a robotic rehabilitation device improve brain plasticity. Architecturally, these devices are designed to offer the patient a progressive adaptation to physical capacities.

The objective of robotic rehabilitation devices to increase the autonomy of patients is recognized in Morocco, and there have been bridges built between specialists in rehabilitation and physical medicine and research laboratories in computer science, robotics, and artificial intelligence. This type of union has enabled the design and development of new robotic rehabilitation tools. In the first example, a Moroccan research team has developed an electric wheelchair control approach based solely on the patient's cerebral electrical signals [30]. This approach is based on controlling the chair's movement direction based on the patient's eye movements. The patient's eye movements are detected by the brain electrical signal measured using electroencephalogram (EEG) sensors. This approach, developed in a laboratory context, has shown a control precision of more than 85% [30]. In a second example of a project from a Moroccan university, a robotic hand exoskeleton was designed to allow rehabilitation of the hemiplegic hand. In this example, several modes of exoskeleton control are proposed. In the first mode, the therapists define the rehabilitation program for the patient through a graphical interface. The graphical interface proposes to select which finger of the hand the practitioner can work. It also offers the option of choosing the duration of the training as well as the speed of rotation of the dedicated servomotors. In the second mode, the developed system offers the patient control of the exoskeleton based on residual muscle electrical signals.

In the context of rehabilitation in general, a team of Moroccan researchers is working on the automation of a process making it possible to recognize the levels of mental effort of an individual during the performance of a cognitive task [31–34]. Using EEG sensors and data analysis algorithms, the proposed approach manages to detect an individual's variations of mental effort at more than 79%. Researchers continue to develop their approach to exploit it as a tool for an objective assessment of functional rehabilitation in patients with mental disorders linked to cerebral palsy.

It is true that for the moment, in Morocco, very few rehabilitation centers and practices use robotic devices for their activities. The ultimate goal would be to make them accessible to everyone who needs it. Faced with this objective, two obstacles exist. First, there is the patient's investment. Often it takes months and years of training to see significant improvements with the robotic device. Such mastery requires significant cognitive expenditure, which often forces the patient to abandon the use of robotic rehabilitation devices. Durability, maintenance of the device, and device care are always a particular struggle in Morocco. In addition to the difficulties of using robotic rehabilitation devices, there is also the difficulty of appropriating adequate devices for their disability, and this is because of their costs, which for the moment remain prohibitive for the Moroccan population. Therefore it is reasonable to focus efforts and direct policies toward the way allowing rehabilitation services to be offered using robotic devices only in rehabilitation centers. Given that with the advent of the Internet of Things, the objective behind the integration of robotics in rehabilitation processes is to enable their use in daily life, and the high prices of these devices (around a few thousand dollars) make them accessible only to a certain category of the Moroccan population.

28.5 Conclusion

The need for any kind of stroke rehabilitation in Morocco is profound. The role of robotic rehabilitation aids holds great potential to decrease the need gap between the low number of therapists and a large number of patients. The Moroccan government needs to increase access of stroke survivors to proper rehabilitation in Morocco to improve their health and recovery and to facilitate the decrease of the level of disability in Morocco.

References

[1] Britannica. Morocco. Available from https://www.britannica.com/place/Morocco (Accessed on 5 May 2022).

[2] United Nations. United Nations Human Rights Treaty Bodies, Acceptance of the inquiry procedure for Morocco. https://tbinternet.ohchr.org/_layouts/15/TreatyBodyExternal/Treaty.aspx?CountryID = 117&Lang = N (Accessed on 7 May 2022).

[3] Institute for Health Metrics and Evaluation. Global Burden of Disease, Morocco Country Profiles. Available from https://www.healthdata.org/morocco (Accessed on 2 April 2022).

[4] World Health Organization. Noncommunicable Diseases (NCD) Country Profiles. Available from https://www.who.int/news-room/fact-sheets/detail/noncommunicable-diseases (Accessed on 2 April 2022).

[5] Chadli S, Taqarort N, El Houate B, Oulkheir S. Epidemiological transition in Morocco (1960−2015). Transition épidémiologique au Maroc (1960−2015). Med Sante Trop 2018;28(2):201−5. Available from: https://doi.org/10.1684/mst.2018.0800.

[6] Ministry of Health of the Kingdom of Morocco and World Health Organization. Results of the STEPwise study: National survey of common risk factors for non-communicable diseases 2017−2018.

[7] Belabbes S. The truth about health in Morocco. Occasional paper series. 2020, Wilson Centre.

[8] Yassine A, Hangouche AJ, El Malhouf N, Maarouf S, Taoufik J. Assessment of the medical expenditure of the basic health insurance in Morocco. Pan Afr Med J 2020;35:115.

[9] Hajjioui A, Fourtassi M, Boujraf S. Spinal cord injury in the Moroccan healthcare system: a country case study. IBRO Neurosci Rep 2021;10:62−5.

[10] Hajjioui A, Abda N, Guenouni R, Nejjari C, Fourtassi M. Prevalence of disability in Morocco: results from a large-scale national survey. J Rehabil Med 2019;51(10):805−12. Available from: https://doi.org/10.2340/16501977-2611.

[11] Hajjioui A, Fourtassi M, Nejjari C. Prevalence of disability and rehabilitation needs amongst adult hospitalized patients in a Moroccan university hospital. J Rehabil Med 2015;47(7):593−8. Available from: https://doi.org/10.2340/16501977-1979.

[12] Bright T, Wallace S, Kuper H. A systematic review of access to rehabilitation for people with disabilities in low- and middle-income countries. Int J Env Res Public Health 2018;15(10):2165. Available from: https://doi.org/10.3390/ijerph15102165.

[13] Ruger JP, Kress D. Health financing and insurance reform in Morocco. Health Aff (Millwood) 2007;26(4):1009−16. Available from: https://doi.org/10.1377/hlthaff.26.4.1009.

[14] Gillen G. Cerebrovascular accident (stroke). In: Pendleton HM, Schultz-Krohn W, editors. Pedretti's occupational therapy: practice skills for physical dysfunction. eighth ed. Elsevier Inc; 2018. p. 809−40.

[15] Hajjioui A, El Hajri Y, Fourtassi M. Outcome and Health-Related Quality of Life after stroke in Morocco. Abstracts of Scientific Papers and Posters Presented at the 13th ISPRM World Congress, Kobe, Japan, June 9−13, 2019. J Int Soc Phys Rehabil Med 2020;3(Suppl S2):489−1157.

[16] Engels T, Baglione Q, Audibert M, Viallefont A, Mourji F, El Alaoui Faris M, et al. Socioeconomic status and stroke prevalence in Morocco: results from the rabat-casablanca study. PLoS One 2014;9(2):e89271.

[17] World Health Organization (2018). Noncommunicable Diseases (NCD) Country Profiles. Retrieved from https://www.who.int/news-room/fact-sheets/detail/noncommunicable-diseases.

[18] Kharbach A, Obtel M, Lahlou L, Aasfara J, Mekaoui N, Razine R. Ischemic stroke in Morocco: a systematic review. BMC Neurol 2019;19(1):349.

[19] American Stroke Association. What is Hemorrhagic Stroke? Available from: https://www. stroke.org/en/about-stroke/types-of-stroke/hemorrhagic-strokes-bleeds (Accessed on 23 October 2022).

[20] Johns Hopkins Medicine. Types of Stroke. Available from. https://www.hopkinsmedicine. org/health/conditions-and-diseases/stroke/types-of-stroke; 2022 (Accessed on 23 October 2022).

[21] De Los Ríos la Rosa F, Khoury J, Kissela BM, Flaherty ML, Alwell K, Moomaw CJ, et al. Eligibility for intravenous recombinant tissue-type plasminogen activator within a population: the effect of the European Cooperative Acute Stroke Study (ECASS) III Trial. Stroke. 2012;43(6):1591−5.

[22] Lees KR, Bluhmki E, von Kummer R, Brott TG, Toni D, Grotta JC, et al. Time to treatment with intravenous alteplase and outcome in stroke: an updated pooled analysis of ECASS, ATLANTIS, NINDS, and EPITHET trials. Lancet. 2010;375(9727):1695−703.

[23] Ministry of Health Morocco. Health in Numbers. Available from: https://www.sante.gov. ma/Documents/2021/12/Sante%20en%20chiffres%202019%20.pdf. 2020.

[24] Hajjioui A, Fourtassi M. Adult rehabilitation after stroke: from injury to inclusion. French book, IpnPub, 2019, p162.

[25] United Nations. United Nations Human Rights Treaty Bodies, Acceptance of the inquiry procedure for Morocco. Available from: https://tbinternet.ohchr.org/_layouts/15/ TreatyBodyExternal/Treaty.aspx?CountryID = 117&Lang = EN; n.d. (Accessed on 23 October 2022).

[26] World Health Organization. The impact of the COVID-19 pandemic on noncommunicable disease resources and services: results of a rapid assessment. Available from: https://apps. who.int/iris/handle/10665/334136 (Accessed on 27 October 2022).

[27] Nafai S, Barlow K, Stevens-Nafai E. OT in Morocco: sustaining service learning trips through telehealth. OT Pract 2017;22(2):20−2.

[28] Kalunga EK, Chevallier S, Rabreau O, Monacelli E. Hybrid interface: integrating BCI in multimodal human-machine interfaces. In Proceeding of the IEEE/ASME International Conference on Advanced Intelligent Mechatronics, AIM, 2014, pp. 530−535, 6878132.

[29] Pila O, Duret C, Laborne F, Gracies J, Bayle N, Hutin E. Pattern of improvement in upper limb pointing task kinematics after a 3-month training program with robotic assistance in stroke. J Neuroeng Rehabilitation 2017;14(1):105. Available from: https://doi.org/10.1186/ s12984-017-0315-1.

[30] Zammouri A, Zerouali S. Prototype of a BCI-Based Autonomous Communicating Robot for Disability Assessment. In Maleh U, Alazab YM, Khan Pathan A-S. Machine Intelligence and Data Analytics for Sustainable Future Smart Cities. in Studies in Computational Intelligence. 2021; pp. 215−227. ISBN: 978-3-030-72064-3.

[31] Zammouri A, Zrigui S, Moussa AA. Evaluating usability of brain signals for content adaptation in intelligent tutoring systems. In Proceeding of Colloquium in Information Science and Technology, CIST; 2016: pp. 517−520.

[32] Zammouri A, Moussa AA. Eye blinks artefacts detection in a single EEG channel. Int J Embedded Syst 2017;9(4):321−7. Available from: https://doi.org/10.1504/IJES.2017.086126.

[33] Zammouri A, Chraa-Mesbahi S, Ait Moussa A, Zerouali S, Sahnoun M, Tairi H, et al. Brain waves-based index for workload estimation and mental effort engagement recognition. J Phys Conf Ser 2017;904(1):012008. Available from: https://doi.org/10.1088/1742-6596/904/1/012008.

[34] Zammouri A, Ait Moussa A, Mebrouk Y. Brain-computer interface for workload estimation: assessment of mental efforts in learning processes. Expert Syst Appl 2018;112:138−47. Available from: https://doi.org/10.1016/j.eswa.2018.06.027.

Part IV

Barriers, best practices, and recommendations for penetration

Chapter 29

Psychosocial dimensions of robotic rehabilitation for stroke survivors

Shovan Saha

Department of Occupational Therapy, Manipal College of Health Professions, MAHE, Manipal, Karnataka, India

Learning objectives

At the end of this chapter, the reader will be able to:

1. Identify the psychosocial dimensions of stroke.
2. Understand neurorehabilitation for patients with stroke.
3. Explain the psychosocial elements of robotic rehabilitation for patients with stroke.

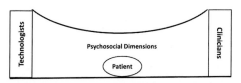

Developing technology that is embedded in the psychosocial contours of the patients may help to transform the concave arch to a convex one.

Shovan Saha

29.1 Introduction

There has been a quantum growth in technical knowledge in recent decades, making it possible to find technological solutions to long-term healthcare problems. These technologies are redefining the landscape in an unprecedented way, opening up new opportunities, thus giving rise to technological optimism and a belief that these advances could solve some of the most compelling challenges. Technological innovations have far-reaching social implications, as echoed by the public reaction, the erosion of privacy, the

Rehabilitation Robots for Neurorehabilitation in High-, Low-, and Middle-Income Countries.
DOI: https://doi.org/10.1016/B978-0-323-91931-9.00007-4

shortening of innovation cycles, and the rapid diffusion of new technology. It is possible for humanity to devise solutions that previously existed only in the realms of imagination [1]. Two important factors that must be considered are social responses to technological innovations and whether the technology is engaging and occupying.

29.1.1   Social responses to technological innovations

Jutai and Day [2] referred to "psychosocial" as "factors within the person and factors attributable to the environment that interact and affect the psychological adjustment of individuals who have a disability." Psychosocial adaptation to disability has been viewed as being composed of both global- and disability-specific indicators.

Advocates of new technologies have tended to focus largely on scientific and technical dimensions. However, the adoption of new technologies is predominantly a process of public education and social learning. Public perceptions about the risks and benefits of novel technologies are difficult to fully understand without knowing the intuitive aspects of the human mind. There is growing evidence that instinctual expectations may render the human mind vulnerable to misinterpretations about new technologies [1].

There are many concerns about the use of technology in rehabilitation, especially in the domain of stroke rehabilitation. Therefore there is a need for careful deliberation by clinicians, technologists, and scientists for improved outcomes.

29.1.2   Technology should be engaging and occupying. So, what is "occupying"?

The term "occupation" can be analyzed using three major concepts, as described by Breines: egocentric, exocentric, and consensual [3].

1. To be "occupied" represents the egocentric aspect. One is engaged in both mind and body when occupied, sometimes alternatively and sometimes simultaneously. Mind and body integrate when performing a task, and the body performs when the mind is occupied.
2. To "occupy" represents the exocentric aspect. When engaged in tasks, one occupies time and space and interacts with the elements of the environment; also these elements themselves interact. That is how an individual understands how to adapt to the world, and in turn, the environment adapts to the individual.
3. "Occupation" represents the consensual aspect and is described in terms of vocation or work. It also reflects endeavors in which one collaborates, contests, or otherwise engages with or for others socially.

Occupation happens through participating in goal-directed activities, assuming intention and purpose on the part of the individual. The activity's motor requirements, the cognitive ability needed to perform the task, and the inherent affective qualities are all part of the picture. It is also necessary to know the tangible components of any activity, as well as its interactive effects because human beings do not act in isolation. They need tools, environments, and others with or for whom they act [3].

The mind and body interact to make choices and perform occupations with skill, ease, and effectiveness. In addition, when these performances meet expectations in a culturally relevant environment, then it results in satisfaction. On the other hand, when therapists' suggestions and patients' lifestyles are in conflict, it may be difficult to resolve the cognitive discord and may produce dissatisfaction. Therefore it is important to recognize the meaning of activity in the individual's life and understand the therapist's role in aiding a person toward a healthy activity.

29.2   The psychosocial dimensions of stroke

Globally, stroke is the third major cause of death and disability, and in 2019, there were 12.2 million cases reported with 101 million prevalent cases [4]. The experience of stroke has far-reaching consequences on the social, physical, and emotional dynamics [5]. The sheer scale of the stroke population globally and the number of people directly or indirectly involved speaks to the enormity of the problem, leading to very complex interactions both at the individual and societal levels. The matter is complicated due to its chronic nature.

Approximately, 85% of stroke survivors recover partially, while 35% are left with serious disabilities [6]. Hemiparesis or hemiplegia is the predominating outcome, leading to movement limitations, specific muscle weakness, tonal abnormality, abnormal postural adjustments, synergistic movement, lack of coordination, and loss of sensitivity [7]. The disabilities affect various aspects of life including gross and fine motor ability, walking, activities of daily living (ADLs), cognition, and speech [8].

29.2.1   Stroke and emotional challenges

Stroke is a traumatic event and is likely to induce an emotional reaction, ranging from a state of grief to more serious emotional disorders. Depression and anxiety seem to be the most common psychological problems following a stroke, with approximately 30%−40% affected. Emotions experienced include shock, disbelief, anger, and frustration [9]. Fatigue is another common complaint following stroke and presents as increased physical and mental fatigability, irritability, poor stress tolerance, and concentration and memory deficits [10]. These symptoms affect both the stroke survivors and their caregivers, and negatively impact the quality of life (QoL) and rehabilitation outcomes [11].

29.2.2   Stroke and psychosocial challenges

The psychosocial challenges of stroke are extensive and complex, including the considerable long-lasting impact on social relationships, mood, identity, QoL, and return to work. Patients and their families struggle to come to terms with their life situations; the physical symptoms are unfamiliar and confusing. The emotional responses and functional progress are hard to predict. They express surprise at the time that it takes to produce small changes despite their strong commitment. Their concerns gradually shift from bodily recovery to struggling to reestablish a functioning and meaningful everyday life, a life in which they seek to assume previous roles, valued activities, and significant relationships, and a life to negotiate their self-understanding and reestablishing their identity in light of the changes brought about by the stroke. The patients also seemed vulnerable to psychological stress during periods of transition from acute care hospitals and attempting to resume new roles. They also experience psychosocial challenges when attempting activities outside the home and when acknowledging that it would probably have a lasting impact on their life [12].

Individuals with stroke also experience major concerns about well-being, leading to a sense of disarray, lack of control, and a sense of incoherence in life. Psychosocial health and well-being are interconnected concepts [13]. Psychosocial well-being has four dimensions: (1) basic mood of joy, pleasure, and well-being, (2) participation and engagement in meaningful activities beyond oneself, (3) good social relations and a feeling of love and being loved in a mutual relationship(s), and (4) a self-concept characterized by self-acceptance, usefulness, and a belief in one's own abilities.

The experience of stroke and its consequences has been shown to cause sudden disruption of the survivor's perceived world. It often leads to a range of complex and multifaceted issues like difficulty in coping, adapting, and adjusting; experiencing the loss of valued roles and pursuing leisure activities; loss of confidence; a profound loss of self; a feeling of continued separation from their pre- and poststroke lives; jeopardized family relationships; loss of income; and reduced social contact [5].

It is often assumed that formal rehabilitation is considered complete when a functional plateau is reached, after which little or no recovery occurs. Stroke survivors usually do not receive any assessment or management of their psychosocial functioning. Therefore the diversity and complexity of problems faced by these individuals demand a shift in service provision from being largely focused on early rehabilitation stages of recovery to taking cognizance of the psychosocial consequences on quality of life in the longer term [5].

29.3   Neurorehabilitation

Neurological rehabilitation is a process that optimizes a stroke survivor's social participation and develops a sense of well-being. In neurorehabilitation, the emphasis is on the patient as a unified whole. The focus is also on social functioning, health, and well-being and is not only restricted to patients who may partially or completely recover but also applies to all living with long-term consequences [14]. It emphasizes the deficits, the limitation of activities, and participation restriction and constitutes a holistic approach [15].

The rehabilitation process can be described under three key approaches: (1) reducing disability; (2) acquiring new skills and strategies to maximize activities; and (3) altering the physical and social environment, so that there is minimal consequent impairment. It is a patient-centered approach involving goals set by the patients based on their own understanding of problems. This approach facilitates greater self-determination to control and enhances the person's potential for active participation [15].

29.3.1   Perspectives of motor learning

Motor learning is "a set of processes associated with practice or experience that leads to relatively permanent changes in the ability to produce skilled action." It is experience-dependent neural plasticity and is influenced by various factors such as specificity of a task, intensity, repetition, and timing [8]. There is increasing evidence that following stroke the motor system still remains plastic and can be modulated by motor training. Neural adaptation leads to more robust motor neuron recruitment and transfer of function to unaffected adjacent or correlated areas. It also leads to the formation of new synapses and strengthening of redundant synapses, increased dendritic sprouting and myelination, and reorganization of cortical and non-cortical representations [7]. Thus there is a need for precise training methods implementing intense multimodal stimulation to induce neural adaptations for enhanced motor and functional recovery.

Among different factors, feedback is one of the key components influencing the acquisition of motor skills. It is the information that is received because of performance. Feedback can be intrinsic or extrinsic. Intrinsic feedback is experienced directly by the individual (e.g., sensory, visual, auditory), and extrinsic (augmented) feedback is provided by an external source, such as a therapist providing verbal or physical guidance. Extrinsic feedback informs the individual about the success and failure of a task or about the quality of performance [8].

29.3.2   Role of motivation in stroke rehabilitation

Motivation is the process whereby people are driven to achieve their own goals and is considered a key element to stroke rehabilitation. Patients who have high motivation can achieve better outcomes as compared to those who have low motivation. It is a multi-construct phenomenon involving various key components including patient characteristics (anxiety, age, personality traits, socioeconomic status), social factors (quality of the therapist and patient–therapist interaction), and an engaging rehabilitation environment [16]. According to goal-setting theory, motivation can be enhanced by setting small, realistic, manageable, and well-specified goals or targets for individual patients. However, they also need to be of the just-right challenge for the patient to be engaged. Some of the other main contributors to positive motivation include acceptance of the conditions, individualized care, control of one's actions, overcoming uncertainty, and receiving timely feedback [17].

29.3.3 Feasibility challenges in neurorehabilitation

Neurorehabilitation remains one of the most important phases of poststroke management [18], but is faced with several barriers. An inadequate amount of rehabilitation therapy, in terms of intensity and repetition, is the key challenge of conventional rehabilitation programs. The clinical manpower for poststroke rehabilitation is predominantly concentrated at the inpatient facility, compared with that in the long-term service for patients with chronic stroke. There is insufficient manpower even in developed countries with rising stroke populations [16]. Traditional rehabilitation methods for stroke rely on laborious manual procedures carried out by therapists. It is often time-consuming and requires constant supervision, thus increasing their economic cost and leading to reduction in rehabilitation time [19]. The presence of cognitive decline, low mood, impaired communication abilities, sensory impairments, visual deficits, and perceptual disorders may impact participation in rehabilitation [18].

29.3.4 Psychosocial challenges in stroke rehabilitation

Rehabilitation of stroke survivors is mainly focused on the recovery of impaired movements and functional deficits in an effort to reduce disability. Many non-motor deficits also result in considerable poststroke disability, negatively impact participation in everyday activities, and influence varying degrees of stroke recovery [18]. Unique challenges are posed when stroke occurs in younger patients, who have responsibilities as the breadwinners of their households. The life situations of younger stroke patients are intrinsically different from those of older stroke survivors. Yet despite minimal physical impairments, only a few of the young stroke patients return to work. The financial impact from years of rehabilitation and the difficulty in performing normal productive activities are stressors for both the patient and family [5]. It is essential to explore the stroke patients' psychosocial aspects that increase motivation toward rehabilitation and thereby enhance their

QoL. This is further challenged by the fact that stroke rehabilitation is often prolonged and extended. It makes it difficult to identify meaningful and motivating treatment tasks that may be adapted and graded to frame the rehabilitation program [10]. Creating an appropriate physical activity intervention for stroke survivors that addresses a broader range of important psychological factors has always been a challenge.

29.3.5 Challenges in upper-extremity stroke rehabilitation

Restoring upper-extremity function is one of the key priorities in poststroke management. As compared to the lower extremity, upper-extremity impairments are more likely to result in activity restrictions. The tasks that involve the arm and hand often require coordinated and precise motor control [8]. Approximately, 80% of stroke survivors report upper-extremity limitation in ADLs. However, only less than 25% of them achieve limited recovery of their affected limb post-rehabilitation [16]. Importantly, there also is paucity of rehabilitation strategies for individualized dissociated finger movements. The lack of recovery in the affected upper extremity after stroke is often due to learned nonuse, typical posturing, tonal abnormality, weakness of muscles, imbalance or muscle tightness, joint stiffness or deformity, diminished circulation, and pain [10].

The rehabilitation aim in poststroke patients is to facilitate functional recovery, restore independence, and early reintegration into premorbid social and domestic life. Hence, there is an imminent need for novel technologies to improve the effectiveness of stroke rehabilitation outcomes.

29.4   Robotic stroke rehabilitation

With the growing demands of physical rehabilitation services globally due to improved stroke survival rates, the role of technology as a therapeutic medium is becoming significant. Robotic systems are increasingly indicated to address unmet needs, ease therapists' burden, and address budgetary constraints while improving patients' QoL. The shortage of therapists and caregivers assisting individuals with physical disabilities at home is another emerging concern. Robotic devices create the possibilities to address these problems, as noted in findings of scientific work over the last couple of decades.

29.4.1   Psychosocial perspective of conventional therapy and structured robotic therapy

As covered in detail in the first section of the book, especially in Chapter 5, robotic therapy achieved similar motor improvements comparable to those of the conventional treatment. Poststroke robotic therapy could be an economical

alternative to conventional rehabilitation services, especially in case of insufficient manpower. However, stroke patients who received therapy in clinical settings had higher motivation compared to those who were in research setups
[16]. The approach in a typical clinical setting is often more realistic and
human-centric and has flexibility in the intervention approaches.

Patients in the clinic achieved better improvements in ADLs although
having lower training frequency per week, although it is possible that the
patients in the clinic setup performed daily practice by themselves besides
the treatment in the clinic. In the clinical setting, the therapists as part of
their practice culture always suggest and encourage stroke patients to practice movements and ADLs, such as dressing, self-feeding, and bathing by
using their affected limb. In contrast, however, the research staff as a routine
do not make any practice suggestions. Voluntary exercises were found to be
an effective training method in enhancing motor recovery, compared to the
forced exercise group, where they achieved the least recovery [16].

Patients in robotic research groups have been optimistic about having specialized robotic therapists. In their presence, patients did not show any concerns about their safety and felt that they were benefiting from the robotic
therapy. However, they preferred a human therapist if they had an option [20].
Majority of the positive findings on robot-assisted studies were based on
research-oriented clinical trials and not necessarily in a real clinical setup. The
translation of rehabilitation effectiveness from structured research setups to
more flexible clinical services has not yet been rigorously evaluated [16].

29.4.2 Psychosocial aspects in robotic rehabilitation

Several robotic technologies from low-cost home rehabilitation devices to
expensive robots with multiple actuated degrees of freedom have been developed for motor rehabilitation. To take full advantage of rehabilitation robots,
it should deliver equivalent clinical therapy doses. Participating in longer
and more intensive therapy is often based on patients' desire to participate in
it. The lack of motivation is a problem, particularly in-home rehabilitation.
For better compliance and higher exercise intensity, it is important to find
ways to increase patient motivation [21]. Many patients have persistent neurological impairments despite intensive rehabilitation limiting activities and
restricting social participation. It has been observed that rehabilitation robots
used to rehabilitate motor impairments lead to improvement in social participation associated with improvement in motor control and manual ability.

The main rehabilitation goal for the lower limb after stroke is becoming
independent in walking. Among the available supporting gait training
machines, the role of electromechanical devices is significant. In contrast
with the effort of one physical or occupational therapist alone, these devices
for the lower limb can provide nonambulatory patients with intensive, high
repetition practice of complex gait cycles. These robotic devices can reduce

the effort applied by the therapists; in fact, as a result of these robots, therapists no longer need to set the paretic limbs or assist trunk movements [7]. Regarding the lower limb, there is evidence of the beneficial effects of electromechanical devices for gait training after stroke, but their relatively high cost limits their usage. Approximately, 35% of survivors with initial paralysis of the leg do not regain useful function, and 20%−25% of all survivors are unable to walk without full physical assistance [22].

The conventional rehabilitation of the lower limbs often would require at least two therapists to train a patient to walk, and the pace and pattern of walking may not be consistent. It is also physically strenuous for the therapists to sustain the exercise over extended periods, thus affecting the rehabilitation progress of the patient. The labor-intensive nature of conventional therapy places a great physical and emotional strain on therapists. Coupled with the requirements of stroke patients for medical care and intensive rehabilitation exercises, providing an optimal rehabilitation program would place a huge strain on time and the organizational budget. Therefore, it is anticipated that with robotic assistive devices, better rehabilitation progress can be achieved for patients, together with the alleviation of time and physical demands on therapists. One study found that robotic training had a comparable treatment effect to conventional training, but for severely impaired lower-limb patients, a statistically significant difference favoring the experimental group was found [23].

Resistance to new technologies is often frowned upon as a temporary phenomenon that is inevitably overcome by technological progress. When society perceives that the risks are likely to occur in the short run and the benefits will only accrue in the long run, then the new technology is likely to face opposition. Technologies may elicit negative responses if they appear to challenge the perceptive view of the natural world. There may be a cultural concern that the new technology is impure or dangerous because it does not fit into accepted social or ecological patterns [1].

The acceptability of a device by users is defined as "demonstrable availability" to use technology and the way people accept, perceive, and adopt technology use. The barriers to the acceptability of a novel technology-based service in healthcare can be classified in terms of technological, behavioral, organizational, and economical [24]. The technological barrier consists of an unwillingness to learn the procedure needed. The human behavior barrier comprises fear of innovation, distrust, and concern about personal data privacy. The organizational barrier is represented by patients' resistance to change. Finally, the economic barriers are represented by the costs of the procedure and the inability to demonstrate cost-saving or clinical benefits to the patient.

The aim of rehabilitation training is to maximize the patient's attention and effort and not only to increase the number of repetitions. Exercises that are monotonous provide worse retention of skill, as compared to adaptive therapy, and assistance as needed and provide better results than fixed pattern therapy. Robotic therapy can possibly reduce recovery if it encourages "slacking" since the patient may reduce attention and effort due to the use of adaptive algorithms [25].

The Usability, Social acceptance, User experience, and Societal impact (USUS) Evaluation Framework for human−robot interaction provides a promising, comprehensive view of the psychosocial aspects of human−machine interaction by incorporating the following categories of evaluation factors: usability, social acceptance, user experience, and societal impact. The sense of usability of robotic device is felt by experiencing the feeling of effectiveness and efficiency of the device, the ease of learning, and the flexibility of the operability. The sense of social acceptance is experienced when the device is consistent with performance expectancy, overall positive attitude, and developing a sense of attachment to the device. The user experience is facilitated with a sense of embodiment and security, an emotional connect, and human-oriented perception. The sense of societal impact is prompted by the introduction of robotic agents' consequences on the social life in the context of a given community [26]. Several barriers have been identified, for rehabilitation robotics. The first barrier is the lack of effective communication between engineers and therapists in the planning stage of designing robotics aids. Second, often devices are incredibly complicated, from both an engineering and a usability point of view. In fact, "simple-to-use" devices are more likely to be accepted by clinicians than those that have long setup times or require multiple therapists. Another important consideration is to promote adherence; therefore it is necessary to provide adequate and motivating feedback [15].

The therapeutic application of robotics in a clinical setup is an immense challenge. In addition, there is a need to describe the appropriate populations for robotics and to develop evidence-based treatment protocols to maximize their use.

29.4.3 Clinician's perspective

Neurorehabilitation is often based on therapists' expertise, with stiff competition among different robotic centers, generating substantial uncertainty about what exactly a neurorehabilitation robot is expected to do [15]. The cultural gap between technology providers, therapists, and end users is becoming smaller, due to the gradual rise in information sharing among users [7].

The acceptability of robotics by clinicians is a significant barrier to their use. Therapists are trained to assess patients' motor dysfunction by touch and to adjust treatment forces based on feedback. Although haptic robotics may be able to perform some of the same functions, the effectiveness of robotic systems has not been compared to a trained, experienced therapist in this area. Therapists also liked some of the treatment programs available but wanted the ability to program customized movement patterns to use in the treatment for specific individuals [20].

Although the objective of the robotic intervention is to effectively maximize the benefit of the limited therapist resources, there might be concern among some therapists that robotic-based rehabilitation may be a threat to their jobs. There are concerns that therapists may stop using devices if setup takes more than 5 minutes [25]. Thus newly developed devices for physical training should be intuitive, easy to use, fast to set up, and have been reasonably priced. Many rehabilitation professionals mistakenly expect a significant increase in muscle hyperactivity and shoulder pain due to intensive training [27]. There are concerns that clinicians and neuroscientists are not in agreement on how to communicate with the engineers about what they want the robotic machines to do [28].

29.5   Conclusion

In conclusion, more research and innovation are needed to establish whether a functional transition is happening from robotic laboratory to service settings and to have a robust understanding of whether ADL tasks are truly enhanced by robotic training. Solutions that improve the interaction between engineers, clinicians, and patients in the robotic field need to be spelled out in a more explicit way. Moreover, the successful clinical implementation of this promising field should be able to negate the opposition from the concerns raised about the possibility that robots could "dehumanize" patient rehabilitation or replace the human workforce [7].

Acknowledgment

I appreciate the help of Ms. Dola Saha (Asst. Prof., Health Information Management, MCHP) for proofreading and reviewing the manuscript.

Conflict of interest

The author has no conflict of interest whatsoever.

References

[1]   Juma C. Innovation and its enemies: why people resist new technologies. Oxford University Press; 2016.
[2]   Jutai J, Day H. Psychosocial impact of assistive devices scale (PIADS). Technol Disabil 2002;14(3):107−11.
[3]   Breines EB. Occupational therapy: activities for practice and teaching. 1st ed. Whurr Publishers Ltd; 2004.

[4] Feigin VL, Stark BA, Johnson CO, Roth GA, Bisignano C, Abady GG, et al. Global, regional, and national burden of stroke and its risk factors, 1990—2019: a systematic analysis for the Global Burden of Disease Study 2019. Lancet Neurol 2021;20(10):795—820.

[5] Thompson HS, Ryan A. A review of the psychosocial consequences of stroke and their impact on spousal relationships. Br J Neurosci Nurs 2008;4(4):177—84.

[6] Truelsen T, Piechowski-Józwiak B, Bonita R, Mathers C, Bogousslavsky J, Boysen G. Stroke incidence and prevalence in Europe: a review of available data. Eur J Neurol 2006;13(6):581—98.

[7] Poli P, Morone G, Rosati G, Masiero S. Robotic technologies and rehabilitation: new tools for stroke patients' therapy. BioMed Res Int 2013;2013. Available from: https://doi.org/10.1155/2013/153872 Oct.

[8] Liu LY, Li Y, Lamontagne A. The effects of error-augmentation versus error-reduction paradigms in robotic therapy to enhance upper extremity performance and recovery poststroke: a systematic review. J Neuroeng Rehabil 2018;15(1):1—25.

[9] Kneebone II, Lincoln NB. Psychological problems after stroke and their management: state of knowledge. Neurosci Med 2012;3(01):83—9. Available from: https://doi.org/10.4236/nm.2012.31013.

[10] Abd El-Kafy EM, Alshehri MA, El-Fiky AA, Guermazi MA. The effect of virtual reality-based therapy on improving upper limb functions in individuals with stroke: a randomized control trial. Front Aging Neurosci 2021;13. Available from: https://doi.org/10.3389/fnagi.2021.731343 Nov.

[11] Kim JS. Post-stroke mood and emotional disturbances: pharmacological therapy based on mechanisms. J Stroke 2016;18(3):244.

[12] Kirkevold M, Bronken BA, Martinsen R, Kvigne K. Promoting psychosocial well-being following a stroke: developing a theoretically and empirically sound complex intervention. Int J Nurs Stud 2012;49(4):386—97.

[13] Tengland PA. Health promotion or disease prevention: a real difference for public health practice? Health Care Anal 2010;18(3):203—21.

[14] Barnes MP. Principles of neurological rehabilitation. J Neurol Neurosurg Psychiatry 2003;74(suppl 4) iv3—7.

[15] Oña ED, Cano-de La Cuerda R, Sánchez-Herrera P, Balaguer C, Jardón A. A review of robotics in neurorehabilitation: towards an automated process for upper limb. J Healthc Eng 2018;2018. Available from: https://doi.org/10.1155/2018/9758939 Apr 1.

[16] Huang Y, Lai WP, Qian Q, Hu X, Tam EW, Zheng Y. Translation of robot-assisted rehabilitation to clinical service: a comparison of the rehabilitation effectiveness of EMG-driven robot hand assisted upper limb training in practical clinical service and in clinical trial with laboratory configuration for chronic stroke. Biomed Eng Online 2018;17:91. Available from: https://doi.org/10.1186/s12938-018-0516-2.

[17] Vourganas I, Stankovic V, Stankovic L, Kerr A. Factors that contribute to the use of stroke self-rehabilitation technologies: a review. JMIR Biomed Eng 2019;4(1):e13732.

[18] Brewer L, Horgan F, Hickey A, Williams D. Stroke rehabilitation: recent advances and future therapies. QJM: An Int J Med 2013;106(1):11—25.

[19] Meng Q, Xie Q, Yu H. Upper-limb rehabilitation robot: state of the art and existing problems. In: Proceedings of the 12th International Convention on Rehabilitation Engineering and Assistive Technology 2018 Jul 13 (pp. 155—158). Available from: https://www.researchgate.net/publication/338037336.

[20] Brewer BR, McDowell SK, Worthen-Chaudhari LC. Poststroke upper extremity rehabilitation: a review of robotic systems and clinical results. Top Stroke Rehabil 2007;14(6):22—44.

[21] Novak D. Promoting motivation during robot-assisted rehabilitation. In: Colombo R, Sanguineti V, editors. Rehabilitation robotics: technology and application. Academic Press; 2018. p. 149−58.

[22] Dobkin BH. Rehabilitation after stroke. N Engl J Med 2005;352(16):1677−84. Available from: https://doi.org/10.1056/NEJMcp043511.

[23] Lo K, Stephenson M, Lockwood C. Effectiveness of robotic assisted rehabilitation for mobility and functional ability in adult stroke patients: a systematic review. JBI Evid Synth 2017;15(12):3049−91.

[24] Mazzoleni S, Turchetti G, Palla I, Posteraro F, Dario P. Acceptability of robotic technology in neuro-rehabilitation: preliminary results on chronic stroke patients. Computer Methods Prog Biomed 2014;116(2):116−22.

[25] Maciejasz P, Eschweiler J, Gerlach-Hahn K, Jansen-Troy A, Leonhardt S. A survey on robotic devices for upper limb rehabilitation. J Neuroeng Rehabil 2014;11(1):1−29.

[26] Koumpouros Y. A systematic review on existing measures for the subjective assessment of rehabilitation and assistive robot devices. J Healthc Eng 2016;2016. Available from: https://doi.org/10.1155/2016/1048964 Oct.

[27] Duret C, Grosmaire AG, Krebs HI. Robot-assisted therapy in upper extremity hemiparesis: overview of an evidence-based approach. Front Neurol 2019;10:412. Available from: https://doi.org/10.3389/fneur.2019.00412 Apr 24.

[28] Patton J, Small SL, Zev Rymer W. Functional restoration for the stroke survivor: informing the efforts of engineers. Top Stroke Rehabil 2008;15(6):521−41.

Chapter 30

Human-centered design for acceptability and usability

Kevin Doan-Khang Bui[1], Michelle J. Johnson[2] and Rochelle J. Mendonca[3]

[1]*Selera Medical, Inc., Portola Valley, CA, United States,* [2]*Departments of Physical Medicine and Rehabilitation, Bioengineering, and Mechanical Engineering and Applied Mechanics, University of Pennsylvania, Philadelphia, PA, United States,* [3]*Department of Rehabilitation and Regenerative Medicine, Programs in Occupational Therapy, Columbia University, New York, NY, United States*

Learning objectives (minimum 2)

At the end of this chapter, the reader will be able to:

1. Identify barriers to the adoption of rehabilitation robotics in HICs and LMICs,
2. Apply human-centered design and biodesign health technology innovation methodologies to their own work.

30.1 Introduction

According to the World Health Organization, 74% of the total burden of disability (measured in years lived with a disability) are linked to conditions which could benefit from rehabilitation [1,2]. Although most disability is experienced in low- and middle-income countries (LMICs), access to rehabilitation services is a global challenge. As the global population continues to age and the prevalence of chronic conditions increases, the global need for rehabilitation will continue to rise.

Through the course of each of the countries spotlighted in this textbook, you hopefully have developed a deeper appreciation for how the landscapes and challenges related to rehabilitation and technology-assisted rehabilitation vary greatly by context. This then raises the question: how do we generate solutions that will be successfully implemented and sustained to improve health outcomes given the various factors that differ across the board? Without careful consideration of how these solutions are designed and implemented—particularly around

Rehabilitation Robots for Neurorehabilitation in High-, Low-, and Middle-Income Countries.
DOI: https://doi.org/10.1016/B978-0-323-91931-9.00025-6
457

their acceptability and usability in different contexts—these solutions will fail to gain traction. This chapter examines the design and deployment of robotic rehabilitation solutions through the lens of human-centered design (HCD), which is the approach of involving the perspective of all stakeholders in each stage of the problem-solving process. Stakeholders include anyone who may have a role— either directly or indirectly—in the adoption of solutions. These can include patients, care partners, clinicians, engineers, policymakers, and patient advocacy groups, just to name a few.

While a true HCD approach would not assume that rehabilitation robotics is the right solution straight away, there are many mindsets and methods that can still be utilized to understand the underlying need more fully. HCD is important to consider because it is a way to drive solutions that are desirable, feasible, usable, and viable. Additionally, the focus goes beyond the solution itself and includes strategies to improve its widespread use. This implementation part of the process is often severely overlooked or underestimated. An understanding of the systems and infrastructure needed for a given solution is important if the goal is to move it to clinical care.

30.1.1 Barriers to the adoption of rehabilitation robotics

30.1.1.1 Barriers in high-income countries

In high-income countries (HICs) such as the United States, many of the barriers to adoption are related to the real-world implementation of rehabilitation robotics. While the technical feasibility has often been demonstrated, adoption of the solution is slow. Without sufficient adoption, the viability of the solution is at risk. While there may be appropriate use cases for robotics to advance rehabilitation research, there are additional considerations in the broader health technology landscape that must be considered the moment one attempts to use these devices as clinical tools.

At a minimum, new health technologies must demonstrate similar or better efficacy at improving patient outcomes at lower costs compared to conventional strategies to be adopted. However, this consideration is often distinctly lacking in this field. A recent systematic review examined studies that attempted to evaluate the cost of robot-assisted neurorehabilitation against standard dose-matched conventional therapy from a hospital perspective [3]. They looked for studies that assessed cost minimization, cost-effectiveness, cost−utility, and cost−benefit. They defined these terms as follows:

- Cost minimization—the cost of providing robotic rehabilitation compared to the cost of conventional physiotherapy.
- Cost-effectiveness—cost minimization is done, but the outcome is presented as relative costs to achieve a unit of effect in terms of therapy benefits such as motor gains.

- Cost−utility—cost minimization is done, but the outcome is presented as relative costs to achieve a unit of utility measured in quality-adjusted life years.
- Cost−benefit—cost minimization is done, but the outcome is presented as relative costs to achieve a unit of monetary benefit such as therapist time and device costs.

They report that only five studies involving 213 patients looked at this cost activity with moderate methodological quality with the studies looking only at cost minimization and cost-effectiveness. None of the five studies were systematic in examining cost−utility or cost−benefit. Despite this, the review suggests that, overall, there is a cost−benefit to robotic rehabilitation.

A standard health economics way to measure the efficacy of robotic rehabilitation strategies compared to standard conventional strategies is by calculating what is called an incremental cost-effectiveness ratio (ICER), which reflects the health economic value of a new solution. An ICER is calculated by measuring both the difference in costs of a new solution compared to the standard of care, as well as the difference in outcome, often measured in quality-adjusted life years (QALYs). QALYs are a measure of disease burden that includes the quality and quantity of life lived, with a value of 1 representing perfect health and a value of 0 representing death [4]. While there is no universally accepted ICER that guarantees adoption, a range of $50,000−$75,000 USD per QALY added is often accepted as cost-effective in countries that take QALYs into account. Disability-adjusted life years (DALYs) are similar in concept to QALYs but are a measure of the total disease burden; they capture the number of years lived with a disability and years of life lost due to dying early. The WHO defines a cost-effective solution as somewhere between 1 and 3x the gross domestic product (GDP) per capita of a given country [5]. Given that there are established ways to measure QALYs and DALYs, this provides a method to measure the health economic value of rehabilitation robotics by using these metrics to calculate the ICER relative to a country's standard of care.

However, it is important to note that the health economic value of a solution may not always reflect its commercial value. Incentives across stakeholders may not always be aligned; thus the willingness to pay may not be there even if a solution has a compelling health economic value. For example, individual rehabilitation sessions are reimbursed by health insurance at a higher rate compared to group therapy, and thus a well-resourced rehabilitation provider will be more incentivized to prioritize individual rehabilitation over group therapy. Given this, an important factor to consider is who is going to pay for a solution, and why they should be compelled to pay for it (i.e., the value proposition). This will often be determined by how the solution fits into the economics of healthcare. Consequently, there is a need to clearly understand how the particular part of care that the solution addresses is coded, covered, and paid for.

Another overlooked factor is how the solution fits into the clinical workflow. This is often a barrier that is realized much later than it should be. If a solution is so novel that it requires specialized training, constant maintenance, a different workflow, or significant added time, this can severely hinder the adoption of a solution. Clinical adoption can also be tied to evidence and what the guidelines recommend.

30.1.1.2 Barriers in low- and middle-income countries

Disability leads to poorer health outcomes, lower education achievements, and less economic participation, which results in a perpetuating cycle of disability and poverty [6]. Despite evidence that rehabilitation is highly effective in improving clinical outcomes and quality of life, people residing in LMICs face specific barriers to accessing rehabilitation.

Barriers to the wider use of rehabilitation include inadequate policies and standards, negative attitudes toward disability, problems with service delivery, lack of accessibility, and a lack of data and evidence. Additionally, there is a lack of rehabilitation professionals to provide rehabilitation services. These professionals include physical and rehabilitation medicine doctors (physiatrists), physiotherapists, occupational therapists, speech and language pathologists, prosthetists, and orthotists. There are fewer than 10 physiotherapists per 1 million residents in many LMICs, whereas high-income countries often have several times more rehabilitation professionals [7]. Addressing these barriers to rehabilitation can provide a positive societal impact by building human capacity, improving the quality and affordability of services, and achieving the Sustainable Development Goal of ensuring healthy lives and promoting well-being for all [6].

However, most solutions are only available in high-income countries, with high costs often associated with limited market penetration of rehabilitation robotic systems. In addition to the high costs of these systems, other barriers to adoption in LMICs include a lack of training (for maintaining these systems), high duty or import taxes, and the lack of studies establishing the feasibility and cost-effectiveness of therapy in relevant settings.

30.1.1.3 Using human-centered design to address these barriers

At first glance, many of the barriers to adoption in both HICs and LMICs seem difficult to address. However, human-centered design (HCD) provides a framework to address these barriers. It can be used to identify clear problems, populations, and outcomes to target. When broken down in this way, solutions do not need to solve everything at once. Instead, by taking an intentional approach and targeting the most compelling areas, HCD can uncover key insights that can lead to powerful new solutions.

30.2   Human-centered design

Much of the material in this chapter will draw heavily from two frameworks around HCD and health technology innovation. The first is IDEO's framework for HCD [8]. The second is the biodesign health technology innovation process [9]. The objective here is to highlight key mindsets and methods from both that can be applied in the rehabilitation engineering space. However, this is not an exhaustive overview, and additional resources are provided at the end of the chapter for further reference. While these frameworks may read as a linear process, that is almost never the case when applied in the real world. Instead, these should be viewed as tools to utilize in an iterative process.

30.2.1   Mindsets for human-centered design

30.2.1.1   Creative confidence [10]

Many people often feel like they have never been creative or that it is an innate talent, but this is a common misconception. The creative confidence mindset reframes creativity in such a way that it is possible to bring an intentional approach to creative problem-solving.

30.2.1.2   Make it

The sooner an idea becomes tangible, the sooner it can be used to answer key questions. No matter the fidelity, something can always be learned, or additional questions raised. This also makes it easier to solicit feedback from various stakeholders, a key aspect of HCD.

30.2.1.3   Learn from failure

The chances of getting something right on the first try are slim. Thus being able to extract new insights, revisit assumptions, and revise the plan are keys to developing innovative solutions. Oftentimes, particularly in the field of rehabilitation robotics, failures are not examined closely enough, leading to many different attempts that look similar but have not unlocked a new insight that allows for a different way of thinking.

30.2.1.4   Empathy

Empathy is the emotional driver behind HCD. It focuses on the end user rather than the problem-solver. It requires a deep understanding of other people's lives, including their aspirations and worries. Every interaction is an opportunity to build empathy, and sharing stories is one key way in which empathy and connection can be built.

30.2.1.5   Embrace ambiguity

The assumption that there is a "right" way to do something limits the set of solutions available. However, not knowing the right answer allows for a more creative approach to problem-solving. Sometimes this can feel like taking a step back, but it also frees you from constraints that may have been baked in and gives you an opportunity to question certain assumptions.

30.2.1.6   Iterate

Many engineers are familiar with the design-build-test iterative cycle. However, this iterative process often stays in the lab or machine shop. A missing piece of the iterative cycle is the validation of the solution with key stakeholders throughout the process, not just at the end.

30.2.1.7   Humility

Particularly when working with a multidisciplinary team, with a patient population with a lived experience different from your own, or in a setting whose cultural values are different, different types of humility are needed. Intellectual humility gives you the ability to adapt to new information and is key in working on teams [11]. Confident humility is the belief that while you may not know the right solution, or even be working on the right problem, that you have the capability to solve it [12].

30.2.2   Methods for human-centered design

30.2.2.1   Inspiration/needs finding

The first step of HCD aims to build empathy with the various stakeholders to identify a compelling problem to target. This involves building a team, defining the need, and identifying the criteria needed to represent a meaningful solution.

- *Building a team*: Given the variety of problems that must be tackled, a variety of skills, experiences, and perspectives are key to building out a versatile team. The ideals of having a mixed, interdisciplinary team must be balanced by the realities of who is available. It is important to recognize the capabilities, strengths, and weaknesses of team members, as well as to share individual motivations. An example core team could consist of a local clinical expert, local engineering expert, and an external international partner with relevant experience. The population you are designing rehabilitation robotic solutions for should also be included as a core team member. While this may start as the core team, additional members can always be added when needed.
- *Need statement:* While the team may have a clear internal understanding of their project's objectives, it is important to be able to communicate this

to others concisely. One way to do this is through crafting a need state-ment. A need statement is a structured way to present the problem, rele-vant population, and desired outcome. It follows a template (a way to address [problem] in [population] that [outcome]) that conveys the purpose of the project in one sentence [9]. A solution should not be embedded in the need statement. The scope of need statements can vary greatly, ranging from incremental improvements to trying to address ambitious problems. Ultimately, the scope is up to the team, but the team should consider if and how people unfamiliar with the project are able to grasp the need based on reading the need statement.

- *Need criteria:* After a need statement has been crafted, the next step is to define what objectives must be achieved to solve the problem [9]. These are the needs criteria and generally encompass the performance, usability, safety, and value targets that will determine if a solution is successful or not. Like the need statement, the needs criteria should be agnostic of a specific solution. This ensures that the work is driven by the need rather than a solution. A helpful way to construct needs criteria is to use current solutions or the standard of care as a benchmark and decide whether a new solution must be better (and by how much) or as good as the bench-mark based on the team's research. The "must-have" criteria define what a potential solution must achieve to justify pursuing it. The needs criteria should clearly convey how a new solution improves on the existing para-digm. When developing solutions for LMICs, factors that may play an important role in the needs criteria may include cost, local production, durability to environmental conditions, functionality despite lack of infra-structure, training requirements, and maintenance of the solution. Needs criteria can also include "nice to have" criteria, which if achieved, add fur-ther value to the solution.

30.2.2.2   Ideation

After the need has been clearly defined, the next phase of HCD involves generating potential ways to address the need. The methods in this section highlight transitioning the need statement and criteria to brainstorming prompts, a framework for brainstorming, concept selection, and testing these concepts through prototyping and rapid validation testing.

- *"How Might We" Questions:* To go from the need statement and criteria to the ideation phase, it is helpful to convert them into a set of "How Might We" (HMW) questions that target the different parts of the need. HMW questions reframe barriers, challenges, and problems as a chance to design solutions. To generate a good set of HMW questions, an under-standing of the anatomy, physiology, pathophysiology, and standards of care is needed by the team. Opportunities for further HMW questions can stem from visible or invisible features, static and dynamic conditions, and other environmental variables.

- *The A.C.E. Framework to Brainstorming:* For an in-depth guide to brainstorming, the IDEO guide is highly recommended. The objective of brainstorming is to generate as many ideas as possible without consideration for their feasibility or ability to solve the problem. Often, the first kinds of ideas to come up during a brainstorm will be broadly focused on the approach level. This is a natural place to start and is generally more abstract in nature. For example, various approaches can be brainstormed by trying to prevent, divert, block, or accept a certain stimulus. A single approach will lend itself to further ideation by thinking about different concept categories. These concept categories can be organized around a technology platform (i.e., chemical, electrical, mechanical, biological), disease mechanism of action, cycle of care, stakeholder, or business model. Within these concept categories, specific concepts can be drawn out. To further refine a concept, quantifying or characterizing different aspects of the concept (i.e., materials, size, mechanism, etc.) will lead to an embodiment. This Approach, Concept, and Embodiment progression represents the A.C.E. framework. The output of a brainstorm can be captured using a mind map, with each layer representing more detail.

- *Concept selection:* While brainstorming is a hallmark method of design thinking, its primary objective is to be generative and creative. Moving all the ideas forward is not feasible, so strategies are needed to select a few. Concepts should first be screened for whether they are likely to meet the need statement and criteria. If further screening is necessary, the technical feasibility of the concept can be evaluated as well. This process will narrow down a large number of ideas to a smaller set of potential solutions that meet the need and are technically feasible.

- *Prototyping:* With the short list of lead concepts, the next step is to build prototypes. Prototypes come in various levels of fidelity, such as looks-like, feels-like, and works-like prototypes. It can also be valuable to prototype the problem. The value of building different kinds of prototypes is to quickly learn something about the solution or problem without investing significant resources or time into developing the actual solution. Simple prototypes can be built out of paper, cardboard, tape, or any other readily available material.

- *Rapid validation testing*: With these prototypes, it becomes possible to get more detailed feedback from stakeholders, generate more questions, and iterate on concepts through rapid validation testing. Rapid validation testing is extensively used in the consumer product space. While healthcare is a more constrained space to operate in, there are certain methods that can still prove useful in addressing key risks and assumptions that are part of a possible solution. The risks and assumptions that pose both the biggest uncertainty and importance should be addressed first. There are various methods that can be used for rapid validation testing. One example is to design a vapor test, which can measure interest in a solution before the solution is built. This can consist of building a landing page or website and tracking user traffic. Another example is to deploy the prototype as a fake front end or fake back end. A fake front end can be used to test whether a user behaves as expected to demonstrate feasibility, without coupling their actions to anything else. On the other hand, a fake back end is meant to replicate a function but with fewer resources (such as testing an "automated" feature by having someone performing the task manually out of the user's sight). The information gained from rapid validation testing can be used to drive further iterations on the solution.

30.2.2.3 Implementation

Bringing a health technology solution to realization is different from any other product given all the milestones that must be achieved. Regulatory approval needs to be secured before a health technology solution is ready to enter the market. Additionally, there is a need to demonstrate some level of clinical efficacy, establish manufacturing partners, and protect intellectual property. In the rehabilitation robotics field, there are a few examples of solutions that have made it all the way to commercialization. Therefore it is important to put thought early on into how a solution will be implemented. Successful implementation of rehabilitation robotic solutions requires partnerships, a funding strategy, and ultimately sustainable revenue.

- *Partnerships:* Identifying key partners and building relationships with them early on is key to implementation. On the clinical side, this can be local champions who will advocate for the solution or key opinion leaders whose words carry great influence in the field. Other key partnerships include the Ministries of Health, universities, and nongovernment organizations. Each partnership offers something different, and understanding what you might be able to ask from the partnership will facilitate progress early in the implementation phase.
- *Funding strategy:* While partnerships may provide an avenue to funding, they likely will not be sufficient to cover a long-term solution. Thus it is important to consider how to bridge short-term and long-term finances. Local or international grants can provide short-term funding to cover potential development costs and staffing the project. Funding can also come from other sources such as crowdfunding, incubators, foundations, and impact investors. In any scenario, key milestones and the funding needed to achieve them should be clearly outlined.
- *Sustaining for the long term:* While it is possible to make significant headway with grants or donations, these sources of funding are inherently unsustainable. Additional resources are needed to scale up operations and

ensure the long-term success of a solution. This starts to introduce commercial market factors primarily driven by the willingness of a stakeholder to pay for a solution for a given value proposition. If a solution is highly desirable over a long period, the revenue generated will sustain the product. It should be noted that market-based and grant-based approaches to funding are not mutually exclusive.

There are also other strategies to implementing a solution that do not rely on building and sustaining a business. One approach is to license the product to another group that has the existing resources to implement the solution more readily. This approach can be used when the team is lacking the technical capabilities, experience, or money needed to develop the solution. However, this comes at the cost of losing control over the project. If the team still seeks to retain or share control, an alternative approach is to partner with another entity to split responsibility for the development and commercialization. A final approach is to give up the rights to the solution through a sale or acquisition. An acquisition can be a strategic decision by a larger company to expand its product portfolio or to enter a new area without having to develop it themselves. In each scenario, a more experienced entity, likely with existing resources in place, takes responsibility for the implementation. This can increase the likelihood that a solution grounded in a HCD approach actually makes it to the patient.

30.2.3 Universal themes

30.2.3.1 Get immersed in the local environment

Particularly when designing for a new context, it is extremely important to understand the local dynamics at play. Assumptions and views do not always translate across different contexts, so it is important to get immersed in the environment to truly understand what is happening. This will provide an opportunity to both test out the assumptions and engage local stakeholders. Getting immersed in the local environment and practicing empathy allow for opportunities to continue to build partnerships and test out solutions.

30.2.3.2 Let the need drive the solution, not the technology

By taking a human-centered and biodesign approach to problem-solving, you may uncover that the best solution is not purely technology-focused. The right mix of factors is needed to implement a solution, and this will heavily influence the technology that can be used in any solution. However, being needs-driven by starting with the needs statement and criteria ensures a higher likelihood that the solution will have a positive impact on patients.

30.2.4   Examples

30.2.4.1   Solution for postpartum hemorrhage in South Africa

Postpartum hemorrhage is a treatable condition that is responsible for about 25% of maternal deaths worldwide—or around 112,000 deaths per year—with most deaths occurring in low-resource areas [13]. In high-income environments, the problem of postpartum hemorrhage is successfully treated with a uterine balloon tamponade device. However, the cost of this device, in the hundreds of US dollars (USD), is not accessible for low-resource settings.

A workaround that had been developed in low-resource settings involved assembling a makeshift condom catheter and filling it with water while inserted into the mother's uterus to stop the bleeding. A partnership between a South African biomedical engineering firm called Sinapi Medical, a global health innovation nonprofit organization called PATH, and the South African Medical Research Council eventually led to the design, manufacturing, and successful commercialization of an accessible and appropriate solution called the Ellavi device sold for under 10 USD per unit. The product has since received regulatory approval in Ghana and Kenya and is projected to save the lives of 169,000 women by 2030 [13].

30.2.4.2   Discovering an untapped key stakeholder in Botswana (an example from the authors)

During our time working in Botswana, we spent a lot of time engaging local stakeholders and were energized by the excitement that our work was generating. However, a big challenge we encountered was in developing the local capacity and interest in biomedical engineering to support our work. We had lots of discussions with the local university about curriculum development, research opportunities, and other ways to build out the next generation of biomedical engineers that were equipped to solve complex healthcare challenges, but this would not be a quick fix.

We set up operations in a lab space in a newly built teaching hospital and had the fortune of connecting with the head of biomedical engineering through a teaching workshop we organized. He had a leadership position at both the hospital and university and was immediately drawn to the mission of our work. It was through this relationship that we learned of a game-changing insight—there was a workforce of biomedical engineers within Botswana who trained abroad, only for them to return and be funnelled into jobs where their skills were underutilized. The mismatch resulted in a population of biomedical engineers, many already trained specifically in rehabilitation engineering, that could be the key to solving one of the big challenges facing rehabilitation in Botswana.

This was an exciting development that was only uncovered by immersing ourselves in the local environment and engaging with local stakeholders.

While the implications meant that we probably would have to rethink a lot of what we were doing given the presence of a new stakeholder, it also opened a lot more possibilities.

30.3 Conclusion

Increasing access to rehabilitation services is a huge global need that requires a variety of solutions to address the challenges and problems in each context. There is great value and benefit in solving these problems to improve the quality of life of people living with disabilities around the world. Specifically in the rehabilitation robotics space, there are considerations outside of the technology itself that are important to address—particularly around usability, acceptability, and sustainability. While it does not guarantee success, grounding the approach in human-centered design and the biodesign process can help lead to meaningful solutions.

30.4 Conflict of interest

No conflicts of interest.

Additional resources

IDEO. The Field Guide to Human-Centered Design. https://www.designkit.org/resources/1.

Stanford Byers Center for Biodesign Resources. https://biodesign.stanford.edu/resources/learning.html.

References

[1] Gimigliano F, Negrini S. The World Health Organization "Rehabilitation 2030: a call for action Eur J Phys Rehabil Med [Internet] 2017;53(2)May [cited 2022 Mar 21]. Available from: https://www.minervamedica.it/index2.php?show = R33Y2017N02A0155.

[2] Roth G. Global Burden of Disease Study 2017 (GBD 2017) Results. [Internet]. Institute for Health Metrics and Evaluation (IHME). Lancet 2018;. Available from: https://www.healthdata.org/sites/default/files/files/policy_report/2019/GBD_2017_Booklet.pdf.

[3] Lo K, Stephenson M, Lockwood C. The economic cost of robotic rehabilitation for adult stroke patients: a systematic review. JBI Evid Synth 2019;17(4):520−47.

[4] Weinstein MC, Torrance G, McGuire A. QALYs: the basics. Value Health 2009;12:S5−9 Mar.

[5] Sachs J. Weltgesundheitsorganisation, editors. *Macroeconomics and health: investing in health for economic development; report of the Commission on Macroeconomics and Health*. Geneva: World Health Organization; 2001. 202 p.

[6] Bickenbach J. The world report on disability. Disabil Soc 2011;26(5):655−8.

[7] World Health Organization. The need to scale up rehabilitation [Internet]. Available from: https://apps.who.int/iris/bitstream/handle/10665/331210/WHO-NMH-NVI-17.1-eng.pdf?sequence = 1& isAllowed = y, 2017.

[8]   IDEO, editor. *The field guide to human-centered design: design kit.* 1st. ed. San Francisco, Calif.: Design Kit; 2015. 189 p.

[9]   Yock PG, editor. Biodesign: the process of innovating medical technologies. 2nd edition Cambridge; New York: Cambridge University Press; 2015. p. 839.

[10] Kelley D, Kelley T. Creative confidence: unleashing the creative potential within us all. New York: Crown Business; 2013. p. 288.

[11] Krumrei-Mancuso EJ, Rouse SV. The development and validation of the comprehensive intellectual humility scale. J Personality Assess 2016;98(2):209−21.

[12] Grant AM. *Think again: the power of knowing what you don't know.* New York, NY: Viking; 2021. 307 p.

[13] Sinapi Medical [Internet]. Available from: https://ellavi.com/. (cited 21 March 2022).

Chapter 31

Toward inclusive rehabilitation robots

Michelle J. Johnson[1], Shafagh Keyvanian[2] and Rochelle J. Mendonca[3]

[1]*Departments of Physical Medicine and Rehabilitation, Bioengineering, and Mechanical Engineering and Applied Mechanics, University of Pennsylvania, Philadelphia, PA, United States,* [2]*Department of Mechanical Engineering and Applied Mechanics, University of Pennsylvania, Philadelphia, PA, United States,* [3]*Department of Rehabilitation and Regenerative Medicine, Programs in Occupational Therapy, Columbia University, New York, NY, United States*

Learning objectives

At the end of the chapter, the reader will be able to:

1. Review general and ethical concerns underlying the use of robots in low-resource settings.
2. Define and discuss issues around increasing access to and inclusiveness of rehabilitation robotic technologies.
3. Describe a potential practice framework for increasing inclusivity in the design and implementation of rehabilitation robotic technologies.

31.1 Introduction

By the World Health Organization's definition, "health is a state of complete physical, mental, and social well-being and not merely the absence of disease or infirmity" [1]. Today our health is being threatened by both communicable diseases (CDs) such as the coronavirus disease (COVID-19) and noncommunicable diseases (NCDs) such as stroke. The threat from NCDs has been growing rapidly and in 2022 was responsible for 74% of all deaths globally, about 41 million people each year with 77% of these deaths occurring in developing countries defined as low- and middle-income countries (LMICs) [2]. Those who survive these CDs or NCDs often face long-term disability, which by the World Health Organization's definition is "a complex interaction between features of a person's body and features of the environment and society in which he or she lives" [3,4].

Rehabilitation Robots for Neurorehabilitation in High-, Low-, and Middle-Income Countries.
DOI: https://doi.org/10.1016/B978-0-323-91931-9.00032-3

By this definition, the determination of one's disability will differ globally and will depend on the country where one resides.

Cardiovascular diseases, which include strokes, account for approximately 43.6% of these NCD deaths, which is about 17.9 million people and 36% attributable to stroke [2,4]. The Global Burden of Disease Study reported that in 2019, there were about 12.2 million incident cases of stroke, 101 million prevalent cases of stroke, and 143 million disability-adjusted life years (DALYs) due to stroke [5]. In 2022, the World Stroke Global Factsheet reported that stroke is still the third-leading cause of death and disability combined in the world [6]. These reports also indicated that more than 85% of the burden of stroke was in LMICs, specifically 86% of deaths and 89% of DALYs [5,6].

Given the disparities in stroke prevention, care, and rehabilitation, we should expect that functional outcomes after a stroke for someone in an HIC such as in the United States will differ greatly from outcomes in LMICs, since rehabilitation access, technology availability, and resources also differ [7−9]. These disparities in stroke care are expected to worsen in HICs over time. For example, it is predicted that by 2060, about 1 in 5 Americans will be over 65 (about 23%), which means that 1 in 5 Americans will have a greater risk for a stroke [10,11]; the related prevalence of stroke is predicted to increase to 14.5 million [11]. At the same time, it predicted that by 2030 there will already be a shortage of health professionals, rehabilitation therapists, nurses and doctors, and people capable of taking care of the growing numbers of people with disabilities [12−14]. So, in the not-too-distant US future, there will be a large gap in rehabilitation care. Unfortunately, this gap in care already exists in LMICs and is exacerbated by challenges such as poverty, lack of resources, and insufficient healthcare dollars [7−9].

Throughout this book, the focus has been on examining how rehabilitation robotic technologies, called therapeutic and assistive robots, are being used globally to support neurorehabilitation after stroke and bridge rehabilitation care gaps. These are a subclass of assistive technologies (ATs), and two examples can be seen in the activities of daily living exercise robot (ADLER) system which uses the HapticMaster as the therapy robot to provide task-oriented therapy using common activities of daily living (ADL) tasks or the BAXTER used as an assistive social robot to monitor and support social exercises (Fig. 31.1). Our collective goal is for rehabilitation clinicians to leverage robotic technology and associated sensors to assess and treat impairment after stroke (refer to Chapter 4 for details on these robots).

Most of these robotic systems, because of their expense [15], are in HICs and are often limited to large private hospitals, hospital-affiliated outpatient clinics, and supervised settings. From 2003 to 2022, we see a rise in the number of review articles trying to understand the cost-effectiveness, benefits, and evidence for the use of rehabilitation robots in healthcare. Most of the randomized control trials have been done in HICs using expensive robotic rehab systems [16]. The evidence suggests that robots are useful as

(A)   (B)

FIGURE 31.1 Examples of rehabilitation robots. A) Therapy robot system—ADLER—and B) assistive robot—BAXTER.

assistants that can co-collaborate with the therapists and other clinicians to administer quantitative assessment tasks, provide objective metrics of recovery, and deliver high-intensity, repeatable, and consistent exercises [16−18]. Despite this, there are still questions as well as a lack of consensus about the type of patient that would benefit as well as on training timing, frequency, session duration, and dosage [18]. When we consider the evidence with respect to the international classification of function (ICF) [4], we can surmise that rehabilitation robots can impact body-level function and impairment such as increasing muscle strength and increasing motor control, impact activity for ADLs, and, more recently, through the exoskeleton and wearable systems such as soft robots that can be attached to the human user, participation; these impact statements are most true for robots targeting the upper limb than the lower limb. Additional nuances about the evidence for these robots are provided in Chapter 5.

The global rehabilitation robotics market has been rapidly growing. One market study reported a compound annual growth rate (CAGR) of 20.1% growth from 2021 to 2022 with expected growth to reach $20.8 billion in 2026 [19]. Most companies making these robots are in North America, Europe, and a few countries in Asia; we do not see many in Latin and South America, or in the Middle East and Africa [19−21] (Fig. 31.2). Table 31.1 lists the most prominent companies with their common commercial rehabilitation robot products. Consequently, there is a disparity in terms of who are benefiting from these systems. The high cost, complexity, and size of most rehab robotic systems also limit their impact on the average stroke survivor with impairment in rural spaces and community-based spaces in HICs and most LMICs. So as a result, the impact of rehabilitation robotics is not inclusive, and access to their potential benefits is unequal. Solutions that are appropriate for LMICs and rural- or community-based settings in HICs are rare (see Table 31.2). Given the potential of these technological solutions to bridge healthcare resource gaps, there is a need to consider innovative ways to increase their inclusivity.

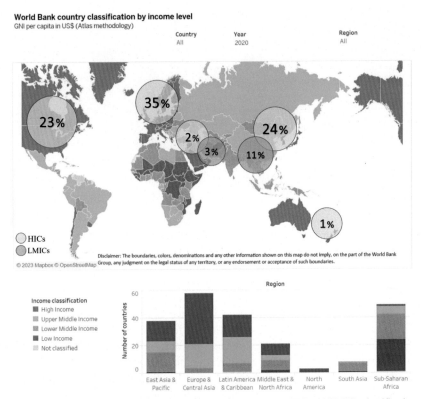

FIGURE 31.2 Location of rehabilitation robot developers across HIC and LMICs classification 2020 [19−21]. The companies listed are those in Tables 31.1 and 31.2. *HICs*, high-income countries; *LMICs*, low- and middle-income countries.

As robots enter rehabilitation spaces [15], documented benefits are offset by increased risks due to increases in the number of human−robot interactions. Some notable concerns that have arisen revolve around access, acceptance, work displacement and social connectedness, empathy, safety, autonomy, legal responsibility for injury, distributive justice, and privacy. These issues are interconnected, and proposed solutions and guidelines may often depend on the type of robot and the medical space. This chapter describes general and ethical issues around access to rehabilitation robot technologies and factors that might influence efforts to increase their inclusivity—a term that encompasses affordability and other common issues that may justify limiting or increasing their use in low-resource settings. Inclusivity can be increased if we address barriers to use by a diverse population. Section 31.2 briefly discusses these barriers from the lens of access, ethics—justice, privacy, safety, and autonomy; affordability; and finally, feasibility, usability, and acceptance. Section 31.3 details a potential practice framework for increasing inclusivity.

TABLE 31.1 Noted rehab robotics companies, location, and products in HICs (high-income countries).

Continent	Location	Company	Rehab robot products
Europe	Austria	Tyromotion GmbH	Diego, Amadeo, Myro, Tyrostation, Pablo
	France	Cutii	Cutii
	France	Technalia	ArmAssit
	France	ezyGain	ema, amy
	Germany	Reha-Stim	Bi-manu Trak
	Germany	Medica Medizintechnik GmbH	TheraTrainer Lyra
	Germany	RECK Technik	MOTOmed therapy devices
	Germany	Hasomed GmbH	RehaMove, RehaCom
	Germany	h/p/cosmos	mercury, pulsar, locomotion, pluto, h/p/cosmos pulsar
	Germany	Kuka	Robert
	Italy	Gloreha	Sinfonia robotic glove, MAESTRO
	Italy	HumanWare	Motore, Ultra
	Luxembourg	ExoAtlet	ExoRehabilitation
	Netherlands	Assistive Innovations	iEat Robot, iFloat NG Dynamic, iArm
	Poland	EGZO Tech	Luna EMG
	Spain	Able Human Motion	Able Exoskeleton
	Spain	Instead Technologies Ltd.	RoboTherapist 3D, RoboTherapist 2D
	Spain	Marsi Bionics	Atlas, MAK
	Spain	Technaid	Ankle H3, Exo-H2 and Exo-H3
	Spain	iDRhA	Helium, Rubidium
	Spain	Cyber Human Systems	ALDAK, BESK G
	Spain	Tcnicalia	Ismore, Rehand, Armassist
	Spain	Adamo Robot	Adamo Robot

(Continued)

TABLE 31.1 (Continued)

Continent	Location	Company	Rehab robot products
	Spain	Gogoa	HANK, BELK
	Sweden	BioServo	Carbonhand
	Switzerland	Hocoma AG	Armeo Power, Lokomat, Erigo, Andago, Senso
	Switzerland	Reha Technologies AG	Armotion, G-EO series
	Switzerland	Motek Medical B.V.	Hero Solution
	Switzerland	Gorbel	SafeGait Active
North America	The United States	Myomo Inc.	MyoPro orthosis series
	The United States	Bioness (Bioventus)	Bioness L300, H300
	The United States	Biodex Medical Systems, Inc.	Biostep, MedBike, FreeStep, Upperbody Cycle
	The United States	Bionik Laboratories Corp.	InMotion Arm/Hand
	The United States	Barret Technologies	Burt
	The United States	Motus Nova	Motus Hand, Motus Foot
	The United States	ReWalk Robotics Ltd.	Rewalk Personal 6.0 system, Re-store soft exosuit
	The United States	Parker Hannifin	Indego
	The United States	Quadriciser Corp.	tQuadriciser Robot
	The United States	Restorative Therapies	iFES system — Rt300
	The United States	AlterG Inc.	AlterG
	The United States	Aretech LLC	ZeroG
	The United States	Flint Rehab	FitMi, MusicGlove
	The United States	Ekso Bionics	EksoNR, Ekso Indego, Ekso Indego Personal, Ekso Health

(Continued)

TABLE 31.1 (Continued)

Continent	Location	Company	Rehab robot products
	The United States	Harmonic Bionics	Harmony SHR
	The United States	Woodway	Loko help, Loko station, Kineassist
	Canada	Kinova	JACO arm, O110/O540
	Canada	BKIN Technologies Ltd.	Kinarm
	Canada	B-TEMIA, Inc.	Keeogo
Oceania	New Zealand	Rex Bionics Ltd.	Rex
Middle East	Israel	Motorika	Reo-Ambulator, ReoGo
	Israel	BioXtreme	deXtreme
Asia	Japan	Cyberdyne Inc.	HAL
	Japan	Softbank Robotics	Pepper, Nao
	Japan	AIST	PARO
	Japan	Toyota Motor Corp.	New Welwalk WW-2000
	Japan	H-Robotics	Rebles pro, rebles planar
	Japan	Honda	Walking Assist
	Japan	Yasukawa	Assist Walking Device, CoCoroe AR2
	Japan	Teijin Pharma	ReoGo-J: customized version of ReoGo (Motorika)
	Singapore	Fourier Intelligence International Pte Ltd.	RehabHub, ArmMotus, WristMotus, HandyRehab, ExoMotus
	Singapore	Articares Pte Ltd.	H-Man
	Singapore	Roceso Technologies	EsoGlove
	South Korea	Neofect	Rapael Smart Glove, Smart Board
	South Korea	Man&Tel Co. Ltd.	3D Rehab Robot trainers
	South Korea	P&S Mechanics	Walkbot
	South Korea	Curexo	Morning Walk

(Continued)

TABLE 31.1 (Continued)

Continent	Location	Company	Rehab robot products
	South Korea	HMH	Exowalk
	South Korea	Hyundai Motor Company	H-MEX, H-LEX, RearMEX
	South Korea	Angel Robotics	Angel Leg
	South Korea	H.Robotics	Rebless
	Taiwan	Free Bionics	Free Walk

TABLE 31.2 Noted rehab robotics companies, location, and products in low- and middle-income countries [19].

Continent	Location	Company	Rehab robot products
Middle East	Iran	AmbuReR	Domestic versions of Lokomat
	Iran	Pedasys	Exoped
	Turkey	BAMA Technology	Robogait, Free gait, Visio gate
Asia	India	Bionic Yantra Ltd.	REARS, WRE
	India	Beable Health	Armable robot
	India	Rymo Technologies	Mobi-L
	India	Gridbots Technologies	Physio
	China	Rehab-robotics Company Ltd.	Hand of Hope, SmarTable, Kineto
	China	Beijing Ai-Robotics Technology	AiLeg, AiWalker
	China	Hangzhou RoboCT Technology	UGO
	Malaysia	Techcare Innovation	CR2 Haptic, HR-30, FIBOD
	Malaysia	Roceso	EsoGLOVE

31.2   Inclusivity

The World Health Organization in "Rehabilitation 2030: A Call for Action [9]" describes a mismatch between the global need for rehabilitation, the growing numbers of patients that need care, and the availability of resources. Currently, the impact of rehabilitation robotics has not been inclusive and a policy of providing equal access to the potential benefits of technology-assisted rehabilitation is missing. As defined, inclusivity is the practice or policy of providing equal access to opportunities and resources for people who might otherwise be excluded or marginalized, such as those having physical or mental disabilities or belonging to other minority groups. [defn. 1 [22]].

Barriers to achieving inclusivity in rehabilitation robotics in low-resource settings are many. For example, many LMICs have limited healthcare budgets and as a result often focus more on the curative model of healthcare service delivery leaving marginal resources for rehabilitation services, personnel, and related assistive technical products. Achieving inclusivity in LMICs means facing the dual challenge of simultaneously building capacity for rehabilitation services and people as well as creating local infrastructure to design and develop assistive technology (AT) solutions. Some challenges identified are very similar across many LMICs including the developing countries in this book. The challenges noted in Chapters 14−28 include shortage of rehabilitation staff; poor insurance coverage for patients; low reimbursement for rehabilitation with long wait times; lack of local rehab manufacturers and distributors that limit the ability to source assistive technology equipment of all types; lack of local places to develop rehabilitation specialists whether therapists or physiatrists; lack of capital funding and investment sources to expand services; and lack of affordable gym space and equipment. Interestingly, extending the reach of rehabilitation robots to low-resource settings in HICs faces similar challenges (refer to Chapters 6−13). Rehabilitation taking place in nursing homes, community-based rehab centers, and at home describes reimbursement challenges due to the high portion of underinsured patients; shortage of rehabilitation staff; lack of gym space for large systems; lack of funds to afford expensive rehabilitation devices; and the need to use technology that is not disease-specific, i.e., able to treat a diverse population of patients.

Clearly, diverse actions are needed to address these barriers to inclusivity. Such actions include addressing ethical concerns related to human subjects; developing affordable devices using innovative design strategies; establishing device feasibility, acceptability, and cost benefits; building rehabilitation research, training, and technological capacity; and working with WHO, United Nations (UN), and other related policy-making organizations to ensure that these actions are equitable and their impact is standardized [22−24].

31.2.1   Extend access to patients with complex stroke

Rehabilitation access is limited in low-resource settings in HICs and LMICs, which puts persons living with residual impairments due to a stroke at a disadvantage. A stroke is classified as a non-traumatic brain injury that can impact all three international classification of function (ICF) levels—body-level impairment, activity, and participation. Typical impairments can be seen in cognitive domains such as executive function, memory, attention, and language and in motor control in upper and lower extremities such as reduced range of motion, reduced dexterity, spasticity, and muscle weakness. One study reported that of 1259 patients more than 60% had six or more impairments with reported common impairments of weakness in upper limb (77.4%), urinary incontinence (48.2%), impaired consciousness (44.7%), dysphagia (44.7%), and impaired cognition (43.9%) [25]. These impairments often result in activity and participation related limits in performing basic ADLs such as self-care, in performing instrumental ADLs such as banking, and in mobility affecting community ambulation [26,27]. Persons with strokes in LMICs are often younger, have poor functional outcomes, and have more complex medical presentations [7,28].

Inclusivity may be examined from the lens of designing rehabilitation robots that can serve persons with complex medical presentations. In the early 1990s when the field began, there was a focus on therapy robots aimed at treating stroke survivors primarily presenting with varying levels of motor impairments in the upper and/or lower extremities only—those with cognitive impairments were typically excluded [26,29,30] (Fig. 31.3). There is an increased need to treat patients living with motor impairments alongside mild to severe cognitive impairments who evidence performance disruptions in ADLs impacting a wider spectrum of their daily lives [28,30,31]; this group would not only include people living with stroke but also people living with dementia, traumatic brain injury (TBI), and human immunodeficiency

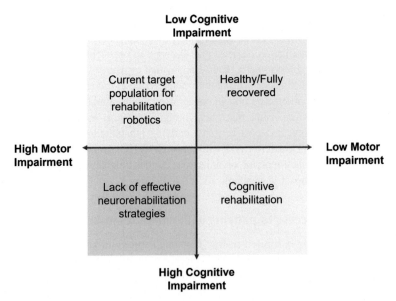

FIGURE 31.3 Complex stroke example motor and cognitive impairment.

virus (HIV) [31−35]. Now it is not uncommon to discuss the need to examine the impact and use of rehab robots with stroke survivors with cognitive impairment [34] and with HIV [35]. Future rehabilitation robots should be personalizable to support the functional retraining of stroke survivors with impairments that affect motor, cognitive, aural and/or visual functions.

31.2.2   Ethical implications of increased access

Human−robot interaction (HRI) considerations are important when designing robots to meet complex stroke needs in HICs or LMICs. The impact of rehabilitation robots applied to stroke rehabilitation heavily depends on the impairment level of the humans expected to interact with the robot—none, mild, moderate, or severe impairments in body function; on the clinical treatment goal of robot—therapeutic, assistive, or diagnostic; on the autonomy of the robot—fully autonomous, semi or passive; and on the amount of energy exchange and contact expected between robot and human—constant, intermittent, or none. HRI designs can be guided by the ethical codes set down in Asimov's laws for robots [36]; assistive roboethics and socialized roboethics guidelines applied to rehabilitation therapy and assistive robots and socially assistive robots, respectively [37]; and more recently artificial intelligence (AI) ethics [38]. Fundamentally, these ethical principles are built on the concept of justice, non-maleficence, beneficence, and autonomy. The robot technologies used should be designed to not intentionally harm a human user, be appropriate, be affordable, help the user only as needed, be effective, reliable, and cost-effective, maximize the user's safety, and respect the freedom and privacy of the person. In addition, the data must be secured to protect personal health data, maximize justice and responsibility by ensuring that the robots are trained on a multiculturally diverse dataset, and maintain human−human relations. When the robot has artificial intelligence and levels of autonomous behavior to sense, think, and act to comprehend its environment and users and to exhibit social behavior and communicate with users, additional emotional and behavioral issues that impact ethical use should be considered [36−39]. Because the action of more socially assistive robots may be mistaken as "real actions—human-like or pet-like," users living with mental health impairments may develop a misplaced attachment, feel deceived when they discover the robot is not real, and be confused about the extent of the robot's authority, thus losing their ability to choose to initiate with or break away from an interactive session with it. Given this, designers should minimize attachment and opportunities for deception by managing exposure to the robot companion, developing robot control strategies that enable the clinician to direct how much authority and autonomy the robot should be given, and how much the clinician should retain. In Fig. 31.4, Johnson and colleagues surmise that a robot can function in all phases of a task-oriented stroke rehabilitation session: as a demonstrator, as an observer,

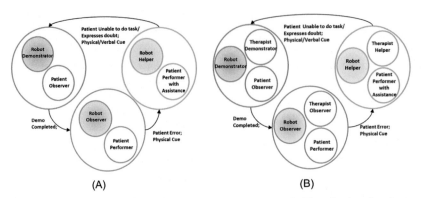

(A)   (B)

FIGURE 31.4 (A) The robot is functioning with full autonomy. (B) The robot has shared control in task-oriented stroke rehabilitation [40]. The robot moves from co-helper with low autonomy and low authority to helper with high levels of autonomy and authority.

and as a helper [40]. The ideal scenario would be the robot in Fig. 31.4B versus the robot in Fig. 31.4A. The robot in Fig. 31.4B has a variable shared control that allows for low and high levels of autonomy. In this scenario, the robot works alongside the clinician as a supportive co-helper versus in Fig. 31.4A where it has full autonomy and function as the clinician [40].

The ethical issue will change as autonomy increases. With this increased autonomy arises the challenge of developing control strategies that can enable robots to make safe and moral decisions [37,41]. Wallach and colleagues [42,43] suggest three levels of morality to apply as the robot "intelligence" and "ability to operate autonomously" increase. The lowest level of morality would be operational morality where the robot designer is responsible for the actions of the system. When the robot designers can only partially predict the robot's actions and their consequences, then we reach functional morality, which would then progress to full morality when the AI robot is able to learn and fully choose its actions without the robot designer's input.

It seems obvious that issues of safety and ethics of use should be considered when developing rehabilitation robots to interact with persons living with disabilities in general and complex stroke specifically, especially when motor impairment coexists with cognitive impairments that affect the users' decision making ability. The consequences of rehabilitation robots' actions when issues of safety and morality are violated should be the responsibility of a country's organizations with research oversight and societal oversight. The ability to deal with the ethical consequences of human subject research violations due to rehabilitation robots is developed in HICs and LMICs, especially when examined with respect to the existence of institutional research oversight boards. On the other hand, societal oversight is still developing, and legal guidelines are nascent in most HICs and many LMICs.

When practically extending the reach of rehabilitation robots to low-resource settings in HICs and LMICs, it becomes clear that most settings lack funds to buy expensive rehabilitation technologies that may only target one type of population. By designing rehabilitation robots to encompass more complex stroke and as such other persons living with multiple types of impairments, we address both the financial and people resource barriers and move closer to the provision of robot technologies that are not disease-specific, i.e., able to treat a diverse population of patients.

31.2.3 Affordable rehab robots for low-resource settings

Another way to improve the inclusiveness of rehabilitation robotics is to reduce costs and make them more affordable. Affordability is defined as "the extent to which something is affordable, as measured by its cost relative to the amount that the purchaser can pay" [44]. This definition suggests a policy of providing neurorehabilitation robots that meet a setting's environmental cost constraints while maintaining effectiveness for assessing and treating disease-related impairments.

> *Definition: Affordability is the extent to which something is affordable, as measured by its cost relative to the amount that the purchaser can pay.*
>
> [defn 2 [44]]

In a recent review [45], Ekechukwu and colleagues suggest that rehabilitation robots are not a pragmatic solution for LMICs, "despite their high cost … a likely advantage is that automated interventions like robotic therapies require minimal input from rehabilitation professionals in terms of time and efforts." If we reduce the overall cost of the design and manufacture of stroke therapy robots, we reduce the price point and enable purchasing and implementation of these robots in low-resource settings; thus we invariably allow clinicians in such settings to treat more patients despite their limited numbers and shortage of time and efforts. Specifically, for people living with residual impairments after stroke, the use of therapy robots and assistive robots, cost-effective and appropriately designed for their environments, could bridge healthcare gaps and support the delivery of more frequent therapy, more nuanced assessments, and long-term assistance to patients with stroke.

In a recent review [45], Ekechukwu and colleagues suggest that rehabilitation robots are not a pragmatic solution for LMICs, "despite their high cost … a likely advantage is that automated interventions like robotic therapies require minimal input from rehabilitation professionals in terms of time and efforts." If we reduce the overall cost of the design and manufacture of stroke therapy robots, we reduce the price point and enable purchasing and implementation of these robots in low-resource settings; thus we invariably allow clinicians in such settings to treat more patients despite their limited numbers and shortage of time and efforts. Specifically, for people living with residual impairments after stroke, the use of therapy robots and assistive robots, cost-effective and appropriately designed for their environments, could bridge healthcare gaps and support the delivery of more frequent therapy, more nuanced assessments, and long-term assistance to patients with stroke.

Affordability is dependent on the wealth of the purchaser. So essentially, defining a rehabilitation technology device as affordable will depend on context. What is an affordable AT in HICs will be different from that in LMICs. The WHO has defined a cost-effective threshold for intervention strategies based on a country's income (defined as GDP or GNI) per capita that is worth using as a guide [46,47]. If we assume that a rehabilitation robot's effectiveness and thus benefit will be the same regardless of country, then the variable of cost will be the primary factor that affects the appropriateness of an intervention or device for a country. An intervention will be considered highly cost-effective if it is less than one country's income per capita—*minimum threshold*, cost-effective if it is between the minimum threshold and three times the threshold—*maximum threshold*, and not reasonably cost-effective if beyond this maximum. Based on this guide, ideal rehabilitation robot interventions should be implemented to achieve effectiveness below a country's maximum threshold (Table 31.3 and Table 31.4). Tables 31.1 and 31.2 defines the minimum thresholds for the HICs and LMICs discussed in this book as well as indicators such as population total, population aged 65 and older, stroke deaths, % health expenditures and the equity index. For the HICs, the minimum thresholds ranged from $29,816 to $65,135 (USD) with an average of $43,492, and for LMICs, from $1502 to $12238 (USD) with an average of $6621 (USD) [46]. So, we loosely argue that affordability is maximized for low-resource settings if, for both HICs and LMICs, rehabilitation robot cost targets 6.6 K (in USD), which would still be unreasonable for 3 of the 23 countries covered in this book, specifically India, Cameroon, and Ghana. To further maximize the adoption of these robots into low-resource settings in LMICs, we may consider aiming for the 1.5 K (in USD), which is the minimum threshold for India. Clearly, although the minimum threshold is thought to approximate the average income of a person in the country, reality informs us that income distribution across a country's population is rarely normal and is often skewed left so there may be lots of people making a lot less than the approximate average, which is further influenced by percentage of the population living in rural versus urban spaces. This income inequality is reflected in the Gini index—gauges economic inequality by measuring income distribution—and the metric in the Gini coefficient, which ranges from 0 to 100 when expressed as a percentage—0% is perfect equality, and 100% would be the maximum possible inequality [46]. On average,

TABLE 31.3 Overview of their demographic and economic information for countries in this book.

	Country	Population (million) *2021	Pop. aged 65 + (%) *2020	Stroke deaths/million *2020	GDP/capita (USD) *2021 minimum threshold	Expend on health (% of GDP) *2021	Gini index (%) *2023
HIC	The United States	332.92	16.63%	2.227	65,134	16.9	41.4
	Canada	38.07	18.10%	1.512	46,550	10.8	33.3
	Italy	60.37	23.30%	2.515	33,090	8.7	35.9
	Spain	46.75	19.98%	1.837	29,816	9	34.7
	Japan	126.05	28.40%	2.321	40,063	11	32.9
	Australia	25.79	16.21%	1.713	54,763	9.3	34.4
	Republic of Korea	51.31	15.79%	2.548	32,143	7.6	31.4
	Israel	8.79	12.41%	1.679	46,376	7.5	39

HIC, High-income countries; LMIC, low- and middle-income countries.

TABLE 31.4 Overview of their demographic and economic information for countries in this book.

	Country	Population (million) *2021	Pop. aged 65 + (%) *2020	Stroke deaths/ million *2020	GDP/capita (USD) *2021 minimum threshold	Expend on health (% of GDP) *2021	Gini index (%) *2023
LMIC	Mexico	130.26	7.62%	3.208	9849	5.4	45.4
	Costa Rica	5.14	10.25%	2.444	12,238	7.6	48.2
	Colombia	51.27	9.06%	2.957	6432	7.6	51.3
	Ecuador	17.89	7.59%	3.247	6184	8.1	45.7
	Serbia	8.7	19.06%	8.33	7359	8.5	36.2
	India	1393.41	6.57%	6.497	2116	3.5	35.7
	Malaysia	32.78	7.18%	8.165	11,414	3.8	41.1
	China	1444.22	11.97%	11.08	10,004	5.4	38.5
	Iran	85.03	6.56%	5.721	7282	8.7	42
	Turkey	85.04	8.98%	4.906	9127	4.1	41.9
	Nigeria	211.4	2.74%	8.868	2361	3.9	35.1
	Botswana	2.4	4.51%	13.33	7961	5.8	53.3
	Ghana	31.73	3.14%	12.402	2203	3.5	43.5
	Cameroon	27.22	2.72%	11.887	1502	3.5	46.6
	Morocco	37.34	7.61%	10.153	3282	5.3	39.5

HIC, High-income countries; LMIC, low- and middle-income countries.

HIC countries have lower Gini Indices indicating more income inequality than LMIC countries. HIC countries have higher GDP per capita, which often allow them to spend higher % of the GDP on healthcare.

In [47], Demofonti and colleagues conducted a scoping review of affordable rehabilitation robots and concluded that many were not many commercially available, and if they were available, they were limited to university-based research efforts. Many manufacturers of rehabilitation robots will have trouble meeting the 1.5 K and below price point, hence the current disparity (Table 31.3). From the examples mentioned in Tables 31.1 and 31.2, most of the lower-cost robots proposed for rehabilitation in homes in HICs such as Motus Hand (Motus Nova: $10 K), Rapael Smart Glove (Neofect: $16 K), CR-2 Haptic (Technocare Innovations: $5 K), Armeo Senso (Hocoma AG: $4.5 K), and MOTOmed viva 2 (Rech Technik: $6 K) are still too costly for LMICs and are beyond the ideal cost of 1.5 K or less. A few mechatronic systems such as FitMi from Flint Rehab have been able to get below $500 and are beginning home-based systems that can be used now in LMICs. The higher-end systems such as the Lokomat from Hocoma AG cost upward of $250 K. Improving the penetration of rehabilitation robots into low-resource settings will require innovation in the design and development of rehabilitation robots to minimize cost and risk while maintaining effectiveness and maximizing benefits.

The review in [47] offered design and manufacturing strategies to increase affordability while maintaining effectiveness. One key strategy proposed is to use a group of simpler low-cost robot/mechatronic devices where each robot is made to be of lower degrees of freedom (DOF)—1-DOF or 2-DOF—and the collective system of robots are used either in a circuit training setting or in a community setting as a shared resource [48–50]. The idea of using robots in circuit training is not new. Buschfort, Hesse, and colleagues [48,49] showed that a suite of four lower DOF robotic devices—ARM Studio by Reha-STIM—can provide effective seated "hands-on" therapy to acute and subacute patients inside the Charite' Rehabilitation Hospital in Germany. The gym consisted of three bilateral training workstations for proximal and distal training of the upper limb and one-finger training workstation.

In [50], Valles and colleagues showed the beneficial impacts of using group therapy with multiple low-cost rehabilitation technologies in a robot/computer-based therapy gym that consisted of six workstations under the supervision of one therapist. Each station used a custom or commercial rehabilitation technology with a total having a cumulative cost of 29 K (USD); these were custom TheraDrive system, two functional electrical stimulation systems called L300 (Bioness) for drop foot and gait practice and H200 (Bioness) for hand and reach and grasping practice, two MOTOmed viva 2 (Reck Technik) for upper- and lower-limb training, and a cognitive station (Captains Log). The study showed that in addition to increasing motivation, group-based, circuit training with robot/computer-based therapy versus one-on-one standard therapy had comparable results for upper-limb recovery and superior results for lower extremity. This work has led to the development of Rehab Community-based Affordable Robot Exercise System (CARES) [51] and the initiation of ongoing work in Botswana (refer to Chapter 25). An advantage of the above in LMICs is that it enables one clinician plus a technician to oversee the activities on all the stations. Both these studies resulted in cost savings compared to standard of care; in Mexico, they found that this method led to a 2.7 reduction in cost per patient session and in Germany a 2.4 reduction in cost per patient session. These results suggest the viability of this approach in LMICs and for low-resource settings in HICs. Lo and colleagues [52] evaluated the economic cost of rehabilitation robotics to hospitals and identified five articles that compared the cost of providing robot intervention against the cost of providing dose-match standard therapy from a hospital benefit prospective. Their review identified [48] and [49] as two of the five studies and three studies using more complex and expensive robotic devices. Overall, [52] determined that from a cost minimization standpoint, the economics favored robot-assisted intervention especially when applied to acute stroke patients, chronic stroke patients, and severely impaired stroke patients. Another similar strategy mentioned in [47] is to design robots with reconfigurability in mind. So, instead of building a multidegree-of-freedom robot, use low-cost single-unit systems that are one degree of freedom, that can be reconfigured quickly and put together as a unit.

The use of a low-cost technology suite as an affordability strategy is advantageous in low-resource settings plagued by limitations in cost, space, and skilled clinical and engineering people resources. Having multiple robots increases the versatility of the suite and allows it to be used with complex stroke as well as with other CDs and NCDs that may result in impairments. In addition, the focus on using a suite in a group training session emphasizes a community-based mindset and the idea that this system does not need to be in someone's home—an often unrealistic for most patients in some LMICs—but could be in a primary care clinic that has the infrastructure to house the suite. Here, the therapy gym becomes a shared resource for the rural or urban setting. Success here would rely on building capacity in the community. For example, educational programs ensure the following: (1) the community recognizes the potential of the suite of rehabilitation technologies; (2) one or more clinicians are adequately trained to use and supervise the technology suite; (3) one or more engineers and/or technicians are trained to respond to usability issues with the suite and help maintain the robots; and (4) a pipeline is developed to adequately train local engineers in design and development of new robots and ATs that can respond to changing rehabilitation needs of persons living in the community.

A final strategy is to rely very carefully on local manufacturing and resources with a country. So essentially the aim should be avoiding the import of robot systems and limiting the import to critical items that are not available in the country or the region. If this is not possible, then the need for capacity building of engineers able to design and build the devices within the country becomes urgent. The advantages of relying on local resources are seen in lower labor costs, increased local technical experts, and increased local pride in ownership and development as well as seeding the buildup of a local rehabilitation market. The focus is on the use of robots made using local resources and in part or in whole using emerging materials and manufacturing tools. For example, (1) novel materials could be created into low-cost products with 3D printers [53]; (2) soft materials could be created using silicone and local resins [54]; (3) cost-effective smart materials such as shape memory alloys (SMAs) could be used that can change shape and hold a deformed shape when energy is applied [55]; (4) found local objects such as sticks that are discarded or recycled [56]; and (5) local objects that are in abundance within the local environment.

Gardner and colleagues [57] suggest that the best solutions for improving access in global contexts are to meld technology, social, and adaptive innovation. Innovative technology-based solutions that are also cost-effective and affordable have the potential to increase the availability of products, while innovative social and adaptive solutions have the potential to ensure the distribution of goods and the appropriate adoption of new goods into local communities. These social and adaptive solutions augment the delivery of rehabilitation care in the face of limited human resources and information by seeking to develop new ways of organizing human resources, information, and decision-making. So, making robots affordable for low-resource settings demands the field to reorient our perspectives on the design of these robots, the method of using them, and the method of deploying them as a clinical resource and tool in these low-resource settings.

Rehabilitation robots for low-resource settings in developing and developed countries should be affordable, effective, multipurpose, community-based, culturally relevant, appropriate, and sustainable. As a recommendation, we identify the following design guidelines for an affordable technology suite for neurorehabilitation in low-resource settings [47,50–52]:

- Provide upper- and lower-extremity training with simple devices that may be reconfigurable and controlled to permit versatility in training arm, hand, or gait.
- Incorporate cognitive training alongside motor training to ensure patients with complex presentations of stroke and other diseases are challenged.
- Leverage country-relevant serious games or virtual reality environments to increase motivation for training.
- Use transportable and, if possible, portable devices that can be shared and implemented in urban or rural settings as a community resource.
- Incorporate disease-specific assessment metrics that provide clinical users with quantitative information on recovery progression.

- Ensure robot devices are safe to use with minimum expert oversight and minimum maintenance. Ideally, the suite can be supervised by a clinician and a trained technician.
- Target a cost-effective intervention that is within the maximum threshold, which is no more than three times a country's income/per capita threshold.
- Track cost-effectiveness using Eq. (31.1) to calculate the cost/patient session. To do so, keep track of the annual cost of the suite, the maintenance cost of the suite, the depreciation period of the suite, and the annual cost of the therapist [50,51].

$$\frac{\text{Therapy cost}}{\text{session}} = \frac{\text{Equipment and maintenance costs} + \text{therapist's annual salary}}{\text{No. of patients treated in a year}}$$

$$(31.1)$$

- Maximize the use of local resources. Where possible the devices should be constructed from parts and materials manufactured in the country or in the region to support the local economy and build capacity for technological and social innovation.
- Leverage other affordable technologies [44] deemed pragmatic for LMICs that can coexist with a robot technology suite. Transcranial direct current stimulation (t-DCS), virtual reality (VR), and functional electrical stimulation (FES) are cost-effective and have proven augmentative effects.
- Leverage mobile and cellular technologies that are widely available in LMICs to provide additional reporting or monitoring capabilities of patients [58].
- Endeavor to still use good robot-assisted rehabilitation therapy practices. Bejarano and colleagues [59] emphasize designing rehabilitation therapy robots that can deliver large dosage, task-specific training, maintain optimal challenge during a session, provide augmented feedback using audio, visual, and/or haptic cuing techniques, and if possible consider implementing controllers that can deliver techniques that further enhance motor learning such as error augmentation, mirror therapy, action observation, and reinforcement learning.

These guidelines mainly pertain to a therapy robot suite. The use of assistive robots and social robots in LMICs is also rare. Although similar issues in terms of cost-effectiveness apply, other drivers may guide the social robot development in LMICs. There are several low-cost do-it-yourself (DIY) guides that are available for creating social robots that can be leveraged here. One of the most notable is the Ono and Opsoro robot kits by Vandevelde and colleagues [60].

31.3 Next steps toward achieving inclusive rehabilitation robotics

Implementing the design guidelines and strategies discussed in Section 31.2 requires significant investment in building successful partnerships and in developing capacity in clinical and engineering spaces to sustain rehabilitation robots and promote continual technological and social innovation. Fig. 31.5 captures the major partnerships that should be formed and considered.

The Rehabilitation 2030 report, acknowledging the unmet need for rehabilitation worldwide, offers several areas for action [9]. While not specific to rehabilitation robotics, these suggested action areas are presented and interpreted in such a way that provides a framework for achieving Inclusive Rehabilitation Robotics, which is defined as achieving equity of access to technology-assisted rehabilitation. The 10 action goals for Rehab2030 are presented (Table 31.5). At least five of these actions are dependent on macro-systems within a country that heavily rely on government action and leadership. Most of the changes will need to occur through external pressures from WHO and UN. For example, Article 26 of the UN's Convention on the Rights of Persons with Disabilities establishes rehabilitation as a human right, stating that "the availability of accessible and affordable rehabilitation plays a fundamental role in achieving Sustainable Development Goal (SDG) 3—Ensure healthy lives and promote well-being for all at all ages" [23,24]. The ratification by member countries of Article 26 and other associated articles around aging and assistive products will help to increase awareness and pressure to realize increased finances aimed at rehabilitation, increased insurance coverage, and rehabilitation woven into all levels of health response.

A focus on *service delivery* is needed. Unless rehabilitation is seen as a critical aspect of the healthcare model, services will continue to be underfunded and little priority given to the development of a local infrastructure that can sustain in country development of these rehabilitation technologies. As it stands, many LMICs must import rehabilitation technology products at high markup costs. The high costs of these products along with the lack of

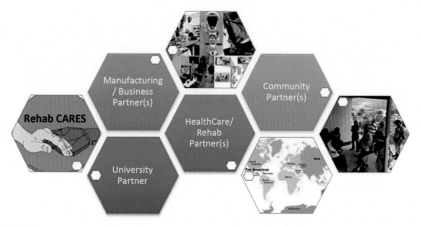

FIGURE 31.5   Essential partnerships for capacity building.

TABLE 31.5 Rehabilitation 2030: call to action [9].

	Actions items
Macro-systemic	Create strong leadership and political support for rehabilitation at subnational, national, and global levels
	Strengthen rehabilitation planning and implementation at national and subnational levels, including within emergency preparedness and response
	Improve integration of rehabilitation into the health sector and strengthen inter-sectoral links to effectively and efficiently meet population needs
	Incorporate rehabilitation into Universal Health Coverage
	Expand financing for rehabilitation through appropriate mechanisms
Service delivery	Build comprehensive rehabilitation service delivery models to progressively achieve equitable access to quality services, including assistive products, for all the population, including those in rural and remote areas
Capacity building	Develop a strong multidisciplinary rehabilitation workforce that is suitable for country context and promote rehabilitation concepts across all health workforce education
	Build research capacity and expand the availability of robust evidence for rehabilitation
Evidence	Collect information relevant to rehabilitation to enhance health information systems including system-level rehabilitation data and information on functioning utilizing the International Classification of Functioning, Disability and Health (ICF)
Partnerships	Establish and strengthen networks and partnerships in rehabilitation, particularly between low-, middle-, and high-income countries

sufficient health insurance per capita make the prescription of standard AT much less rehabilitation robots by clinicians rare and their use by patients a luxury. The push to increase access to rehabilitation robotics must be undertaken with the push to increase access to assistive technology products in general.

A focus on *partnerships* is needed. Successful development and deployment of rehabilitation robots in LMICs are feasible and are clearly easier to do in middle-income countries than in low-income countries. The development, validation, and subsequent commercialization of CR-2 Haptic robot by Techcare Innovation in Malaysia is an example of this possibility. The robot was created and tested initially via academic course work at the University Teknologi Malaysia with funding support from a university-based research grant, a Lab2Market commercialization grant, and a Collaborative Research in Engineering, Science and Teknologi Center Research and Development grant [61−63]. Khor and colleagues acknowledge that the funding via the academic partnership was critical for the success of this robot. In addition, the testing of the robot with patients required hospital collaboration with several hospitals such as Hospital Sultanah Aminah and Hospital Serdang, the Ministry of Higher Education in Malaysia, and the National Stroke Association of Malaysia. The authors of the LMIC chapters detailed here that discussed some rehabilitation robot developments describe a similar process as seen in Malaysia. They often describe the formation of critical relationships between the local universities, public or private hospitals, and the ministries of health. Another common method for the successful development of local rehab robotics infrastructure is through partnership between universities in HICs and those in LMICs. The initial activities for testing affordable robots in Mexico [50] were supported by such collaborations— Marquette University (USA) and ITESM, Chihuahua (Mexico); this is described in Chapter 14. This collaboration model was leveraged to deploy affordable robots in Botswana with funds from the National Institutes of Health facilitating academic partnership between the University of Pennsylvania (the United States) and the University of Botswana (Botswana) (refer to Chapter 25). Many universities support the lab to commercialization process by providing funds for start-up companies to move prototypes to market.

A focus on *capacity building and evidence* is needed. Design, development, and use of rehabilitation technologies in general and rehabilitation robots specifically require multidisciplinary capacity-building efforts that include engineers of all types, medical doctors, nurses, and therapists. Many workforce development initiatives have been made possible through international academic partnerships where universities in HICs form partnerships with universities in LMICs to train engineers, therapists, students, and other allied workers. Rehabilitation robotics can learn from recommendations for increasing clinical engineering and managing healthcare technologies in developing countries [64]. Some key recommendations are to (1) create stronger connections between engineers in order to increase their presence in hospitals and health centers both urban and rural; (2) since many LMICs have a "geographical imbalance" to urban capital, efforts should be made to connect engineers to remote/rural clinics and provide training of local technical staff using telemedicine and mHealth methods; and (3) immerse engineers into the rehabilitation and medical communities and into the local patient community so that they learn to design medical technology that is not only affordable but appropriate with a high probability for acceptance and uptake.

It is not unusual for many people needing rehabilitation in LMICs to be discharged home without it; the need to provide evidence and support for a comprehensive rehabilitation delivery model is an opportunity for rehabilitation robots [7,65,66]. The narrative that rehabilitation robot technologies can bridge gaps in the care given limited clinical resources and provide objective metrics to quantify impairment could help persuade ministries of health in LMICs. For sure, the success of rehabilitation robots in low-resource settings will require research studies to be completed on the robots developed and evidence gathered in the local country context. Engineers, clinicians, and other researchers must be trained and given the requisite skills to safely use and test the efficacy of devices with local populations. Research capacity building is also important and has been boosted through cooperative global action such as the US actions taken to address the global NCD burden. Malekzadeh and colleagues [67] describe the Fogarty International Center of the US National Institutes of Health programs to build individual and institutional research capacity in LMICs using innovative funding mechanisms and partnership-building approaches. Train-the-trainer models were used by colleagues in Botswana to train occupational therapists and engineers to conduct research in Botswana; this was facilitated through an international academic partnership of the University of Pennsylvania, Columbia University, and the University of Botswana (Chapter 25). In summary, Table 31.5 is in itself a guide to achieving inclusive rehabilitation robotics.

Acknowledgments

We thank the authors who contributed to write the HIC and LMIC chapters in this book. Their chapters provided support for the insights presented here.

Conflict of interest

Dr. Michelle Johnson and Dr. Rochelle Mendonca are cofounders of Recupero Robotics, LLC, a company that licenses the Rehab CARES from the University of Pennsylvania. Dr. Johnson is funded by the National Institute of Health to conduct several studies using haptic TheraDrive for stroke and HIV.

References

[1]   WHO. Constitution [Internet]. World Health Organization. Available from: <https://www.who.int/about/governance/constitution>; 2023 [cited 01.03.23].

[2]   WHO. Noncommunicable diseases [Internet]. World Health Organization. Available from: <https://www.who.int/news-room/fact-sheets/detail/noncommunicable-diseases>; 2023 [cited 01.03.23].

[3]   WHO. Disabilities [Internet]. World Health Organization. Available from: <https://www.afro.who.int/health-topics/disabilities >; 2023 [cited 01.03.23].

[4]   Stucki G, Cieza A, Melvin J. The International Classification of Functioning, Disability and Health (ICF): a unifying model for the conceptual description of the rehabilitation strategy. J Rehabil Med 2007;39:279−85, Jun 1.

[5]   GBD 2019 Stroke Collaborators. Global, regional, and national burden of stroke and its risk factors, 1990−2019: a systematic analysis for the Global Burden of Disease Study 2019. Lancet Neurol 2021;20(10):795−820.

[6]   Feigin VL, Brainin M, Norrving B, Martins S, Sacco RL, Hacke W, et al. World Stroke Organization (WSO): Global Stroke Fact Sheet 2022. Int J Stroke 2022;17(1):18−29.

[7]   Bernhardt J, Urimubenshi G, Gandhi DBC, Eng JJ. Stroke rehabilitation in low-income and middle-income countries: a call to action. Lancet Lond Engl 2020;396(10260):1452−62.

[8]   Pandian JD, Kalkonde Y, Sebastian IA, Felix C, Urimubenshi G, Bosch J. Stroke systems of care in low-income and middle-income countries: challenges and opportunities. Lancet Lond Engl 2020;396(10260):1443−51.

[9]   World Health Organization. Rehabilitation 2030: A call to action. 2017.

[10] Fact Sheet: Aging in the United States [Internet]. Population Reference Bureau. Available from: <https://www.prb.org/resources/fact-sheet-aging-in-the-united-states/>; 2023 [cited 01.03.23].

[11] Mohebi R, Chen C, Ibrahim NE, McCarthy CP, Gaggin HK, Singer DE, et al. Cardiovascular disease projections in the United States Based on the 2020 census estimates. J Am Coll Cardiol 2022;80(6):565−78.

[12] Zimbelman JL, Juraschek SP, Zhang X, Lin VWH. Physical therapy workforce in the United States: forecasting nationwide shortages. PM&R. 2010;2(11):1021−9.

[13] Lin V, Zhang X, Dixon P. Occupational therapy workforce in the United States: forecasting nationwide shortages. PM&R. 2015;7(9):946−54.

[14] Zhang X, Lin D, Pforsich H, Lin VW. Physician workforce in the United States of America: forecasting nationwide shortages. Hum Resour Health 2020;18(1):8.

[15] Johnson MJ, Bui K, Rahimi N. Medical and assistive robotics in global health. Handbook of global health. 1st ed. Spring Cham; 2021. p. 1815−60.

[16] Mehrholz J, Pohl M, Platz T, Kugler J, Elsner B. Electromechanical and robot-assisted arm training for improving activities of daily living, arm function, and arm muscle strength after stroke. Cochrane Database Syst Rev 2018;2018(9). Available from: http://www.embase.com/search/results?subaction = viewrecord&from = export&id = L623714198.

[17] Veerbeek JM, Langbroek-Amersfoort AC, van Wegen EEH, Meskers CGM, Kwakkel G. Effects of robot-assisted therapy for the upper limb after stroke. Neurorehabil Neural Repair 2017;31(2):107−21.

[18] Calabrò RS, Sorrentino G, Cassio A, Mazzoli D, Andrenelli E, Bizzarini E, et al. Robotic-assisted gait rehabilitation following stroke: a systematic review of current guidelines and practical clinical recommendations. Eur J Phys Rehabil Med 2021;57(3):460−71.

[19] Research and Markets. Global Rehabilitation Robots Market 2023−2027 [Internet]. Research and MarketsL The World's largest Market Research Store. Available from: <https://www.researchandmarkets.com/reports/5017650/global-rehabilitation-robots-market-2023-2027>; 2023 [cited 05.10.22].

[20] World Bank. New World Bank country classifications by income level: 2022−2023 [Internet]. World Bank Blogs. Available from: <https://blogs.worldbank.org/opendata/new-world-bank-country-classifications-income-level-2022-2023>; 2023 [cited 25.03.23].

[21] World Bank. World Bank Country and Lending Groups [Internet]. The World Bank Data. Available from: <https://datahelpdesk.worldbank.org/knowledgebase/articles/906519-world-bank-country-and-lending-groups; 2023 [cited 25.03.23].

[22] Cambridge Dictionary. Inclusivity [Internet]. Cambridge Dictionary. Available from: <https://dictionary.cambridge.org/us/dictionary/english/inclusivity; 2023 [cited 25.03.23].

[23] United Nations. Article 26 — Habilitation and rehabilitation [Internet]. United Nations: Department of Economic and Social Affairs. Available from: <https://www.un.org/development/desa/disabilities/convention-on-the-rights-of-persons-with-disabilities/article-26-habilitation-and-rehabilitation.html; 2023 [cited 25.03.23].

[24] World Health Organization. Assistive technology [Internet]. World Health Organization. Available from: <https://www.who.int/news-room/fact-sheets/detail/assistive-technology; 2023 [cited 25.03.23].

[25] Lawrence ES, Coshall C, Dundas R, Stewart J, Rudd AG, Howard R, et al. Estimates of the prevalence of acute stroke impairments and disability in a multiethnic population. Stroke. 2001;32(6):1279–84.

[26] Everard GJ, Ajana K, Dehem SB, Stoquart GG, Edwards MG, Lejeune TM. Is cognition considered in post-stroke upper limb robot-assisted therapy trials? A brief systematic review. Int J Rehabil Res 2020;43(3):195–8.

[27] Lee PH, Yeh TT, Yen HY, Hsu WL, Chiu VJY, Lee SC. Impacts of stroke and cognitive impairment on activities of daily living in the Taiwan longitudinal study on aging. Sci Rep 2021;11(1):12199.

[28] Prvu Bettger J, Liu C, Gandhi DBC, Sylaja PN, Jayaram N, Pandian JD. Emerging areas of stroke rehabilitation research in low- and middle-income countries: a scoping review. Stroke. 2019;50(11):3307–13.

[29] Fasoli SE. A paradigm shift in rehabilitation robotics: moving toward more functional outcomes after stroke? Neurorehabil Neural Repair 2018;32(4–5) 389–389.

[30] Bui KD, Lyn B, Roland M, Wamsley CA, Mendonca R, Johnson MJ. The impact of cognitive impairment on robot-based upper-limb motor assessment in chronic stroke. Neurorehabil Neural Repair 2022;36(9):587–95.

[31] Rusch M, Nixon S, Schilder A, Braitstein P, Chan K, Hogg RS. Impairments, activity limitations and participation restrictions: prevalence and associations among persons living with HIV/AIDS in British Columbia. Health Qual Life Outcomes 2004;2:46, Sep 6.

[32] Logan LM, Semrau JA, Debert CT, Kenzie JM, Scott SH, Dukelow SP. Using robotics to quantify impairments in sensorimotor ability, visuospatial attention, working memory, and executive function after traumatic brain injury. J Head Trauma Rehabil 2018;33(4): E61–73.

[33] Chang YL, Luo DH, Huang TR, Goh JOS, Yeh SL, Fu LC. Identifying mild cognitive impairment by using human-robot interactions. J Alzheimers Dis 2022;85(3):1129–42.

[34] Bourke TC, Lowrey CR, Dukelow SP, Bagg SD, Norman KE, Scott SH. A robot-based behavioural task to quantify impairments in rapid motor decisions and actions after stroke. J Neuroeng Rehabil 2016;13(1):91.

[35] Bui KD, Johnson MJ. Robot-based measures of upper limb cognitive-motor interference across the HIV-stroke spectrum. IEEE Int Conf Rehabil Robot 2019;2019:530–5.

[36] Asimov I. Runaround. Astounding science fiction. New York: Street and Smith Pub.; 1942.

[37] RESNA. Code of Ethics [Internet]. RESNA: Rehabilitation Engineering and Assistive Technology Society of North America. Available from: <https://www.resna.org/Portals/0/RESNA%20Code%20of%20Ethics%20and%20Standards%20of%20Practice%20_%20FINAL.pdf?ver = mcmYEkPAonprJKGVFRlWdA%3d%3d; 2023 [cited 25.03.23].

[38] Tzafestas SG. Roboethics: a navigating overview. 1st ed. Springer; 2015. p. 1–204.

[39] Prem E. From ethical AI frameworks to tools: a review of approaches. AI Ethics [Internet]. Available from: <https://doi.org/10.1007/s43681-023-00258-9; 2023 [cited 09.02.23].

[40] Johnson M, Mohan M, Mendonca R. Therapist-patient interactions in task-oriented stroke therapy can guide robot-patient interactions. Int J Soc Robot 2022;14:1–20, Jun 10.

[41] Powell D, O'Malley MK. The task-dependent efficacy of shared-control haptic guidance paradigms. IEEE Trans Haptics 2012;5(3):208–19.

[42] Wallach W, Allen C, Smit I. Machine morality: bottom-up and top-down approaches for modelling human moral faculties. AI Soc 2008;22(4):565–82.

[43] Wallach W, Allen C. Moral machines: teaching robots right from wrong. Oxford: Oxford University Press; 2009. p. 1–288.

[44] Cambridge Dictionary. Affordabilty [Internet]. Cambridge Dictionary. Available from: <https://dictionary.cambridge.org/us/dictionary/english/affordability; 2023 [cited 25.03.23].

[45] Ekechukwu END, Olowoyo P, Nwankwo KO, Olaleye OA, Ogbodo VE, Hamzat TK, et al. Pragmatic solutions for stroke recovery and improved quality of life in low- and middle-income countries-a systematic review. Front Neurol 2020;11:337.

[46] World Health Organization. New cost-effectiveness updates from WHO-CHOICE [Internet]. World Health Organization. Available from: <https://www.who.int/news-room/feature-stories/detail/new-cost-effectiveness-updates-from-who-choice; 2023 [cited 25.03.23].

[47] Demofonti A, Carpino G, Zollo L, Johnson MJ. Affordable robotics for upper limb stroke rehabilitation in developing countries: a systematic review. IEEE Trans Med Robot Bionics 2021;3(1):11–20.

[48] Buschfort R, Brocke J, Hess A, Werner C, Waldner A, Hesse S. Arm studio to intensify the upper limb rehabilitation after stroke: concept, acceptance, utilization and preliminary clinical results. J Rehabil Med 2010;42(4):310–14.

[49] Hesse S, Heß A, Werner C, Kabbert N, Buschfort R. Effect on arm function and cost of robot-assisted group therapy in subacute patients with stroke and a moderately to severely affected arm: a randomized controlled trial. Clin Rehabil 2014;28(7):637–47.

[50] Valles KB, Montes S, de Jesus Madrigal M, Burciaga A, Martínez ME, Johnson MJ. Technology-assisted stroke rehabilitation in Mexico: a pilot randomized trial comparing traditional therapy to circuit training in a Robot/technology-assisted therapy gym. J Neuroeng Rehabil 2016;13(1) 83–83.

[51] Johnson MJ, Rai R, Barathi S, Mendonca R, Bustamante-Valles K. Affordable stroke therapy in high-, low- and middle-income countries: from theradrive to rehab CARES, a compact robot gym. J Rehabil Assist Technol Eng 2017;4, May.

[52] Lo K, Stephenson M, Lockwood C. The economic cost of robotic rehabilitation for adult stroke patients: a systematic review. JBI Database Syst Rev Implement Rep 2019;17(4):520–47.

[53] Hussaini A, Kyberd P, Mulindwa B, Ssekitoleko R, Keeble W, Kenney L, et al. 3D printing in LMICs: functional design for upper limb prosthetics in Uganda. Prosthesis 2023;5(1):130–47.

[54] Vo QNT, Huynh TTM, Dao SVT. Applying soft actuator technology for hand rehabilitation. In: Van Toi V, Nguyen TH, Long VB, Huong HTT, editors. 8th International conference on the development of biomedical engineering in Vietnam. Cham: Springer International Publishing; 2022. p. 123–32.

[55] Alaneme KK, Okotete EA. Reconciling viability and cost-effective shape memory alloy options – A review of copper and iron based shape memory metallic systems. Eng Sci Technol Int J 2016;19(3):1582–92.

[56] Carroll D, Yim M. Truss optimization with random piecewise linear beams. 2017. V011T15A009.

[57] Gardner CA, Acharya T, Yach D. Technological and social innovation: a unifying new paradigm for global health. Health Aff 2007;26(4):1052−61.

[58] Beratarrechea A, Lee AG, Willner JM, Jahangir E, Ciapponi A, Rubinstein A. The impact of mobile health interventions on chronic disease outcomes in developing countries: a systematic review. Telemed J E Health 2014;20(1):75−82.

[59] Chia Bejarano N, Maggioni S, De Rijcke L, Cifuentes GC, Reinkensmeyer D. Robot-assisted rehabilitation therapy: recovery mechanisms and their implications for machine design. Emerging Therapies in Neurorehabilitation II. Springer; 2016. p. 197−223.

[60] Vandevelde C, Wyffels F, Vanderborght B, Saldien J. Do-it-yourself design for social robots: an open-source hardware platform to encourage innovation. IEEE Robot Autom Mag 2017;24(1):86−94.

[61] Khor KX, Chin PJH, Hisyam AR, Yeong CF, Narayanan ALT, Su ELM. Development of CR2-Haptic: A compact and portable rehabilitation robot for wrist and forearm training. In: 2014 IEEE Conference on Biomedical Engineering and Sciences (IECBES); 2014. pp. 424−9.

[62] Khor KX, Chin PJH, Yeong CF, Su ELM, Narayanan ALT, Abdul Rahman H, et al. Portable and reconfigurable wrist robot improves hand function for post-stroke subjects. IEEE Trans Neural Syst Rehabil Eng 2017;25(10):1864−73.

[63] Khor K, Chin P, Fai Y, Su E, Narayanan L, Abdul Rahman H, et al. Evaluation of CR2-haptic movement quality assessment module for hand. In: 3rd International Conference for Innovation in Biomedical Engineering and Life Sciences; 2021. pp. 95−102.

[64] Secca MF. Clinical engineering in developing countries. Clinical engineering handbook. Elsevier; 2004.

[65] Owolabi MO, Platz T, Good D, Dobkin BH, Ekechukwu END, Li L. Editorial: translating innovations in stroke rehabilitation to improve recovery and quality of life across the globe. Front Neurol 2020;11:630830.

[66] Aguiar Noury G, Walmsley A, Jones RB, Gaudl SE. The barriers of the assistive robotics market-what inhibits health innovation? Sensors. 2021;21(9) Apr 29.

[67] Malekzadeh A, Michels K, Wolfman C, Anand N, Sturke R. Strengthening research capacity in LMICs to address the global NCD burden. Glob Health Action 2020;13 (1):1846904.

Chapter 32

Toward global use of rehabilitation robots and future considerations

Sumudu Perera Kimmatudawage[1,2,3], Rakesh Srivastava[4], Kavita Kachroo[5], Suman Badhal[6] and Santosh Balivada[7]

[1]Department of Medicine, @AgeMelbourne, Royal Melbourne Hospital (Melbourne Health), University of Melbourne, Melbourne, VIC, Australia, [2]Western Health, Melbourne, VIC, Australia, [3]London School of Tropical Medicine, London, United Kingdom, [4]WISH Foundation, Indian Council of Medical Research, Medical Council of India, NBE, Ministry of Health and Family Welfare, New Delhi, Delhi, India, [5]Kalam Institute of Health Technology (KIIT), Visakhapatnam, Andhra Pradesh, India, [6]Vardhman Mahavir Medical College, Safdarjung Hospital, New Delhi, Delhi, India, [7]Andhra Pradesh Med Tech Zone, Visakhapatnam, Andhra Pradesh, India

Learning objectives

At the end of this chapter, the reader will be able to:

1. Describe the emerging sector of robotic rehabilitation.
2. Understand the future growth potential of robotic rehabilitation with joint research between technology and medical experts for product development.

32.1 Introduction

Much interest has been raised over the preceding decade about the benefits of the implementation of rehabilitation robotics within standard rehabilitation practice. In 2020 alone, the global market for rehabilitation robotics was valued at US $1.05 billion, with projected growth of 14.8% annually by 2026 [1]. Three major categories of robotic rehabilitation exist—operational machines, wearable robotics [2], and social robotics [3]. While a vast majority of this valuation comes from exoskeleton systems and trials, the market encompasses any number of existing or prototype automated therapy machines. Rehabilitation robotics have been proposed both to automate and augment traditional rehabilitation

Rehabilitation Robots for Neurorehabilitation in High-, Low-, and Middle-Income Countries.
DOI: https://doi.org/10.1016/B978-0-323-91931-9.00002-5

services, thus improving a patient's autonomy, self-control, functional independence, and mood and reducing future morbidity and mortality, while hopefully reducing long-term costs of rehabilitation [4]. Despite these advances in robotic rehabilitation, however, a number of key barriers exist presently for global/regional/national adoption, resulting in high cost, poor access, inadequate provision, and penetration in rehabilitation institutions and robotic market.

The current ongoing COVID-19 pandemic has shown both a global necessity for technologies such as rehabilitation robotics (which may allow for care provision remotely, in a time-efficient manner), while also demonstrating global supply issues that may impede global accessibility to such therapy. Current issues with disruptions to the global supply chain and global chip shortage, projected to possibly continue into 2023 [5], demonstrate the possibly tenuous nature by which global access to rehabilitation robotics may be impeded.

While the above barriers are significant, much work is being conducted in breaking them to allow for clinically beneficial advances in rehabilitation robotics. Significant work is being considered and conducted in policy reform, regulatory reform, pricing reform, and marketing reform to enable equitability of care and access. Standards, protocols, and research are being conducted to advance the field, while production advances, mass production capability, and creative financing are proposed to improve cost-effectiveness of therapy and care.

There is a need to explore at depth both existing barriers, current workarounds, and future solution possibilities to enable global use of robotics, while also discussing future considerations for ongoing robotic rehabilitation care.

32.1.1 Need for robotics in rehabilitation

As per the WHO, about 2.4 billion population are presently dealing with a health issue that could benefit from rehabilitation globally. With changes taking place in both the health and characteristics of the population worldwide, this estimated need for rehabilitation is only going to increase in the coming years [6]. Neurorehabilitation will be occupying central space in the coming years in rehabilitation services.

Several neurological disorders such as stroke require institutional neurorehabilitation; despite timely and comprehensive multidisciplinary management, patients often have some residual permanent disability. Comprehensive rehabilitation management varies by patient in terms of its duration and processes. However, the goals of neurorehabilitation are that it should be multidisciplinary, improve quality of life with minimum residual deficits and disability, and maximize capacity for quality performance of tasks in day-to-day life. The growing need for technological solutions to improve functional outcomes after rehabilitation has led to many assistive technology solutions, including the application of robotics in neurorehabilitation.

Availability of rehabilitation institutions—more so neurorehabilitation institutions—the shortage of qualified neurorehabilitation professionals, and the neurorehabilitation robotic advancement are surrogate indicators to estimate this demand and supply gap. The WHO Global Atlas of the Health Workforce demonstrates that both the general health workforce and rehabilitation human resources are below par; i.e., it is often absent or, when available, is deficient or splintered [7]. This dearth of globally disaggregated data makes it more arduous to evolve guidelines and further recommend policies which are required for addressing unmet rehabilitation needs and barriers to access. The data presented reveal that there is a global deficiency of qualified rehab professionals, which leads to unmet needs for rehabilitation [8,9]. Across the globe, the levels of provision of institutional rehabilitation services may be taken as a marker for the number of health workforce for rehabilitation management. In light of that, it can be said that institutional rehabilitation demand is much higher than the supply [9]. Especially in middle- and low-income countries, the density of skilled rehabilitation professionals is often less than 10 per 1 million when compared to greater than 10,000 per 1 million for those in high-income countries (Fig. 32.1).

Other health professionals like physiotherapists who are able to give rehabilitation services are also meager in low-income countries (Fig. 32.2). Figs. 32.1–32.3 provide an overview of the differences in rehabilitation service capacity across the world's income spectrum [9].

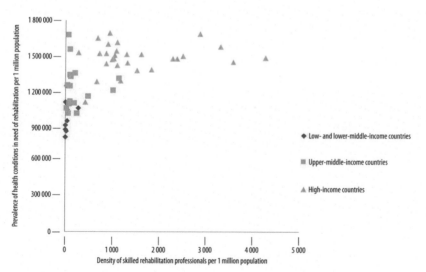

FIGURE 32.1   Prevalence of rehabilitation-relevant health conditions compared to the density of skilled rehabilitation professionals in 12 low- and lower-middle-income countries, 16 upper-middle-income countries, and 31 high-income countries per 1 million population.

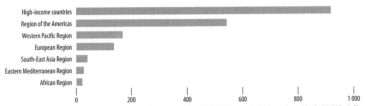

Note: High-income countries are those with a gross national income per capita of US$12 475 or more in 2015, as estimated by the World Bank. The remaining data in this figure are from low- and middle-income countries only.
Source: WCPT, and OECD

FIGURE 32.2   Density of physiotherapists per 1 million populations by region (107 countries).

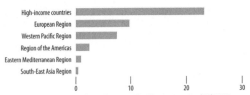

Note: High-income countries are those with a gross national income per capita of US$ 12 475 or more in 2015, as estimated by the World Bank. The remaining data in this figure are from low- and middle-income countries only. No data are available from the African Region.
Source: International Society of Physical and Rehabilitation Medicine (ISPRM)

FIGURE 32.3   Density of physical and rehabilitation medicine doctors per 1 million populations by region (data from 48 countries).

Low-income countries often have the lowest density of therapists (Fig. 32.3). There are less than 10 physiotherapists per million population in many countries in regions in Africa and Southeast Asia. Data from several countries are still not available.

A medically qualified physiatrist is needed to lead the multidisciplinary neurorehabilitation care. The density of physiatrists (physical and rehabilitation medicine doctors) is below 1 per million in East Mediterranean and Southeast Asian regions. The estimates on the number of physiatrists in many countries like those in Africa are not clear, but an article in the South African Medical Association revealed that there were six physicians in the country who specialized in PM&R, whereas there are no PM&R doctors residing in East and West Africa [10].

Additionally, occupational therapists, speech therapists, psychologists, and orthotists are part of the core neurorehabilitation team. There are no global reliable data on their numbers. It is estimated that they will all be in short supply in all regions of the world.

32.2   Application areas for robots in rehabilitation and health

Robotics can also be classified according to the broad areas of its application within comprehensive rehabilitation and health programs. The following section briefly describes some of these areas.

32.2.1   Therapeutic robots

Whenever a robotic product is used as complementary to therapies within a rehabilitation program, it is called a therapeutic robot. In neurorehabilitation, where it is been used for repetitive exercise therapy, typically done by a therapist, the therapeutic robotic device is built on the principle of machine learning and can perform specific movements for a defined time/protocol. There are two types of robotic devices: end effectors which adjust to any body type and exoskeleton devices, which have to be customized to body types. The advantages of these robotic devices are improved efficiency and output, the ability to target a large patient load, data storage, and memory for better comparison, and provision of biofeedback for the patient. More information on these devices can be found in Chapter 4 of this book.

32.2.2   Robots for clinical training

Medical training requires a simulation lab and simulators for practical learning. Initially, mannequins and crash test dummies were used. Now, robots can be used for clinical training. These training robots are practical models or devices for simulation that can enormously enhance the healthcare professional's practical and clinical knowledge. One recent advance in the field of training robots is the pediatric robot, HAL [11]. This is an excellent example of robotics with artificial intelligence (AI) that can emulate the expressions of a pediatric patient and clinical symptoms of cardiac conditions like arrhythmia, arrest, and rapid breathing. Further, it can also express a variety of facial expressions, like pain and anxiety, thus creating a realistic situation of dealing with a patient and enhancing the quality of learning.

32.2.3   Surgical robots

The surgical field is a potential and promising area for the applications of robots in the past few years and is gaining popularity. Surgical robots can show a high-definition, magnified, and three-dimensional view of the surgical field. Robotic surgeries are gaining momentum because these are "minimally invasive" as compared to conventional surgeries and have less blood loss, are more time-efficient, comparatively error-free, and the operative procedures can be easily done with a very small incision. These robots are used for urology surgeries, prostatic surgery, transplant surgeries, orthopedics (Robodoc) [12] for hip and knee replacement, gynecology for hysterectomy, and neurosurgery (robotic invasive surgical assistant [ROSA]). From the patient's perspective, robotic surgery means fewer complications and reduced chances of infection because of shorter hospital stays and quicker recovery. It reduces the clinician or surgeon load.

32.2.4   Personal-use robots

Personal-use robots, often called social or service robots, are most often used in the care of older adults and persons with disabilities. Prescription and medication dispensing robots are service robots that are being developed to help older adults take medications more efficiently and with better precision. Robots to help caretakers are another new trend. Caretaking is demanding, more so with patients with neurological disabilities who have difficulty in home/community living, the aging population, and individuals with dementia. The role of social robotics in this segment is increasing significantly. These robots help patients in their day-to-day activities, e.g., helping with mobility, transport, routine checkups, memory updates, and reminders.

32.2.5   Hazardous task-performing robots

As the world is facing COVID-19 and other infectious diseases, robots can be deployed for hospital sanitization. There are robots which are specialized to scan and disinfect the hospital environment and air using UV for removing harmful organisms. Siemens IRbot, Skytro, Bioquell, and Intellibot are some of the companies at the global level involved in sanitization and disinfection robots [13].

32.2.6   Teleservices robots

Telemedicine robots connect a healthcare provider to a patient/specialist/diagnostic lab/pharmacy for making consultations more comprehensive and proving end-to-end solutions. It provides needs-based solutions for at-home and community care by using tele-consult platforms and wearable robotics for basic consultation, specialist consultation, e-diagnostic, e-pharmacy, repair/replacement of assistive devices, mobility ADL technology provision or service in the home, etc. VGo and PadBotB3 are a few examples.

32.2.7   Barriers to the adoption of rehabilitation robotics

A number of key barriers exist to the global/regional/national adoption of rehabilitation robotics. They can be grouped as clinical, technological, psychological, and financial barriers.

32.2.7.1   Clinical barriers

Robots are used for baseline assessments before initiating automated therapy. The major clinical barrier is in the human−machine relationship, which often results in the therapist choosing manual over automated machine assessment. Lack of training and infrastructure is another issue, especially in the developing world [14]. The universal application of automated robotics assessments still has a long way to go, but with systematic training, advocacy, scientific evidence, and cost reduction, it can become an integral part of assessment for neurorehabilitation.

Deploying robotic therapy in health care can expedite therapy; however, there are several barriers including chances of errors and mechanical failure with these robots, removal of the flexibility of operation for the therapist, and mechanical malfunction can cause injury or even cost human lives. As we know, the human neural network is a very complex system, which facilitates and reenforces learning, however creating a machine learning-based robotic therapy solution that is as sound, precise, flexible, and accurate as the human brain is a complex process. The scope of medicine is vast, and it is very difficult to incorporate everything in a machine from initial assessment to therapy and surgery. We may need multiple robots for these tasks which increases the costs, although a few examples of humanoid robots like Sophia [15] have been used for nursing tasks. The missing link for robotic treatment is intelligent adaptation and flexibility to situations that are unique to an individual patient. Social factors also play a role, and lack of exposure and acceptability to robots is a big barrier in integrating in robotic neurorehabilitation. A comprehensive and perfect decision support system in neurorobotic rehabilitation, which can complement all facets of a multidisciplinary rehabilitation protocol, is yet to come.

32.2.7.2 Financial barriers

Robotic therapy devices, wearables, and social robotic solutions are expensive. Additionally, infrastructure development expenses are also high for the initial setup, training, and maintenance for both rehabilitation institutions and for patients in home/community care. Financial factors usually outweigh technological advantages, both while planning investment at the institutional level and when including it in health insurance for home/community care. The infrastructure, logistics, identification of qualified and trained personnel for robotic therapy/surgery, coordination and collaboration among various specialties, appropriate patient selection, and dos and don'ts list to monitor and record the outcomes of the robotic intervention are essential and need to be planned. There also needs to be meticulous, lifelong, continuous training requirements in neurorehabilitation institutes to stay updated with technological advances. The cost of bringing robotics to healthcare settings is the biggest barrier today. Wearable and social robotics are yet not covered by any insurance system. If the increased cost is covered by a health scheme or is subsidized, it may lead to better integration of robotic rehabilitation both at the institutional and individual level, especially in developing countries, where it can address the shortage of physiatrist and therapists.

32.2.8 Psychological barriers

The psychological barriers revolve around a patient with already compromised health issues, denial of disabling consequence of diseases/injury, high possibility of residual disability, lack of motivation, gradual erosion of friends/familial support system, loss of job/earning, gradual depletion of the financial reserve, all leading to endogenous depression. Patients do not typically opt for robotic rehabilitation without professional counseling. Active participation by patients at both physical and mental levels during therapy is pivotal for recovery. This can be promoted through adaptive assistance by the therapist or the robot; however, the robot cannot deal with psychological issues. Last, a major barrier for healthcare workers is their fear of losing jobs with the employment of more robots, which should be addressed.

32.2.8.1 Technological barriers

The major barrier to current robotic solutions as per OSHA is that "many robot accidents do not occur under normal operating conditions but instead, during programming, program touch-up or refinement, maintenance, repair, testing, setup, or adjustment." It said there are seven main hazards with robots [16]. With the improvement in rehabilitation robotics, there is potential for universal integration once we eliminate all known accidents.

- Robots require computing speeds close to the human brain, which puts tremendous strain on the silicon chip. Therefore we need other solutions for withstanding the speed and preventing breakdown.
- Lack of exposure and access to this newer technology is another barrier, especially in the developing countries.
- A challenge of current robotic technology is its inflexibility in terms of customization of robotic care, and one program does not necessarily fit everyone.
- Technical errors, errors in the control system, or software errors due to faulty parts are other barriers, which can result in the failure of the robot.
- The environment may also create electromagnetic or radio frequency interference that can influence the robot function or cause glitches in power systems. This may cause electric shock or short circuits and fire, which need to be eliminated if it is used in developing countries.
- Inappropriate installations are frustrating and totally avoidable. Incorrectly installed robots lead to many potential hazards, even costing human lives (Fig. 32.4).

32.3 The future robotic rehabilitation technologies

Advanced robotic technology has great potential for rehabilitation clinicians, especially physiatrists and therapists. Robots can be programmable, globally standardized as products/parts, and can be maneuverable for local application in rehabilitation. Different mechanical, electronic, and mechatronic devices from artificial limbs to robotic devices can be integrated through the Internet of Things (IoT) into a single robotic application device for rehabilitation robotics [17].

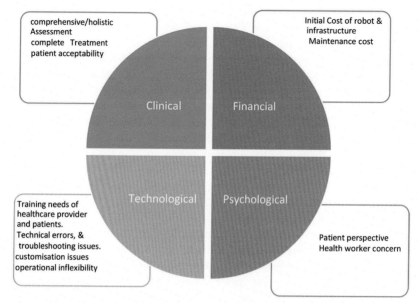

FIGURE 32.4   Barriers and challenges of robotics rehabilitation in various domains.

Demand has been increasing for physical medicine and for therapy across the world. Robotic rehabilitation and assistive technologies have the capacity to mitigate both times for recovery and expenses on human resources for a comprehensive, multidisciplinary rehabilitation program. Currently, rehabilitation robots can move a certain body part, but not the entire body. Similarly, cloud robotics has come out as a collective technology which links cloud computing with service robotics through developments in wireless networking, broadband storage, communication techniques, wearable sensor systems, and the existence of Internet sources [18].

Despite the number of rehabilitation robotics increasing globally, they are limited in scope due to technical gaps in the algorithms, mechanical designs, sensor selection, reduced battery life, mismatch in human—robotic interface, and proper materials selection. The algorithms used for systems include dynamic Bayesian network, support vector machine, linear-based artificial neural network [19], a long short-term memory algorithm and deep neural network, multilayer perception neural network, etc. By using these algorithms, the intelligence of the robot is enhanced to accommodate rehabilitation with proper coding [19]. The utilization of Industry 4.0 technologies like AI algorithms has facilitated recovery by using a complete estimation of a patient's responsiveness to rehabilitation when interaction occurs with robots. However, mechanical design might be a gap which can be addressed by additive manufacturing, 3D printing with a structured framework, or the innovative use of soft robots leveraging smart materials. Sensors like force sensors, encoders, EMG sensors, or foot pressure sensors can also be used. The battery life depends upon the use of the machine. Battery power can be increased by using motors and rechargeable lithium-ion battery with battery backup.

Material selection is an increasingly important criterion and can include polylactic acid, acrylonitrile butadiene styrene, copolymers (modified) such as polyethylene terephthalate glycol, or other materials such as aluminum, titanium, or polytetrafluoroethylene. These are low-cost and effective to meet required mechanical parameters while satisfying the FDA medical standards [20]. Consideration of these factors with respect to the human–robotic interface may solve current issues.

The human–robotic interface differs in the design of upper-limb rehabilitation robots and lower-limb rehabilitation robots. There is an increasing move to provide personalized and wearable exoskeleton robots that persons with disabilities can use inside and outside hospital environments and in the community as both therapeutic and assistive systems. To realize this dream, upper- and lower-limb rehabilitation robots need a more lightweight framework design to enhance safety and comfort. There is a need for compact actuators, long-lasting power supplies, and the use of flexible and lighter materials in building robotic devices [20]. Safe systems can enable users to train independently, without a highly skilled rehabilitation clinician. There is also a need to continue to examine the intersection of these systems with virtual reality (VR), augmented reality, and AI in order to provide more motivation and real-life practice scenarios for therapy. For example, studies show that in VR-based robotic intervention, tasks can be accomplished according to the patient's need even through video games, so as to enhance the patient's ability to imagine movements, reengage the motor circuits, and improve upper-limb motor functions. Providing biofeedback and/or interfacing robot movements with the biological brain signals can help to create brain–computer interfaces that support more natural learning or relearning paradigms after brain injury. Unfortunately, there is still a technology gap here in that we do not yet understand how to best enable complete interfacing between the biological brain and AI. Solving this technology gap requires a good theoretical understanding of energy cost, neural alteration, and biomechanical function after stroke.

32.3.1 Robotics in neurorehabilitation

Robotic-mediated neurorehabilitation is a rapidly emerging research discipline that combines robotics, neuroscience, and rehabilitation to reimagine how neurological illnesses are treated. Research on robot-mediated treatment is widespread and is not just confined to neurorehabilitation. The outcome of this research can be integrated across all robotic applications in health care; therefore it is investment attractive. The two primary goals of robotic neurorehabilitation are to regain the functionality of the upper limb and to assist in gait reeducation. Multiple clinical trials have been conducted regarding the efficacy of these robotic technologies. However, the time has come to address the global need for a comprehensive systems-based approach toward robotic-mediated neurorehabilitation for a large patient population. This implies the adoption of universal design, compliance with global standards both in processes and products, intellectual property right and patent compliance, good manufacturing and marketing practices, and transparent pricing systems. Investment in research in neurorehabilitation is both a public and private responsibility. It can lead to universal access with assured health outcomes.

Research is being conducted to further advance technological gaps existing in rehabilitation robotics. Such advances include increasing the range of motion and obtaining highly sophisticated sensory input from the users for AI. Additional areas for improvement include the development of exercise programs that use real objects, and the development of robotics not only as training devices but also as assistive devices for daily routines in the home.

In the coming years, robotic systems may be accepted as a specific rehabilitation tool; however, this requires specific and large-sample studies of comparisons between robot-assisted therapy and conventional therapy. Due to COVID-19, there was a significant need to develop alternative methods and modes of providing rehabilitation, specifically in home environments. Hence, the effective use of rehabilitation robots for telerehabilitation is needed.

32.3.2 Future neurorehabilitation market

The global neurorehabilitation market was estimated to be $1.05 million in 2020 and is anticipated to reach $2.45 million by 2030, representing an 8.3% compound annual growth rate (CAGR) from 2021 to 2030 [18]. The demand for neurorehabilitation is increasing with the rise in the prevalence of neurological disorders. It is anticipated that this will also drive the growth of the neurorehabilitation device market [21]. According to the WHO, over 6 million die each year due to stroke. It is also predicted that the number of people affected by neurological diseases including stroke will increase from 95 million in 2015 to 103 million in 2030 [21]. These statistics highlight the increasing need for innovative interventions, ultimately increasing the demand for neurorehabilitation devices, including robotics.

32.3.3 Market segmentation

The global neurorehabilitation market is segmented into four different parts [22]:

1. Type: Neurorobotic devices, noninvasive stimulators, and brain−computer interfaces.

2. Operation: Stroke, Parkinson's disease, multiple sclerosis, spinal cord injury, traumatic brain injury, and cerebral palsy.
3. End user: Rehabilitation centers, hospitals, and conventions.
4. Disease severity: Level of impairment and disability.

The global neurorehabilitation devices market by region is $950 million for the United States, $890 million for Europe, $800 million for Asia Pacific, and $400 million for Middle East and Africa [23] (Fig. 32.5).

Owing to the ongoing technological advancements and the rising prevalence of neurological illnesses, the market for neurorehabilitation devices is also growing at an exponential rate. Other drivers for market growth include the availability of well-established healthcare infrastructure and increased awareness among providers and end users about the benefits of neurorehabilitation devices.

Because of favorable government initiatives, the European market for neurorehabilitation devices is expected to grow at a considerable CAGR for a certain period. In addition, the expansion of the healthcare system as a result of the improving economy will almost certainly fuel demand and growth in the future years. Germany will be the leading market among all countries, contributing significantly to the ongoing market growth. In the Asia Pacific region, the market for neurorehabilitation devices has primarily become an economic one, owing to its population size and the number of end users. The increasing number of openings provided by the region's ever-evolving healthcare structure has done wonders for the market's overall growth in recent years; India, Japan, and China are the topmost countries in terms of espousing neurorehabilitation devices. Furthermore, the region is

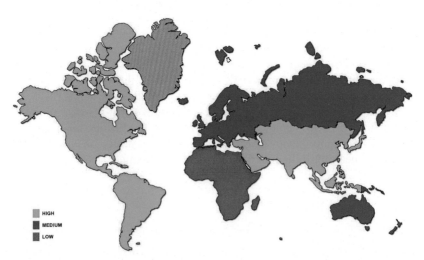

FIGURE 32.5   Neurorehabilitation devices market growth rate by region.

defined by a demographic characteristic that has a positive impact on the market, as elderly individuals are more susceptible to numerous neurological ailments. Apart from that, the neurorehabilitation devices market in China is expected to rise at a fast pace, owing to the rising prevalence of stroke [23].

32.3.4 Key players

The present neurorehabilitation devices market is highly competitive, with a large number of key companies acting as lead players. The following are a few of the companies that are now dominating the market:

1. Hocoma AG, Switzerland.
2. Medtronic Plc, USA.
3. Bionik Labs, Canada.
4. St. Jude Medical Inc. (Abbott Laboratories), USA.
5. Ectron Ltd., California.

32.3.5 Robot-assisted telerehabilitation

Objective motor assessment which is performed by collecting data via rehabilitation robots or robotic devices can provide consistent and reliable data [24]. These assessments are automatic assessments of the motor function of the patient and are free of subjective bias from a therapist. This is unlike the traditional assessment systems that require one-on-one interaction. In systems for home-based rehabilitation, the assessments and evaluation are done with remote supervision by a physiatrist and therapist. This can be achieved by combining rehabilitation devices with information technology. The assessment and performance data can be transmitted to the physiatrist and therapist over the Internet. This creates a scenario where a single person or a small team can treat numerous patients at the same time, which in turn can significantly improve access on the one hand and save the time of the physiatrist and therapist, thus increasing their efficiency. The transition from a hospital-based rehabilitation program to a home-based program and providing assisted control in the hands of the end user or his/her family member or community care provider in itself has great potential in lowering the total expenses of rehabilitation programs, making it more affordable and accessible (Table 32.1).

Telerehabilitation, in general, employs telecommunication, computing technology, and remote sensing to enable the delivery and assessment of rehabilitation services from a distance. While traditional telerehabilitation often demands a one-on-one connection with a physiatrist and therapist, albeit from afar, a robotics-based approach offers more flexibility in the frequency and durations of sessions. Robot-mediated telerehabilitation research is an emerging trend, resulting in a range of relevant services, that allow a

TABLE 32.1 FDA standards [23].

Category	Standard
Electrical	
Software (entire life cycle)	IEC 62304 Ed. 1.1 2015-06
EMC/EMI	AAMI/ANSI/IEC 60601-1-2:2014
Electrical safety testing	IEC 60601-1:2005 (ReWalk^{TM}) ANSI/AAMI ES 60601-1-:2005(R)2012 (Indego)
Medical electrical equipment (home use)	ANSI/AAMI HA 60601-1-11:2015
Mechanical	
Durability testing (used in prosthetics)	ISO 10328:2006
Cycling loading testing (used in prosthetics)	ISO 22675:2006
Particle ingress	ANSI IEC 60529:2004
General	
Risk management	ISO 14971:2007
Quality management	ISO 13485:2003
Labeling	ISO 15223-1:2012
Biocompatibility	ISO 10993-1:2009
Human factors engineering	AAMI ANSI HE 75:2009/(R)2013
Training	AAMI TIR 49:2013
Application of usability	AAMI ANSI IEC 62366-1:2015

qualified expert to remotely verify that actions are carried out with the required intensity, precision, and posture, while simultaneously supervising many robots and patients [24]. This area has been the subject of comprehensive assessment from its inception until 2020 (Fig. 32.6).

32.4   Regulatory aspects

32.4.1   Challenges of the neurorehabilitation market

- Developing research guidelines for robotic trials, with clear definitions and names of the robotic devices [25,26].
- Trails should be capable of determining the specific patient population that benefits from the particular robotic device.

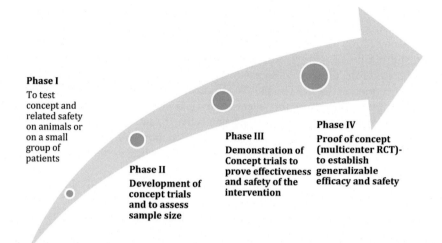

FIGURE 32.6 Clinical trial phases in motor rehabilitation.

- Clinical guidelines can be developed after overcoming the first two challenges.
- Maximizing the full range of motion.
- Use less virtual reality and instead more real objects to practice daily living activities.
- Shift from hospital/clinic-centered to home-based rehabilitation.
- Development of robots not as a therapeutic training device but as an assistive device.
- Use of cost-effective approaches in developing the technologies in order to improve the cost−benefit ratio.

32.5   Conclusions

The world is moving toward universal health coverage since the introduction of sustainable development goals in 2016, where the targets can only be achieved by 2030 through convergence of technology, health care, and social justice.

Globally technological development through cooperation and partnerships at regional and country levels is building, which has been boosted by the COVID-19 pandemic, thus establishing the value of technology application in healthcare delivery in high-income countries, middle-income countries, and low-income countries. WHO reports a higher disease burden of noncommunicable diseases globally. Neurological impairments are a high priority, needing technology-driven solutions from governments, pharmaceutical industries, technology companies, global bodies like the World Health Organization

(WHO), General Agreement on Trade in Services (GATS), and Trade Related Intellectual Property Rights (TRIPs). It has been established beyond doubt that the market of robotic rehabilitation is growing fast; therefore investment shortage will not be an issue; however, the issue of global convergence of R&D, acceptance of the universal design, agreement for benchmark standards, easing out of restriction in GATS and World Intellectual Property Organization (WIPO), global processes for patents, universal single pricing mechanisms, acceptable data security systems, and innovative financing for making these robotic rehabilitation products available for all countries in future will make United Healthcare (UHC) a reality by 2030 and more so through robotic neurorehabilitation.

Finally, robotic-mediated neurorehabilitation could potentially boost the efficiency and accessibility of therapy by supporting physiatrists and therapists in providing therapy for prolonged periods and collecting data to monitor progress. Through telerehabilitation, automation of therapy may allow many patients to be treated at the same time and maybe even remotely from their homes. Many consumers have accepted low-cost, off-the-shelf technologies for use in therapy sessions in recent years, along with strategies for enhancing motivation and participation.

Robot-mediated neurorehabilitation is promising for the future and can be attained through calibrated training and practice. They can also develop confidence among patients. Improvement of precision and accuracy flexibility in the mechanical movement of prosthetics and the development of personalized solutions are some of the areas that need to be addressed in order to bring about advancements in the field. Both engineering and clinical practice have been challenged in the field of machine-mediated neurorehabilitation. More integrated solutions are required on the engineering side [25,26].

References

[1] Rehabilitation Robots Market: Global Industry Trends, Share, Size, Growth, Opportunity and Forecast 2021−2026 [Internet]. Available from: https://www.researchandmarkets.com/reports/5441961/rehabilitation-robots-market-global-industry (Cited 7 January 2022).

[2] Guglielmelli E, Johnson MJ, Shibata T. Guest editorial special issue on rehabilitation robotics. Trans Robot 2009;3(25):477−80. Available from: https://www.infona.pl//resource/bwmeta1.element.ieee-art-000004982753.

[3] Giansanti D. The social robot in rehabilitation and assistance: what is the future? Healthcare 2021;9(3) /pmc/articles/PMC7996596/.

[4] Winstein CJ, Stein J, Arena R, Bates B, Cherney LR, Cramer SC, et al. Guidelines for adult stroke rehabilitation and recovery: a guideline for healthcare professionals from the American Heart Association/American Stroke Association. Stroke 2016;47(6):e98−e169. Available from: https://pubmed.ncbi.nlm.nih.gov/27145936/.

[5] Intel CEO. Warns chip shortage to last until 2023 as demand soars - Nikkei Asia [Internet]. Available from: https://asia.nikkei.com/Spotlight/Supply-Chain/Intel-CEO-warns-chip-shortage-to-last-until-2023-as-demand-soars (Cited January 2022).

[6] Rehabilitation [Internet]. Available from: https://www.who.int/news-room/fact-sheets/detail/rehabilitation (Cited 7 January 2022).

[7] World Health Organization (WHO). WHO global disability action plan 2014−2021: better health for all people World Heal Organ 2015;1−25[cited 2022 Jan 7]. Available from: https://apps.who.int/iris/bitstream/handle/10665/199544/?sequence = .

[8] Understanding the consequences of access barriers to health care: experiences of adults with disabilities PubMed [Internet]. Available from: https://pubmed.ncbi.nlm.nih.gov/12554383/ (Cited 7 January 2022).

[9] World Health Organization. The need to scale up rehabilitation. World Health Organization; 2017. Available from: <https://apps.who.int/iris/handle/10665/331210> License: CC BY-NC-SA3.0 IGO.

[10] Haig AJ, Im J, Adewole A, Nelson VS, Krabak B. The practice of physical medicine and rehabilitation in Subsaharan Africa and Antarctica: a white paper or a black mark? PM&R. 2009;1(5):421.

[11] Pediatric H.A.L. S2225 - Advanced Pediatric Patient Simulator - Gaumard [Internet]. Gaumard.com. Available from: https://www.gaumard.com/s2225/#: ~ :text = Pediatric%20HAL%C2%AE%20is%20the,expressions%2C%20movement%2C%20and%20speech; 2018 (Cited 17 August 2022).

[12] Stiehl J. Computer navigation in primary total knee arthroplasty. J Knee Surg 2010;20 (02):158−64.

[13] Robots In Healthcare | Benefit, Disadvantages and Future of Medical Robot [Internet]. Available from: https://www.delveinsight.com/blog/robotics-in-healthcare (Cited 7 January 2022).

[14] Demofonti A, Carpino G, Zollo L, Johnson M. Affordable robotics for upper limb stroke rehabilitation in developing countries: a systematic review. IEEE Trans Med Robot Bionics 2021;3(1):11−20.

[15] Stephanie HWC, Joanne Y. The future of service: the power of emotion in human-robot interaction. J Retail Consum Serv 2021;61:102551. Available from: https://doi.org/10.1016/j.jretconser.2021.102551 ISSN 0969-6989.

[16] A Robot May Not Injure a Worker: Working safely with robots | Blogs | CDC [Internet]. Blogs.cdc.gov. Available from: https://blogs.cdc.gov/niosh-science-blog/2015/11/20/working-with-robots/; 2021 (Cited 27 December 2021).

[17] Truelsen T, Piechowski-Jóźwiak B, Bonita R, Mathers C, Bogousslavsky J, Boysen G. Stroke incidence and prevalence in Europe: a review of available data. Eur J Neurol 2006;13(6):581−98. Available from: https://pubmed.ncbi.nlm.nih.gov/16796582/.

[18] Di Pino G, Pellegrino G, Assenza G, Capone F, Ferreri F, Formica D, et al. Modulation of brain plasticity in stroke: a novel model for neurorehabilitation. Nat Rev Neurol 2014;10 (10):597−608. Available from: https://pubmed.ncbi.nlm.nih.gov/25201238/.

[19] Kolaghassi R, Al-Hares M, Sirlantzis K. Systematic review of intelligent algorithms in gait analysis and prediction for lower limb robotic systems. IEEE Access 2021;9:113788−812.

[20] Kumar A, Choudhary A, Tiwari A, James C, Kumar H, Kumar Arora P, et al. An investigation on wear characteristics of additive manufacturing material. Mater Today Proc 2021;47:3654−60.

[21] Neurorehabilitation Market Size | Growth Prediction - 2030 [Internet]. Available from: https://www.alliedmarketresearch.com/neurorehabilitation-market-A10461 (Cited 7 January 2022).

[22] Neurological Disorders public health challenges WHO Library Cataloguing-in-Publication Data; 2006.

[23] Neurorehabilitation Devices Market | 2021 - 26 | Industry Share, Size, Growth - Mordor Intelligence [Internet]. Available from: https://www.mordorintelligence.com/industry-reports/neurorehabilitation-devices-market (Cited 7 January 2022).

[24] Johnson MJ, Schmidt H. Robot assisted neurological rehabilitation at home: motivational aspects and concepts for tele-rehabilitation Public Heal Forum 2009;17(4)8. e1−8. e4. Available from: https://www.researchgate.net/publication/43656685_Robot_Assisted_Neurological_Rehabilitation_a_Home_Motivational_Aspects_and_Concepts_for_Tele-Rehabilitation.

[25] Harwin WS, Patton JL, Edgerton VR. Challenges and opportunities for robot-mediated neurorehabilitation. Proc IEEE 2006;94(9):1717−26. Available from: https://www.researchgate.net/publication/2998102_Challenges_and_Opportunities_for_Robot-Mediated_Neurorehabilitation.

[26] Iandolo R, Marini F, Semprini M, Laffranchi M, Mugnosso M, Cherif A, et al. Perspectives and challenges in robotic neurorehabilitation. Appl Sci 2019;9(15). Available from: https://www.researchgate.net/publication/334976616_Perspectives_and_Challenges_in_Robotic_Neurorehabilitation.

Further reading

Veneman JF. On rehabilitation robotics safety, benchmarking, and standards: safety of robots in the field of neurorehabilitation—context and developments Human−Robot Interact 2019;91−102Apr 12 [cited 2022 Jan 7]. Available from: https://www.taylorfrancis.com/chapters/edit/10.1201/9781315213781-7/rehabilitation-robotics-safety-benchmarking-standards-jan-veneman.

Index

Note: Page numbers followed by "*f*" and "*t*" refer to figures and tables, respectively.

Printed in the United States
by Baker & Taylor Publisher Services